Behavioural Diversity in Chimpanzees and Bonobos

Chimpanzees (*Pan troglodytes*) and bonobos (*Pan paniscus*), otherwise known as pygmy chimpanzees, are the only two species of the genus *Pan*. As they are our nearest relatives, there has been much research devoted to investigating the similarities and differences between them. This book offers an extensive review of the most recent observations to come from field studies on the diversity of *Pan* social behaviour, with contributions from many of the world's leading experts in this field. A wide range of social behaviours is discussed including tool use, hunting, reproductive strategies and conflict management, as well as demographic variables and ecological constraints. In addition to interspecies behavioural diversity, this text describes exciting new research into variations between different populations of the same species. Researchers and students working in the fields of primatology, anthropology and zoology will find this a fascinating read.

CHRISTOPHE BOESCH is a Scientific Director of the Max-Planck-Institute for Evolutionary Anthropology in Leipzig, Germany, and is also Professor of Primatology at the University of Leipzig. His long-term research into the behaviour of chimpanzees from the Taï National Park in the Côte d'Ivorie has earned him numerous academic accolades.

GOTTFRIED HOHMANN is a research assistant at the Max-Planck-Institute for Evolutionary Anthropology in Leipzig, Germany. For the past ten years, he has conducted field research into the social behaviour of bonobos from Lomako in the Democratic Republic of Congo.

LINDA F. MARCHANT is Professor of Biological Anthropology in the Department of Anthropology at Miami University, Ohio, USA. Her research has concentrated on the study of chimpanzee behaviour in the Gombe and Mahale Mountains National Parks in Tanzania and at Assirik in Senegal. Her video, *Chimpanzee Grooming as Social Custom* won Best Short Film, 2000, from the Society for Visual Anthropology. She is co-editor of the book *Great Ape Societies* (1996, ISBN 0 521 55494 2 (hardback) / 0 521 5536 1 (paperback)).

Behavioural Diversity in Chimpanzees and Bonobos

Edited by

Christophe Boesch
Gottfried Hohmann
Linda F. Marchant

PUBLISHED BY THE PRESS SYNDICATE OF THE UNIVERSITY OF CAMBRIDGE
The Pitt Building, Trumpington Street, Cambridge, United Kingdom

CAMBRIDGE UNIVERSITY PRESS
The Edinburgh Building, Cambridge CB2 2RU, UK
40 West 20th Street, New York, NY 10011-4211, USA
477 Williamstown Road, Port Melbourne, VIC 3207, Australia
Ruiz de Alarcón 13, 28014 Madrid, Spain
Dock House, The Waterfront, Cape Town 8001, South Africa

http://www.cambridge.org

First published 2002

Printed in the United Kingdom at the University Press, Cambridge

Typeface Ehrhardt 9/12pt. *System* QuarkXPress® [SE]

A catalogue record for this book is available from the British Library

ISBN 0 521 80354 3 hardback
ISBN 0 521 00613 9 paperback

Contents

Contributors

ANDERSON, DEAN P.
Department of Zoology, University of Wisconsin-Madison, 430 Lincoln Drive, Madison, Wisconsin 53706, USA

BRADLEY, BRENDA J.
Interdepartmental Program in Anthropological Sciences, Stony Brook, SUNY at Stony Brook, NY 11794–4364, USA, and Department of Primatology, Max-Planck-Institute for Evolutionary Anthropology, Inselstr. 22, 04103 Leipzig, Germany

BOESCH, CHRISTOPHE
Department of Primatology, Max-Planck-Institute for Evolutionary Anthropology, Inselstr. 22, 04103 Leipzig, Germany

DORAN, DIANE M.
Department of Anthropology, SUNY at Stony Brook, Stony Brook, New York 11794, USA

FLEAGLE, JOHN G.
Department of Anatomical Sciences, SUNY at Stony Brook, Stony Brook, New York 11794, USA

FRUTH, BARBARA
Max-Planck-Institute for Behavioural Physiology, Seewiesen, P.O. Box 1564, 82305 Starnberg, Germany

FURUICHI, TAKESHI
Laboratory of Biology, Meiji-Gakuin University, Kamikurata, Totsuka, Yokoyama 244–8539, Japan

HASHIMOTO, CHIE
Primate Research Institute, Kyoto University, Kanrin, Inuyama, Aichi 484–8506, Japan

HEESY, CHRISTOPHER P.
Interdepartmental Program in Anthropological Sciences, SUNY at Stony Brook, Stony Brook, New York 11794, USA

HOHMANN, GOTTFRIED
Department of Primatology, Max-Planck-Institute for Evolutionary Anthropology, Inselstr. 22, 04103 Leipzig, Germany

HUNT, KEVIN D.
Department of Anthropology, Indiana University, Bloomington, IN 47405, USA

IHOBE, HIROSHI
School of Human Sciences, Sugiyama Jogakuen University, 37–234, Takenoyama, Iwasaki, Nisshin, Aichi 470–0131, Japan

JUNGERS, WILLIAM L.
Department of Anatomical Sciences, SUNY at Stony Brook, Stony Brook, New York 11794, USA

LANJOUW, ANNETTE
International Gorilla Conservation Programme, P.O. Box 48177, Nairobi, Kenya

LIU, HSIEN-YANG
Department of Computer Science, University of Minnesota, 200 Union Street S.E., Minneapolis, MN 55455, USA

LWANGA, JEREMIAH S.
Makerere University Biological Field Station, P.O. Box 409, Fort Portal, Uganda

MARCHANT, LINDA F.
Department of Anthropology, Miami University, Oxford, OH 45056, USA

MATSUMOTO-ODA, AKIKO
Human Evolution Studies, Department of Zoology, Division of Biological Sciences, Kyoto University, Kitashirakawa Oiwake-cho, Sakyo-ku 606–8502, Japan

MATSUZAWA, TETSURO
Primate Research Center, Kyoto University, Inuyama, Aichi, 484–8506, Japan

MCGREW, WILLIAM C.
Department of Zoology and Department of Anthropology, Miami University, Oxford, OH 45056, USA

MITANI, JOHN C.
Department of Anthropology, University of Michigan, Ann Arbor, MI 48109–1382, USA

MOERMOND, TIMOTHY C.
Department of Zoology, University of Wisconsin-Madison, 430 Lincoln Drive, Madison, WI 53706, USA

MULLER, MARTIN N.
Department of Anthropology, Harvard University, 11 Divinity Avenue, Cambridge, MA 02138, USA

MYERS THOMPSON, JO A.
Lukuru Wildlife Research Project, c/o P.O. Box 5064, Snowmass Village, CO 81615–5064, USA

NAKAMURA, MICHIO
Japan Monkey Centre, 26 Kanrin, Inuyama 484–0081, Japan

NEWTON-FISHER, NICHOLAS E.
Department of Biological Anthropology, University of Cambridge, Downing Street, Cambridge, CB2 3DZ, UK

NORDHEIM, ERIK V.
Department of Forest Ecology and Management, University of Wisconsin-Madison, 1630 Linden Drive, Madison, WI 53706, USA

PUSEY, ANNE E.
Department of Ecology, Evolution and Behavior, University of Minnesota, 1987 Upper Buford Circle, Saint Paul, MN 55108, USA

REYNOLDS, VERNON
Institute of Biological Anthropology, Oxford University, 58 Banbury Road, Oxford, OX2 6QS, UK

SMALL, MEREDITH F.
Department of Anthropology, Cornell University, Ithaca, NY 14850, USA

SUGIYAMA, YUKIMARU
Tokai-Gakuen University, Nakahira 2–901, Tenpaku, Nagoya, 468–8514, Japan

UEHARA, SHIGEO
Primate Research Institute Kanrin, Kyoto University, Inuyama, Aichi 484–8506, Japan

VIGILANT, LINDA
Department of Primatology, Max-Planck-Institute for Evolutionary Primatology, Inselstr. 22, 04103 Leipzig, Germany

WALLIS, JANETTE
Department of Psychiatry and Behavioral Science, Health Science Center, University of Oklahoma, P.O. Box 26901, Oklahoma City, OK 73190, USA

WATTS, DAVID P.
Department of Anthropology, Yale University, New Haven, CT 06520–8277, USA

WILLIAMS, JENNIFER M.
The Jane Goodall Institute's Center for Primate Studies, 202 Lower Street, Shelburne Falls, MA 01370, USA

WRANGHAM, RICHARD
Department of Anthropology, Peabody Museum, Harvard University, 11 Divinity Avenue, Cambridge, MA 02138, USA

Preface

This book comes from a conference on 'Behavioural Diversity in Chimpanzees and Bonobos', held 11–17 June, 2000 at the monastery of Seeon in southern Bavaria, Germany. Organized by Gottfried Hohmann and Christophe Boesch, and monitored by Linda F. Marchant, the cloistered meeting heard reports from 30 researchers on 13 different populations of these African apes. Instead of publishing proceedings of the conference, we decided to organize a collection of integrated findings on key topics. Thus, the chapters in this book deviate from the contributions to the Seeon conference in several ways. First, most chapters were greatly edited in order to draw direct comparisons across sites and between species. Second, for the same purpose, some authors who gave separate presentations at the meeting agreed to prepare combined chapters. Third, to enlarge the scope of the book, we asked others who were not present at the meeting to contribute.

This book is the result, and it is devoted only to research on wild populations of *Pan,* a hominoid genus with two sister species, chimpanzee (*P. troglodytes*) and bonobo (*P. paniscus*). Although many research facilities offer excellent housing and stimulating social and non-social environments, when compared with field data, the steady flow of resources, reduced fluctuation in health, safety from predators, and constrained movements limit the ecological validity of their findings.

Behavioural diversity is not a new research topic. There is a rich tradition in comparative ethology of examining given behaviours across species (Lorenz 1950; Hinde & Tinbergen 1958; Wickler 1967; Eibl-Eibesfeldt 1975). These studies focused on species-typical behavioural patterns by comparing closely related taxa. On another frontier, researchers explored geographic variation in behaviour across populations of the same species (Emlen 1971; Huntingford 1982; for reviews see Foster & Cameron 1996 and Foster & Endler 1999). The aim of this book is to investigate behavioural diversity from both perspectives: within and between species.

In history, structure and content, with *Pan* as its sole subject, this book has several influential predecessors that have marked the progress of research on great apes: *The Great Apes* (Hamburg & McCown 1979), *Understanding Chimpanzees* (Heltne & Marquardt 1989), *Chimpanzee Cultures* (Wrangham *et al.* 1994) and *Great Ape Societies* (McGrew *et al.* 1996). New findings demand a successor.

Since the 1990s, field research on chimpanzees has advanced in several ways. First, several projects have lasted long enough to document whole life histories of known individuals. Second, new study sites have been established with niches of 'unusual' environmental features. Third, at several sites, studies now incorporate neighbouring communities, so that intrapopulation variability can be explored. Fourth, analyses of the genetic relationships among group members reveal the history of group members and the fitness consequences of social and reproductive strategies. Finally, a new generation of researchers has entered the arena, tackling old themes with new theories and techniques.

Field research on the bonobo still lags behind the work done on chimpanzees. However, the advantage of being in the early stages of its development is that the slope of the knowledge growth curve is still steep. Data from known individuals in long-term research have accumulated, genetic relationships within communities have been identified and, against long odds, new sites have been founded in areas outside the equatorial belt of closed lowland forest. In spite of the copious attention the media has focused on the natural history of *Pan*, the notion that each species lives in a different social world remains widespread. As the chapters of this book will show, the images of the two *Pan* species are rich in diversity and are still being revealed.

ACKNOWLEDGEMENTS

The conference at Seeon was supported by grants from The Wenner-Gren Foundation for Anthropological Research; Deutsche Forschungsgemeinschaft; Bayerisches Staatsministerium für Wissenschaft, Forschung, und Kunst; Max-Planck-Gesellschaft; and a private donor. The editors thank the following people for their careful and substantive reviews: Todd Disotell, Tony L. Goldberg, Peter M. Kappeler, William C. McGrew, Jim Moore, Carel P. van Schaik and David P. Watts. We also thank Tracey Sanderson from Cambridge University Press for her editorial guidance and helpful comments on earlier drafts. Michael Seres kindly donated the wonderful photographs as a source for the book's cover. This book would have never appeared without the combined efforts and skills of many people: Sylvio Tüpke, Anja Kösler, André Reissaus and Bela Dornon gave technical advice and helped to prepare the graphics of the book. Special thanks go to Paula Ross who edited the text, and co-ordinated the entire process of editing with great competence and infinite patience.

REFERENCES

Eibl-Eibesfeldt, I. (1975). *Ethology, the Biology of Behavior*. New York: Holt, Rinehart & Winston.

Emlen, S. T. (1971). Geographic variation in indigo bunting song (*Passerina cyanea*). *Animal Behaviour* **19**, 407–8.

Foster, S. A. & Cameron, S. A. (1996). Geographic variation in behavior: a phylogenetic framework for comparative studies. In: *Phylogenies and the Comparative Method in Animal Behavior*, ed. E. Martins, pp. 138–65. Oxford: Oxford University Press.

Forster, S. A. & Endler, J. A. (eds) (1999). Geographic variation in behavior: perspectives on evolutionary mechanisms. New York & Oxford: Oxford University Press.

Hamburg, D. A. & McCown, E. R. (eds) (1979). *The Great Apes*. Menlo Park, CA: Benjamin/Cummings.

Heltne, P. G. & Marquardt, L. A. (eds) (1989). *Understanding Chimpanzees*. Cambridge, MA: Harvard University Press.

Hinde, R. A. & Tinbergen, N. (1958). The comparative study of species-specific behavior. In: *Behavior and Evolution*, ed. A. Roe & G. G. Simpson, pp. 251–68. New Haven, CT: Yale University Press.

Huntingford F. A. (1985). Do inter- and intra-specific aggression vary in relation to predation pressure in sticklebacks? *Animal Behaviour* **30**, 909–16.

Lorenz, K. (1950). The comparative method in studying innate behavior patterns. *Symposium of the Society for Experimental Biology* **4**, 221–68.

McGrew, W. C., Marchant, L. F. & Nishida, T. (eds) (1996). *Great Ape Societies*. Cambridge: Cambridge University Press.

Wickler, W. (1967). Vergleichende Verhaltensforschung und Phylogenetik. In: *Die Evolution der Organismen*, ed. G. Heberer, pp. 420–508, Stuttgart: Fischer.

Wrangham, R. W., McGrew, W. C., de Waal, F. B. M. & Heltne, P. G. (eds) (1994). *Chimpanzee Cultures*. Cambridge, MA: Harvard University Press.

Gottfried Hohmann, Christophe Boesch &
Linda F. Marchant

Behavioural Diversity in *Pan*

CHRISTOPHE BOESCH

A BRIEF HISTORY

In the late 1950s, the great anthropologist Louis Leakey searched for a student to study wild chimpanzees in their natural habitat; he was convinced that this kind of investigation would provide important information about the behaviour of our early ancestors. His quest eventually ended with Jane Goodall, who began her research on chimpanzees in Gombe in the early 1960s, near the shore of Lake Tanganyika in Tanzania, East Africa.

During the course of the following years, our knowledge of wild chimpanzees relied almost exclusively on observations collected from two communities – Goodall's work at Gombe, and the work of Toshisada Nishida and his colleagues, conducted in Mahale Mountains National Park, some 200 km south of Gombe. Both populations, located on the eastern side of Lake Tanganyika, live in relatively similar environments, characterised by a mosaic of savanna, woodland and dense shrub habitats. The last three decades of the 20th century saw the success of these two studies turn the Tanzanian chimpanzee into something of a prototype for 'The Wild Chimpanzee', and behavioural diversity was restricted to the differences between these two populations. Anthropological and psychological literature and textbooks are still strongly biased in assuming that Tanzanian chimpanzee behaviour represents all chimpanzees.

The chimpanzee: variations revealed

Meanwhile, new populations of chimpanzees have been studied in different types of environments throughout their distribution in Africa. West African chimpanzees have been the focus of two studies that began in the late 1970s, one in Guinea in the small Bossou forest and the second one in the Taï National Park, Côte d'Ivoire. It quickly became apparent that our knowledge about Tanzanian chimpanzees could not simply be assumed to apply to all chimpanzee populations. For example, the tool use repertoire was discovered to differ markedly for each of the studied populations (Sugiyama & Koman 1979; Boesch & Boesch 1990; McGrew 1992), and the hunting behaviour seemed to be more organised in West Africa (Boesch & Boesch 1989). Rapidly, new studies of other chimpanzee populations in East Africa started as well, for example in the Budongo and Kibale forests (Ghiglieri 1984; Newton-Fisher, Chapter 9). The picture emerging from all these studies was that each wild population presents many different behaviours, not only in the domain of tools and hunting, but in core, basic social interactions as well. A recent review of nine chimpanzee populations revealed that cultural differences are systematic, and that it is more precise to talk about the 'Gombe chimpanzee', the 'Mahale chimpanzee' or the 'Taï chimpanzee' than 'chimpanzees' in general. The current volume extends this approach by presenting new and important aspects of chimpanzee behaviour from known populations, and also by providing data on new populations about which little has been known.

The 'pygmy chimpanzees': bonobos coming into their own

Looking to the chimpanzees' 'sister species', studies of bonobos, originally called pygmy chimpanzees, started much later than was the case for chimpanzees. Consequently, we know relatively less about them. The questions we address are can we speak of a 'prototype bonobo', or are we also dealing with population-specific behaviour patterns here? These questions remain more difficult to answer, and the contributors to this collection can help us only up to a point. This is primarily because we have detailed observations from fewer populations of bonobos, so it is simply too early to answer such a question fully and properly. Bonobos at Lomako and Wamba have been studied for many years, but the two populations live near each other and within very similar forest types (Hohmann & Fruth, Chapter 10; Furuichi & Hashimoto, Chapter 11). These populations represent only a small proportion of the habitat inhabited by this species. The new observations of the Lukuru bonobos, situated in a drier forest further south than the sites at which earlier research on bonobos was conducted, are therefore of special interest (Myers-Thompson, Chapter 4).

WHEN SHOULD WE EXPECT BEHAVIOURAL DIVERSITY?

Chimpanzees

Modern evolutionary thinking predicts that populations of the same species living in different habitats will need to adopt different strategies in order to survive and reproduce because of the different challenges they face daily (Ridley 1995; Futuyma 1998). In animal species, the response to environmental conditions can have a genetic base, or be behaviourally based on extended learning abilities. Learning is expected to be more flexible than genetics, and species that learn how to respond flexibly are expected to exhibit high levels of behavioural diversity. Primates, in general, possess more learning abilities than other mammals (Byrne 1995; Tomasello & Call 1997). Chimpanzees and bonobos in particular have demonstrated extended learning abilities under different conditions, whether in captivity or in their natural habitat (Goodall 1986; Byrne 1995; Boesch & Boesch-Achermann 2000). In addition, it has been known for a long time that chimpanzees and bonobos live in diverse environments, ranging from tropical rainforest to very dry savanna regions that contain some gallery forests (Suzuki 1969; Kano 1971; McGrew *et al.* 1979; Kortlandt 1983; Moore 1996). Therefore, we should expect chimpanzee populations to show greater behavioural diversity than species living under more similar conditions.

The main problem we face in expanding our knowledge about these populations is the need for more detailed behavioural observations in order to answer questions about behavioural diversity in this species. More informed answers can come only if more scientists go into the field and study more wild chimpanzee populations living under different ecological conditions. What might we expect to be the outcome of such research? The more divergent the habitats, the more diverse the behavioural strategies adopted by chimpanzees. New research projects currently under way that are investigating as yet little-known populations will eventually advance our knowledge and convincingly demonstrate that this knowledge is refined and increased with the number of populations examined. For example, a new study of chimpanzees in the Kalinzu Forest, Uganda, revealed the first instances in this region of the use of tools to dip for driver ants (Hashimoto *et al.* 2000). Some observations of the Tenkere chimpanzees in Sierra Leone, revealed the use of small branches as 'stepping-sticks' and 'seat-sticks' (Alp 1997). These observations remain unique for all known chimpanzee populations. Similarly, new observations of tool use in the Lossi chimpanzees in Congo revealed the existence of a tool-set consisting of three components, which they use in sequence – a stout chisel, a bodkin, and a dip-stick – to extract honey from melipone bee nests (Bermejo & Illera 1999). Lossi chimpanzees are the first known wild populations using such a complex tool-set. Finally, Ndakan chimpanzees in the Dzanga-Sangha region in Congo are the only ones known to pound nests of Melipone bees with large pieces of wood to gain access to the honey (Fay & Carroll 1994).

This collection of articles presents detailed data on some new populations of chimpanzees, some of which live in especially dry conditions compared to the classical image of the chimpanzee as living in a humid forest. The Tongo chimpanzees in the Democratic Republic of Congo and the Semliki chimpanzees of Uganda nicely illustrate the extent to which we still might be underestimating behavioural diversity in this species. Both populations live in areas where, at certain times of the year, water is an important limiting factor, and both populations have invented unique ways to obtain access to water during these periods. Tongo chimpanzees live in a forest that grows on volcanic soil and where water rapidly infiltrates the ground. When water is limited, Tongo chimpanzees dig rather deep holes into the soil to reach large tubers that contain plenty of water; they then suck the tubers, extracting as much of the moisture as possible (Lanjouw, Chapter 3). Semliki chimpanzees live in a much more open habitat, cut by the gallery forest and following small rivulets. When the rivulet beds are dry, Semliki chimpanzees were seen digging holes in the dry beds to reach the underground water. These holes functioned like wells, providing the chimpanzees with clear water (Hunt & McGrew, Chapter 2). Never before had chimpanzees been seen digging wells in order to access underground water sources! It has been suggested that exploitation of underground resources was a niche that our human ancestors discovered, and we see here that chimpanzees in dry environments also seem to use such resources. These two examples show just how inventive chimpanzees can be and how much more we might learn about chimpanzee behaviour if more populations are studied.

Bonobos

To date, bonobos have only been studied in the deep forest of the Congo basin, and they were thought to be restricted to such tropical rainforests. However, their precise distribution within the Congo Basin remains unclear and questions have been raised about how far south they occur and in what kinds of environments they can survive. New observations presented in this volume about the bonobos of the Lukuru region in the south of the Democratic Republic of Congo, show that they can live in a mosaic forest with large savanna areas. They have been observed entering deep into the savanna to feed on grass, as well as venturing into swamp areas, walking bipedally into waist-deep water of streams or pools (Myers Thompson, Chapter 4). This behaviour demonstrates unexpected flexibility in bonobos, seeming to indicate greater behavioural diversity than previously assumed. Here as well, the type of habitat inhabited seems crucial in explaining behavioural diversity. Thus, part of the information presented here points out that, if we want to know the real spectrum of behavioural diversity in bonobos, then we need to be patient – more data on new populations are required. Hopefully, both chimpanzees and bonobos will survive long enough in Africa, despite the present threats to their well-being, so that we can more fully address and answer questions about the behavioural diversity that exists in these species.

HOW DO ENVIRONMENTAL DIFFERENCES AFFECT BEHAVIOUR?

Dating from the time of the great trophy hunters, long before researchers appeared on the scene, we have known about the differences in geographic distribution of the two *Pan* species: chimpanzees occur in many different parts of tropical Africa, while bonobos were restricted to the Democratic Republic of Congo (formerly Zaïre). And these observations later came to be supported by scientific data. Chimpanzees have always lived in diverse environments (Suzuki 1969; Kano 1971; McGrew *et al.* 1979). The distribution of bonobos is still a matter of debate, but it seems clear that the tropical rainforest of Central Africa is their main habitat.

Traditionally, socio-ecological models have placed heavy emphasis on the importance of food resources in influencing social structure and grouping patterns (Wrangham 1980; Dunbar 1988; Begon *et al.* 1990), although others have suggested predation to be the most important in explaining grouping patterns (van Schaik & van Hooff 1983; Dunbar 1988). The large body size of the great apes and the relatively weak evidence of predation against them was taken as confirmation that, as an explanation of the differences in social behavior between chimpanzees and bonobos, food is a more important influence on social structure and grouping patterns (Wrangham 1986; White & Wrangham 1988; Stanford *et al.* 1994; Doran 1997; but see Boesch 1991). More precisely, chimpanzees' greater dependence on large fruit patches and bonobos' more extensive reliance on terrestrial herbal vegetation (TVP) were suggested as an explanation of the supposed important differences in social domain between the two species (White & Wrangham 1988). Alternatively, proposals positing that additional factors, for example the presence of estrous females, were important in explaining social grouping in chimpanzees (Goodall 1986; Boesch 1996; Boesch & Boesch-Achermann 2000) have also been put forth. This debate between an unifactorial explanation of social grouping patterns versus a multifactorial one has been limited by the quality of data available on food distribution and production in the different populations compared.

Food availability, sexual opportunity and party size

A number of researchers in this book present precise quantifications of food production and distribution in different populations, allowing us for the first time to test the influences these different factors may exert on social grouping patterns in chimpanzees (Anderson *et al.*, Chapter 6; Mitani *et al.*, Chapter 7). Based on painstaking monitoring of fruit production of many trees of the most important food species over the whole year and estimation of the density of these species, a precise quantification of fruit availability was obtained, making a test of the relative importance of food versus sexual opportunities possible. The results are fascinating. In Ngogo chimpanzees, both food availability and number of estrous females present in the party explain the largest part of the variation seen in party sizes (Mitani *et al.*, Chapter 7), and the influence of each of these factors seems to be independent of the other. In the Taï forest, the number of estrous females present in a party was shown to be the only factor affecting party size – the more estrous females present, the larger the party. Food availability played a clear role only if no estrous female was present in the party

(Anderson *et al.*, Chapter 6). In this case, the more food available, the larger the party size. Thus, the only two studies that have been able to test the respective roles of food availability and sexual opportunities on party size in chimpanzees concur in granting sexual opportunity at least as much importance as food availability.

Although direct comparisons between the sites are still difficult, greater food availability in Taï might explain the differences between the two sites, in the sense that the presence of estrous females may be a more important factor than food availability in explaining group size in chimpanzees. If food is generally less abundant in the habitat, then it will play a more important role, as it does in the Ngogo chimpanzee population. Food might be even more limited in Gombe, as suggested by their smaller body size, and food competition might dominate, as suggested by the results presented and discussed in Chapter 14 (Williams *et al.*). An analysis of rainfall, often proposed as a good indicator of an area's food productivity, shows that in both Budongo and Gombe, the number of estrous females and party size have a very high correlation. However, in Gombe party size decreases before the dry season, while in Budongo party size decreases after the dry season (Wallis, Chapter 13). Since a precise quantification of food production is still missing for both sites, these results are difficult to interpret, but they do at least show that party size fluctuates more directly with the number of estrous females than with the amount of rainfall. We might, therefore, be witnessing a general pattern in chimpanzees, where party size is driven much more by sexual opportunity than food.

Hunting behaviour

Hunting behaviour in chimpanzees is another domain where the role of the environment has been proposed as an important one (Boesch 1994; Boesch & Boesch-Achermann 2000): forest structure has been suggested as a key agent since it directly affects the ways in which hunters can achieve a capture. Alternatively, it has been proposed that the number of hunters or the number of males in the party is the key factor in explaining the occurrence of hunting and its success (Stanford *et al.* 1994; Mitani & Watts 1999). A direct test of these hypotheses shows that, independent of party size, when Ngogo chimpanzees hunt for red colobus in a continuous forest, their success is much lower than when they hunt them in an interrupted forest (Watts & Mitani, Chapter 18). Interestingly, if the hunting takes place in continuous forest, then success of the Ngogo chimpanzees is the same as that of the Taï chimpanzees, although the former have three times more males in the community. Thus, forest structure seems to be the key factor affecting hunting success in chimpanzees, and demographic parameters, for example number of males, may play a more limited role in hunters' compensating for a lower organisation during the hunt. Ngogo chimpanzees reach surprising success with lower level of organisation (Watts & Mitani, Chapter 18). Similarly, a comparison of the interactions between hunters and red colobus prey across the Mahale, Gombe and Taï sites revealed that the structure of the forest where the hunt takes place plays a crucial role (Boesch *et al.*, Chapter 16).

BONOBOS AND CHIMPANZEES: DIFFERENCES AND SIMILARITIES

Morphological differences between bonobos and chimpanzees were first thought to be rather clear, and led to the name 'pygmy chimpanzee' for *Pan paniscus*, thought to be smaller than the chimpanzee (*Pan troglodytes*). However, measurements made in the late 1970s using a larger sample of individuals showed that, while bonobos were not smaller than chimpanzees (most anatomical measurements overlapped between the two species), body proportions were distinct: the bonobo has shorter upper limbs and longer lower limbs (Zihlmann 1996). Molecular analysis of both nuclear and mitochondrial DNA confirms that bonobos and chimpanzees are very closely related (Bradley & Vigilant, Chapter 19). These results suggest a separation between the two species somewhere between 1.2 and 2.7 million years ago, while the divergence between chimpanzee and human is proposed to have occurred 5–8 million years ago (Kaessmann *et al.* 1999). However, despite these similarities in morphology and genetics, many aspects of bonobo and chimpanzee sexual and social behaviour are quite different. The bonobo has been portrayed as overly sexual, with regular homosexual interactions between the females, and as exhibiting very cohesive social grouping patterns, with females dominant over males (Kano 1992). Chimpanzees, in contrast, have been presented as generally less cohesive, with smaller social groups where males are more social, and clearly dominant over females (Wrangham 1986; Nishida & Hiraiwa-Hasegawa 1987). Initially, these differences were considered to be interspecies differences (Wrangham & Peterson 1996; de Waal & Lanting 1997). However, new data from forest chimpanzees emphasise the intraspecies variability, attributed to living under different ecological conditions: the differences between chimpanzees and bonobos could be related more to the ecological conditions prevailing in the populations under consideration than to interspecies differences (Boesch 1996; Boesch & Boesch-Achermann 2000). Recently, it has also been suggested that the differences between the two species are more apparent than real, simply because there are fewer data available on bonobos (Stanford 1998).

Many contributions to this current collection add important information to this debate over interspecies differences. Furthermore, they confirm that given that behavioural diversity is a function of the different types of environments a species inhabits, we need to take ecological differences into consideration when comparing the two species. For example, sexual behaviour differs less between the two species than initially thought (Takahata *et al.* 1996; Furuichi & Hashimoto, Chapter 11), to the point that both species could be directly included in the same framework, showing that strategies used by female bonobos and by females of some chimpanzee populations (Taï) are very similar and could be explained by the higher cohesiveness in the social grouping typical for both populations (Wrangham, Chapter 15). Similarly, when looking at social dynamics from the point of view of feeding competition, the pattern that emerges is that females react in the same way in the two species, and if conditions permit, bonobo and chimpanzee females may adopt very similar social grouping patterns (Hohmann & Fruth, Chapter 10; Matsumoto-Oda, Chapter 12; Williams *et al.*, Chapter 14). The image emerging is that both species, due to their long and common phylogenetic history, are likely to respond very similarly to changes in local conditions. In a multivariate analysis of differences between the bonobo and chimpanzee, it has been shown that the analysed differences correlated most powerfully with the number of dry months per year, with rainforest chimpanzees occupying a position most similar to rainforest bonobos (Doran *et al.*, Chapter 1).

The answer (or answers) to questions of how different and how similar bonobos and chimpanzees are from and to each other will affect the way we understand the evolutionary forces located at the origin of the divergence between the two species of *Pan*. Based only on the distribution of the two species, many scenarios have been proposed to explain how different the two species *should* be. In this sense, the information about the Lukuru bonobos is very important because it indicates that both species could survive outside the dense forest during the more dry periods in the past, and that differences in this respect are not enough to explain the divergence that led to the two species. Similarly, the fact that both species seem to make similar adaptations to the tropical rainforest environment indicates that it is not only the shift from open to closed forest that led to the divergence between the two species.

Our models for the evolution of both species need to be refined by including some of the evidence presented here. The classical proposition is that an important change in the ecological conditions in East Africa produced a split, one leading to chimpanzees and one leading to modern humans. This savanna model argued that climatic changes East of the Rift Valley changed the forest environment into a savanna, and that the ancestors trapped there had to adapt in order to survive (Dart 1925; Leakey 1980; Johanson & Edey 1981). However, after comparing forest and savanna chimpanzees, it has been suggested that the ancestors common to chimpanzees, bonobos and humans were living in a more forested environment than normally proposed by anthropologists (Boesch & Boesch 1989). Recent discoveries of three possible early ancestors of the human evolutionary lineage seem to support this possibility completely: the environment in which *Ardipithecus* sp. and *Orrorin tugenensis* lived between 6 and 3.5 million years ago seems to have been predominantly a woodland and not a drier savanna habitat (WoldeGabriel *et al.* 2001; Leakey *et al.* 2001; Pickford & Senut 2001), a habitat that is still inhabited today by a few bonobo populations and a number of chimpanzee populations. This, combined with further evidence that more recent ancestors, for example the *Australopithecus* (Rayner *et al.* 1993; Brunet *et al.* 1995), occupied a wooded habitat suggests that it was a woodland habitat where the common ancestors of humans, chimpanzees and bonobos could be found, and that it was also the habitat in which the split occurred. *Orrorin tugenensis* lived some 6 million years ago, a time that is within the range of the last common ancestor between humans and chimpanzees (Senut *et al.* 2001). If the dating of *Orrorin tugenensis* is accurate, then, contrary to the savanna model, which suggests that an important shift in ecological conditions is responsible for the divergence between human and chimpanzee ancestors, the recent discoveries of fossils, as well as new evidence included in the chapters that follow on the distribution and behaviour of chimpanzees and bonobos living in relatively dry habitats, suggest that a woodland habitat was part of the living environment for much of the line from the last common ancestor to contemporary populations of chimpanzees and bonobos. The cause of the divergence in those lines may either be found in subtle ecological differences within the 'woodland' framework, or in changes in life-history traits due to the selection pressure that comes from increasing predation.

ARE ALL BEHAVIOURAL DIFFERENCES RELATED TO ECOLOGICAL DIFFERENCES?

This question has been the centre of a lively debate because it is directly related to the question of culture in animals. When the behavioural repertoire differs between two populations of the same species, but the environmental conditions are the same, then the differences might be cultural, that is learned from other group members (Kummer 1971; Bonner 1980; Boesch 1996). Excluding environmental factors is more easily said than done, however, since the influence might be more subtle than expected (Tomasello 1990). Nevertheless, consensus is growing that chimpanzees have cultural abilities, and that many of the differences we observe between populations cannot all simply be explained by ecological differences (Boesch & Boesch 1990; McGrew 1992; Boesch *et al.* 1994; Boesch 1996; Whiten *et al.* 1999; Whiten & Boesch 2001). Yet this is, by no means, the last word on the breadth of cultural abilities in chimpanzees. For example, as demonstrated by Nakamura (Chapter 5), cultural variation might develop in the social domain through interactions between individuals, and an analysis of social grooming patterns has revealed differences between Mahale and other populations that are apart from any ecological explanations.

In addition to chimpanzees, the possession of cultural abilities has also been suggested in whales, dolphins and killer whales, as well as in orangutans and in some birds (Bonner 1980; van Schaik & Knott 2001; Rendell & Whitehead 2001). The conservative expectation is that we will find similar abilities in bonobos but, to test whether or not this expectation will be fulfilled, we need more data on more bonobo populations. If bonobos do have similar cultural abilities, then it seems safe to assume that such abilities were common within the chimpanzee–bonobo–human clade, and this assumption will require using new eyes when studying the human cultural abilities (e.g. Boesch & Tomasello 1998).

The extent to which behavioural diversity exists has been one of the important lessons learned from the growing number of observations of more and more wild populations of chimpanzees and bonobos. The contributions made to this area of primatology by this volume indicate that we have still more to learn about the full degree of diversity in both species. Yet even at this point in time, we can already say that it is larger than we ever suspected.

REFERENCES

Alp, R. (1997). 'Stepping-sticks' and 'seat-sticks': new types of tools used by wild chimpanzees (*Pan troglodytes*) in Sierra Leone. *American Journal of Primatology*, 41, 45–52.

Begon, M., Harper, J. L. & Townsend, C. R. (1990). *Ecology: Individuals, Populations and Communities* (2nd Edition). Oxford: Blackwell Scientific Publications.

Bermejo, M. & Illera, G. (1999). Tool-set for termite-fishing and honey extraction by wild chimpanzees in the Lossi forest, Congo. *Primates*, 40(4), 619–27.

Boesch, C. (1991). The effect of leopard predation on grouping patterns in forest chimpanzees. *Behaviour*, 117(3–4), 220–42.

Boesch, C. (1994). Chimpanzees – red colobus: a predator–prey system. *Animal Behaviour*, 47, 1135–48.

Boesch, C. (1996). Social grouping in Taï chimpanzees. In *Great Ape Societies*, ed. W. C. McGrew, L. Marchant & T. Nishida, pp. 101–13. Cambridge: Cambridge University Press.

Boesch, C. & Boesch, H. (1989). Hunting behavior of wild chimpanzees in the Taï National Park. *American Journal of Physical Anthropology*, 78, 547–73.

Boesch, C. & Boesch, H. (1990). Tool use and tool making in wild chimpanzees. *Folia Primatologica*, 54, 86–99.

Boesch, C. & Boesch-Achermann, H. (2000). *The Chimpanzees of the Taï Forest: Behavioural Ecology and Evolution*. Oxford: Oxford University Press.

Boesch, C. & Tomasello, M. (1998). Chimpanzee and human cultures. *Current Anthropology*, 39(5), 591–614.

Boesch, C., Marchesi, P., Marchesi, N., Fruth, B. & Joulian, F. (1994). Is nut cracking in wild chimpanzees a cultural behaviour? *Journal of Human Evolution*, 26, 325–38.

Bonner, J. (1980). *The Evolution of Culture in Animals*. Princeton: Princeton University Press.

Brunet, M., Beauvilain, A., Coppens, Y., Heintz, E., Moutaye, H. & Pilbeam, D. (1995). The first australopithecine 2500 kilometres west of the Rift Valley (Chad). *Nature*, 378, 273–4.

Byrne, R. (1995). *The Thinking Ape*. Oxford: Oxford University Press.

Dart, R. (1925). *Australopithecus africanus*, the man-ape of South Africa. *Nature*, 115, 195–9.

de Waal, F. & Lanting, F. (1997). *Bonobo: The Forgotten Ape*. Berkeley: University of California Press.

Doran, D. (1997). Influence of seasonality on activity patterns, feeding behavior, ranging and grouping patterns in Taï chimpanzees. *International Journal of Primatology*, 18(2), 183–206.

Dunbar, R. (1988). *Primate Social Systems*. New York: Cornell University Press.

Fay, M. & Carroll, R. (1994). Chimpanzee tool use for honey and termite extraction in Central Africa. *American Journal of Primatology*, 34, 309–17.

Futuyma, D. (1998). *Evolutionary Biology*, 3rd Edition. Sunderland, MA: Sinauer Associates.

Ghiglieri, M. P. (1984). *The Chimpanzees of the Kibale Forest*. New York: Columbia University Press.

Goodall, J. (1986). *The Chimpanzees of Gombe: Patterns of Behavior*. Cambridge, MA: The Belknap Press of Harvard University Press.

Hashimoto, C., Furuichi, T. & Tashiro, Y. (2000). Ant dipping and meat eating by wild chimpanzees in the Kalinzu Forest, Uganda. *Primates*, 41(1), 103–8.

Johanson, D. & Edey, M. (1981). *Lucy: The Beginnings of Humankind*. New York: Simon and Schuster.

Kaessmann, H., Wiebe,V. & Pääbo, S. (1999). Extensive nuclear DNA sequence diversity among chimpanzees. *Science*, 286, 1159–62.

Kano, T. (1971). The chimpanzees of Filabanga, western Tanzania. *Primates*, 12, 229–46.

Kano, T. (1992). *The Last Ape: Pygmy Chimpanzee Behavior and Ecology*. Stanford, CA: Stanford University Press.

Kortlandt, A. (1983). Marginal habitats of chimpanzees. *Journal of Human Evolution*, 12, 231–78.

Kummer, H. (1971). *Primate Societies: Group Techniques of Ecological Adaptation*. Chicago: Aldine.

Leakey, M., Spoor, F., Brown, F., Gathogo, P., Kiarie, C., Leakey, L. & McDougall, I. (2001). New homonin genus from eastern Africa shows diverse middle Pliocene lineages. *Nature*, 410, 433–40.

Leakey, R. (1980). *The Making of Mankind*. London: Book Club Associates.

McGrew, W. C. (1992). *Chimpanzee Material Culture: Implications for Human Evolution*. Cambridge: Cambridge University Press.

McGrew, W. C., Baldwin, P. J. & Tutin, C. (1979). Chimpanzees, tools and termites: cross-cultural comparisons of Senegal, Tanzania and Rio Muni. *Man*, 14, 185–214.

Mitani, J. & Watts, D. (1999). Demographic influences on the hunting behavior of chimpanzees. *American Journal of Physical Anthropology*, 109, 439–54.

Moore, J. (1996). Savanna chimpanzees, referential models and the last common ancestor. In *Great Ape Societies*, ed. W. C. McGrew, L. Marchant & T. Nishida, pp. 275–92. Cambridge: Cambridge University Press.

Nishida, T. & Hiraiwa-Hasegawa, M. (1987). Chimpanzees and bonobos: cooperative relationships among males. In *Primate Societies*, ed. B. Smuts, D. Cheney, R. Seyfarth, R. Wrangham & T. Struhsaker, pp. 165–77. Chicago: Chicago University Press.

Pickford, M. & Senut, B. (2001). The geological and faunal context of Late Miocene hominid remains from Lukeino, Kenya. *Comptes Rendus de l'Académie des Sciences, Serie IIa*, 332, 145–52.

Rayner, R. J., Moon, B. P. & Masters, J. C. (1993). The Makapansgat australopithecine environment. *Journal of Human Evolution*, 24, 219–31.

Rendell, L. & Whitehead, H. (2001). Culture in whales and dolphins. *Behavioral and Brain Sciences*, 24, 309–82.

Ridley, M. (1995). *Evolution*. London: Blackwell

Senut, B., Pickford, M., Gommery, D., Mein, P., Cheboi, K. & Coppens, Y. (2001). First hominid from the Miocene (Lukeino Formation, Kenya). *Comptes Rendus de l'Académie des Sciences, Serie Iia*, 332, 137–144.

Stanford, C. (1998). *Chimpanzee and Red Colobus: The Ecology of Predator and Prey*. Cambridge, MA: Harvard University Press.

Stanford, C., Wallis, J., Mpongo, E. & Goodall, J. (1994). Hunting decisions in wild chimpanzees. *Behaviour*, 131, 1–20.

Sugiyama, Y. & Koman J. (1979). Tool-using and -making behavior in wild chimpanzees at Bossou, Guinea. *Primates*, 20, 513–24.

Suzuki, A. (1969). An ecological study of chimpanzees in a savanna woodland. *Primates*, 10, 103–48.

Takahata, Y., Ihobe, H. & Idani, G. (1996). Comparing copulations of chimpanzees and bonobos: do females exhibit proceptivity or receptivity? In *Great Ape Societies*, ed. W. C. McGrew, L. Marchant & T. Nishida, pp. 146–55. Cambridge: Cambridge University Press.

Tomasello, M. (1990). Cultural transmission in tool use and communicatory signaling of chimpanzees? In *Comparative Developmental Psychology of Language and Intelligence in Primates*, ed. S. Parker & K. Gibson, pp. 274–377. Cambridge: Cambridge University Press.

Tomasello, M. & Call, J. (1997). *Primate Cognition*. Oxford: Oxford University Press.

van Schaik, C. P. & van Hooff. J. (1983). On the ultimate causes of primate social systems. *Behaviour*, 85, 91–117.

van Schaik, C. P. & Knott, C. (2001). Geographic variation in tool use on *Neesia* fruits in Orangutans. *American Journal of Physical Anthropology*, 114, 331–42.

White, F. J. & Wrangham, R. (1988). Feeding competition and patch size in the chimpanzee species *Pan paniscus* and *Pan troglodytes*. *Behaviour*, 105(1–2), 148–64.

Whiten, A. & Boesch, C. (2001). The cultures of chimpanzees. *Scientific American*, 284(1), 48–55.

Whiten, A., Goodall, J., McGrew, W., Nishida, T., Reynolds, V.,

Sugiyama, Y., Tutin, C., Wrangham, R. & Boesch, C. (1999). Cultures in chimpanzee. *Nature*, 399, 682–5.

WoldeGabriel, G., Haile-Selassie, Y., Renne, P., Hart, W., Ambrose, S., Asfaw, B., Heiken, G. & White, T. (2001). Geology and palaeontology of the Late Miocene Middle Awash valley, Afar rift, Ethiopia. *Nature*, 412, 175–8.

Wrangham, R. (1980). An ecological model of female-bonded primates. *Behaviour*, 75, 262–300.

Wrangham, R. (1986). Ecology and social relationships in two species of chimpanzee. In *Ecological Aspects of Social Evolution: Birds and Mammals*, ed. D. I. Rubenstein & R. Wrangham, pp: 352–78. Princeton: Princeton University Press.

Wrangham, R. & Peterson, D. (1996). *Demonic Males: Apes and the Origins of Human Violence*. Boston: Houghton Mifflin Co.

Zihlmann, A. (1996). Reconstructions reconsidered: chimpanzee models and human evolution. In *Great Ape Societies*, ed. W. C. McGrew, L. Marchant & T. Nishida, pp. 293–304. Cambridge: Cambridge University Press.

Part I
Behavioural flexibility

Introduction

TETSURO MATSUZAWA

Chimpanzees and bonobos are our closest living relatives. However, many aspects of their lives in the natural environment – which ranges from the moist primary forest to dry open habitat – still elude us. This book, particularly the section on behavioural flexibility, outlines some of the most recent advances in our knowledge of the two species. People inhabiting the tropical forests of Africa have long coexisted with the two extant species of *Pan*, chimpanzees and bonobos, while sharing partly overlapping niches. Take, for example, the 20 or so chimpanzees living at Bossou, Guinea, West Africa, who have been studied continuously for more than two decades. More than 600 plant species (664, to be precise) have been identified as comprising the flora of Bossou (Sugiyama & Koman 1992). Chimpanzees utilise 200 of these for food and at least two species for medicinal use, while humans use 76 species for food and 81 as traditional medicine, in addition to those utilised in the construction of houses, furniture and for other purposes.

There are only a handful of tribes whose totem beliefs prohibit the hunting of chimpanzees. In others, hunters have been killing chimpanzees for meat and using parts of the body such as the skull for medicinal and animistic religious practices. This kind of coexistence may have been common practice throughout Africa for thousands of years.

The first chimpanzees known to have been brought to Europe came from Angola and were presented to the Prince of Orange in 1640 (Yerkes & Yerkes 1929). Over the next three centuries, white people continued to shoot chimpanzees, not for meat, but as specimens to be transported back to Europe and studied for their anatomy and morphology, all in the name of 'natural history'. Large numbers of chimpanzee skulls and bones are still kept in museums and universities. Demand for individual chimpanzees to be trapped and shipped to the West also came from various zoological gardens.

With the opportunity to observe them in captivity, people began to realise that chimpanzees were in many respects much like us: they could use tools, and even build tools to obtain food out of reach by joining sticks together. However, realising the necessity for studying the behavior of chimpanzees in their natural habitat is relatively recent. After some earlier attempts by others, Jane Goodall, a young British woman, working under the auspices of the visionary palaeontologist Louis Leakey (1903–72), arrived in Gombe, Tanzania, on 14 July 1960. At about the same time, Adriaan Kortlandt of the Netherlands was setting up a short field study in Congo, Kinji Imanishi (1902–92) and Junichiro Itani (1925–2001) and their students from Japan had just begun field surveys in Tanzania, and Vernon and Frances Reynolds arrived in Uganda for a 9-month study in the Budongo Forest.

Knowledge about wild chimpanzees' behavioural diversity has accumulated chiefly through research carried out at the six main sites where long-term projects have continued for more than two decades: Gombe, Mahale, Budongo, and Kibale for Eastern (*P. troglodytes schweinfurthii*) chimpanzees, and Bossou and Taï for Western (*P. t. verus*) chimpanzees. For an overview of the behavioural diversity of chimpanzees, three original books are especially useful: for Gombe see Goodall (1986), for Mahale see Nishida (1990), and for Taï see Boesch & Boesch-Achermann (2000).

Until recently, in contrast to chimpanzees, our knowledge of wild bonobos was scarce. Takayoshi Kano, one of Imanishi's last students, carried out an extensive survey in the Congo Basin in 1973. He rode his bicycle from village to village, questioning locals. He compiled information from 103 villages, and confirmed traces of bonobo feeding and nesting in 30. He finally settled at Wamba, in what was then Zaïre, to begin his direct observation of the last unkown ape in the wild (Kano 1992).

The last two decades of the 20th century saw a great deal of effort devoted to learning more about chimpanzees and bonobos in their natural habitat. Many people struggled with binoculars and field notes in the humid forests and dry, hot savannas of Africa. Guns were replaced, first by still cameras, then by video equipment. A genuine coexistence between humans and their evolutionary neighbours started to take shape.

Two previous international meetings have focused on chimpanzees and bonobos: one held in 1986 (Heltne & Marquardt 1989), another in 1991 (Wrangham *et al.* 1994), both supported by the Chicago Academy of Sciences. Two further meetings, in 1974 (Hamburg & McCown 1979) and in 1994 (McGrew *et al.* 1996), supported by Wenner-Gren Foundation, dealt with the Great Apes from a larger perspective. Most recently, the Max Plank Institute hosted a meeting in Seeon, Germany in the summer of 2000. This gathering provided a forum for the presentation and discussion of the most up-to-date information on diversity in chimpanzees and bonobos, especially in terms of their behavioural flexibility in adapting to a variety of ecological environments.

As one of the participants, I enjoyed all of the talks and was excited to learn of new discoveries from various field sites. Diane Doran and her coauthors (Chapter 1) proposed a multivariate approach to understanding the behavioural diversity of chimpanzees and bonobos. Concentrating on long-term research sites, they developed a list of 57 characters at each site, including, among others, habitat, demography, and social behaviour within and between groups. Results showed that phylogeny plays a significant role in explaining behavioural diversity in *Pan*. Overall, more closely related groups tend to be more similar in behaviour. Chimpanzees and bonobos are very distinct when overall behaviour is the basis of comparison, while within the chimpanzee lineage, the distinction is greatest between Eastern and Western subspecies. Although further detailed reports from various sites are definitely necessary, Doran *et al.* clearly show how analysing a wide range of information can lead to uncovering important aspects of the origins of behavioural diversity.

It was William McGrew, Caroline Tutin and Pamela Baldwin who first observed chimpanzees inhabiting a dry open area – one of the findings of a study that began in 1976 in Mount Assirik, in the Niokolo Koba National Park, Senegal. Kevin Hunt, with McGrew as coauthor (Chapter 2), reported the behavioural adaptation of the chimpanzees to the dry habitats in Semliki Wildlife Reserve, Uganda, as well as Mount Assirik. Semliki chimpanzees often dig drinking holes, or 'wells', in dry sandy riverbeds. Although this behaviour itself has been seen elsewhere, at Semliki the drinking hole or 'well' digging is not limited to locations where there is no standing water. Chimpanzees dig holes in the sand less than a metre from the clear, running river. Although the occurrence of the behaviour is negatively correlated with rainfall, the ultimate reason for its existence needs further investigation.

Annette Lanjouw (Chapter 3) surprised us with her observations in Tongo, ex-Zaïre, where the chimpanzees' practice of eating tubers comprises an adaptation that has not been reported for any other population. Tuber collection occurs only during the dry season, when other sources of water are no longer available inside the forest. The chimpanzees dig into the soil, at times inserting their entire arms into the holes. It takes several minutes (up to 15) to dig up the tubers. They then chew on the unearthed tubers, sucking out all the moisture. Video records clearly show that the chimpanzees even occasionally 'carry' the water away, stored in the tuber just like in a water-bottle.

Information about bonobos has come mainly from two research sites, Wamba and Lomako. The two populations, separated by only about 175 km, are both located within uniformly hot, wet and flat ground sheltered by closed canopy with moist, evergreen, lowland forest vegetation. In contrast to these, Jo Myers Thompson (Chapter 4) reported on the Lukuru Wildlife Research Project, where, far from the other two sites, bonobos live in dry forest/savanna mosaic habitat at the southern periphery of their current species distribution. This report revised our existing view of the two species, and convincingly demonstrated that bonobos adapt to the dry open habitat as well as chimpanzees.

Since the uncovering in the 1990s of evidence of the behavioural diversity of chimpanzees, 'culture' has emerged as a hot topic of discussion. A review by Yamakoshi (2001) listed 54 cases from 14 study sites of reported tool use in a feeding context. However, the regional differences in chimpanzee behavioural repertoire are not – and should not be – limited to feeding contexts. Michio Nakamura (Chapter 5) reported on his study of Grooming-Hand-Clasp (GHC) behavior as a greeting among Mahale chimpanzees. In contrast to feeding, there is no explicit reason for the existence of diversity in greeting behaviors – why, for instance, do people shake hands in the West and bow in Japan? The origins of behavioural diversity in social interaction are not clear. Nakamura's detailed report on GHC behaviour was based on information accumulated over 36 years of research in Mahale.

In summary, the following section on behavioural flexibility contains much useful information about chimpanzees and bonobos in the wild. Readers will be impressed by the behavioural flexibility of *Pan* in adapting to various ecological environments. Thus, studies of *Pan* illuminate the evolutionary process in hominids who extended their niche from the moist forest to dry open habitats. Ironically, by doing so, humans have emerged as a species so overpowering that they ultimately threaten the very existence of *Pan*.

REFERENCES

Boesch, C. & Boesch-Achermann, H. (2000). *The Chimpanzees of the Taï Forest: Behavioural Ecology and Evolution*. Oxford: Oxford University Press.

Goodall, J. (1986). *The Chimpanzees of Gombe: Patterns of Behaviour.* Cambridge, MA: The Belknap Press of Harvard University Press.

Heltne, P. & Marquardt, L. (1989). *Understanding Chimpanzees.* Cambridge, MA: Harvard University Press.

Hamburg, D. & McCown, E. (ed.) (1979). *The Great Apes*. Menlo Park: Benjamin/Cummings.

Kano, T. (1992). *The Last Ape: Pygmy Chimpanzee Behavior and Ecology*. Stanford, CA: Stanford University Press.

McGrew, W., Marchant, L. & Nishida, T. (1996). *Great Ape Societies*. Cambridge: Cambridge University Press.

Nishida, T. (1990). *The Chimpanzees of the Mahale Mountains: Sexual and Life History Strategies*. Tokyo: University of Tokyo Press.

Sugiyama, Y. & Koman, J. (1992). The flora of Bossou: its utilization by chimpanzees and humans. *African Study Monographs*, 13 (3), 127–69.

Wrangham, R., McGrew, W., de Waal, F. & Heltne, P. (1994). *Chimpanzee Cultures*. Cambridge, MA: Harvard University Press.

Yamakoshi, G. (2001). Ecology of tool use in wild chimpanzees: toward reconstruction of early hominid evolution. In *Primate Origins of Human Cognition and Behavior*, ed. T. Matsuzawa, pp. 557–74. Tokyo: Springer.

Yerkes, R. & Yerkes, A. (1929). *The Great Apes: A Study of Anthropoid Life*. New Haven, CT: Yale University.

1 • Multivariate and phylogenetic approaches to understanding chimpanzee and bonobo behavioral diversity

DIANE M. DORAN, WILLIAM L. JUNGERS, YUKIMARU SUGIYAMA, JOHN G. FLEAGLE & CHRISTOPHER P. HEESY

INTRODUCTION

Primates exhibit considerable diversity in their social systems (Smuts *et al.* 1987), a phenomenon that is thought to have evolved through the interaction of many factors. These include: (1) ecological variables, particularly predation pressure and the abundance and distribution of food (Alexander 1974; Wrangham 1979, 1980, 1987; van Schaik 1983, 1989, 1996; Sterck *et al.* 1997); (2) social factors, primarily sexual selection and the potential risk of infanticide (Wrangham 1979; Watts 1989; van Schaik 1996); (3) demographic and life history variables (DeRousseau 1990; Ross 1998); and (4) phylogenetic constraints (Wilson 1975; DiFiore & Rendall 1994). Generally, tests of models of the effect of these variables on behavior have been made through broad comparisons of many taxa, usually across genera (Wrangham 1980; van Schaik 1989; DiFiore & Rendall 1994; Sterck *et al.* 1997). There have been fewer attempts to consider the influence of these factors on variability in social organization within and between closely related taxa, largely as a result of a dearth of species for which such data are available (but see Mitchell *et al.* 1991; Koenig *et al.* 1998; Boinski 1999; Barton 2000). In addition, most tests have focused intensively on a single class of traits, and their proposed influence on behavior (e.g. the influence of ecology on behavior, van Schaik 1989; but see Nunn & van Schaik 2000), rather than the role of all proposed factors. To date, no study has quantitatively examined the combined influence of ecology, habitat, demography, and phylogeny on behavior.

Chimpanzees (*Pan troglodytes*) and bonobos (*Pan paniscus*) provide a unique opportunity to address the variability of social behavior in relation to these factors within closely related taxa because they have been studied at several different sites for long (greater than 15 years) periods of time (e.g. chimpanzees: Goodall 1986; Nishida 1990b; Sugiyama 1999; Boesch & Boesch-Achermann 2000; bonobos: Malenky 1990; Kano 1992; White 1996b; Fruth & Hohmann 1999). Currently, authorities recognize one species of bonobo, *Pan paniscus* (which is restricted in its distribution to the Democratic Republic of Congo) and three geographically distinct subspecies of chimpanzees, including *P. troglodytes verus* in West Africa, *P. t. troglodytes* in Central Africa, and *P. t. schweinfurthii* in East Africa (Fleagle 1999; but see Gonder *et al.* 1997 and Gagneux *et al.* 1999 for discussion of a potential fourth subspecies of chimpanzees, *P. t. vellerosus*, in Nigeria and Cameroon). Results from mitochondrial DNA studies have been used to estimate dates of divergence of chimpanzees from bonobos at approximately 2.5 million years ago (MYA), and between western (*P. t. verus* and *P. t. vellerosus*) and eastern/central chimpanzees (*P. t. schweinfurthii* and *P. t. troglodytes*) at 1.6 MYA (Morin *et al.* 1994; Gagneux *et al.* 1999).

There are currently eight habituated communities of chimpanzees (two communities each at Taï and Kibale, one each at Gombe, Mahale Mountains, Budongo, and Bossou), making chimpanzees the best studied of all nonhuman primates (Reynolds & Reynolds 1965; Sugiyama & Koman 1979; Goodall 1986; Nishida 1990a; Boesch 1996a; Wrangham *et al.* 1996; reviewed by McGrew 1992). Although chimpanzees have many features in common across sites, such as a primarily frugivorous diet and fluid fission–fusion, male-bonded social organization (but see Boesch 1996b), there is also considerable intersite diversity in chimpanzee behavior (reviewed in Wrangham *et al.* 1994). This intersite variation includes differences in both foraging strategies (including hunting, nut cracking, insect eating, and types of fallback resources) and social behavior (including party sizes, type and extent of association, the degree of seasonal influences on sociality, frequency of female transfer, and possibly male–female affiliation patterns). The most striking differences in behavior are between western (Taï and Bossou) and eastern (Gombe, Mahale, and Kibale) chimpanzees. Taï chimpanzees have been argued to be bisexually-bonded, rather than male-bonded (Boesch 1996b; but see Doran 1997), and to differ in their hunting strategies, relying on more cooperation than eastern chimpanzees (Boesch 1994).

Bonobos are less well-studied than chimpanzees, with

only two long-term study sites, Wamba and Lomako. Like chimpanzees, bonobos are highly frugivorous and are characterized by a fission–fusion society with female dispersal. However, they are reported to differ from chimpanzees in the nature of social relations within and between the sexes (with more frequent male–female and female–female association), in their reduced levels of aggression within and between communities, and in their more frequent and varied sexuality (e.g. de Waal 1989; Wrangham & Peterson 1996; Nishida 1997). Intersite differences in bonobo behavior have also been reported (White 1992; White 1996b).

Until recently, chimpanzees and bonobos have been viewed as morphologically (e.g. Susman 1984), genetically (e.g. Ruvolo *et al.* 1994), and behaviorally (e.g. de Waal 1989; Wrangham & Peterson 1996; Nishida 1997) distinct species. During the last decade, the number of long-term chimpanzee studies has increased, and researchers have emphasized the diversity in chimpanzee behavior across sites. One result of this is that the dichotomy between chimpanzee and bonobo behavior has been challenged (Boesch 1996b; Fruth 1998; McGrew 1998; Stanford 1998). Stanford (1998) suggested that 'the dichotomy currently drawn between the social systems of chimpanzees and bonobos may not accord well with field data' (p. 406). In a commentary following his article, several authors support his position that many of the previously perceived distinctions between the two species were a result of unequal sampling (Fruth 1998; McGrew 1998); others disagree and see clear behavioral differences (de Waal 1998; Kano 1998; Parish 1998).

One reason for these disparate opinions is that comparative studies have not assessed overall similarity and dissimilarity between species, but rather have focused on one or, at most, a few aspect(s) of behavior (e.g. reproductive behavior: Takahata *et al.* 1996; hunting: Boesch 1991; Stanford *et al.* 1994; Mitani & Watts 1999; grouping patterns: Chapman *et al.* 1994; Sakura 1994; Boesch, 1996b; Doran, 1997; cultural diversity: McGrew 1992; Whiten *et al.* 1999). To date, no study has quantified the behavioral variation in chimpanzees and bonobos based on a wide range of traits from many sites. Furthermore, although several hypotheses have been offered to explain the evolution of specific behavioral traits, such as increased male–female affiliation as a female counter strategy to reduce the risk of infanticide (van Schaik 1996), or the role of herb consumption in bonobo sociality (Wrangham 1986a), few studies have specifically examined whether proposed behaviors and causal factors vary in a predictable manner across taxa.

The overall objectives of this study are to examine the range of variation in behavior within and between chimpanzees and bonobos, and to identify the factors associated with this variability. We did this through four steps. First, we constructed a data matrix of 82 characters, chosen specifically to represent the key components of *Pan* social structure and behavior and the factors (ecological, habitat, and demography) considered potentially important in the evolution of sociality. Second, we used multivariate analyses (Sneath & Sokal 1973; Flury & Riedwyl 1988) to examine overall behavioral similarity within and between species of *Pan*, and to identify which behavioral characters are important in distinguishing taxa. Third, we used multivariate analyses to consider the extent to which differences in habitat, ecology, demography, and phylogeny were associated with the observed patterns of behavior in African apes. Finally, we used phylogenetic analyses (Nunn & van Schaik 2000; Borgerhoff *et al.* 2001) to place the similarities and differences among the study populations in an explicitly evolutionary context, by reconstructing the evolution of behavioral, ecological, and demographic characters onto the accepted phylogeny of *Pan*. We tested whether changes in specific factors, argued to be important in the evolutionary history of *Pan*, such as (1) increased herbivory or (2) seasonality of rainfall, are associated with predictable behavioral differences, such as (1) more stable groups or (2) altered patterns of association. This combination of multivariate and phylogenetic analysis enabled us to formulate new hypotheses based on the identification of novel patterns of association of traits.

METHODS

Data matrix

A data matrix of 90 variables for six distinct chimpanzee and bonobo studies, referred to here as taxa (sensu stricto 'operational taxonomic unit') was compiled primarily from the published literature, with additional original data contributed by one coauthor (YS). Characters were chosen to describe key components of social structure and behavior as well as those factors considered potentially important in the evolution of *Pan*. Invariant traits were culled, resulting in a reduced data set of 82 variables (data and complete list of references are available in an electronic appendix[1]).

Variables

The 82 variables, which include both continuous and coded data, can be divided into the following four data subsets:

I. Behavior Subset (n = 57)

This describes *Pan* social structure and behavior and includes the following five subsets. *A. Social Behavior Within Communities* documents male–female, female–female and male–male social relationships. *B. Social Behavior Between Communities* documents territoriality, the nature of interactions between communities, and the patterns of dispersal between them. *C. Reproductive Strategies* documents mating strategies and behavior, sexual selection, and infanticidal behavior. *D. Social Structure* assesses fission-fusion nature of the community by describing party size and composition, the extent of male–female association, and time spent as solitary individual. *E. Cultural Behavior* is limited to cultural behaviors that were well studied at several sites (so that absence of a behavior cannot be ascribed to differences between observers, or in the lengths of studies) and, when present, are common among the majority of individuals in a community (or some clear and predictable subset). Since data describing the frequencies of occurrence (both within and between communities) of many recently described behaviors (Whiten *et al.* 1999) are not currently available, these characters were not included in this study.

A. Social Behavior Within Communities (*n* = 20)
Adult patterns of grooming: Who grooms most frequently?

1. Grooming dyads (0 = male/male; 1 = male/female; 2 = female/female)

Adult patterns of association (0 = absent; 1 = rare; 2 = common): What is the degree of

2. Male–male association
3. Male–anestrous female association
4. Female–female association

Dominance hierarchies (0 = none; 1 = linear; 2 = high, middle and low rank detectable):

5. Female dominance hierarchy
6. Male dominance hierarchy
7. Male–female dominance? (0 = no dominance; 1 = males dominant to females; 2 = females dominant to males)

Cooperation (0 = never; 1 = rare; 2 = common):

8. Male–male coalitions for home range defense
9. Male–male coalitions for rank acquisition
10. Male–male coalitions in committing infanticide

Food sharing:

11. Food most commonly shared among adults (0 = meat; 1 = fruit)

Percentage of sharing among adults when:

12. Females share food with males
13. Females share food with females
14. Males share food with males
15. Males share food with females

Miscellaneous

16. Presence of genital–genital (G–G) rubbing (0 = absent; 1 = rare; 2 = common)
17. Presence of rump contact (0 = absent; 1 = rare; 2 = common)
18. Female response to immigrants (0 = welcome; 1 = neutral; 2 = aggressive)
19. Immigrants associate with (0 = males; 1 = females)
20. Percentage of infanticide victims who are cannibalized

B. Social Behavior Between Communities (*n* = 8)

21. Territorial (0 = no; 1 = yes)
22. Observed extra-group mating (0 = never; 1 = rare; 2 = common)
23. Degree of female transfer from natal community (0 = absent; 1 = <25%; 2 = 25–50%; 3 = 50–75%; 4 = >75%)
24. Degree of male transfer from natal community (0 = absent; 1 = <25%; 2 = 25–50%; 3 = 50–75%; 4 = >75%)

During inter-group encounters, there is:

25. Peaceful intermingling (0 = absent; 1 = rare; 2 = common)
26. Mating between members of adjacent communities (0 = absent; 1 = rare; 2 = common)
27. Female G–G rubbing (0 = absent; 1 = rare; 2 = common)
28. Most common interaction during inter-group encounters (0 = very aggressive; 1 = somewhat aggressive; 2 = aggressive leading into peaceful; 3 = nonaggressive)

C. Reproductive Strategies (*n* = 14)

29. Percentage of copulations that occur in maximal swelling
30. Mean length of postpartum amenorrhea (in months)

Percentage of matings that:

31. Are opportunistic
32. Occur in consortship
33. Are possessive
34. Have a dorsal–ventral mating position
35. Have a ventral–ventral mating position
36. Are interrupted by other males
37. Are initiated by males
38. Are obtained by the alpha male

Infanticide:

39. Occurs (0 = absent; 1 = present)
40. Percentage of infanticides committed by males
41. Percentage of infanticidal events occurring within a community
42. Percentage of infanticide victims who are male

D. Social structure (*n* = 9)

Percentage of parties that are:

43. Mixed sex and age classes
44. All-male
45. Mothers and dependent offspring only
46. Adult males and females with no offspring present
47. Solitary individuals
48. Mean party size
49. What percentage of time do adult females (with dependent offspring) spend alone
50. Size of female core area relative to that of males (in percent of male home range)
51. Is there a sex difference in day range? (0 = no; 1 = males travel farther)

E. Cultural Behavior (*n* = 6)

Tool use:

52. Occurrence of nut cracking (0 = absent; 1 = present)
53. Occurrence of ant or termite fishing (0 = absent; 1 = present)
54. Use of tools for food acquisition (0 = absent; 1 = present).
55. Occurrence of hunting or mammal eating (0 = absent or very rare (less than 1 per year); 1 = occasional; 2 = common)
56. Percentage of mammalian prey that are red colobus monkeys (0 = >50% ; 1 = 25–50%; 2 = <25%)
57. Individuals who obtain prey most frequently (0 = lone individuals; 1 = group of males).

II. Ecology Subset (*n* = 8)

Resource density and distribution are hypothesized as influencing female competitive regimes and resultant social behavior (e.g. van Schaik 1989). Increased reliance on herbs (with resulting decreased competition) has been argued to be a key factor in the evolution of sociality of bonobos (Wrangham 1986a). Differences in ripe fruit availability have been suggested as playing an important role in the relative sociality of female chimpanzees across sites (reviewed in Doran 1997). Since measures of resource availability and seasonal variation in diet (including fallback resources use) are not available from all sites, we use diet as a proxy. Ecological variables represent diet, ranging behavior, and home range overlap, and include:

Percentage of time spent feeding on:

58. Fruits
59. Herbs and leaves
60. Insects
61. Nuts

62. The average percentage of time spent feeding per day
63. Home range size (km^2)
64. Amount of home range overlap (0 = absent; 1 = <25%; 2 = 25–50%; 3 = 50–75%)
65. Average day range (m).

III. Habitat Subset (*n* = 9)

These variables represent the physical characteristics of the environment, as well as relative predation and competition risk at the study site.

66. Elevation (m)
67. Average annual rainfall (mm)
68. Seasonal variation in rainfall (average number of months per year in which rainfall is less than 50 mm)
69. Mean minimum temperature (degrees Celsius)
70. Mean maximum temperature (degrees Celsius)
71. Degree of variation in mean monthly temperature (variation in monthly mean maximum or minimum temperature in 1 year)
72. Latitude (absolute number of degrees north or south of equator)
73. Number of sympatric anthropoid species present at site
74. Number of potential predators present at site.

IV. Demography Subset ($n = 8$)
These variables represent the demographic and life history variables at a study site.

75. Total community (unit-group) size
76. Adult sex ratio (female/males) in the community
77. Sex ratio (females/males) at birth
78. Mean length of maximum swelling (in months)
79. Percentage of estrus cycle spent in maximum swelling
80. Age at first pregnancy (years)
81. Average interbirth interval (months)
82. Number of adult males in community (unit-group).

Taxa

The taxa include two samples each of *P. t. schweinfurthii* (Gombe and Mahale), *P. t. verus* (Bossou and Taï), and *P. paniscus* (Lomako and Wamba), and are referred to by study-site name. Sites were selected because they provided the only existing studies from which data were available for a minimum of 20% of the 82 variables. We are not suggesting that these six sites represent the entire potential range of chimpanzee and/or bonobo behavior. Regrettably, no long-term study of habituated *P. t. troglodytes* is available for comparison.

At a few study sites, more than one community has been studied, although none had complete data for more than one community. Therefore, all variables were based on a single community when possible (Gombe, Kasakela; Mahale, M ; Taï, North; Lomako, Eyengo; and Wamba, E 1). Since many of these studies have been ongoing for 20 years or more, and since behavior and demographic variables can alter through time, demographic data used in each study were taken, when possible, from the same time period as the behavioral data. For example, the demographic data from Bossou were taken from the same time period as Sakura's (1994) behavioral data.

The Bossou chimpanzee community is considered 'unique' by many ape researchers because it is a small isolated community (with one to two adult males), which has been relatively stable in size and composition for over 20 years. Although the presence of so few males at Bossou is unique for data presented here, this demographic makeup is not unique among all currently known communities (e.g. only one to two adult males have been present in the Taï (North) community since 1995; Boesch & Boesch-Achermann 2000). Although data from Taï, Gombe, and Mahale are based on the presence of neighboring contiguous communities, the relative isolation of the Bossou community is also no longer unique to Bossou. The chimpanzees at Gombe have become increasingly isolated from other chimpanzee communities through time. Thus, Bossou provides an important case for evaluating how changing ecological, demographic, and environmental factors influence behavior in *Pan*.

Broad genetic sampling of chimpanzee mitochondrial DNA haplotypes across Africa has provided an understanding of the phylogenetic relationships of *Pan* subspecies included in this study (Bradley & Vigilant, Chapter 19). While there is no clear subdivision within chimpanzee subspecies, the two subspecies considered here (*P. t. verus* and *P. t. schweinfurthii*) form clear monophyletic clades, which cluster together to the exclusion of the bonobos. Therefore, all chimpanzees are more closely related to each other than they are to bonobos, and genetic similarity is greater among individuals within a chimpanzee subspecies than across subspecies.

Multivariate analysis

We use multivariate analysis to assess overall phenetic similarity, or overall similarity without any consideration of phylogeny. This method does not distinguish between features shared through inheritance from a common ancestor versus those that are acquired uniquely in a particular set of related taxa.

Distance statistics are used to summarize chimpanzee and bonobo affinities based on each of the four data sets (behavior, demography, ecology, and habitat). More specifically, average Euclidean distances (or average taxonomic distances, or ATDs) are calculated among taxa within each data set using data standardized to Z-scores for each of the 82 variables (Sneath & Sokal 1973; Reyment *et al.* 1984). This approach makes no parametric assumptions about homogeneity of dispersion matrices, normality, and so on, and some missing data can be accommodated without bias. Because raw variables are in very different scales and dimensionality, conversion of variables to Z-scores serves to weight them equally and render them commensurate and dimensionless (e.g. virtually all will take on values between -2.0 and $+2.0$). Tabular results of ATD matrices discussed in text are available in electronic appendices.

The information in the ATD matrices is then summarized and presented graphically via clustering and ordination. We used the UPGMA algorithm (unweighted pair group method with arithmetic mean) for clustering (Sneath

& Sokal 1973), and principal coordinates for ordination to reduce the information into two or three orthogonal axes of variation (Gower 1966; Rohlf 1972). Minimum spanning trees are superimposed on the ordination to help identify the group(s) most similar overall to another group(s); this also helps to disclose any distortion of the total ATD in the reduced dimensional space. We also examine the correlations between the original standardized data and their principal coordinates scores. This helps to identify especially influential variables associated with each axis of variation, and can thereby reveal 'contrast vectors' among groups in many cases (Corruccini 1978; Jungers 1988).

The ATD matrices, produced from each data subset, are then compared directly to each other via matrix correlations (or the standardized Mantel statistic) in order to discover predictable patterns of covariation (or the lack thereof). Significant correlations among such matrices were determined using a permutation approach (e.g. 5000 random permutations of one of the matrices in a pair-wise comparison; one-tailed probabilities are reported). The magnitude of the matrix correlation is less important than its level of significance. All calculations were performed using NTSYS-pc, version 1.80 (Rohlf 1993).

Phylogenetic analysis

The evolution of behavioral, ecological, and demographic characters was reconstructed onto the accepted phylogeny of the study populations sensu stricto Brooks & McLennan (1991; see also Wrangham 1986b; Ghiglieri 1987; Kappeler 1999), using MacClade 4.0 (Maddison & Maddison 2000), with the mountain gorilla as an outgroup. The theoretical and practical applicability of phylogenetic methods to the study of individual populations within a species is an intensively debated issue (see discussion in Borgerhoff *et al.* 2001). It is worth emphasizing that we did not use these data to generate a phylogeny for African apes, but rather mapped individual features onto an accepted molecular phylogeny. As noted below, this phylogeny is not, in fact, the most parsimonious resolution of this data set. However, our goal was not to generate a phylogeny of African apes based on behavioral characters. Rather it was to use phylogenetic methods to reconstruct the probable history of character evolution among a group of variously related populations of apes.

For this aspect of the study, continuous characters were recoded into a discontinuous format using visual gap coding. All characters were entered as unordered and unpolarized, that is we placed no limitation on how a character can change through time. We did not examine the habitat characteristics of the study sites for this part of the study. In addition, two characters from the behavior subset were eliminated because of extensive missing data (frequency of possessive mating and frequency of interrupted mating) and a third character (frequency of ventral–ventral mating) was eliminated because it was the reciprocal of another character and therefore contained no additional information. Using this tree with the data set mapped onto the accepted phylogeny of the populations studied, we calculated consistency indices (CI) and retention indices (RI) for the entire tree, for individual characters, and for subsets (or ensembles) of characters. These are two widely used measures of the extent to which the data indicate that there has been parallel evolution in the characters. The CI indicates how many changes in a character take place in a tree, although it does not indicate where changes occur in the tree. The RI indicates to what extent characters define nodes (for examples, see discussion in Kitching *et al.* 1998). We also examined the distribution of features throughout the tree, reconstructed ancestral states and character evolution using maximum parsimony and tabulated the changes reconstructed along each branch, using the 'almost all changes' options of MacClade 4 (Maddison & Maddison 2000).

RESULTS

Multivariate analysis

BEHAVIORAL AFFINITIES

Overall behavioral affinities

The average taxonomic distance for all taxa, based on the 57 behavioral variables, and summarized by UPGMA clustering (Figure 1.1(*a*)), indicates that chimpanzee and bonobo behavioral affinities are similar to those reflective of phylogeny. The phenogram displays two clusters, (1) bonobos (Wamba and Lomako) and (2) chimpanzees. However, within the chimpanzee cluster, the *P. t. schweinfurthii* taxa (Gombe and Mahale) cluster together, as would be expected on the basis of phylogeny, unlike the *P. t. verus* taxa (Taï and Bossou). Bossou is distant from all other common chimpanzees, and does not show special affinity to the other *P. t. verus* taxon (Taï), as would be predicted on the basis of phylogeny alone.

This distinction is even more apparent in the principal coordinates ordination of the same behavior distance matrix (Figure 1.1(*b*)). The first two principal coordinates axes

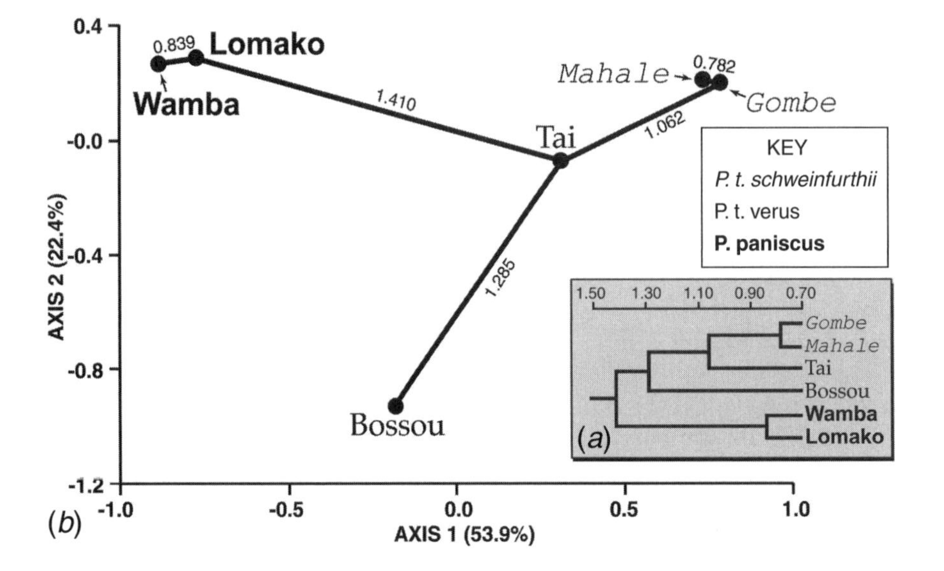

Fig. 1.1. Summary of average taxonomic distances generated from BEHAVIOR data subset ($n = 6$ taxa, 57 variables) using (*a*) UPGMA clustering and (*b*) principal coordinate analysis. In principal coordinate summary, taxa are joined to nearest neighbor by minimum spanning trees; average taxonomic distances between taxa are indicated. Contrast vector is itemized separately in Table 1.1.

account for 76% of the variation in the data, with the first axis alone accounting for more than half of total variation. Along the first axis there is a tight clustering of the two bonobo taxa (Wamba and Lomako), and an even tighter clustering of the two *P. t. schweinfurthii* taxa (Mahale and Gombe), with the *P. t. verus* taxa intermediate between them. Taï groups most closely with Mahale and Gombe; Bossou falls mid-way along the axis, although closest to Taï in overall behavioral similarity. Along the second axis, Bossou (and Taï to a much lesser extent) is the major outlier.

What variables are influential in driving the behavioral affinities?

Thirty-three of the original 57 behavioral variables are significantly correlated with the first axis, and are thus influential in distinguishing taxa along it, primarily serving to separate bonobos from chimpanzees (Table 1.1(*a*)). Twenty-nine of these 33 (significant) variables have complete or nearly complete data (data missing from 0–1 sites). The 29 variables include 'bonobo' traits, which occur commonly in bonobos, and are absent or rare in chimpanzees (strong male–female and female–female bonds, less disparity in male and female ranging behavior, less violent and more varied inter-group encounters), and 'chimpanzee' traits, which are common in chimpanzees and absent, or greatly reduced, in bonobos (strong male bonds, infanticide, territoriality, frequent hunting of monkeys, and tool use) (Table 1.1(a)).

Of these significant variables for which data is complete for *all* sites ($n = 21$), Bossou chimpanzees share (1) 38% with all chimpanzees to the exclusion of bonobos (including tool use, mating style, male dominance to females and longer postpartum amenorrhea), (2) 38% with bonobos to the exclusion of all other chimpanzees (including the absence of (a) hunting, (b) territoriality and (c) differentiated male–male relationships), and (3) 24% with Taï chimpanzees and bonobos, to the exclusion of *P. t. schweinfurthii* (including the absence of infanticide and more frequent association between males and females) (Table 1.1(a)).

Five variables are influential in distinguishing taxa along the second axis, and thus Bossou and, to a lesser extent, Taï, from all other taxa. These include nut cracking, a behavior shared exclusively by Taï and Bossou, as well as behaviors that are unique or exaggerated in frequency at Bossou compared to every other site, including male transfer and heightened male–female and female–female social relations (Table 1.1(b)).

Are the results from each subset of behavior similar?

A comparison of the ATD matrices from each of the five subsets of behavior (within-group, between-group, social

Table 1.1. *Contrast vector among variables for behavior subset*
(a) *Contrast vector among variables (Axis 1) for behavior subset*

Character: 'Bonobo' Traits	Load	Bossou	Character: 'Chimpanzee' Traits	Load	Bossou
I. No missing data			*I. No missing data*		
Lack of preference for red colobus	[a]−0.917	2	Degree of male–male association	[a]0.917	2
Fruit (versus meat) is most commonly shared food item	[a]−0.917	2	Frequency of hunting or mammal meat-eating	[a]0.917	2
Intergroup encounter – G–G rubbing common	[a]−0.880	1	Tendency for groups of males to hunt more frequently than solitary males	[a]0.917	2
Immigrants associate with females (versus males)	[a]−0.880	1	Territorial	[a]0.917	2
Intergroup encounter – mating common	[a]−0.872	1	Male–male coalition: home range defense	[a]0.917	2
Frequency of ventral–ventral mating	[b]−0.789	1	Tool use: ant or termite fish	[a]0.880	1
Degree of male–female association	[b]−0.806	3	Tool use: for food acquisition	[a]0.880	1
Degree of female–female association	[b]−0.806	3	Males dominant to females	[a]0.880	1
			Frequency of dorso-ventral mating	[b]0.789	1
			Occurrence of infanticide	[b]0.806	3
			Percentage of infanticides that are within group	[b]0.784	3
			Male–male coalition: for rank	[b]0.775	2
			Time spent alone by females	[b]0.762	3
II. Data missing from one site			*II. Data missing from one site*		
Nonaggressive intergroup encounters	[a]−0.973		Males share food with males	[a]0.876	
Male–male rump contact	[a]−0.833		Mean length of postpartum amenorrhea	[a]0.893	
Female core area size relative to male's	[a]−0.847		Percentage lone individual	[a]0.853	
Females share food with males	[b]0.765	1	Male–male cooperation for infanticide	[b]0.791	1

(b) *Contrast vector among variables (Axis 2) for behavior subset*

Character	Load	Bossou
I. No data missing		
Percentage of parties–mothers	[a]−0.890	Unique
Degree of male transfer	[a]−0.830	Unique
Tool use – nut crack	[a]−0.812	Shares with Taï
II. Data Missing From one Site		
Males share food with females	[a]−0.904	
Adult grooming – who grooms who most?	[b]−0.804	

Notes:
Load is the correlation between original variable and the summary variable; $n = 6$ taxa; 4 degrees of freedom; [a] data significant at $p < 0.05$ (i.e. $r = \pm 0.81$); [b] marginally significant ($0.05 < p < 0.1$). Similarity of Bossou to other sites is indicated by: 1, similar to all other chimpanzee sites; 2, similar to bonobos and distinct from all other chimpanzee sites; and 3, similar to Taï chimpanzees and bonobos and distinct from *P. t. schweinfurthii*.

structure, sex, and culture) indicates that all matrices are highly correlated with each other, except the sex matrix, which is not significantly correlated with any matrix, and the social structure matrix, which is only significantly correlated with within-group and cultural matrices (data in the electronic appendix). UPGMA clustering (of ATD data) from each behavioral data subset (except sex) indicates a clear segregation of chimpanzees and bonobos. Results from the different data subsets vary primarily in the relationship of Bossou to other taxa.

Affinities based on reproductive strategies (sex) indicate a clustering of taxa that is independent of phylogeny, with Lomako, Bossou, Taï and to a lesser extent, Wamba, clustering together versus Gombe and Mahale chimpanzees. As a result of greater than average missing data (20%), these results should be viewed with caution. The presence or absence of infanticide drives clustering patterns in this subset.

DEMOGRAPHIC AFFINITIES

The average taxonomic distances for all taxa, summarized by UPGMA clustering (Figure 1.2(*a*)), indicate that chimpanzee and bonobo demographic affinities show little similarity to either the known phylogenetic relationship or behavioral affinities of the taxa. Mahale and Taï cluster together versus all other taxa. Bonobos cluster together, but they are also similar to Gombe chimpanzees. Demographically, Bossou does not show strong overall similarity to any other taxa.

Principal coordinate analysis illustrates this more clearly: there are three distinct clusters on the basis of demographic variables (Figure 1.2(*b*)). The first two principal coordinate axes account for 86.2% of the variation in the data, with the first axis accounting for 59.4%. Traits that distinguish taxa along the first axis and, thus, Mahale and Taï from other taxa, include larger community size, higher adult female/male sex ratio, and longer interbirth interval. Traits that are correlated with the second axis and distinguish Bossou from other taxa include fewer adult males in the community and a decreased age of first pregnancy.

BEHAVIORAL ECOLOGY AFFINITIES

The average taxonomic distances for all taxa, summarized by UPGMA clustering (Figure 1.3(*a*)), indicate that chimpanzee and bonobo affinities based on diet and ranging behavior (ecology) show almost no similarity to the known phylogenetic relationships of the taxa. Neither the bonobo taxa nor any chimpanzee subspecies cluster together. There is much less variance in ATDs between taxa in this data subset than in any other. In the principal coordinate ordination, taxa space themselves evenly, with 61.1% of the total variation explained nearly equally by the first two axes (Figure 1.3(*b*)). The percentage of fruit in the diet is a significant variable in the separation of taxa along the first axis; bonobos and Taï chimpanzees have a tendency towards greater frugivory than the other taxa. The amount of herbs and leaves in the diet does not distinguish taxa.

HABITAT AFFINITIES

The average taxonomic distances for all taxa, summarized by UPGMA clustering (Figure 1.4(*a*)), indicate that chimpanzee and bonobo affinities based solely on physical characteristics of the environment (habitat) show a pattern suggestive of phylogeny, but with two obvious exceptions. Contrary to phylogenetic predictions, (1) Taï clusters with the bonobos and (2) Bossou is distant from all other taxa. The first two principal coordinates axes account for 77.4% of the variation in the data. Bossou separates from the other taxa on the basis of its greater rainfall and higher average maximum temperature; bonobos and Taï, to a lesser extent, are distinguished from (other) chimpanzees primarily by reduced seasonality of rainfall in their habitats (Figure 1.4(*b*)).

COMPARISONS OF RESULTS FROM THE FOUR DATA SETS

Having considered chimpanzee and bonobo affinities based on four independent data sets, we next ask which, if any, data sets give similar results. The behavior and habitat ATD matrices are highly correlated (electronic appendix). This indicates that patterns of association determined from 57 variables that summarize social behavior and nine variables describing the physical habitat are similar.

AFFINITIES BASED ON TOTAL DATA SET

When the entire data set of 82 variables is combined, results are similar to those produced from the behavior data set (Figure 1.5(*a*) and (*b*)). This is hardly surprising given that variables are weighted equally and behavioral variables make up the vast majority of the total data set. Figure 1.5(*b*) lists the variables that are significant, in addition to behavior variables discussed previously (Table 1.1), and those that are significantly correlated with the first axis, and thus serve to distinguish chimpanzees and bonobos. The variable most highly correlated to the first axis (including all behavioral variables), is the average number of months with less than 50 mm of rainfall, with the bonobo locations showing

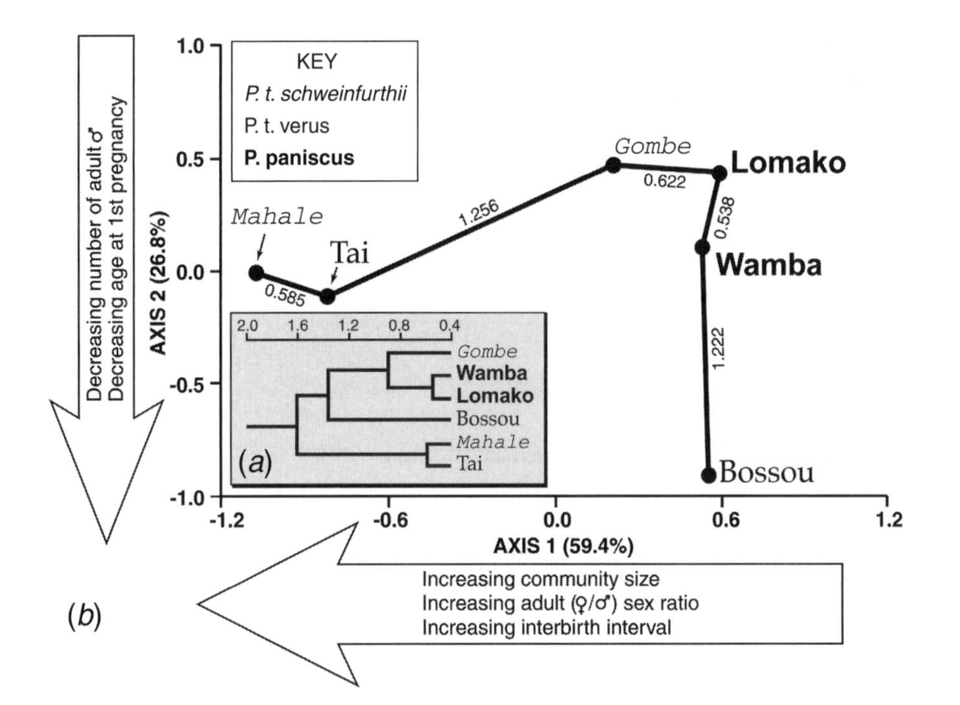

Fig. 1.2. Summary of average taxonomic distances generated from DEMOGRAPHY/LIFE HISTORY data subset ($n = 6$ taxa, 8 variables) using (*a*) UPGMA clustering and (*b*) principal coordinate analysis. In principal coordinate summary, taxa are joined to nearest neighbor by minimum spanning trees; average taxonomic distances between taxa are indicated.

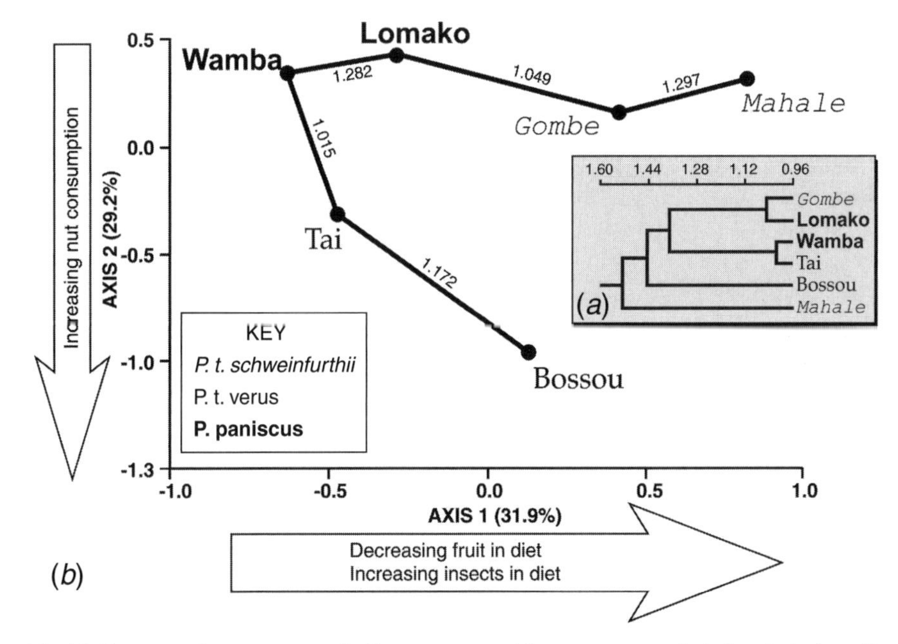

Fig. 1.3. Summary of average taxonomic distances generated from ECOLOGY (diet and ranging behavior) data subset ($n = 6$ taxa, 8 variables) using (*a*) UPGMA clustering and (*b*) principal coordinate analysis. In principal coordinate summary, taxa are joined to nearest neighbor by minimum spanning trees; average taxonomic distances between taxa are indicated.

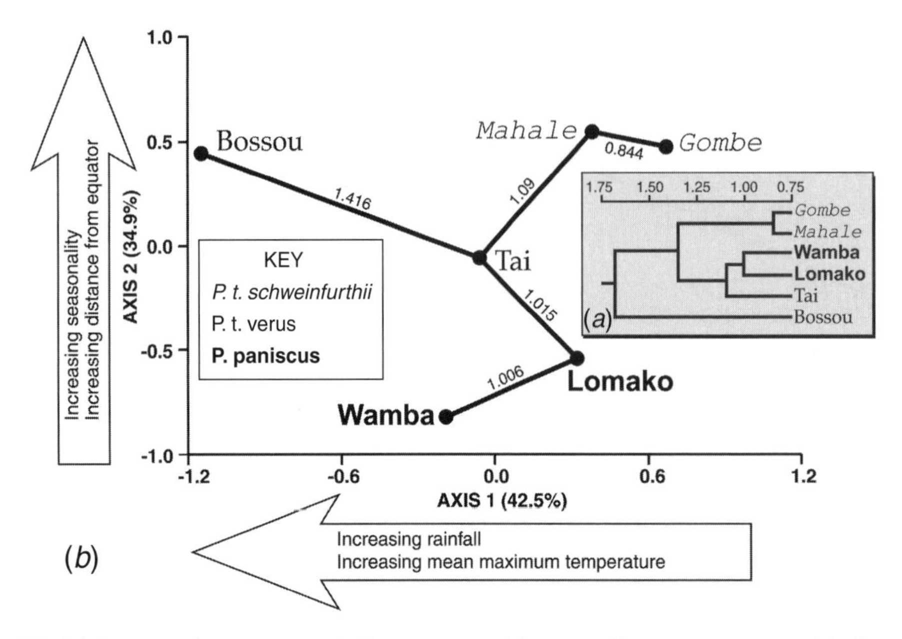

Fig. 1.4. Summary of average taxonomic distances generated from HABITAT data subset ($n = 6$ taxa, 9 variables) using (*a*) UPGMA clustering and (*b*) principal coordinate analysis. In principal coordinate summary, taxa are joined to nearest neighbor by minimum spanning trees; average taxonomic distances between taxa are indicated.

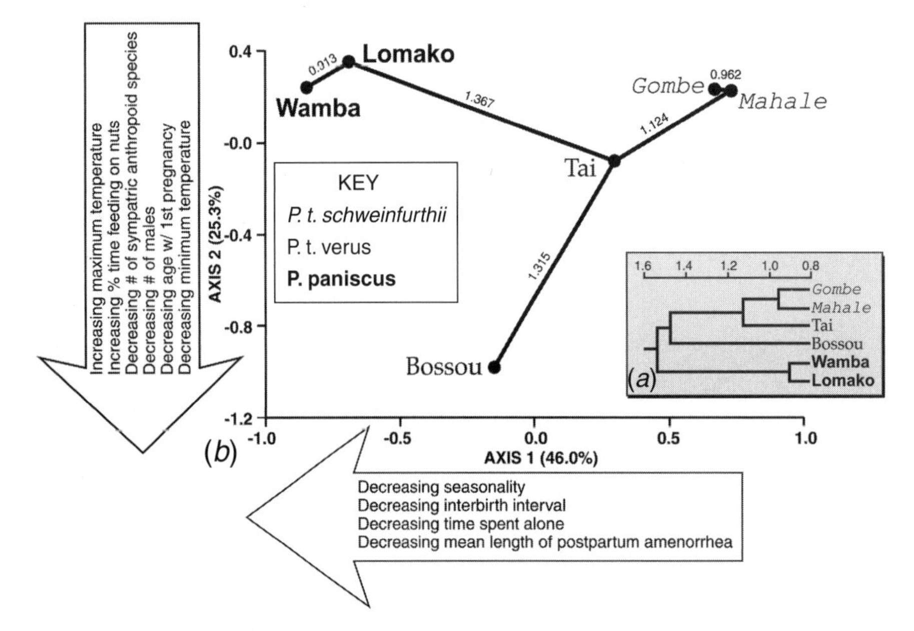

Fig. 1.5. Summary of average taxonomic distances generated from ENTIRE chimpanzee and bonobo data subset ($n = 6$ taxa, 82 variables) using (*a*) UPGMA clustering and (*b*) principal coordinate analysis. In principal coordinate summary, taxa are joined to nearest neighbor by minimum spanning trees; average taxonomic distances between taxa are indicated. Note that the contrast vectors show only significant nonbehavioral variables; the behavioral variable loadings are essentially the same as those seen in Table 1.1.

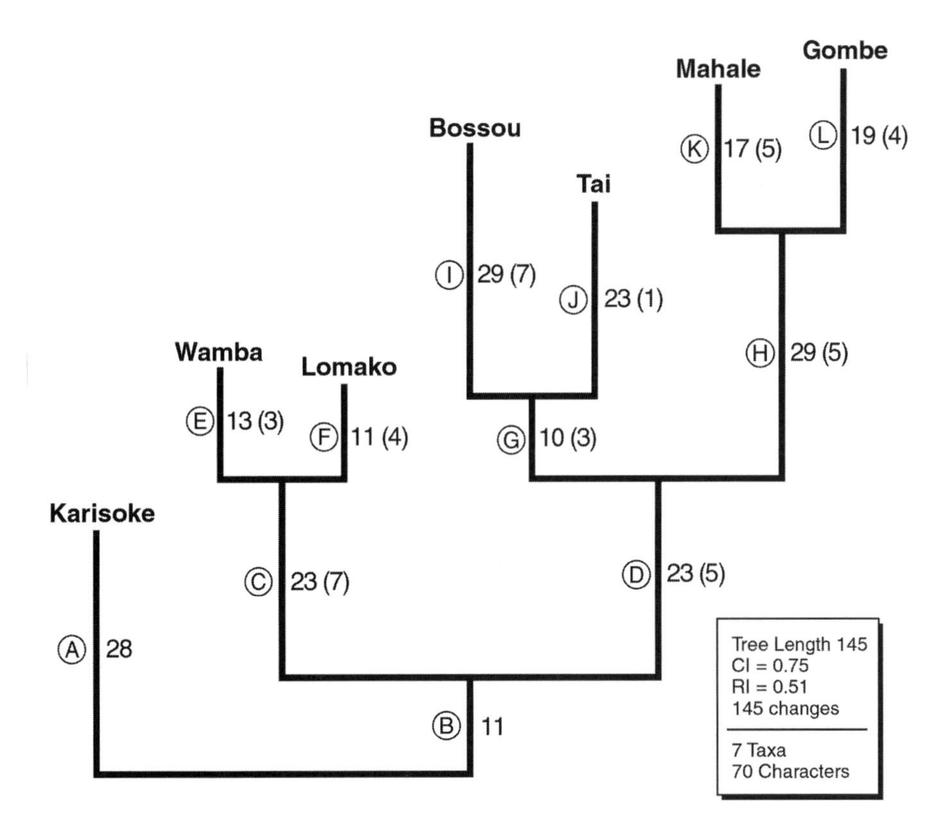

Fig. 1.6. Molecular phylogeny of study populations used in this study with the number of character changes along each branch indicated. The higher number indicates total changes (both ambiguous and unambiguous); the number in parentheses is the number of unambiguous changes (discussed in the text). The character changes along each of the lettered branches are listed in the electronic appendix.

reduced seasonality of rainfall. Additional variables that are significantly correlated include infanticidal behavior (absent in bonobos) and interbirth intervals (shorter in bonobos).

Phylogenetic analysis

The reconstruction of trait evolution accords well with and complements the results of the multivariate analyses by placing the similarities and differences among the study populations in an explicitly evolutionary perspective. Overall, these data (excluding the habitat features) mapped onto the chimpanzee phylogeny have a consistency index of 0.75 and a retention index of 0.51. The lower retention index reflects the fact than many of the features included in this analysis are characteristic of individual study populations and hence do not contribute to resolution of nodes.

Among the different subsets of the data, the large behavior subset has a consistency index of 0.76, similar to that for the overall tree, and a retention index of 0.58. The smaller number of ecological features has a slightly higher consistency index (0.82) but a substantially lower retention index (0.40), reflecting the fact that the ecological changes tend to be characteristic of individual sites or studies rather than groups of sites. The demographic and life history characters have a lower consistency index (0.62) and a very low retention index (0.11), reflecting the fact that the data on life history and demography largely record site-specific differences, and these features are rarely shared by related populations.

Figure 1.6 shows the number of character state changes occurring along each of the branches in the accepted phylogeny of the study populations. Because of the homoplasy (parallel evolution of similar features) in the data and the lack of additional outgroups, there were several equally parsimonious reconstructions of character change for many traits on this tree and, hence, many changes that cannot be reconstructed precisely using parsimony. The reconstructed changes are referred to as ambiguous changes when there are several possible reconstructions and thus the beginning and/or ending character states are impossible to reconstruct precisely, and as unambiguous changes when the character change along the branch node can be reconstructed precisely. In our tabulations, we have shown both the total number of (ambiguous + unambiguous) changes and the number of unambiguous changes along each branch. The overall pattern is the same for both sets of changes (for individual characters changing along each branch, see the electronic appendix).

The node uniting bonobos and chimpanzees is characterized by 11 character changes (Figure 1.6). These are, by necessity, all ambiguous as we did not root the tree or have any additional outgroup except the mountain gorillas. If the condition in the mountain gorillas is taken as the appropriate outgroup (for lack of any comparable data on any other gorilla taxon), then the ancestral chimpanzee lineage was characterized by several differences in ecology, demography, and behavior. These include a decrease in folivory, an increased day range, an increase in both the time and the percentage of the cycle of maximum sexual swelling, an increase in the age of first pregnancy, an increase in the frequency of female transfer, a decrease in the frequency of male transfer, an increase in the amount of mating initiated by males, changes in group cohesion, including an increase in time spent as lone individuals (rather than in a party), particularly for females, as well as an increase in the frequency of nursery parties (mothers and offspring to the exclusion of males). Greater knowledge of the ecological and behavioral features of lowland gorillas, a more concerted effort at identifying features distinguishing chimpanzees from other apes, and a consideration of the likely ancestral conditions for African apes would undoubtedly modify this list.

The individual branch leading to the bonobo populations and that uniting all of the chimpanzee groups are each well-supported. The lineage leading to bonobos (Wamba and Lomako; *Pan paniscus*) as a species is characterized by 23 ambiguous and seven unambiguous features. The unambiguous features are mostly related to intergroup encounters and reproduction, including decreased aggression, more frequent intergroup mating, occurrence of ventral–ventral mating and G–G rubbing, lack of male dominance to females, and by the tendency of immigrant females to associate with females rather than males. Each of the bonobo populations is characterized by a much smaller number of site-specific features.

The common node uniting western (Taï and Bossou; *Pan troglodytes verus*) and eastern populations (Gombe and Mahale; *Pan troglodytes schweinfurthii*) is supported by a total of 23 changes, 5 of which are unequivocal. The unambiguous changes uniting the different populations of *Pan troglodytes* include using tools to acquire food, termite fishing with tools, longer interbirth interval, decreased frequency of mixed-sex parties, and increased frequency of male–male coalitions.

The patterns of character evolution in the two subspecies of *Pan troglodytes* are strikingly different. The eastern chimpanzees, *Pan troglodytes schweinfurthii*, have many more shared features than site-specific characteristics, while the western chimpanzees, *Pan troglodytes verus*, have few shared features and many site-specific features. Thus, the common branch leading to the Gombe and Mahale populations is characterized by a total of 29 changes of which 5 are unambiguous. In contrast, the number of characters that are unique to each of these sites is much smaller – 17 for Mahale and 19 for Gombe. These findings accord well with the multivariate results that found that these two populations clustered tightly in all analyses, but were usually well-separated from the (western) chimpanzees. The unambiguous changes on the Mahale–Gombe branch include increased number of male parties, lone individuals and lone females, increased cooperative male infanticide and post-infanticidal cannibalism, and reduced female sharing.

Compared with the eastern (Mahale and Gombe) populations, the western chimpanzees from Bossou and Taï are striking for their lack of similarity. The node linking them is characterized by a total of only ten changes of which three are unambiguous (nuts in their diet, nut-related tool use, and higher degree of association between males and anestrus females).

In contrast, the Taï chimpanzees show 23 site-specific characteristics (one unambiguous) and the Bossou chimps have diverged more from their last common ancestor (with the Taï chimpanzees) than any population in the study, with 29 changes, of which 7 are unambiguous. The unambiguous change unique to the Taï population is a dramatic increase in the frequency of male parties. The Bossou population is characterized by a smaller home range, decreased age at first pregnancy, increased frequency of parties including mothers and offspring, and several features related to the presence of only one male, including greater frequency of alpha male mating, decreased number of male–male coalitions, increased male transfer, and increased male sharing with females.

The phylogenetic differences in the patterning of changes in eastern and western chimpanzees accords well with the results of the multivariate analyses, in which the geographically close Mahale and Gombe always clustered together and are distinct from the western chimpanzees, whereas the two western sites –Taï and Bossou – were as distant from one another as they were from the eastern chimpanzees. Indeed, the Bossou population was the most widely separated chimpanzee population in all of the multivariate analyses, and also has the greatest number of site-specific changes in the analysis of trait (character) evolution.

DISCUSSION

Distinctive features of African apes

CHIMPANZEE–BONOBO DICHOTOMY

A major goal of this study was to examine the extent of similarities and differences between bonobos and chimpanzees, an issue that has recently been brought to the forefront by Stanford's (1998) stimulating review of this topic. In the multivariate analyses, bonobos and chimpanzees segregate on the basis of overall behavior (based on 57 variables), as well as in nearly every subset of behavioral data, including social relationships and social structure within the community, intercommunity relationships and in culture related to tool use and hunting. Traits that distinguish bonobos from chimpanzees in this study include: (1) greater female sociality, as indicated by greater male–female and female–female association, reduced tendency for females to be found alone, and less disparity in male and female ranging behavior; (2) absence of male dominance and a greater tendency for females to possess and be responsible for distribution of resources; (3) more varied intergroup encounters, as indicated by G–G rubbing and mating between communities during intergroup encounters; and (4) different mechanisms by which female immigrants transfer into and become established in a new community, as indicated by their associating with other females versus males. Traits that are common in chimpanzees and absent, or greatly reduced, in bonobos include: (1) strong male–male bonds, as indicated by a high degree of male–male association and the frequent formation of male–male coalitions to establish and maintain rank, defend territories, and while engaging in infanticidal activities; (2) male dominance to females; (3) decreased female sociality; (4) territoriality and aggressive defense of home range; (5) frequent hunting of monkeys; and (6) tool use for food acquisition.

The large number of character changes that are reconstructed along both the branch leading to bonobos and that leading to chimpanzees in the phylogenetic analysis are in accord with the results of the multivariate analyses. These two species are each distinguished by numerous character changes; moreover, the unique evolutionary changes reconstructed in the evolution of the bonobo ancestor and in the chimpanzee ancestor accord with those identified in other studies (Wrangham 1986b; Ghiglieri 1987; White 1996a).

These results are noteworthy in reinforcing the distinction between the two species, at least on the basis of currently available data. In and of itself, this is a not insignificant point given Stanford's (1998) recent claim. Yet in addition, this study provides the first quantitative analysis of similarities and differences between the two species based on a wide range of variables.

CHIMPANZEES: DICHOTOMY BETWEEN EASTERN AND WESTERN POPULATIONS

Results from multivariate analysis indicate similarity in behavior within the eastern chimpanzees (*Pan troglodytes schweinfurthii*) at Mahale and Gombe. However, the western (*Pan troglodytes verus*) chimpanzees at Taï and Bossou are widely separated from one another. Ecologically, behaviorally, and in terms of demography and habitat, the two sites are often more similar to another ape population (chimpanzee, bonobo, or gorilla) than to each other (Figure 1.7). This same pattern is evident in the phylogenetic analyses. The branch leading to the eastern chimpanzees is one of the longest in the tree, reflecting the fact that Mahale and Gombe share many unique features relative to other chimpanzees. In contrast, the two western populations are linked by a very short branch and show many more changes characteristic of the individual sites.

Geographic distances between Mahale and Gombe (200 km) and Taï and Bossou (300 km) are similar (Boesch & Boesch-Achermann 2000), and therefore are unlikely to explain the behavioral differences, particularly since dispersal (as evidenced by maternally transmitted genotypes) has been detected over distances of 900 km in *P. t. verus* (Morin *et al.* 1994). Results of wider variation in western versus eastern chimpanzee behavior are interesting, in part, because they coincide with greater genetic (mtDNA) variability. However, whether this indicates a true difference in the variation in behavior of western versus eastern chimpanzees is unclear, since this finding is based on a very small sample size, with only two communities of each subspecies, one of which (Bossou, discussed below) is characterized by

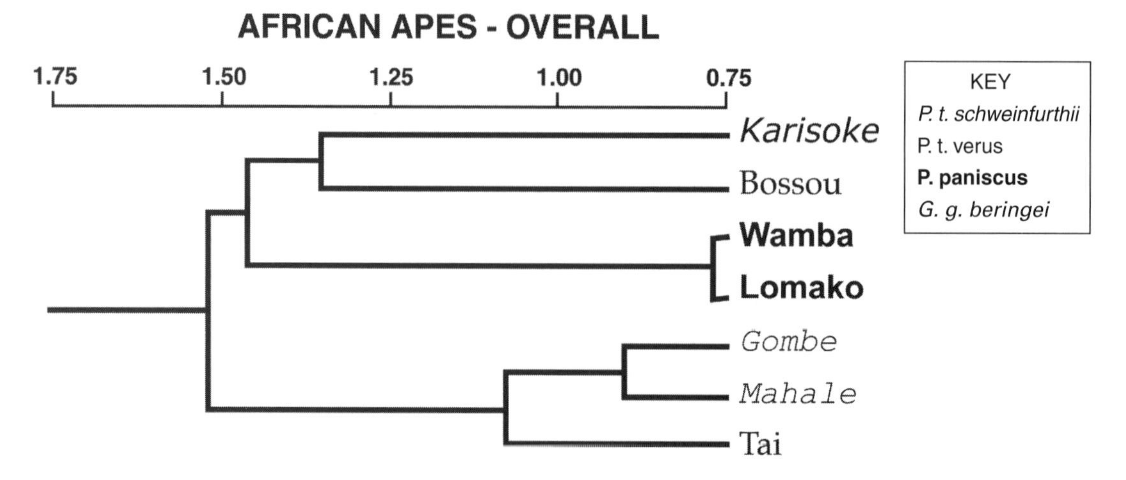

Fig. 1.7. A phenogram of the UPGMA clustering of African apes, including mountain gorillas in addition to bonobos and chimpanzees, based on average taxonomic distances computed from overall data set ($n=7$ taxa, 81 variables).

unusual demographic and environmental conditions. Additional study of chimpanzees from a wider range of their geographic distribution should clarify the issue. Currently, there are three additional ongoing studies of habituated eastern chimpanzees (Kanyawara, Ngogo, and Budongo). As more data on chimpanzee behavior become available from these and other sites, it will be interesting to test whether there remains a behavioral distinction between eastern and western chimpanzees.

Two key traits distinguish western from eastern chimpanzees (considered in this study). First, the presence of nut eating and associated tool use occurs in western (but not eastern) chimpanzees, a trait that has been noted previously (McGrew 1992), and which does not typify all known western chimpanzees (Boesch 1994). Second, western chimpanzees show greater male-(anestrus) female association and affiliation than eastern chimpanzees, where females are less social and spend more time alone with their offspring. This decreased female sociality is coupled with the absence or reduction in frequency of infanticide in western chimpanzees. The prevalence of infanticide has been documented at every eastern site studied to date, including sites not included in this study, such as Kanyawara, Ngogo, and Budongo (Clark-Arcadi & Wrangham 1999). It has not been recorded for western chimpanzees, in spite of studies of longer duration (and thus likelihood of detection) than at Kanyawara, Ngogo, and Budongo.

Decreased female sociality and increased prevalence of infanticide in eastern chimpanzees are associated with differences in habitat. The two eastern sites are characterized by increased seasonality of rainfall, that is, an increased number of dry season months per year, relative to the two western chimpanzee and central African bonobo sites. Increased seasonality of rainfall is likely to result in greater seasonal variance in fruit availability, although phenological data are not currently available to test this hypothesis. Mean chimpanzee party size is known to be limited by the availability of ripe fruit; chimpanzees, and in particular female chimpanzees, respond to reduction in ripe fruit availability by reducing party size (for review see Doran 1997). Permanent female affiliation with males has been hypothesized to be a female counter-strategy to infanticide (van Schaik 1996; van Schaik & Kappeler 1997). Thus, it is possible that the costs (in terms of feeding competition) of female association with males on a more permanent basis may be too great to permit it to serve as an effective counter-strategy to infanticide in habitats with considerable annual variance in fruit production (as measured in this study by rainfall). As a result, in eastern chimpanzees, male–female association occurs less frequently and infanticide occurs (relatively) more often than in more stable environments, such as at Taï and Bossou, where female chimpanzees may benefit from more frequent male–female association with a reduction in infant mortality though infanticide. Thus, we hypothesize that for, chimpanzees in more seasonal environments, seasonality of rainfall results in greater variance in fruit availability which, in turn, necessitates more independent female foraging, leaving females more vulnerable to infanticide. Thus, seasonality of rainfall may have a potentially profound impact on infanticidal behavior in fission–fusion species.

Interrelationship of behavior, ecology, habitat, and life history/demography

ECOLOGICAL FACTORS

Resource density and distribution have been hypothesized as influencing female competitive regimes and resultant social behavior (e.g. van Schaik 1989). Ecological explanations have figured heavily in the discussion of the evolution of chimpanzee and bonobo behavior. Reduced female feeding competition (for bonobos relative to chimpanzees) has been hypothesized as enabling the formation of more stable bonobo parties, which in turn permitted greater female sociality, an important step in the evolution of bonobo sociality (Wrangham 1986a). Among the hypotheses that have been proposed to account for reduced feeding competition in bonobos are bonobos' (1) greater reliance on herbs in diet (Wrangham 1986a); (2) use of larger fruit trees (White 1986; Badrian & Badrian 1988); and (3) more constant source of fruit, as a result of release from fluctuation in fruit availability as a result of their restricted geographical distribution relative to chimpanzees (Malenky 1990).

Results from this study indicate that although there are clear species-specific differences in behavior, there is no species-specific difference in basic diet. Bonobos do not consume, or at least spend more time eating, a greater amount of herbs and leaves compared to chimpanzees. In fact, although all chimpanzees and bonobos are highly frugivorous, bonobos (along with Taï chimpanzees) had the highest percentage of fruit in their diet, implying that herb use does not decrease their reliance on fruit. Thus, it is unlikely that bonobo herb use significantly reduces feeding competition relative to chimpanzees and, therefore, may not be a major factor in the evolution of bonobo party stability (and subsequent sociality).

The most significantly correlated variable distinguishing chimpanzees and bonobos in this study is the average number of dry season months per year, with bonobos having no, or few, dry months per year. This finding is concordant with Malenky's (1990) hypothesis that bonobo sociality may result from their release from the fluctuation in fruit availability that is common at chimpanzee sites, if seasonality of rainfall is considered a proxy for seasonality in fruit production (as discussed above). Phenological data from a wide range of chimpanzee and bonobo sites, which are not currently available, are necessary to make a direct test of this hypothesis. Additionally, to examine this hypothesis more fully it will be important to examine whether variation in fruit availability is greater in a more southern (and seasonal) end of bonobo distribution (see Myers-Thompson, Chapter 4) and, if so, whether its impact on behavior is predictable (smaller parties and more variation in party size).

The phylogenetic analysis offers an additional way to test for consistent causal relationships between ecological and behavioral variables by looking for consistent patterns of associated change between individual characters, for example diet and group size (Eggleton & Vane Wright 1994). In this study, for example, Lomako, but not Wamba bonobos, are characterized by increased herbivory. The typical bonobo patterns of sociality (increased female–female and male–female association and affiliation) actually precede the occurrence of herbivory, appearing on the branch leading to the bonobo lineage (Figure 1.6). Another example relates to causal factors of infanticide. Differences in the degree of home range overlap are not associated with differences in the prevalence of infanticide, since the prevalence of infanticide differs remarkably at Mahale, Gombe, and Taï, in spite of a similar degree of home range overlap.

LIFE HISTORY/DEMOGRAPHY

In this study, demographic factors do not uniformly have an impact on behavior. The behavior and demography ATD matrices were not significantly correlated. In the multivariate analyses, Taï and Mahale clustered together on the basis of demographic features, particularly on the basis of large community size, high adult female to male sex ratio, and relatively longer interbirth intervals. These demographic similarities were not associated with clear behavioral differences. The phylogenetic analysis reinforces the results from the multivariate analysis; there is a lack of clear causality between demographic/life history factors and behavior in this study. One of the more surprising results of the phylogenetic analysis was the relatively low consistency index (CI = 0.62) and retention index (RI = 0.11) of the life history/demography characters compared with the ecological and behavioral characters. Comparative biologists have debated the extent to which life history features are more (Alberch 1990; Vrba 1990) or less (Vrba 1990) likely to show extensive homoplasy (parallel evolution of similar features) relative to other aspects of an animal's biology. This study supports the view that these features are subject to extensive homoplasy, as most of the change in life history/demography characters takes place at the local population level. Of 21 reconstructed changes in demography/life history characters among chimpanzees, 15 were at the level of individual populations, two at the subspecies level, three at the species level, and three at the generic level. It is worth noting, however, that this study, like most others, involved considerable preselection of characters to be analyzed. That is, characters were chosen specifically because they showed variation between species, subspecies, or populations, that is invariant characters were culled. Aspects of life history and demography that were the same for all African apes or all hominoids, such as a tendency for single births, were not included in the analysis.

The role of phylogeny

In recent years, primatologists have become increasingly aware that phylogeny is an important consideration in attempts to evaluate the evolution of adaptations (e.g. Fleagle 1992; DiFiore & Rendall 1994; Lee 1999; Nunn & Barton 2001). However, there is considerable debate as to the proper way to evaluate the role of phylogeny in the evolution of aspects of behavior and ecology, the significance of phylogenetic patterns in the distribution of behavioral and ecological characteristics, and in the appropriateness of even considering behavioral characteristics in a phylogenetic perspective (Robson-Brown 1999). Indeed, among the five coauthors of this paper, we probably embody many of the varied views of enthusiasm, utility, trepidation, skepticism, and confusion that are evoked among primatologists today during the discussion of phylogenetic considerations of behavior. Nevertheless, the results of this study illustrate quite clearly why phylogenetic considerations are both appropriate and valuable for understanding the evolution of behavior and ecology, as well as point to some of its limitations.

A common view of behavioral (and ecological) features is that they are so variable between populations (or within species and genera) that they do not generally reflect phylogenetic relationships. In other words, they show too much parallel evolution, compared to other aspects of an animal's biology (such as tooth structure or ear morphology) commonly used in taxonomic studies. However, the few formal attempts to compare levels of homoplasy among behavioral, cranial, postcranial, and biomolecular features, have not generally found a consistent pattern indicating greater homoplasy of behavioral features (see Lockwood & Fleagle 1999 and references therein). Indeed, among phylogenetic studies of primate behavior and morphology, the behavioral studies show some of the lowest levels of homoplasy – much lower than many studies of morphological or biomolecular evolution.

Features of behavior, ecology, and life history/demography used in this study and mapped onto the commonly accepted phylogeny of African great apes, show a CI of 0.75. This is almost exactly the level of homoplasy expected for seven taxa, according to a formula based on a broad analysis of a very large sample of phylogenetic studies from a wide range of organisms and many types of characters (Sanderson & Donoghue 1989). Thus, behavioral features show no more homoplasy than other aspects of an animal's biology. Indeed the CI of 0.75 shows that, in this study, a parsimonious reconstruction of the evolution of these behavioral features among the taxa being considered accords well with the accepted phylogeny of the taxa.

What is the significance of all this? Essentially, it just documents that, overall, more closely related groups of chimpanzees tend to be more similar in behavior. There are some notable exceptions; there are some extremely divergent characters and some divergent taxa (discussed below). However, this general conclusion is a fairly routine expectation based on the assumption that species, subspecies, and populations have not been created de nova, but have evolved by modification from other chimpanzee species, subspecies, and populations. Thus, as we have discussed in the previous paragraphs, gorilla, bonobos, eastern chimpanzees, and western chimpanzees can each be characterized by clusters of shared features of behavior and ecology. This does not mean that the behavior of chimpanzees is 'constrained' by phylogeny in anything other than a retrospective sense. Rather, only that, by and large, the pattern of changes that are reconstructed to have evolved in these chimpanzee populations generally follow what is regarded as the phylogenetic history of the individual populations based on genetic studies. Indeed, as the comparisons of the different patterns of behavioral similarities and differences among eastern and western populations clearly demonstrate, individual populations show tremendous differences in the extent to which they resemble their closest relatives.

HOW IS BOSSOU UNIQUE, AND WHAT DOES THIS INDICATE ABOUT HOW DIFFERENT FACTORS MAY INFLUENCE BEHAVIOR IN AFRICAN APES?

The chimpanzees of Bossou provide an especially interesting case to consider in regard to the question of phylogeny because, although they share some features in common with Taï chimpanzees, the two populations are, in many ways, behaviorally, environmentally, and demographically distinct. As noted above, the demographic makeup of the chimpanzee community of Bossou, a small community having only one adult male, is considerably different from that of the other three chimpanzee communities in this study, although not unique from all known communities (the Taï community has undergone recent demographic changes; only one adult male is currently present in the north community). Thus, by default, strong male bonds, and the traits associated with them that are generally considered 'chimpanzee traits,' are absent. Generally, male–male cooperation is important for territory defense through boundary patrolling. At some, but not all, other sites, community disintegration resulted when the number of males decreased as a result of active 'warfare' by males of adjacent communities (Goodall 1986, pp. 503–14; but see Nishida 1990b). The long-term stability (greater than 10 years) of a one-male group at Bossou may be due to the population's isolation from any close neighboring communities of chimpanzees, and a distance of at least 7 km of savanna between it and any adjacent chimpanzee dwelling forest (Sugiyama 1999). It would be especially informative to broaden this study by including recent behavioral results from Taï, based on data collected after the number of adult males in the community decreased. Looking at how (a wide variety of) behavior changes within a community through time, while maintaining roughly constant environmental features, would help to elucidate the effect that demographic versus environmental features have on the distinctive position Bossou occupies in this study.

As a result of its demographic distinction, Bossou shares many features in common with mountain gorillas, to the extent that when an outgroup (Karisoke mountain gorillas) is added in the UPGMA clustering of the ATDs based on all 82 variables, Bossou clusters with mountain gorillas, albeit distantly, leaving the underlying pattern of behavioral affinities of the other chimpanzee and bonobo taxa unchanged. Similarly, a phylogeny linking Bossou with gorillas rather than with the Taï chimps provides better resolution of the distribution of characters in this data set, even though it clearly does not correspond to the most probable phylogeny of these apes.

Far more interesting, however, is the evidence that Bossou provides as a test case for the association between different behavioral, demographic, and ecological features. Many of the distinctive features characterizing the Bossou population are, at least in part, causally related to the unusual demography (reduced number of adult males) and isolation of this population. As a one-male group, with the resulting absence of strong male–male bonds and coalitions, Bossou chimpanzees are similar to gorillas and bonobos (and distinct from all other chimpanzees studied to date) in their absence of territoriality, lack of hunting of monkeys, lack of male–male affiliations, and potential occurrence of male dispersal. In spite of these considerable differences in demography, Bossou, in their reduced prevalence of infanticide in conjunction with more stable male–female association, is similar to Taï chimpanzees and unlike eastern chimpanzees. We propose that in seasonal environments, such as those of the eastern chimpanzees considered in this study, greater variation in fruit supply makes more permanent male–female association too costly for females. Finally, in spite of considerable differences in habitat, demography, and ecology, Bossou chimpanzees remain similar to all other chimpanzees (considered in this study) and distinct from bonobos in (1) their use of tools to acquire food; (2) the continued dominance of males over females, in spite of a reduction in the number of males present and an increase in female–female association; (3) a considerably longer time before resumption of estrus after parturition; and (4) the absence of a ventral–ventral mating pattern. Thus, these behavioral traits seem relatively invariant in chimpanzees despite considerable differences in habitat, demography, and ecology.

ACKNOWLEDGEMENTS

We thank Gottfried Hohmann and Christophe Boesch for their invitation to participate in the symposium on chimpanzee behavioral diversity in Seeon, Germany. DD thanks Roman Wittig, Nick Newton-Fisher, Becky Stumpf, and Brenda Bradley for useful discussion on the topic. We thank the editors and Carel van Schaik and Becky Stumpf for their comments on an earlier version of the manuscript. We would also like to thank Toshisada Nishida and Akiko Matsumoto-Oda for unpublished data, Vernon Reynolds for his willingness to participate in this study, and Luci Betti-Nash for producing the figures. Financial support was provided by the National Science Foundation (SBR-9422438 and SBR-9729126).

NOTE

1. Data, statistical analyses and further references relevant to this chapter are available on Diane Doran's web site at http://www.informatics.sunsyb.edu.anatomy/IDPAS/Doran/.

REFERENCES

Alberch, P. (1990). Natural selection and developmental constraints: external versus internal determinants of order in nature. In *Primate Life History and Evolution*, ed. C. J. DeRousseau, pp. 15–35. New York: Wiley-Liss, Inc.

Alexander, R. D. (1974). The evolution of social behavior. *Annual Review of Ecology and Systematics*, 5, 325–83.

Badrian, A. J. & Badrian, N. L. (1988). Group composition and social structure of *Pan paniscus* in the Lomako Forest of Central Zaire. In *The Pygmy Chimpanzee: Evolutionary Biology and Behavior*, ed. R. L. Susman, pp. 173–81. New York: Plenum Press.

Barton, R. A. (2000). Socioecology of baboons: the interaction of male and female strategies. In *Primate Males: Causes and Consequences of Variation in Group Composition*, ed. P. M. Kappeler, pp. 97–107. Cambridge: Cambridge University Press.

Boesch, C. (1991). The effect of leopard predation on grouping patterns of forest chimpanzees. *Behaviour*, 117, 220–42.

Boesch, C. (1994). Hunting strategies of Gombe and Taï chimpanzees. In *Chimpanzee Cultures*, ed. R. W. Wrangham, W. C. McGrew, F. B. M. de Waal & P. G. Heltne, pp. 77–91. Cambridge, MA: Harvard University Press.

Boesch, C. (1996a). Taï site report. In *Great Ape Societies*, ed. W. C. McGrew, L. F. Marchant & T. Nishida, pp. 317–18. Cambridge: Cambridge University Press.

Boesch, C. (1996b). Social grouping of Taï chimpanzees. In *Great Ape Societies*, ed. W. C. McGrew, L. F. Marchant & T. Nishida, pp. 101–13. Cambridge: Cambridge University Press.

Boesch, C. & Boesch-Achermann, H. (2000). *The Chimpanzees of the Taï Forest: Behavioural Ecology and Evolution*. Oxford: Oxford University Press.

Boinski, S. (1999). The social organizations of squirrel monkeys: implications for ecological models of social evolution. *Evolutionary Anthropology*, 8, 101–12.

Borgerhoff Mulder, M. (2001). Using phylogenetically based comparative methods in anthropology: more questions than answers. *Evolutionary Anthropology*; 10, 99–111.

Brooks, D. R. & McLennan, D. (1991). *Phylogeny, Ecology, and Behavior*. Chicago: University of Chicago Press.

Chapman, C. A., White, F. J. & Wrangham, R. W. (1994). Party size in chimpanzees and bonobos: a reevaluation of theory based on two similarly forested sites. In *Chimpanzee Cultures*, ed. R. W. Wrangham, W. C. McGrew, F. B. M. de Waal & P. G. Heltne, pp. 41–57. Cambridge, MA: Harvard University Press.

Clark-Arcadi, A. & Wrangham, R. W. (1999). Infanticide in chimpanzees: review of cases and a new within-group observation from the Kanyawara study group in Kibale National Park. *Primates*, 40, 337–51.

Corruccini, R. S. (1978). Morphometric analysis: uses and abuses. *Yearbook of Physical Anthropology*, 21, 134–50.

de Waal, F. B. M. (1989). *Peacemaking Among the Primates*. Cambridge, MA: Harvard University Press.

de Waal, F. B. M. (1998). Comment on C. B. Stanford, The social behavior of chimpanzees and bonobos. *Current Anthropology*, 39, 407–8.

DeRousseau, C. J. (1990). *Primate Life History and Evolution*. New York: Wiley-Liss.

DiFiore, A. & Rendall, D. (1994). Evolution of social organization: a reappraisal for primates using phylogenetic methods. *Proceedings of the National Academy of Science, USA*, 91, 9941–5.

Doran, D. M. (1997). Influence of seasonality on activity patterns, feeding behavior, ranging and grouping patterns in Taï chimpanzees. *International Journal of Primatology*, 18, 183–206.

Eggleton, P. & Vane Wright, R. (1994). *Phylogenetics and Ecology*. London: Academic Press.

Fleagle, J. G. (1992). Trends in primate evolution and ecology. *Evolutionary Anthropology*, 1, 78–9.

Fleagle, J. G. (1999). *Primate Adaptation and Evolution*. San Diego: Academic Press.

Flury, B. & Riedwyl, H. (1988). *Multivariate Statistics: A Practical Approach*. Cambridge: Cambridge University Press.

Fruth, B. (1998). Comment on C. B. Stanford, The social behavior of chimpanzees and bonobos. *Current Anthropology*, 39, 408–9.

Fruth, B. & Hohmann, G. (1999). A ladies success story. *Max Planck Research*, 1, 14–23.

Gagneux, P., Wills, C., Gerloff, U., Tautz, D., Morin, P. A., Boesch, C., Fruth, B., Hohmann, G., Ryder, O. & Woodruff, D. S. (1999). Mitochondrial sequences show diverse evolutionary histories of African hominoids. *Proceedings of the National Academy of Science, USA*, 96, 5077–82.

Ghiglieri, M. P. (1987). Sociobiology of the great apes and the hominid ancestor. *Journal of Human Evolution*, 16, 319–57.

Gonder, M. K., Oates, J. F., Disotell, T. R., Forstner, M. R. J., Morales, J. C. & Melnick, D. J. (1997). A new west African chimpanzee subspecies? *Nature*, 388, 337.

Goodall, J. (1986). *The Chimpanzees of Gombe*. Cambridge, MA: Harvard University Press.

Gower, J. C. (1966). Some distance properties of latent root and vector analysis. *Biometrika*, 53, 325–38.

Jungers, W. L. (1988). Relative joint size and hominoid locomotor adaptations with implications for the evolution of hominid bipedalism. *Journal of Human Evolution*, 17, 247–65.

Kano, T. (1992). *The Last Ape: Pygmy Chimpanzee Behavior and Ecology*. Stanford, CA: Stanford University Press.

Kano, T. (1998). Comment on C. B. Stanford. The social behavior of chimpanzees and bonobos. *Current Anthropology*, 39, 410–11.

Kappeler, P. M. (1999). Lemur social structure and convergence in primate socioecology. In *Comparative Primate Socioecology*, ed. P. C. Lee, pp. 273–99. Cambridge: Cambridge University Press.

Kitching, I. J., Forey, P. L., Humphries, C. J. & McWilliam, D. (1998). *Cladistics: The Theory and Practice of Parsimony Analysis*. Oxford: Oxford University Press.

Koenig, A., Beise, J., Chalise, M. K. & Ganzhorn, J. (1998). When females should contest for food–testing hypotheses about resource density, distribution, size and quality with Hanuman langurs (*Presbytis entellus*). *Behavioral Ecology and Sociobiology*, 42, 225–37.

Lee, P. C. (1999). *Comparative Primate Socioecology*. Cambridge: Cambridge University Press.

Lockwood, C. A. & Fleagle, J. G. (1999). The recognition and evaluation of homoplasy in primate and human evolution. *Yearbook of Physical Anthropology*, 42, 189–232.

Maddison, D. R. & Maddison, W. P. (2000). *MacClade 4: Analysis of Phylogeny and Character Evolution*. Sunderland, MA: Sinauer Associates.

Malenky, R. K. (1990). Ecological factors affecting food choice and social organization in *Pan paniscus*. PhD thesis, State University of New York at Stony Brook, Stony Brook, New York.

McGrew, W. C. (1992). *Chimpanzee Material Culture*. Cambridge: Cambridge University Press.

McGrew, W. C. (1998). Comment on C. B. Stanford, The social behavior of chimpanzees and bonobos. *Current Anthropology*, 39, 411.

Mitani, J. C. & Watts, D. P. (1999). Demographic influences on the hunting behavior of chimpanzees. *American Journal of Physical Anthropology*, 109, 439–54.

Mitchell, C. L., Boinski, S. & van Schaik, C. P. (1991). Competitive regimes and female bonding in two species of squirrel monkeys (*Saimiri oerstedi* and *S. sciureus*). *Behavioral Ecology and Sociobiology*, 28, 55–60.

Morin, P. A., Moore, J. J., Chakraborty, J. L., Goodall, J. & Woodruff, D. S. (1994). Kin selection, social structure, gene flow, and the evolution of chimpanzees. *Science*, 265, 1193–201.

Nishida, T. (1990a). A quarter century of research in the Mahale Mountains: an overview. In *The Chimpanzees of the Mahale Mountains*, ed. T. Nishida, pp. 3–36. Tokyo: University of Tokyo Press.

Nishida, T. (1990b). *The Chimpanzees of the Mahale Mountains*. Tokyo: University of Tokyo Press.

Nishida, T. (1997). Sexual behavior of adult male chimpanzees of the Mahale Mountains National Park, Tanzania. *Primates*, 38, 379–98.

Nunn, C. L. & Barton, R. A. (2001) Comparative methods for studying primate adaptation and allometry. *Evolutionary Anthropology*, 10, 81–98.

Nunn, C. L. & van Schaik, C. P. (2000). Social evolution in primates: the relative roles of ecology and intersexual conflict. In *Infanticide By Males and Its Implications*, ed. C. P. van Schaik & C. H. Janson, pp. 388–420. Cambridge: Cambridge University Press.

Parish, A. R. (1998). Comment on C. B. Stanford, The social behavior of chimpanzees and bonobos, *Current Anthropology*, 39, 413–14.

Reyment, R. A., Blackith, R. E. & Campell, N. A. (1984). *Multivariate Morphometrics*. London: Academic Press.

Reynolds, V. & Reynolds, F. (1965). Chimpanzees of the Budongo Forest. In *Primate Behavior*, ed. I. De Vore, pp. 368–424. New York: Holt, Rinehart and Winston.

Robson-Brown, K. (1999). Cladistics as a tool in comparative analysis. In *Comparative Primate Socioecology*, ed. P. C. Lee, pp. 23–43. Cambridge: Cambridge University Press.

Rohlf, F. J. (1972). An empirical comparison of three ordination techniques in numerical taxonomy. *Systematic Zoology*, 21, 271–80.

Rohlf, F. J. (1993). Numerical Taxonomy and Multivariate Analysis. New York: Exeter Software.

Ross, C. (1998). Primate life histories. *Evolutionary Anthropology*, 6, 54–63.

Ruvolo, M., Pan, D., Zehr, S., Goldberg, T., Disotell, T. R. & von Dornum, M. (1994). Gene trees and hominoid phylogeny. *Proceedings of the National Academy of Science, USA*, 91, 8900–4.

Sakura, O. (1994). Factors affecting party size and composition of chimpanzees (*Pan troglodytes verus*) at Bossou, Guinea. *International Journal of Primatology*, 15, 167–83.

Sanderson, M. & Donoghue, M. (1989). Patterns of variation in levels of homoplasy. *Evolution*, 43, 1781–95.

Smuts, B. B., Cheney, D. L., Seyfarth, R. M., Wrangham, R. W. & Struhsaker, T. T. (1987). *Primate Societies*. Chicago: University of Chicago Press.

Sneath, P. H. A. & Sokal, R. R. (1973). *Numerical Taxonomy*. San Francisco: W. H. Freeman and Company.

Stanford, C. B. (1998). The social behavior of chimpanzees and bonobos: empirical evidence and shifting assumptions. *Current Anthropology*, 39, 399–420.

Stanford, C. B., Wallis, J., Mpongo, E. & Goodall, J. (1994). Hunting decisions in wild chimpanzees. *Behavior*, 131, 1–18.

Sterck, E., Watts, D. P. & van Schaik, C. P. (1997). The evolution of female social relationships in nonhuman primates. *Behavioral Ecology and Sociobiology*, 41, 291–309.

Sugiyama, Y. (1999). Socioecological factors of male chimpanzee migration at Bossou, Guinea. *Primates*, 40, 61–8.

Sugiyama, Y. & Koman, J. (1979). Social structure and dynamics of wild chimpanzees at Bossou, Guinea. *Primates*, 20, 323–39.

Susman, R. L. (1984). *The Pygmy Chimpanzee*. New York: Plenum Press.

Takahata, Y., Ihobe, H. & Idani, G. (1996). Comparing copulations of chimpanzees and bonobos: do females exhibit proceptivity or receptivity? In *Great Ape Societies*, ed. W. C. McGrew, L. F. Marchant & T. Nishida, pp. 146–55. Cambridge: Cambridge University Press.

van Schaik, C. P. (1983). Why are diurnal primates living in groups? *Behaviour*, 87, 120–44.

van Schaik, C. P. (1989). The ecology of social relationships amongst female primates. In *Comparative Socioecology*, ed. V. Standen & R. A. Foley, pp. 195–218. Oxford: Blackwell.

van Schaik, C. P. (1996). Social evolution in primates: the role of ecological factors and male behavior. *Proceedings of the British Academy*, 88, 9–31.

van Schaik, C. P. & Kappeler, P. M. (1997). Infanticide risk and the evolution of male–female association in primates. *Proceedings of the Royal Society of London B*, 264, 1687–94.

Vrba, E. S. (1990). Life history in relation to life's hierarchy. In *Primate Life History and Evolution*, ed. C. J. DeRousseau, pp. 37–46. New York: Wiley-Liss, Inc.

Watts, D. P. (1989). Infanticide in mountain gorillas: new cases and a reconsideration of the evidence. *Ethology*, 81, 1–18.

White, F. J. (1986). Behavioral ecology of the pygmy chimpanzee. PhD thesis, State University of New York at Stony Brook, Stony Brook, New York.

White, F. J. (1992). Pygmy chimpanzee social organization: variation with party size and between study sites. *American Journal of Primatology*, 26, 203–14.

White, F. J. (1996a). Comparative socio-ecology of *Pan paniscus*. In *Great Ape Societies*, ed. W. C. McGrew, L. F. Marchant & T. Nishida, pp. 29–41. Cambridge: Cambridge University Press.

White, F. J. (1996b). Twenty-three years of field research. *Evolutionary Anthropology*, 5, 11–17.

Whiten, A., Goodall, J., McGrew, W. C., Nishida, T., Reynolds, V., Sugiyama, Y., Tutin, C. E. G., Wrangham, R. W. & Boesch, C. (1999). Culture in chimpanzees. *Nature*, 399, 682–5.

Wilson, E. O. (1975). *Sociobiology*. Cambridge, MA: Belknap Press of Harvard University Press.

Wrangham, R. W. (1979). On the evolution of ape social systems. *Social Sciences Information*, 18, 334–68.

Wrangham, R. W. (1980). An ecological model of female-bonded primate groups. *Behaviour*, 75, 262–300.

Wrangham, R. W. (1986a). Ecology and social relationships in two species of chimpanzees. In *Ecology and Social Evolution: Birds and Mammals*, ed. D. I. Rubenstein & R. W. Wrangham, pp. 352–78. Princeton: Princeton University Press.

Wrangham, R. W. (1986b). The significance of African apes for reconstructing human social evolution. In *The Evolution of Human Behavior: Primate Models*, ed. W. G. Kinzey, pp. 51–71. New York: SUNY Press.

Wrangham, R. W. (1987). Evolution of social structure. In *Primate Societies*, ed. B. Smuts, D. L. Cheney, R. M. Seyfarth, R. W. Wrangham & T. T. Struhsaker, pp. 282–97. Chicago: University of Chicago Press.

Wrangham, R. W. & Peterson, D. (1996). *Demonic Males*. Boston: Houghton Mifflin Company.

Wrangham, R. W., McGrew, W. C., de Waal, F. B. M. & Heltne, P. G. (1994). *Chimpanzee Cultures*. Cambridge, MA: Harvard University Press.

Wrangham, R. W., Chapman, C. A., Clark-Arcadi, A. & Isabirye-Basuta, G. (1996). Social ecology of Kanyawara chimpanzees: implications for understanding the costs of great ape groups. In *Great Ape Societies*, ed. W. C. McGrew, L. F. Marchant & T. Nishida, pp. 45–57. Cambridge: Cambridge University Press.

NOTE

The publisher has used its best endeavours to ensure that URLs for external websites are correct and active at the time of going to press. However the publisher has no responsibility for the websites and can make no guarantee that a site will remain live or that the content is or will remain appropriate.

2 • Chimpanzees in the dry habitats of Assirik, Senegal and Semliki Wildlife Reserve, Uganda

KEVIN D. HUNT & WILLIAM C. MCGREW

INTRODUCTION

Scenarios of ape–human divergence have often represented hominins as savanna-adapted (e.g. Dart 1959; Robinson 1972; Coppens 1988; Wheeler 1992), whereas the chimpanzee (*Pan troglodytes*) has been represented as rainforest-adapted. This dichotomy has eroded from both sides of the divide. Kano (1972) and McGrew *et al.* (1981) noted that the longest-studied chimpanzees, those at Gombe and Mahale, live in a grassland–woodland–forest mosaic, not tropical forest, and that other populations range into even drier, more open habitats (see the 'savanna ape model' in Moore 1996). The habitats of the earliest hominins were at least partly forested (Bonnefille 1984; WoldeGabriel *et al.* 1994, 2001). A less severely drawn contrast between ape and hominin habitats now seems a better fit with the evidence: Chimpanzee habitats range from closed-canopy rainforest to savanna, whereas early hominins likely occupied somewhat drier habitats, ranging from forest to woodland to savanna. This overlap means that primatologists can study the behavior of chimpanzees living in habitats nearly identical to those in which Mio-Pliocene hominins lived.

Study of chimpanzee diet, habitat use, ranging, and positional behavior in peripheral habitats will help establish whether water availability, food availability, food dispersion, vegetative cover, or climate limits the pan-African chimpanzee distribution. Better understanding of the factors that limit chimpanzees may help to identify niche differences between Mio-Pliocene apes and hominins.

Study of chimpanzees at the edges of the species' range may also have conservation implications. Documentation of the minimum requisites of the species is valuable for rehabilitation, relocation, and green corridor projects. Documenting dry-habitat adaptations should extend our knowledge of the capacities of chimpanzees to adapt to their environment through cultural as well as ecological means.

Here we compare the behavioral ecology of chimpanzees at Assirik and Semliki. The Assirik apes were studied in 1976–79 and 2000, while their Semliki counterparts have been studied since 1996.

METHODS

Semliki

Chimpanzees were found most mornings by listening for their calls, movements, or feeding sounds in areas where pant-hoots had been heard the previous evening. When chimpanzees had not been found the day before, we searched for them on foot. On 31 occasions chimpanzees were followed to their night nests, and attempts were made to recontact them before they left the nest the next morning. Chimpanzees were first encountered at distances of more than 100 m, but since mid-1998 we have approached individuals to within 10–20 m. When we found a feeding party, we counted individuals and noted their age and sex, then we followed them for as long as possible. Once the feeding party left the feeding site, we followed them on the ground, though we rarely found them at a second feeding site. As of August, 2000, observations have rarely lasted more than 45 minutes, with 142 minutes being the longest. Observers listen for pant-hoots in the evening about 5 days a week.

Trees were identified according to species using standard reference sources (Eggeling & Dale 1951; Palgrave 1977). Voucher specimens were collected for each item to be keyed, and items that could not be keyed were identified by a specialist (Anthony Katende) at Makerere University. Diet was determined by direct observation of chimpanzees eating ($n = 28$ cases), by screening seeds and food fragments from feces ($n = 4$), and by looking at feeding remnants ($n = 1$). Feces were collected only when a chimpanzee was seen to drop them. Feces were screened through 1-mm wire mesh, and seeds were matched to those collected from plant species at Semliki.

We collected daily minimum and maximum temperature, rainfall, and humidity data (using a wet bulb–dry bulb thermometer) beginning on 1 September, 1997. Climatological instruments were sited in a shaded, protected area near the center of the Mugiri chimpanzee range ('research station' in Figure 2.2, below). Cloud cover was estimated to the nearest 5%.

From July, 1997 we recorded data on fresh nests, including location, height (measured by clinometer), height of tree, and diameter of the nest; proximity to other nests (estimated from ground), number of nests within 15 m, whether it was more than 50% open to the sky or covered by foliage; age (estimate based on ad hoc experience with known-age nests monitored over weeks), season, and species of tree. Nest data were taken when fresh nests were discovered in the course of tracking chimpanzees.

When drinking wells were found, we measured their depth, minimum and maximum diameter at both ground level and bottom; location, minimum distance to standing water, number of wells when clustered, distance to other wells, and presence or absence of leaf sponges.

Assirik

On foot, we searched daily for chimpanzees and most often found them acoustically, either from their calls or the distinctive sound of baobabs cracking (see below). Most calls were at the beginning or end of the day, and whenever possible we followed them to the nest at dusk and returned to the site before dawn. When chimpanzees were found, we watched them for as long as possible, but did not pursue them once they fled. Vigils at fruiting groves also yielded sightings (Tutin *et al.* 1983).

Samples of vegetation were collected and sent to the Royal Botanic Gardens, Kew, for definitive identification. Many specimens were keyed in the field using the standard flora (Hutchinson & Dalziel 1927–1936). Overall, 173 species of vegetation were identified. Diets of chimpanzees were determined by direct observation, collection of feeding remnants, and analysis of fecal specimens. McGrew *et al.* (1988) gave criteria for acceptance of indirect evidence.

Climatological data were collected from March, 1976 to December, 1979, and from February to April, 2000 (McGrew *et al.* 1981; unpublished data). Besides rainfall, we collected daily minimum and maximum temperature; for comparison across microhabitats. We also collected hourly temperature and humidity (using a wet bulb–dry bulb thermometer) at three sites: gallery forest, woodland and plateau. In 2000, we collected temperature data from five microhabitats, using automatic data loggers.

In 1976–79, we collected data on 252 chimpanzee nests, using 19 variables: season, tree species, tree height, tree girth, lowest branch, nearest neighbor tree, chimpanzee food species, age, height, width, breadth, depth, main branches, open or closed, nearest neighbor nest, nests in same tree, arboreal escape route, understory, branches bent or broken. Baldwin *et al.* (1981) analyzed a subset of these variables for comparison with nests from a rainforest site in Equatorial Guinea. In 2000, we collected limited data on the distribution, density, and abundance of 736 nests in Niokolo-Koba (Pruetz *et al.* in press).

RESULTS

Physical and botanical characteristics

SEMLIKI

Gazetted in 1932 as the Toro Game Reserve, then renamed the Semuliki Game Reserve (later shortened to Semliki), the 548 km^2 Semliki Wildlife Reserve (Figure 2.1) is northwest of Fort Portal, near the eastern edge of the western (or Albertine) Great Rift Valley (0°50′ to 1°05′N, 30°20′ to 30°35′E). Most of the surface area of the reserve is undulating plain on the great rift floor, at elevation 620–800 m (Verner & Jenik 1984). The eastern boundary of the reserve runs along the crest of the rift escarpment, at 1500 m and above. Lake Albert and the Ruwenzori Mountains are the northern and southern boundaries.

Semliki is predominantly dry *Combretum* savanna and *Borassus* palm savanna, dominated by *Acacia*, *Albizia*, *Piliostigma*, and *Combretum*. Dominant savanna grass species are *Hyparrhenia* spp., *Themeda triandra*, *Imperata cylindrica*, *Sporoboleyum pyramidalis*, and *Chloris gayana* (Verner & Jenik 1984; Allan *et al.* 1996). In the study community's range, *Themeda* and *Chloris* grasses are common on steep slopes, and *Panicum maximum* and *Echinochloa pyramidalis* are found in moist, flat areas.

The reserve is crosscut and partly bounded by watercourses that support 50 to 250-m-wide strips of riverine forest, some trees of which reach 50 m in height (Allan *et al.* 1996; Hunt unpublished data). The main watercourse is the Wasa River, which runs north from the southern part of the reserve and empties into Lake Albert. The Nyabaroga River constitutes much of the southwest border; it supports a riverine forest and the Nyabaroga community of chimpanzees. The Muzizi River skirts the northeast boundary of the reserve and also supports a chimpanzee community. Research has focused mainly on chimpanzees in gallery forests along the Mugiri River and its tributaries at the eastern edge of the reserve. The Mugiri supports a species-diverse flora grading from woodland to tall gallery forest. Dominant tree species in the riverine habitat are *Celtis africana*, *C. integrifolia*, *C. mildbraedii*, *C. brownii*, *Albizia*

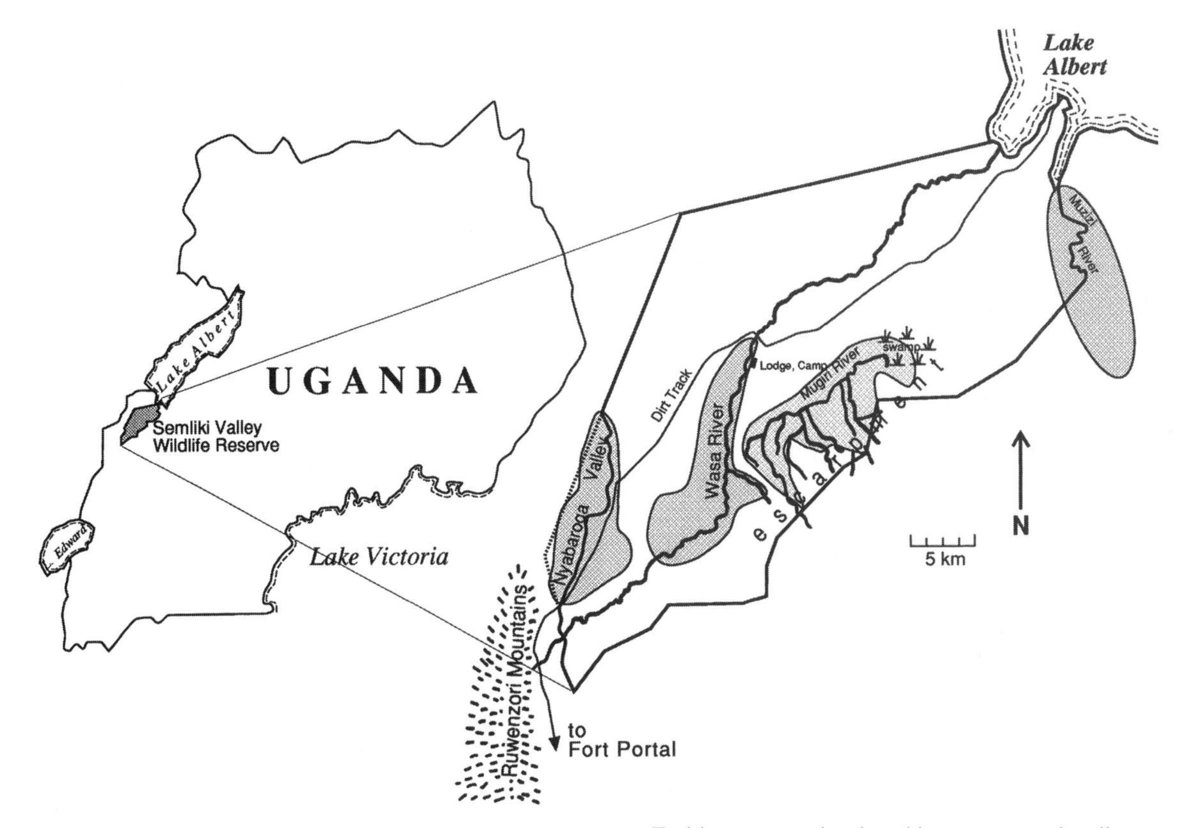

Fig. 2.1. Semliki Wildlife Reserve. Bordered to the south by foothills of the Ruwenzori Mountains, to the north by Lake Albert, and to the east by the rift escarpment. Four communities of chimpanzees (stippled areas) are believed to live in the reserve, at the Muzizi River, the Mugiri River (the study group, lighter stippling), the forested stretch of the Wasa, and the Nyabaroga Valley.

grandibracteata, *A. coriaria*, *Strychnos mitis*, *Chrysophyllum* spp., *Cynometra alexandri*, *Diospyros abyssinica*, *Funtumia africana*, *Trichilia prieuriana* (Verner & Jenik 1984; personal observation), *Phoenix reclinata*, *Beilschmiedia ugandensis*, *Polyscias fulva*, and *Cola gigantea* (Hunt, unpublished data; Katende, personal communication). *Tamarindus indica* are widely dispersed at the savanna–forest ecotone. A handful of small tributaries that also support riverine forest flow into the Mugiri from the escarpment to the east. Between these tributaries, the habitat is woodland. The Mugiri chimpanzees' range is mostly limited to the escarpment slopes. Peasant small-holdings and tea estates border the reserve at the top of the escarpment, and these limit the Mugiri community's range.

Fruiting tree species that chimpanzees use heavily are distinctly distributed. *Beilschmiedia*, *Cynometra*, *Cola*, *Tamarindus*, and *Phoenix* trees are on the rift floor in or within meters of the Mugiri riverine forest. *Securinega* trees are less distinctly distributed, but tend to occur on the lower escarpment slopes. *Grewia* trees grow only on the escarpment and are common only in open woodland and bushland.

ASSIRIK

Protected since 1926, first as a reserve and later as a national park, the Parc National du Niokolo Koba, Republique du Senegal, covers 9130 km^2. Since 1990, it has been combined with the Parc National du Badiar Nord, Republique du Guinea, as the Parc Transfrontalier Niokolo Badiar. It encompasses part of the watershed of one of Africa's major rivers, the Gambia, especially two main tributaries, Kouloumtou and Niokolo Koba. In the southeastern part of the park, Mont Assirik is a flat-topped, low hill of 311-m elevation (the highest point in the park). At 12°50′N, 12°45′W, Assirik is 125 km southeast of the regional capital of Tambacounda and 65 km northwest of the town of Kedougou. The study area within a 4-km radius of Assirik's summit was chosen because it had year-round flowing surface water, thought to be necessary to support a resident population of wild chimpanzees (but see Lanjouw, Chapter 3).

The vegetation of the study area was divided into five types:

1. gallery forest (riverine semi-deciduous lowland forest), 3% of the surface area of 503 ha;
2. woodland (drought-deciduous lowland woodland), 37%;
3. bamboo thicket (flat-leafed tree savanna with isolated palms and deciduous trees), 5%;
4. grassland (narrow-leafed savanna with isolated deciduous trees), 27%;
5. plateau (narrow-leafed savanna with isolated deciduous shrubs), 28%.

The vegetation types are taken from Ellenberg & Müller-Dombois's (1967) world-wide classification.

Gallery forest (mostly evergreen and often closed canopy) grows from steep-sided watercourses in narrow, alluvial valleys cut by erosion through laterite pans. The woody vegetation is a multi-layer mix of lianas, shrubs, and trees; characteristic species are *Saba senegalensis*, *Combretum tormentosum*, *Oncoba spinosa*, *Ficus umbellata*, and *Diospyros mespiliformes*. The tallest emergents are *Ceiba pentandra*, which sometimes reach 40 m in height.

Woodland (open canopy) occurs on gentler, well-drained slopes of valley and hillsides; 83% of trees are 10 m or below high. The understory is herbaceous and sparse, mostly of grass. The most common woody species are *Hexolobus monopetalus*, *Pterocarpus erinaceus*, and *Afzelia africana*.

Bamboo thickets have dense clumps of *Oxytenanthera abyssinica* interspersed with scattered trees, growing on clay-based soils in wide, well-watered depressions. Their thick shade prevents an understory from forming. A few tree species dominate the spaces between the deciduous bamboo clumps: *P. erinaceus*, *Acacia macrostachya*, *Piliostigma thonningii*, and *D. mespiliformes*.

Grasslands comprise tall, dense, often monospecific stands ('elephant grass') growing in seasonally water-logged depressions. Verdant in the wet season, they burn spectacularly in the late dry season. Interspersed are scattered trees and shrubs, especially *Combretum* spp.

Plateaus are wide, flat, impervious sheets of laterite or bauxite upon which little vegetation can thrive, except for a few species of grasses and shrubs, such as *Combretum collineum*, with shallow root systems. The grass (especially *Rhytachne triaristata*) grows in dense, uniform swathes of less than one m in height; it burns completely in the dry season, leaving an almost lunar, scoured rocky surface.

Watercourses radiate from the slopes of Mont Assirik. During the wet season (June–November) these flow constantly, but from December onward more and more of these streams dry up, until by the onset of the new rains in May, only two main valleys have running surface water. Fires sweep through several times from January to April in the dry season, clearing herbaceous undergrowth from all vegetation types. All woody species except those growing in streambeds are fire-resistant; even the leaf litter of the gallery forest burns.

In this mosaic habitat, fruiting tree species that are crucial to the chimpanzees are patchily distributed, even within the five main types of vegetation. For example, *Pseudospondias microcarpa* grows only in permanent watercourses in gallery forest; *Grewia lasiodiscus* is confined to the ecotone of plateau–edge woodland, which is only a few meters wide. Chimpanzee food resources are also highly seasonal, in keeping with the predictable, stark annual cycle. Succulent fruits of *Spondias mombin* (September), *Pseudospondias microcarpa* (April), and *Diospryros mespiliformes* (January) have regular, 1-month-long seasons, while other species fruit regularly but for longer: *Saba senegalensis* (May–July), *Adansonia digitata* (November–January), and *Grewia lasiodiscus* (October–November).

Fauna

SEMLIKI

Grasslands historically supported populations of Uganda kob in numbers estimated as high as 20000, but during the political unrest from 1971 to 1986 numbers were reduced to approximately 2000. Buffalo, reedbuck, waterbuck, bushbuck, and elephant also occur in the reserve, but hartebeest are extinct. Warthog were reported extinct in the late 1970s (Verner & Jenik 1984), but we have seen them almost daily since 1998. Leopard are heard nightly. Other large predators are lion and spotted hyena. Five species of primates are sympatric with the apes: redtail monkey, black-and-white colobus, blue monkey, vervet, and baboon.

Redtail monkey and black-and-white colobus range throughout the rift-floor section of the Mugiri, but only a few hundred meters up the escarpment slopes. The blue monkeys' range begins where the redtails' range finishes, originating on the lower slopes of the escarpment and continuing up to the reserve boundary. Baboons range over the chimpanzee community's whole range. The vervet range overlaps little with that of chimpanzees since these monkeys are found predominantly in the *Hyparrhenia–Themeda* savanna.

Baboon, redtail monkey, and blue monkey are the main competitors for fruits exploited by the apes, all having been seen to eat fruit species that chimpanzees prefer. Baboons forage in the same trees at the same time as chimpanzees.

ASSIRIK

Chimpanzees are sympatric with more potential predators at Assirik than at any other study site in Africa: lion, leopard, spotted hyena, and African hunting dog. All hunt nocturnally, and hyena and leopard are heard almost nightly. Apart from a few extinctions in historical times (giraffe, ostrich, topi), the large vertebrate fauna is intact, although elephant are seriously threatened by poaching. Chimpanzees share plateau water holes with eland, roan antelope, hartebeest, warthog, buffalo, and so on, as well as the more usual bushbuck, duiker, and so on. The remaining fauna is savanna-typical: jackal, oribi, ground squirrel, and so on.

Because of the dry habitat, the primate fauna at Assirik is impoverished by comparison with other chimpanzee study sites. There are no colobines, and the only guenon is the green (vervet) monkey. Baboons thrive, and this may be the only place in Africa where chimpanzee and patas monkey coexist. At least two nocturnal prosimian species are at Assirik: bushbaby and potto (both eaten by chimpanzees); only baboons and green monkeys compete for many of the same food-species that chimpanzees eat (McGrew *et al.* 1988).

Population

SEMLIKI

Since 1996 most of the reserve has been surveyed on foot, and nest and chimpanzee calls have only been found in four areas. The pattern of pant-hooting and the geographic spread of nests suggests that there are four communities of chimpanzees in Semliki (Figure 2.1, stippled areas). Chimpanzees have been heard and nests found far in the north of the reserve, along the forested banks of the Muzizi River, northeast of the Mugiri community. No nests have been found nor calls heard in the more than 10-km-wide stretch of land between the Muzizi and Mugiri rivers, suggesting separate communities. Along the Wasa River we have seen and followed groups of chimpanzees on 12 occasions. We did not recognize individuals, and the apes were clearly more wary than those of the Mugiri community, suggesting that the Wasa community is distinct. In the Nyabaroga Valley, near the reserve's entrance, is another community of chimpanzees, as evidenced by nests and pant-hoots. The southernmost site of chimpanzee nests at Nyabaroga is 27 km from the northernmost observation of Wasa chimpanzees, which suggests that they are also distinct communities.

ASSIRIK

Only one community of chimpanzees appears to live in the Assirik study area, and it is likely to be the only one in the park (Tutin *et al.* 1983). The rest of the area is too dry to support the apes year-round. Only to the south of Assirik are there watercourses that allow access to the Gambia River, at its closest point about 20 km away. In the other directions, drinking water is absent, except for residual pools in the Niokolo Koba River, but no nests were found beyond 10 km north of Assirik (Baldwin *et al.* 1982). It is unclear if the Gambia is a barrier to chimpanzee dispersal; by the end of the dry season it can be crossed at certain points by rock-hopping.

Chimpanzees exist in greater-than-expected numbers outside the park to the south and east, on the south side of the Gambia River (Pruetz *et al.* in press). Whether there is gene and meme flow between these communities and Assirik remains to be seen.

Climate

SEMLIKI

In August, September, October, November, and April, measurable rainfall averaged over 15 days per month, and total rainfall averaged over 100 mm (see Table 2.1). December and March rainfall was over 100 mm as well, though each averaged fewer than 15 days of rain. Thus, Semliki has a main rainy season from August to December, and a season of short rains in March and April.

Previous estimates of annual rainfall at Semliki suggested that it varied from 700 to 1300 mm (Pratt & Gwynne, 1977). We measured 1439 mm from 1 September, 1997 to 31 August, 1998. The next year yielded only 973 mm, dropping the 2-year average to 1206 mm. Annual rainfall of 1206 mm is 300 mm less than all other research sites except Assirik. Even the wet year of 1439 mm places Semliki as drier than all chimpanzee sites except Assirik.

Semliki is hot and humid. The mean daily high temperature was 34 °C, and the mean daily low was 20 °C. Only Assirik was hotter. Relative humidity daily maximums

Table 2.1. *Climatological data from study sites of wild chimpanzees, ranked by annual total of rainfall*

Site (Country)	Annual mean (mm) days of rain	Monthly mean (mm)												Temperature (°C)	
		Jan	Feb	Mar	Apr	May	Jun	Jul	Aug	Sep	Oct	Nov	Dec	Mean max.	Mean min.
Assirik (Senegal)[a]	954	0	0	0	8	54	141	205	215	209	110	10	3	35	23
	84	0	0	0	1	4	11	19	19	19	8	2	1	—	—
Semliki (Uganda)[b]	1206	54	59	103	196	85	34	55	142	106	107	132	127	34	20
	151	10	5	10	21	11	7	7	17	17	18	17	11	—	—
Lopé (Gabon)[c]	1531	—	—	—	—	—	—	—	—	—	—	—	—	—	—
Nouabale-Ndoki (PRC)[c]	1540	—	—	—	—	—	—	—	—	—	—	—	—	—	—
Kibale (Uganda)[c]	1671	—	—	—	—	—	—	—	—	—	—	—	—	23	16
Budongo (Uganda)[d]	1684	—	—	—	—	—	—	—	—	—	—	—	—	28	14
	167	—	—	—	—	—	—	—	—	—	—	—	—	—	—
Gombe (Tanzania)[e]	1775	174	90	183	212	157	17	0	0	0	71	272	241	28	19
	152	24	16	19	25	14	3	0	0	0	7	20	24	—	—
Kahuzi-Biega (DRC)[c]	1800	—	—	—	—	—	—	—	—	—	—	—	—	—	—
Taï (Ivory Coast)[c]	1829	—	—	—	—	—	—	—	—	—	—	—	—	—	—
Mahale (Tanzania)[c]	1836	250	182	271	366	101	2	1	5	34	81	232	238	27	19
	141	—	—	—	—	—	—	—	—	—	—	—	—	—	—
Okorobiko (Mbini)[a]	2112	—	—	—	—	—	—	—	—	—	—	—	—	24	18
	169	—	—	—	—	—	—	—	—	—	—	—	—	—	—
Bossou (Guinea)[e]	2230	—	—	—	—	—	—	—	—	—	—	—	—	—	—

Notes:

[a] McGrew *et al.* (1981); [b] Hunt, unpubl. data; [c] Appendix, McGrew *et al.* (1996); [d] Reynolds, V., pers. comm., 2000; [e] Moore, J., pers. comm., 2000.

averaged 95% in the wet season and 92% in the dry season (Hunt *et al.* 1999).

ASSIRIK

Assirik is hot and dry. Table 2.1 shows a mean annual rainfall of 954 mm, but the median is notably less, at 885 mm. This is because one atypically wet year (1224 mm) pulled the overall average up from 891, 824, and 879 mm. Based on another measure – number of days per annum of recordable rainfall – at 84 days of rainfall on average (versus 151 at Semliki), Assirik stands out as much drier than other sites for which data are available. This makes Assirik by far the driest site of chimpanzee study for which rainfall has been published (McGrew *et al.* 1981).

Not only is rainfall less at Assirik, it is also more delimited by seasonal extremes. With November to mid-May as the dry season, it is the only site of chimpanzee study where more months of the year are dry than wet (7:5, versus 4:8 for Gombe and Mahale).

Assirik is always hot – day or night, but humidity varies dramatically. Mean daily high air temperature in woodland never falls below 30°C, even in the coolest months, which are November–December. Mean daily low air temperature in woodland never falls below 20°C, even during this coolest period. Humidity varies from near 100% in the rainy season in July–October to below 50% in December–February (Baldwin 1979).

Community area, numbers, seasonality, and ranging

SEMLIKI

As of 31 October, 1999, we had found chimpanzees on 600 days. In seeking to determine a community's home range

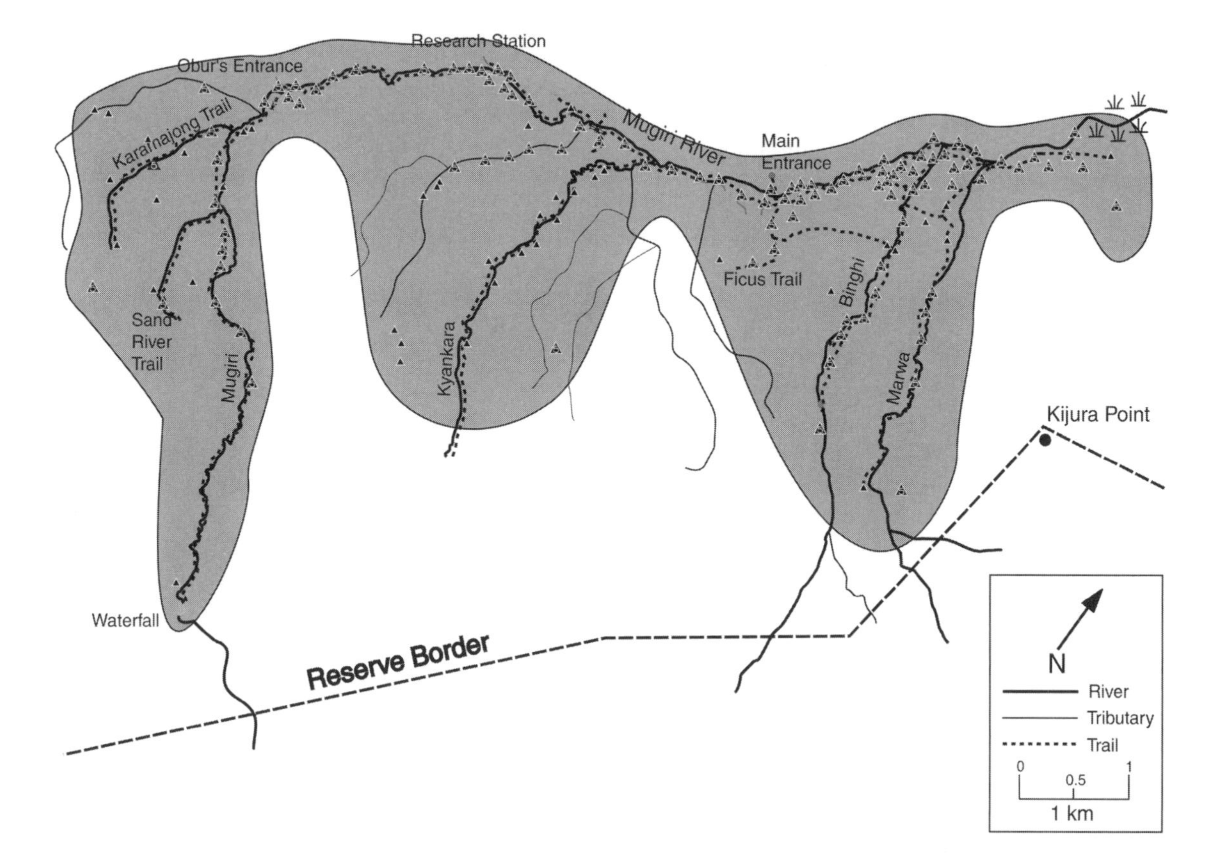

Fig. 2.2. Mugiri River and tributaries. Small triangles indicate the location of our first sighting of day, large triangles indicate five first sightings. Stippling represents a rough approximation of range. The minimum convex polygon area for the range encompassed in first sightings is 38.3 km^2.

when individuals' ranging cannot be seen directly in nest-to-nest follows, indirect measures are necessary (Baldwin *et al.* 1982). The most conservative measure, a minimum convex polygon enclosing first-sighting-of-day locations, yields an area of 38.3 km^2. The minimum convex polygon area for all sightings of individuals believed to be in the Mugiri community is 50.1 km^2. None of these sightings is in the savanna north of the Mugiri, though we have occasionally seen apes there. They commonly use the open *Piliostigma–Acacia* bushland, especially between the Sand and Mugiri Rivers (e.g. 23 first sightings are in these areas; see Figure 2.2). The Mugiri community numbers at least 29, which is the largest party seen so far. Several known individuals were not seen in this large party, suggesting a larger community of perhaps 40–50 members of all ages.

Month-by-month distribution of first-sightings revealed seasonal differences. The Mugiri habitat can be divided into a rift floor zone (<250 m from the rift valley portion of the river) and an escarpment zone (>250 m southeast of the rift valley portion of the Mugiri and the escarpment portion of the river). Chimpanzees spent significantly more time in the more-watered rift floor zone in the dry season (76.5% of first-sightings, $n=294$, compared to only 57% in the wet season, $n=246$; $\chi^2=30.3$, $p<0.001$). This may relate to the distribution of two food resources, *Grewia* and *Beilschmiedia*. Found only on the escarpment in relatively open habitats, *Grewia* fruits are only eaten in the wet season. By contrast, *Beilschmiedia* fruits, found only on the rift floor, are only eaten in the dry season. Another food source, *Cynometra*, occurs only on the rift floor and it is eaten in both seasons, but especially during the short rains in March and April.

Feeding party size averaged 4.8 individuals (SD$=4.1$, $n=384$; numbers include independent juveniles, but not offspring carried by mother), but poor habituation likely renders these data as underestimates. Party size in the wet season versus the dry season did not differ; large parties were eating *Cynometra* pods in February (mean$=5.11\pm4.13$, $n=46$).

Table 2.2. *Characteristics of night nests of wild chimpanzees, compared across study sites*

Site	Nest tree height (m)	Nest tree DBH (cm)	Nest height (m)	Distance to next nest (m)	Nest group size	Nests per tree
Assirik[a]	13	42	11	4	4	2
Semliki[b]	14	45	11	5	4	1
Lopé[c]	—	25	10	6	1	—
Taï[d]	—	33	20	—	—	—
Okorobiko[a]	—	16	10	4	2	1
Sapo (Liberia)[e]	—	52	12	—	1	—

Notes:

[a] Baldwin *et al.* (1981), median; [b] Hunt, unpubl. data, median; [c] Wrogemann (1992), median; [d] Fruth & Hohmann (1994), median; [e] Anderson *et al.* (1983), median.

ASSIRIK

At Assirik, between February 1976 and December 1979, we saw wild chimpanzees 367 times; much more often they were heard but not seen (Tutin *et al.* 1983). The number of identified individuals of all ages was 24, and the calculated likely size of the community was about 28. The overall median party size was four (range: 1–22), which is similar to other populations of wild chimpanzees, but given the smaller-than-usual community size, Assirik apes on average were more sociable. Parties of mixed composition (adults of both sexes, plus immature offspring) were larger in more open (grassland and plateau) habitats than in less open (woodland and forest) ones. Travelling parties were much larger (median = 19 individuals) than were parties engaged in eating (six), resting (seven), or socializing (five). Party size did not differ between wet and dry seasons, despite the highly seasonal nature of the site.

A minimum convex polygon enclosing the area from sightings of identified individuals yields 37.4 km^2. A measure of the continuous distribution of all chimpanzees in the study area amounts to 51.4 km^2. Continuous distribution of nests, plus sightings, totals 72.1 km^2, a range that exceeds the home-range size of all habituated communities studied, that is Bossou, Budongo, Gombe, Kibale, Mahale, and Taï. If density is taken into account, then a projected home range can be calculated and amounts to 278–333 km^2 (for details of analysis, see Baldwin *et al.* 1982).

Nests

SEMLIKI AND ASSIRIK

At Semliki, overnight nests were found in groups averaging 5.0 (±3.23, $n = 348$, max = 12) (Table 2.2). The median group size of four was the same as at Assirik (Baldwin *et al.* 1981), but was larger than at Sapo (Anderson *et al.* 1983) and Lopé (Wrogemann 1992), each at one, and at Okorobiko (Baldwin *et al.* 1981), two. At Semliki, *Cynometra* was the preferred species for nests at 147 out of 181 nests of known species; *Cola* (17 nests), *Phoenix* (nine), *Albizia* (five), *Beilschmiedia* (two) and *Pseudospondias* (one) made up the rest of the sample. Sept (1992) also reported *Cynometra* as the species of choice for nests at Ishasha, southern Uganda.

All Semliki nests were in riverine forest and were equally distributed between the rift floor and the escarpment. Mean and median nest heights at Semliki were 11 m (sd = 5.81, $n = 324$, range 1.5–47.8). These resemble medians of 10 m at Okorobiko, 11 m at Assirik (Baldwin *et al.* 1981), 12 m at Sapo (Anderson *et al.* 1983), and 10 m at Lopé (Wrogemann 1992). Although the range at Taï (4-45 m) was similar to Semliki, the mean (23.2 m) and median (20 m) at Taï were much higher.

Distances between nests averaged 5.4 m at Semliki, versus 4 m at Okorobiko and Assirik (Baldwin *et al.* 1981), and 6 m at Lopé (Wrogemann 1992). The median diameter at breast height (DBH) of nesting trees at Semliki was 45 cm (range: 6–197 cm). This is in the Okorobiko range 13–19 cm and similar to medians at Assirik (42 cm) and Sapo (52 cm; Anderson 1983). Medians were smaller at Taï (33 cm; Fruth 1990; Fruth & Hohmann 1994) and Lopé (25 cm; Wrogemann 1992).

Nest height, tree height, and number of nests per group did not differ seasonally at Semliki, although trends were similar to those seen for chimpanzees elsewhere (Baldwin *et al.* 1981; Wrogemann 1992). However, nest-tree diameter (mean = 54.0 cm dry vs. 45.5 cm wet, T-test: $n = 285$, 120, $p = 0.02$) and distance between nests (5.6 m dry vs. 4.7 m

Table 2.3. *Technology of wild chimpanzees at Assirik*

Type	Prey species	Common name	Frequency[a]	Evidence[b]
Ant fish[c]	*Camponotus* sp.?	Wood-boring ant	H	U R U
Ant peel[d]	*Oecophylla longinoda*	Weaver ant	H	O R F
Ant dip[d]	*Dorylus nigricans*	Driver ant	H	U R F
Baobab crack[d,e]	*Adansonia digitata*	Baobab tree	C	O R F
Branch clasp	—	—	H	O – –
Branch shake[d]	—	—	H	O – –
Buttress beat	—	—	C	O U –
Honey probe[e]	*Apis mellifera*	Honey bee	H	U R F
Leaf napkin[d]	—	—	P	O U –
Nest build[d,f]	—	Woody spp.	C	O R –
Nest line[d,i]	—	—	H	O – U
Play start[i]	—	—	P	O U –
Termite fish[d,g]	*Macrotermes subhyalinus*	Termite	C	U R F
Throw object[i]	—	—	P	O U –
Water dig[h]	—	Surface water	H	U R –

Notes:

[a] P, present; C, customary; H, habitual; [b] O, observation; F, fecal analysis; R, remnant/artifact; –, not applicable; U, unknown (See McGrew *et al.* 1988); [c] McGrew (1983); [d] Baldwin (1979); [e] Bermejo *et al.* (1989); [f] Baldwin *et al.* (1981); [g] McGrew *et al.* (1979), McBeath & McGrew (1982); [h] Galat-Luong & Galat (2000); [i] McGrew *et al.* (unpublished data).

wet, T-test: $n = 120$, 285, $p = 0.04$) were both greater in the dry season.

At Semliki, of 348 nests, 62% were covered, that is, constructed under foliage versus open to the sky. This value resembles those for forested sites: 79% at Sapo, 61.8% at Lopé, and 83% in Guinea (see references above), but is much greater than the 25% recorded at Assirik (Baldwin *et al.* 1981). More nests were covered in the dry season (188 of 266) than in the wet season (44 of 86) ($\chi^2 = 42.5$, $p < 0.001$, df = 1), a trend observed elsewhere (e.g. Baldwin *et al.* 1981).

Behavior

SEMLIKI

Tool use at Semliki was seen rarely, perhaps because habituation is still incomplete. The chimpanzees make and use leaf sponges for gathering water, even when water is easily accessible without a tool. A female sitting on a stone in the center of the Binghi tributary, surrounded by water, was seen to use a wadded leaf to drink water. No evidence of ant or termite fishing has emerged; no fishing tools have been found, despite thorough searches in likely areas, nor have ant or termite body-parts been seen in feces.

ASSIRIK

Fifteen patterns of technology (i.e., material culture) have been described for the chimpanzees of Assirik, but there are observational data for only ten (see Table 2.3). Other than *Nest build* (ubiquitous for great apes, treated above) three other patterns are known to be customary (i.e., shown normatively, at least by appropriate age and sex classes).

Baobab crack is cracking open the hard-shelled fruit of this arid-country tree by pounding it on an anvil of stone, root, or bough. *Termite fish* is inserting a flexible probe of vegetation into the earthen mound of a large species of macrotermitinine termite, then extracting the insects for eating (McGrew *et al.* 1979; McBeath & McGrew 1982). *Buttress beat* is drumming on the buttress roots of large trees, using palms or soles, to produce a characteristic, long-distance signal.

Of the other 11 types of technology, there are enough data for eight of them to be termed habitual (i.e., repeated instances by several individuals). *Ant dip* is the use of long, straight wands of vegetation to harvest driver ants from their surface nest (McGrew 1983). The excavated nests and abandoned tools are obvious when present, as are fragments of the ants in the apes' fecces. *Ant peel* is a food-processing

Table 2.4. *Plant parts in diets of wild chimpanzees compared across study sites*

	Plant parts eaten[a]							
Site	Fruit	Flower	Seed	Leaf	Stem	Bark	Root	Total
Assirik[b]	34	6	6	6	2	4	2	60
	57%	10%	10%	10%	3%	7%	3%	
Semliki[c]	13	1	5	10	3	1	0	33
	39%	3%	15%	30%	9%	3%	—	
Lopé[d]	106	5	17	21	9	3	0	161
	66%	3%	10%	13%	6%	2%	—	
Gombe[e]	86	19	14	54	15	13	0	201
	43%	9%	7%	27%	7%	7%	—	
Mahale[f]	87	21	13	100	36	14	0	271
	32%	8%	5%	37%	13%	5%	—	
Bossou[g]	128	11	16	45	33	7	6	246
	52%	5%	7%	18%	13%	3%	2%	

Notes:

[a] Classification follows Peters & O'Brien (1981): Fruit = fruit, including nuts; Flower = flower + inflorescence; Seed = seed + pod; Leaf = leaf & shoot; Stem = stem + stalk (pith); Bark = bark, cambium, gum, sap, resin; Root = underground storage organ; [b] McGrew et al. (1988, Table IV); [c] Hunt, unpubl. data; [d] Tutin & Fernandez (1993, Table 1); [e] Wrangham (1977, Table II); [f] Nishida & Uehara (1983, Table II); [g] Sugiyama & Koman (1992, Table II).

task in which a silk-bound leafy bundle of weaver ant nest is detached, crushed, and then peeled leaf-by-leaf. The leafy remnants are characteristic, and the ants are commonly found in fecal samples. *Nest line* is the use of detached leafy twigs to line the inner surface of the nest or sleeping platform. *Ant fish* involves flexible probes of vegetation to extract wood-boring ants from their arboreal nest. *Branch shake* is an attention-getting pattern used in courtship, usually from male to female. *Branch clasp* is holding onto vegetation with an arm extended overhead, while engaged in mutual grooming. *Honey probe* yields distinctive, sticky artifacts. Bee remnants are common in feces, but the behavior has not been seen (Bermejo *et al.* 1989). Galat-Luong & Galat (2000) reported that Niokola Koba chimpanzees use sticks as tools to dig wells in riverbeds (*Water dig*). We found wells at Assirik but cannot be sure who dug them.

Finally, three behavioral patterns have been seen only once, *Throw object*, *Leaf napkin*, and *Play start*.

Diet

SEMLIKI

Although new food items may yet be recorded, the declining rate of new dietary additions suggests that the Mugiri chimps have a small dietary repertoire (see Table 2.4). When at least one eating bout was seen on one of 199 days that were part of the more than 3 years of study, and when 72 fecal samples were screened, only 36 types of food-items were recorded, including those that were unidentified. We have not added a new item to the food list since 14 September, 1999.

After only 3 years of study, annual cycles of diet are necessarily preliminary, but one pattern is clear. *Cynometra* pods appeared in January in all 3 years, and they were eaten avidly until depleted in March or April. Large pods were often broken open and only the seeds consumed, but smaller pods were eaten whole. *Cynometra* may be the staple food item at Semliki; at least one feeding bout consisting of *Cynometra* pods was recorded in 11 of the 39 months for which we have data. It may be the Semliki chimps most preferred food. In the early morning, chimpanzees often uttered food-grunts as they approached *Cynometra* trees and as they began to eat; otherwise, these vocalizations have not been heard at Semliki. The second most common food item was *Saba florida* fruit (8 months), which was eaten most often in August but also in May, June, and July. *Tamarindus indica* fruit was a staple in November and December (7 months). *Phoenix reclinata* fruit seems to be a fallback item, since it was ignored when other items were available (6 months). *Cola gigantea* arils were chewed and sucked off seeds, which were then spat out. *Cola* arils seem to be less preferred than many other food items since it was often ignored.

Chimpanzees wadged small fruits and piths, but their treatment of *Phoenix* pith was unusual. When eating this item, chimpanzees removed the outer cortex of palm frond ribs, folded the pith into an accordion shape, placed the resulting bolus in the mouth, and ejected it after sucking only a short time.

During June 1998, we saw chimpanzees stripping bark from *Acacia siberiana* trees and eating the cambium. We found stripped bark with chimpanzee-sized tooth marks in areas chimpanzees frequented on a weekly basis in the following months of July and August. This may be an unusual fallback food since this occurred at the end of the dry season in a very dry year.

We often found chimpanzees near black-and-white colobus and redtail monkeys, and occasionally found the apes among baboons or blue monkeys, but hunting was rare. In 39 months, we have evidence of only seven colobus predation episodes. We saw two colobus killed, and heard three hunts. Colobus hair was found in two of 72 analyzed fecal samples. There was no other evidence of carnivory, but eggshells were found twice.

ASSIRIK

As at Semliki, the dietary repertoire of 60 plant parts eaten by the apes from Assirik is small by comparison with their counterparts elsewhere in Africa (see Table 2.4). As with most populations of wild chimpanzees, fruit was the predominant plant part eaten, followed by leaves (McGrew *et al.* 1988).

Figs (*Ficus* spp.) of at least three (and probably seven) species were the only plant food eaten in every month of the study, as determined by analysis of fecal contents. Second ranking were the fruits and seeds of baobab, a dry country tree; at least some part of this species was eaten from August through May. Assirik seems to be the only study site at which chimpanzees and baobabs coexist. Eaten equally often were the immature seeds and mature arils of *Afzelia africana*, a big-podded leguminous tree. The most-often eaten succulent fruit is *Saba senegalensis*, the tart, juicy matrix of which is avidly consumed. This genus, like *Ficus*, is commonly eaten by chimpanzees across Africa (see above). As at Semliki, *Tamarindus indica* has a short but intensive season in November–December. None of the nut-bearing genera of trees (*Elaeis*, *Coula*, *Panda*, etc.) were present, as the climate is too dry.

Faunivory occurred, but the prey profile was unlike that of any other population of wild chimpanzees. There was no evidence of consumption of monkeys or ungulates, but instead, two species of nocturnal prosimian, *Galago senegalensis* and *Perodicticus potto*, were eaten. These were apparently not actively pursued at night but, rather, were taken from nest holes or sleeping places by day. The apes ate six species of invertebrates, all social insects. One, the fiercely biting and stinging army ant, *Megaponera foetens*, seems to be eaten only at Assirik (McGrew 1983).

Habituation

SEMLIKI

There has been steady progress on habituation (Hunt *et al.* 1999), but the rate of progress has been slow compared with other populations. It seems unlikely that this is due to encountering chimpanzees only rarely since we approached individuals closely enough that they were aware of us 20 days per month. Yet despite this proximity, direct observation of chimpanzees averaged only 3.4 hours per month. In early observations, some Mugiri chimpanzees were wary but curious when they saw us, suggesting that they had had little prior contact with humans. Only 3 months after the study began, a female approached KDH to within 15 m and stared for 90 seconds before slowly moving away. This occurred only 2 km from the nearest human small holding farms. In the first year of study, many females were less wary of us than were all males. In October 1997, KDH stood for an hour 4 m directly below two females feeding in a *Cynometra* tree. In contrast, during 1996 and 1997 males never allowed observers to approach closer than 50 m without fleeing. As at Assirik, chimpanzees in trees were more tolerant of human observers than were those chimpanzees in open areas.

ASSIRIK

With even wider-ranging subjects, progress toward habituation of the chimpanzees was slow: After 44 months of study, the apes tolerated observers at 20–50 m distance, depending on the type of habitat (Tutin *et al.* 1983). The chimpanzees were most tolerant when arboreal in forest and least tolerant when terrestrial in the open. The ratio of hours spent searching to hours spent in contact with the subjects fell from 27:1 in 1976 to 10:1 in 1979, with approximately equal amounts of time spent searching. However, there were marked seasonal differences in the chimpanzees' observability. The best months were the late dry season (March–May) when the apes concentrated their time in gallery forest for food, shade, and nesting. The worst months were during the early wet season (July and August) and early dry season (November and December) when they ranged widely in the non-forested parts of the study area. Few individuals provided enough data to be conclusive, but one adult male, Brown Bear, had a home range of at least 27.6 km, which is bigger than the community ranges of all habituated populations of chimpanzees.

Wells

Dry habitats offer a tough challenge to chimpanzees. When temperature or humidity is high, surface water may be seasonally scarce. When water is unavailable at Semliki, chimpanzees dig drinking holes in dry, sandy riverbeds (Hunt *et al.* 1999; Hunt *et al.* unpublished data), as they do elsewhere (Nishida *et al.* 1999; Galat-Luong & Galat 2000). However, drinking-hole or 'well' digging is not limited to locations that lack standing water. On 12 August, 1997, KDH saw a female chimpanzee dig a hole in the sand less than a meter from the clear, gently flowing Mugiri River. She turned away and plucked a leaf, then used the leaf to dip water out of the newly dug hole. The 'well' was 30 cm deep and had a maximum diameter of 20 cm. After that, chimpanzees were seen digging six wells. We also found holes surrounded by chimp hand- and foot-prints (but no other animal tracks) and whole or wadded leaves (presumably sponges), suggesting that these were also wells. We recorded 132 wells at 36 sites.

All wells were dug in sand. There was no evidence that tools were used to dig the wells, neither by direct observation, nor impressions in sand, nor items near holes that might have been used as digging tools. The wells were shallow, averaging only 15 cm in depth (range: 3–50 cm, sd $=9.8$, $n=96$), though a few were up to 50 cm deep. Wells were sometimes so narrow that apes seemed to have trouble extracting the hand or leaf used to sponge the water. Average well diameter at the surface was 20 cm for the narrowest dimension (range: 7–43 cm, sd $=6.6$, $n=96$) and 28.5 cm for the widest (range: 29–50 cm, sd $=7.4$, $n=96$). Deeper holes were sometimes triangular in cross-section, with evidence of first one hand then the other used for digging. This method of alternate hand use was seen once.

Wells were often dug in clusters, averaging 5.9 wells at each well site. Such wells were tightly packed, averaging only 4.4 cm apart (range: 3–29 cm, sd $=7.1$, $n=79$).

Forty of 127 wells had one or more leaf sponges nearby. We counted sponges as being present at a well site if they were found within a cluster, even if they were not found directly by a particular well.

Low rainfall was a good predictor of well-digging, as 121 wells were dug in the dry season versus only 11 in the wet season. Furthermore, 53 wells were dug during the month with the least rainfall. There was a negative correlation between number of wells in a month and rainfall per month (Pearson 2-tailed, $r=-0.41$, $p=0.04$, $n=26$ months). In short, many of the wells were dug because there was little standing water available for drinking.

Access to water is not the only function of the wells, however, since 78 of 132 wells were dug near, often very near, surface water. The distance from the hole to the water source averaged 8.5 m, and one well was only 15 cm from the nearest water source. Many of the water sources were rivers or streams that looked potable, and some wells were found near the water source used for our drinking water. Other wells were dug near pools that looked to be poor sources of drinking water. Twenty-six wells were dug near algae-choked pools in otherwise dry streambeds, as seems to have been the case in Senegal (Galat-Luong & Galat 1999).

The presence of many wells near standing water suggests that some wells were dug to filter out particulates or other contaminants such as leaf litter, algae, woody detritus, plant parts, suspended soil, or the larvae of intestinal parasites. The only data on wells dug by chimpanzees in Senegal seem to be those of Galat-Luong & Galat (2000).

Injuries

SEMLIKI

Despite finding snares, we have seen no apes with snare injuries. All kinds of injuries, such as ear bite, facial scar, missing digit, and healed wound, are rare at Mugiri. One old male, Bahati, has intact ears and digits. None of the three individuals identified to date at Semliki have damaged ears.

ASSIRIK

We saw no chimpanzees with typical snare injuries to the hands or feet. Almost all other injuries seen were the usual scars of healed wounds, such as notched ear, missing digit joint, and so on. Of the thirteen identified adults, seven (Brown Bald, Brown Bear, Colin, Grizzle, Kojak, Muldoon, One-Eyed Sam) had at least one damaged ear, while six (Burdock, Guy, Ptarmigan, Snoopy, Unnamed 6, Unnamed New Valley) did not. In addition to his chronic blindness in one eye (probably a cataract), One-Eyed Sam was severely attacked at least twice, suffering a ripped scrotum and bitten-off finger. He died shortly after the last attack (Baldwin 1979).

DISCUSSION

The climates at Assirik and Semliki are in stark contrast to those of other chimpanzee research sites. Daily temperature maxima are +6 °C hotter than the next hottest site, and rainfall averages over 300 mm less. Despite these differences, there are no readily apparent adaptations to heat stress, such as relative hairlessness or high surface area to volume ratios, for example longer limbs. Chimpanzees cope with these hot, dry habitats by limiting most of their activity to gallery forests.

At both sites, forested areas are only a fraction of the community's range, yet at both sites most daily activity occurs in forests. That chimpanzees avoid open areas is evidenced in the greater range sizes of dry-habitat chimpanzees. Ranges for dry-habitat chimpanzees are 50 km^2 or larger, compared to 4–24 km^2 at Gombe (Wrangham, 1979; Goodall, 1986) and 7–14 km^2 at Mahale (Hasegawa & Hiraiwa-Hasegawa, 1983). Chimpanzees in dry sites apparently must range farther because forests are a smaller part of their habitat. Although no anatomical study of dry-habitat chimpanzees has been done, we see no evidence of adaptations in body size or proportions to this long-distance ranging.

At Assirik and Semliki, greater cloud cover and lower temperatures in the wet season seemed to reduce heat stress, so that chimpanzees were able to use more open parts of their habitat during those periods (Baldwin *et al.* 1982; Collins & McGrew 1988). There is no evidence of such seasonal patterning at wetter sites (Kano 1971). If early hominins had a similarly seasonal pattern of habitat usage, exploiting forest foods more in the dry season and open habitat resources in the wet season, one might expect to find different artifact accumulations at wet- and dry-season sites. Early hominin archaeological assemblages differ by site, and are sometimes interpreted as being sites of tool manufacture, single-prey butchery, and as loci for socializing and food processing (Sept 1992). The differences between the sites may be due to seasonal differences in resource use – different resources requiring different tools.

Underground storage organs, such as roots, tubers, bulbs, and corms have been suggested as possible early hominin dietary staples (Peters & O'Brien 1981; Wrangham *et al.* 1999). Yet, dry-habitat chimpanzees do not utilize these items at all (but see Lanjouw, Chapter 3). If early hominins specialized in such resources, it would signify a clear niche separation between apes and humans.

At Assirik and Semliki, chimpanzees dig holes to get water in the dry season; at Semliki, apes dig drinking holes even when water is readily available. If early hominins occupied even drier habitats and more often used open canopy zones, then water was probably a limiting resource for them. Therefore, we might expect to find more open-habitat hominins in lake- and river-side environments.

Food-item lists at Assirik and Semliki are much shorter than at other sites. Although it is not clear that there is any systematic difference in dietary characteristics between dry-habitat and wetter-habitat chimpanzees (Table 2.4), a limited food list might encourage physiological or anatomical specializations more readily than would be the case for closed-forest generalists. Fossil hominins 2–3 million years old vary enough to warrant assigning the fossils to more than five different species. This sort of divergence suggests that different populations of hominin adapted to different local resources. Some fossil hominins display unexpected dental specializations (e.g. *Australopithecus garhi*; Asfaw *et al.* 1999), suggesting adaptation to a food–species list that has distinctive physical qualities and mechanical processing requirements. Less selective pressure on dentition, even in species-poor and variable habitats, may have occurred with the emergence of stone tools for food processing at 2.5 million years ago (mya) (Semaw *et al.* 1997). Thus, fossil hominins found in drier habitats are more likely to display distinctive population dental morphology before, but not after, 2.5 mya.

Large predators are common in dry habitats. Assirik and Semliki chimpanzees make their nests in larger groups than at other sites, perhaps in response to the presence of more predators than at other sites. Larger party sizes may also relate to lower population densities, which might make it difficult for individuals to find one another once separated. Great body size dimorphism in early hominins (Leigh & Shea 1995) suggests male–male competition (Plavcan 1999). If early hominins foraged in large parties, as is suggested by party size in these dry-habitat chimpanzees, and if hominins lacked strong female–female social bonds, as is typical of apes (Wrangham 1986), then early hominin social structure may have been much like that of the hamadryas baboon. If so, then principal social bonds were between a breeding male and several unrelated females, and larger groups might contain several such breeding units.

Wide-ranging, dry-habitat chimpanzees are more difficult to habituate than wet-habitat ones. At both sites, chimpanzees are most wary in open habitats, either on the ground or in trees. Chimpanzees are most comfortable with observers when relatively high in trees in closed-canopy forest, where overlapping tree crowns allow escape. Although chimpanzees seem less frightened when in the trees than on the ground, they are fearful if the tree crown is isolated. Perhaps such discomfort in open canopy contributes to their wariness of observers, which in turn, impedes habituation.

Although quantitative data on positional behavior are lacking, we found nothing to suggest that dry habitats encourage more terrestriality. In an earlier study (Hunt 1989), the more closed habitat at Mahale elicited no more climbing than the more open habitat at Gombe, nor were chimpanzees at Gombe more terrestrial than at Mahale. Feeding is mostly an arboreal activity in chimpanzees. Although trees were farther apart in the more open Gombe habitat (Collins & McGrew 1988), chimpanzees did not shift to more terrestrial food items.

Early hominin morphology suggests substantial arboreal positional behavior (Stern & Susman 1993; Hunt 1994). If dry-habitat hominins responded as do chimpanzees to dispersed, low-diversity food resources, then they had day-ranges that were large compared to typical chimpanzees. A large day-range and arboreal gathering might encourage caching of tools, since carrying tools long distances or returning to retrieve them between arboreal feeding bouts would be energy- and time-consuming. However, these two sites do not suggest a link between dry habitats and tool use. Assirik evinces a typical chimpanzee tool kit, while Semliki seems technologically impoverished.

We also saw nothing to suggest that dry habitats elicit more hunting. Chimpanzees hunt more often in the moister habitats of Taï, Gombe, and Mahale (Boesch 1994; Stanford 1999). Preliminary evidence suggests that dry-habitat chimpanzee prey are smaller (Assirik), or that hunting is less frequent (Semliki), or both. While scavenging opportunities may be abundant in some dry habitats (Blumenschine *et al.* 1994), we saw no sign that carcass availability was higher at Assirik or Semliki compared to Gombe, Mahale, or Kibale.

Patterns of aggression appear to differ across habitats. Like Gombe, Mahale, and Kibale, Assirik chimpanzees show damaged ears and fingers attributable to fighting, but Semliki chimpanzees are largely unscarred. Snares cannot account for the differences, as they are absent at Assirik, Gombe, and Mahale, but frequent at Kibale and Semliki.

Although study of dry-habitat chimpanzees is still young, early evidence suggests that savanna chimpanzees show few patterns expected in the savanna ape hypothesis (Moore 1996). Compared with forest chimpanzees, open-country apes are not more terrestrial, do not use tools more intensively, do not focus on underground storage organs, do not hunt more often, and are not more violent. However, some expectations of savanna adaptation may be borne out. Chimpanzees at Assirik and Semliki may be under greater predation pressure and may range farther. Other aspects of dry-habitat chimpanzees are not surprising, but have not been widely considered. Chimpanzees in these habitats have small food-item lists, and they are probably more influenced by seasonality than elsewhere. While Assirik and Semliki chimpanzees share a number of distinctive patterns, they show some differences as well, such as prey choice in faunivory. Although there are some predictable patterns, we cannot yet generalize about The Savanna Chimpanzee as an ecotype, any more than we can generalize about The Chimpanzee as a species.

ACKNOWLEDGEMENTS

At Assirik, we thank: Service des Parcs Nationaux for permission to study in the Parc National du Niokolo Koba; Délégation Générale de Recherche Scientifique et Technique for permission to study in Senegal; Science Research Council, Carnegie Trust for the Universities of Scotland, L.S.B. Leakey Foundation, Wenner-Gren Foundation for Anthropological Research, American Philosophical Society, and Philip and Elaina Hampton Fund for financial aid; many persons who contributed data and help in the field, especially P.J. Baldwin, M.J.S. Harrison, L.F. Marchant, J.D. Pruetz, and C.E.G. Tutin; D.L. Deaton for word processing.

At Semliki, we thank: the Government of Uganda, particularly the National Research Council and the Uganda Wildlife Authority; Green Wilderness Group (U) Ltd for invaluable logistical and financial support; Chief Warden John Makombo and staff at the Semliki Wildlife Reserve for essential assistance; George Tuhairwe, Rachel I. Weiss,

James Latham, Alexander J. M. Cleminson, Esther Bertram, James Fuller and Aisling Wilson for contributions to long-term data collection initiatives and records; National Science Foundation grants BNS 97–11124 and BNS 98–15991, and Indiana University for funding.

We thank the editors for their help and patience.

This chapter is dedicated to Dr Pamela Jane Baldwin (1954–1999) who did the first Ph.D. on open country chimpanzees. She is sorely missed.

REFERENCES

Allan, C., Sivell, D. & Lee, T. (1996). Semuliki (Toro) Game Reserve, Uganda: Results of the Frontier-Uganda Biological Assessment. Report No. 7: Society for Environmental Exploration.

Anderson, J. R., Williamson, E. A. & Carter, J. (1983). Chimpanzees of Sapo Forest, Liberia: Density, nests, tools and meat-eating. *Primates*, 24, 594–601.

Asfaw, B., White, T., Lovejoy, O., Latimer, B., Simpson, S. & Suwa, G. (1999). *Australopithecus garhi*: A new species of early hominid from Ethiopia. *Science*, 284, 629–34.

Baldwin, P. J. (1979). The Natural History of the Chimpanzee (*Pan troglodytes verus*) at Mt. Assirik, Senegal. Ph.D. Thesis, University of Stirling.

Baldwin, P. J., Sabater Pi, J., McGrew, W. C. & Tutin, C. E. G. (1981). Comparison of nests made by different populations of chimpanzees (*Pan troglodytes*). *Primates*, 22, 474–86.

Baldwin, P. J., McGrew, W. C. & Tutin, C. E. G. (1982). Wide-ranging chimpanzees at Mt. Assirik, Senegal. *International Journal of Primatology*, 3, 367–85.

Bermejo, M., Illera, G. & Sabater Pi, J. (1989). New observations on the tool-behavior of the chimpanzees from Mt. Assirik (Senegal, West Africa). *Primates*, 30, 65–73.

Blumenschine, R. J., Cavallo, J. A. & Capaldo, S. D. (1994). Competition for carcasses and early hominid behavioral ecology: a case study and conceptual framework. *Journal of Human Evolution*, 27, 197–213.

Boesch, C. (1994). Hunting strategies of Gombe and Taï chimpanzees. In *Chimpanzee Cultures,* ed. R. W. Wrangham, W. C. McGrew, F. B. M. deWaal & P. G. Heltne, pp. 77–92. Cambridge, MA: Harvard University Press.

Bonnefille, R. (1984). Cenozoic vegetation and environments of early hominids in East Africa. In *The Evolution of the East Asian Environment, Vol. II: Palaeobotany, Palaeozoology and Palaeoanthropology*, ed. R. O. Whyte, pp. 579–612. Hong Kong: University of Hong Kong Press.

Collins, D. A. & McGrew, W. C. (1988). Habitats of 3 groups of chimpanzees (*Pan troglodytes*) in western Tanzania compared. *Journal of Human Evolution*, 17, 553–74.

Coppens, Y. (1988). Hominid evolution and the evolution of the environment. *Ossa*, 14, 157–63.

Dart, R. A. (1959). *Adventures with the Missing Link*. New York: Harper.

Eggeling, W. J. & Dale, I. R. (1951). *The Indigenous Trees of the Uganda Protectorate*, 2nd edn, Entebbe: Government Printer, Uganda.

Ellenberg, H. & Müller-Dombois, D. (1967). Tentative physiognomic-ecological classification of plant formations of the earth. *Beriche des Geobotanischen Institutes der Eidgenossichen Technischen Hochschule, Stiftung Rubel, Zurich*, 37, 20–47.

Fruth, B. (1990). Nussknackplatze, Nester und Populationsdichte von Schimpansen: Untersuchungen zu regionalen Differenzen im Sud-Westen der Elfenbeinkuste. Master's thesis, Ludwig-Maximilians Universität. (München).

Fruth, B. & Hohmann, G. (1994). Comparative analysis of nest building behavior in bonobos and chimpanzees. In *Chimpanzee Cultures*, ed. R. W. Wrangham, W. C. McGrew, F. B. M. de Waal & P. G. Heltne, pp. 109–28. Cambridge, MA: Harvard University Press.

Galat-Luong, A. & Galat, G. (2000). Chimpanzees and baboons drink filtered water. *Folia Primatologica*, 71, 258.

Goodall, J. (1986). *The Chimpanzees of Gombe: Patterns of Behavior*. Cambridge, MA: Harvard University Press.

Hasegawa, T. & Hiraiwa-Hasegawa, M. (1983). Opportunistic and restrictive matings among wild chimpanzees in the Mahale Mountains, Tanzania. *Journal of Ethology*, 1, 75–85.

Hunt, K. D. (1989). Positional behavior in *Pan troglodytes* at the Mahale Mountains and Gombe Stream National Parks, Tanzania. PhD dissertation, University of Michigan.

Hunt, K. D. (1994). The evolution of human bipedality: ecology and functional morphology. *Journal of Human Evolution*, 26, 183–202.

Hunt, K. D., Cleminson, A. J. M., Latham, J., Weiss, R. I. & Grimmond, S. (1999). A partly habituated community of dry-habitat chimpanzees in the Semliki Valley Wildlife Reserve, Uganda. *American Journal of Physical Anthropology*, Suppl. 28, 157.

Hutchinson, J. & Dalziel, J. M. (1927–1936). *Flora of West Tropical Africa*. London: Crown Agents.

Kano, T. (1971). The chimpanzees of Filabanga, western Tanzania. *Primates*, 11, 1–46.

Kano, T. (1972). Distribution and adaptation of the chimpanzee on the eastern shore of Lake Tanganyika. *Kyoto University African Studies*, 7, 37–129.

Leigh, S. R. & Shea, B. T. (1995). Ontogeny and the evolution of adult body-size dimorphism in apes. *American Journal of Primatolology*, 36, 37–60.

McBeath, N. M. & McGrew, W. C. (1982). Tools used by wild chimpanzees to obtain termites at Mt. Assirik, Senegal: The influence of habitat. *Journal of Human Evolution*, 11, 65–72.

McGrew, W. C. (1983). Animal foods in the diets of wild chimpanzees: Why cross-cultural variation? *Journal of Ethology*, 1, 46–61.

McGrew, W. C., Tutin, C. E. G. & Baldwin, P. J. (1979). Chimpanzees, tools and termites: cross-cultural comparisons of Senegal, Tanzania and Rio Muni. *Man*, 14, 185–214.

McGrew, W. C., Baldwin, P. J. & Tutin, C. E. G. (1981). Chimpanzees in a hot, dry and open habitat: Mt. Assirik, Senegal, West Africa. *Journal of Human Evolution*, 10, 227–44.

McGrew, W. C., Baldwin, P. J. & Tutin, C. E. G. (1988). Diet of wild chimpanzees (*Pan troglodytes verus*) at Mt. Assirik, Senegal: I. Composition. *American Journal of Primatology*, 16, 213–26.

McGrew, W. C., Marchant, L. F. & Nishida, T., eds. (1996). *Great Ape Societies*. Cambridge: Cambridge University Press.

Moore, J. (1996). Savanna chimpanzees, referential models and the last common ancestor. In *Great Ape Societies*, ed. W. C. McGrew, L. F. Marchant & T. Nishida, pp. 275–92. Cambridge: Cambridge University Press.

Nishida, T., Kano, T., Goodall, J., McGrew, W. C. & Nakamura, M. (1999). Ethogram and ethnography of Mahale chimpanzees. *Anthropological Science*, 107, 141–88.

Nishida, T. & Uehara, S. (1983). Natural diet of chimpanzees (*Pan troglodytes schweinfurthii*): long-term record from the Mahale Mountains, Tanzania. *African Studies Monographs*, 3, 109–30.

Palgrave, K. C. (1977). *Trees of Southern Africa*. Capetown: C. Struik Publ.

Peters, C. R. & O'Brien, E. M. (1981). The early hominid plant-food niche: insights from an analysis of plant exploitation by *Homo*, *Pan*, and *Papio* in eastern and southern Africa. *Current Anthropology*, 22, 127–40.

Plavcan, J. M. (1999). Mating systems, intrasexual competition and sexual dimorphism in primates. In *Comparative Primate Socioecology*, ed. P. C. Lee, pp. 241–60. Cambridge: Cambridge University Press.

Pratt, D. J. & Gwynne, M. D. (1977). *Rangeland Management and Ecology in East Africa*. London: Hodder and Stoughton.

Pruetz, J. D., Marchant, L. F., Arno, J. & McGrew, W. C. Status of the savanna chimpanzees (*Pan troglodytes verus*) in southeastern Sénégal. *American Journal of Primatology*, in press.

Robinson, J. T. (1972). *Early Hominid Posture and Locomotion*. Chicago: University of Chicago Press.

Semaw, S., Renne, P., Harris, J. W. K., Feibel, C. S., Bernor, R. L., Fesseha, N. & Mowbray, K. (1997). 2.5-million-year-old stone tools from Gona, Ethiopia. *Nature*, 385, 333–6.

Sept, J.M. (1992). Was there no place like home? A new perspective on early hominid archaeological sites from the mapping of chimpanzee nests. *Current Anthropology*, 33, 187–207.

Stanford C. B. (1999). *The Hunting Apes: Meat Eating and the Origins of Human Behavior*. Princeton: Princeton University Press.

Stern, J. T. Jr. & Susman, R. L. (1983). The locomotor anatomy of *Australopithecus afarensis*. *American Journal of Physical Anthropology*, 60, 279–317.

Sugiyama, Y. & Koman, J. (1992). The flora of Bossou: Its utilization by chimpanzees and humans. *African Studies Monographs*, 13, 127–69.

Tutin, C. E. G. & Fernandez, M. (1993). Composition of the diet of chimpanzees and comparisons with that of sympatric gorillas in the Lopé Reserve, Gabon. *American Journal of Primatology*, 30, 195–211.

Tutin, C. E. G., McGrew, W. C. & Baldwin, P. J. (1983). Social organization of savanna-dwelling chimpanzees, *Pan troglodytes verus*, at Mt. Assirik, Senegal. *Primates*, 24, 154–73.

Verner, P. H. & Jenik, J. (1984). Ecological study of Toro Game Reserve (Uganda) with special references to Uganda Kob. *Rozpravy Ceskoslovenké Akademi Ved, Rada Matematickych A Prírodních, Tovnik 94–Sesit 4*. Czechoslovakia: Praha.

Wheeler, P. E. (1992). The thermoregulatory advantages of large body size for hominids foraging in savannah environments. *Journal of Human Evolution*, 23, 351–62.

WoldeGabriel, G., White, T. D., Suwa, G., Renne, P., de Heinzelin, J., Hart, W. K. & Heiken, G. (1994). Ecological and temporal placement of early Pliocene hominids at Aramis, Ethiopia. *Nature*, 371, 330–3.

WoldeGabriel, G., Haile-Selassie, Y., Renne, P., Hart, W. K., Ambrose, S. H., Asfaw, B., Heiken, G. & White, T. D. (2001). Geology and palaeontology of the late Miocene Middle Awash Valley, Ethiopia. *Nature*, 412, 175–8.

Wrangham, R. W. (1977). Feeding behavior of chimpanzees in Gombe National Park, Tanzania. In *Primate Ecology*, ed. T. H. Clutton-Brock, pp. 503–38. London: Academic Press.

Wrangham, R. W. (1979). Sex differences in chimpanzee dispersion. In *The Great Apes*, ed. D. A. Hamburg & E. R. McCown, pp. 481–89. Menlo Park: Benjamin/Cummings.

Wrangham, R. W. (1986). The significance of African apes for reconstructing human social evolution. In: *Primate Models for the Evolution of Human Behavior*, ed. W. G. Kinzey, pp. 51–71. Albany, New York: SUNY Press.

Wrangham, R. W., Jones, J. H., Laden, G., Pilbeam, D. & Conklin-Brittain, N. (1999). The raw and the stolen. *Current Anthropology*, 40, 567–94.

Wrogemann, D. (1992). Wild Chimpanzees in Lope, Gabon: Census Method and Habitat Use. PhD dissertation, University of Bremen, Germany.

3 • Behavioural adaptations to water scarcity in Tongo chimpanzees

ANNETTE LANJOUW

INTRODUCTION

Chimpanzees, *Pan troglodytes*, occur over a wide range of habitats in Africa and their behavioural ecology is adapted to the specific conditions at each site. Numerous long-term studies of chimpanzees have been conducted across Africa, in habitats ranging from woodland forests in Tanzania (Goodall 1986; Collins & McGrew 1988), moist forests in Gabon (Tutin & Fernandez 1993), and dry open habitats in Senegal (McGrew *et al.* 1981; Bermejo *et al.* 1989) and Guinea (Sugiyama & Koman 1992). Behaviours related to the choice and collection of foods, processing of food items, and collection of water have been observed to vary from site to site. These differences have been attributed to cultural differences between chimpanzee populations and are a function both of adaptations to the ecology of the site and of learning (McGrew 1983; Boesch 1991; Whiten *et al.* 1999).

Similar constraints facing chimpanzees across different sites have resulted in typical, site-specific adaptations. Limited access to water in dry, open habitats such as Mt Assirik, Senegal (McGrew *et al.* 1981), Semliki Wildlife Reserve, Uganda (Hunt *et al.* 1999), and Tongo, eastern Democratic Republic of Congo, have led to the evolution of both similar and different methods for obtaining moisture. In sites where access to water can be seasonally limited, the chimpanzees have developed similar techniques for obtaining moisture, for example using tools like leaf sponges or cups.

The Tongo Chimpanzee Conservation Project of the Frankfurt Zoological Society focused on the conservation of a community of chimpanzees in the Southern Sector of the Virunga National Park, in the Democratic Republic of Congo (formerly Zaïre). The community was habituated, monitored, and observed over a period of three consecutive years (1987–90), with the objectives of (1) conserving a threatened community of chimpanzees in the Virunga National Park; (2) eradicating a growing charcoal production industry inside the Virunga National Park; and (3) developing a sustainable source of funding, through tourism, for the protected area authority in the Congo, the Institut Congolais pour la Conservation de la Nature (ICCN).

This chapter examines some of the specific adaptations the chimpanzees at Tongo have developed in order to obtain moisture in a forest that is very dry during much of the year. The Tongo chimpanzees were studied without interruption from 1987 to 1990.

Historical background

Beginning with the identification of a community of chimpanzees in the Southern Sector of the Virunga National Park, the Tongo Chimpanzee Conservation Project was founded in October 1987 in order to develop a carefully controlled chimpanzee-tourism programme. The habitual range was defined, and a comprehensive network of trails was developed in the forest. Teams of trackers attempted to locate and follow the chimpanzees twice a day. Over a period of 2 years of intensive effort, a community of chimpanzees numbering approximately 50 individuals was sufficiently habituated to permit tourism to open. The habituation process took 1 year, followed by a 12-month trial period for tourism. Tourism to the chimpanzees of Tongo was officially opened in December 1989. At that time, there was no other area where chimpanzees had been habituated for tourism without the use of artificial inducements (food, reproduced chimpanzee calls or other sounds).

From 1989 to 1992, tourism grew steadily, with a regular flow of revenue coming to the ICCN for park management. Attention from other conservation and rural development programmes helped the small rural community of Tongo flourish and develop. Little hotels and restaurants opened and local people started to develop small enterprises linked to the tourism industry. The pressures on the park decreased and charcoal production inside the park around Tongo halted. The complete destruction of this industry, which at its peak involved ten trucks fully laden with charcoal (the equivalent of at least four times that amount in fresh wood) per day, indicated a considerable conservation impact.

The political problems in Zaïre came to a peak in 1993; a period of political and social conflict followed, and it continues up to the present time (2001). Thousands of people have been killed and infrastructures have been destroyed. Areas of the forest have been encroached upon for agriculture and much forest has been destroyed for the production of charcoal. Large numbers of wildlife have been poached to feed the warring parties in the region. The ICCN park staff has stayed in Tongo throughout this period, although regular monitoring of the park has not been possible due to security constraints. The monitoring that has occurred indicates that the chimpanzees have not been targeted or harmed. Those chimpanzees that were habituated and identified for tourism, with names and clearly distinguishable features, have all been located. The level of habituation has decreased, and it is difficult for the park staff to follow the animals on the ground. Once security permits, however, it will be possible to resume the monitoring work, with a minimum investment in the habituation of the animals and the infrastructure within the park.

Geography

The Southern Sector of the Virunga National Park is dominated by the Virunga Volcanoes, whose most recent eruption dates to the late 1940s. The Virungas, or Bufumbiro chain of volcanoes, form an arc straddling the divide between the Eastern and Western Rift Valleys. The volcanoes form a barrier between the chain of lakes in the Western Rift, barring the northward flow of water from Lake Kivu to Lake Edward. At the west are the Nyiragongo and Nyamulagira volcanoes. These two volcanoes are active, with the most recent eruption of the Nyiragongo dating to 1977 and the most recent eruption of the Nyamulagira dating to early 2000. The lava flows from these two volcanoes are slowly filling one of the arms of the Western Rift Valley, dividing the Central Congo basin and the Nile basin. Both the Nyiragongo and Nyamulagira volcanoes have been described as 'Hawaii-type' volcanoes (Tazieff 1951), which produce very liquid lava. Distinctive 'chaotic' lava fields result, typified by large rough boulders and crevasses still showing the solidified flow of the liquid lava. The lava fields of both volcanoes extend towards the south, in what has been labelled the Low Lava Plains (Basse Plaine de Lave), and to the north, known as the High Lava Plains (Haute Plaine de Lave) (Lebrun 1960).

The Tongo forest lies between 1°20′ and 1°15′ latitude south and 29°05′ and 29°10′ longitude east, on one of the lava flows from the Nyamulagira. The early vegetation colonizing the lava flows are lichens, ferns and *Rumex* spp. These slowly break down the rock to liberate more minerals and capture more humidity so that other types of vegetation can grow. This process of colonization of the lava flows takes decades, even centuries. By comparing the Tongo forest with other areas of determined age (the earliest noted lava flows in the region date to the 1930s), it is estimated that the lava flow on which the Tongo forest is growing is over 300 years old.

Forest cover

Apart from a small number of old craters containing little ponds or swamps, or temporary stands of water, the High and Low Lava Plains are devoid of any standing or running water. The relatively high rainfall immediately drains and filters through the porous lava plains, down to a considerable depth, collecting only along the original substrate to form a network of waterways. No water remains at or near the surface. As a consequence, despite the estimated age of the lava flow on which the forest in Tongo grows, it is a very dry forest, dominated by plants adapted to low soil humidity. The soil layer is thin and the underlying rock is exposed in most areas. Dominating trees are *Cussonia holstii* and *Olea chrysophylla*. The relative density of *Ficus* spp. in Tongo appears very high (Gauthier Hion 1989, personal communication).

The forest is surrounded by lava flows of relatively recent origin, some dating from as late as 1958, and thus showing only the very early stages of vegetation colonisation (lichen and small clusters of *Rumex* and pteridophytes in crevasses and the shade of large boulders). The forest 'island' of Tongo, covering approximately 10 km^2, could therefore be described as an area of well-developed forest surrounded by much drier scrub and exposed rock, thus encouraging the chimpanzees to remain in the relatively small forest patch.

Only a small area in the Tongo forest is not covered by lava flows. This area is a series of four hills, emerging from below the level of the lava flow and exposing the original substrate. On these hills, the soil is loose and sandy and the forest is more open with a less dense lower storey. A number of lianas that are rare on the rocky lava plain can be found on these hills. The hills are only exposed by between 20 m and 100 m above the lava plain and are called Rugomba, Katwa I, Katwa II, and Kanyanbundu.

Fauna

The forest in Tongo supports a variety of other non-human primates, such as the redtailed monkey (*Cercopithecus ascanius*), the black and white colobus (*Colobus guereza*), the blue monkey (*Cercopithecus mitis doggetti*), and the baboon (*Papio anubis*). Predators are rare in the forest growing on the chaotic lava rocks, although leopard (*Panthera pardus*) were heard regularly and lions (*Panthera leo*) had been seen moving through the forest. Large ungulates were not seen, although a number of forest antelopes (including *Tragelaphus scriptus*) and Giant Forest Hogs (*Hylochoerus meinertzhageni*) were observed in the forest.

The chimpanzees were observed feeding in the same trees as colobus monkeys and redtailed monkeys. The chimpanzees at Tongo also successfully hunted both species.

Water availability

Despite a high and regular level of rainfall, averaging 1753 mm per year (ICCN 1990), due to the high level of solar radiation and the porosity of the volcanic substrate, the vegetation covering the lava plains is adapted to very low humidity. The peak of the rainy season is September to November, with a small rainy season from March to April. The long dry season generally extends from May to August. From December to March, occasional rains keep the region relatively humid (ICCN 1990).

The Tongo forest, as described above, is practically devoid of any standing water. The only permanent water source in the vicinity of the forest is the Mulindi Springs. Approximately 5 km beyond the northern edge of the forest, the water flowing beneath the lava plains filters through to the surface to emerge into a series of crystal clear pools, forming a chain and then a river that flows into the Rwindi Plain. During the study period, the chimpanzees were only observed ranging to these pools on one occasion. It is clear, therefore, that most of their needs for water were being met through other sources in the Tongo forest.

In the forest, the only standing water available was what collected in the hollows of trees or forked branches, or was absorbed in decomposing vegetation. The forest itself, as described earlier, is relatively sclerophyllous and adapted to periods without water, despite the abundant rain. The fruit of the trees are therefore small and not high in moisture content.

METHODOLOGY

The Tongo forest, in the Virunga National Park, was selected as the site for the conservation programme. In the first phase of the project, the forest island of 10 km^2 was covered in a network of trails along quadrats every 200 m. The second phase of the project, the habituation process, involved two teams of trackers who were in radio communication with each other, localising the animals every morning at 5:30. These teams followed the animals for as long as possible each day until they lost the animals. Each team recorded the location, how they were found (monitoring a food tree, nesting the previous night, or following calls), group size (and changes), composition (and changes), behaviour (including feeding behaviour and food source), duration of contact, and weather. Group composition and behaviour were also recorded using scan sampling, with 5-minute intervals recorded on datasheets in notebooks.

Individuals were classed into age/sex categories according to guidelines provided by the National Research Council (Committee on Nonhuman Primates 1981), and Goodall (1986): Infants: 0–5 years; Juveniles: 5-7 years; Subadults: 8–15 years; Adult: 15 years and more. A party of chimpanzees was defined as a group including subadult and/or adult animals. Infants and juveniles were not counted in a party.

A database was kept of all individuals encountered during the duration of the project, with data on each individual chimpanzee. These included physical data enabling identification of the animal, but also behavioural data including who the animal associated with, where it was observed and what it was doing (feeding, resting, travelling, grooming, etc.). Each card in the database included a sketch of the animal to assist identification. Within the first year of habituation most of the adult male chimpanzees were individually identified. Some of the more frequently encountered females and subadult animals were also individually identified.

RESULTS

Over a period of 28 months, this observer spent almost 400 hours with the chimpanzees. A total of 338 contacts were

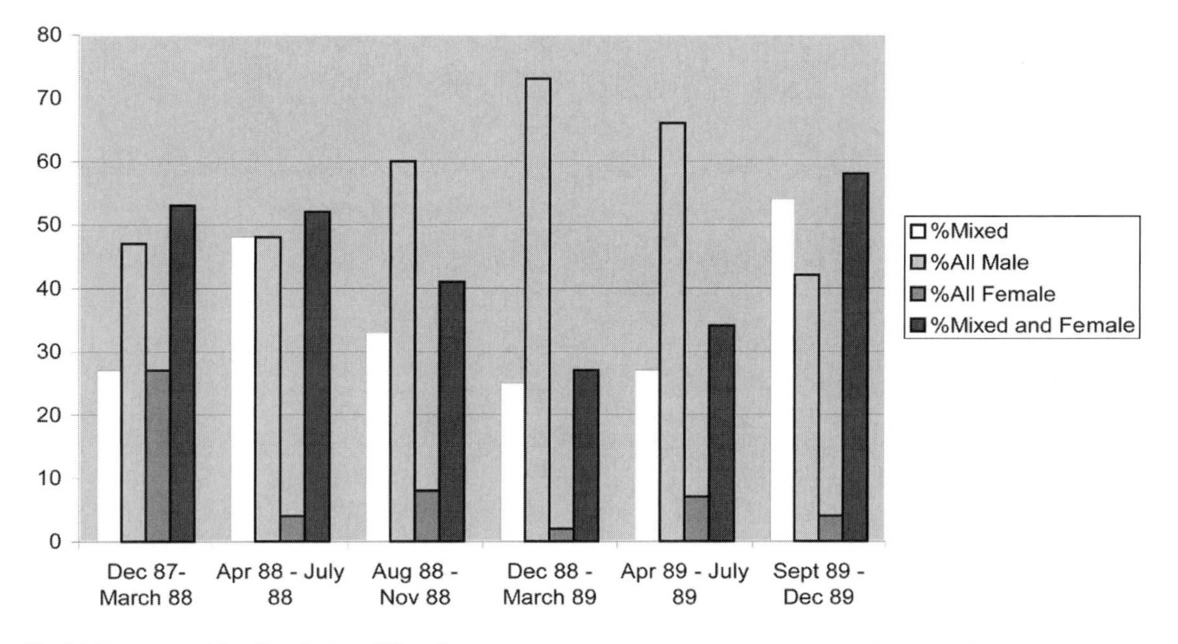

Fig. 3.1. Party composition. Sample size = 300 parties.

made, primarily between December 1987 and January 1990. During the last year of the project, February 1990 to November 1990, the tourism programme opened and other priorities took precedence over the collection of behavioural data on the chimpanzees. However, less systematic, qualitative observations were still made and these complement the database.

Chimpanzee population

It is felt that only one community of chimpanzees inhabited the 10 km² forest island described above. Although not all the females and subadult animals had been identified and habituated at the end of the 3-year project, 20 adult males had been identified and approximately 15 females had been identified. It is estimated that the size of the population inhabiting the Tongo forest island totalled approximately 50 individuals.

In the forest, on the southern side of the road cutting from the east of the Rift to the western escarpment, another community of chimpanzees could be found. Also, in the forests outside of the Virunga National Park, following the ridges and waterways along the escarpment, chimpanzee calls were regularly heard, indicating the presence of other communities.

Party size and composition

A party was defined as a group of adult and/or subadult chimpanzees. An all-male party or all-female party included only adult and/or subadult males or females. A mixed party included both males and females, respectively. The clustering of mixed and all-female groups in Figure 3.1 allows comparison of groups including at least one female with groups including no females. The quality of observations varied greatly throughout the study period. In the early phases of habituation, complete counts and composition of parties could not be obtained. As the animals became more habituated, more accurate data could be recorded on the size and composition of the parties, and more individuals could be recognised. Comparison over time therefore includes minimum counts, thus presenting a probable underestimate of the average size of parties. With time, the party size increased (fewer animals left the party as a consequence of the arrival of human observers), and the accuracy of counts increased (Figure 3.2). Party composition was probably also affected by the habituation process. The relatively high number of solitary and all-female parties in the early phases of habituation results from all parties of unknown composition being discarded from the analysis. In the early phases of habituation, the larger parties of males and mixed parties were often not included in the analysis since an accurate count of the composition of the party was not possible (Figure 3.1).

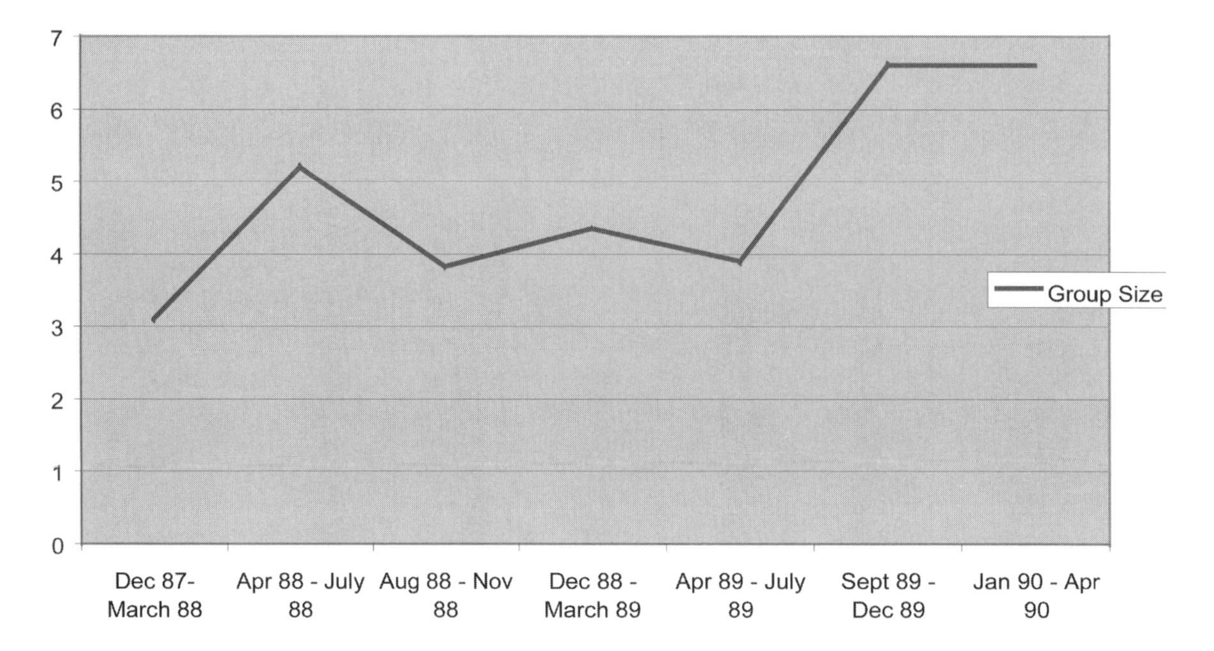

Fig. 3.2. Average party size. Sample size = 338 parties encountered.

Chimpanzees were found by following vocalisations (primarily in the early morning when leaving nests) or waiting at large fruit trees where they were likely to feed (presence of previous feeding signs indicated that they had fed at a tree and were likely to return). Occasionally the animals were followed from the nest when their nesting sites had been located the previous evening. Most observations were based on following their vocalisations, and this may have led to a bias in observations of all-male parties (Figure 3.1).

Other chimpanzee communities

The Tongo forest is a patch or 'island' of mature forest with abundant food sources surrounded by dry and less well-developed vegetation on more recent lava flows. Other patches of mature forest were found in the vicinity of Tongo, on the escarpment outside the park, and on other ancient lava flows within the park. Although the chimpanzees of the Tongo community could, and frequently did leave the forest to visit other areas, including other patches of 'good' forest, they tended not to leave the Tongo forest 'island' for long. Chimpanzees were frequently heard in nearby forest patches within the park, as well as on remaining forest patches on the escarpment outside the park. Approximately seven observations of unknown chimpanzees in these other forest patches were recorded during the study period. On three occasions, small bands of males from the Tongo community were followed while travelling to a forest patch containing chimpanzees south of Tongo. These parties were not followed for long distances, however, due to the lack of trails and difficulty of the terrain. Although the males travelled silently during these occasions, it is impossible to determine if their behaviour was similar to observations of chimpanzees on male patrols in other sites since they were not observed for sufficiently long periods of time.

Behavioural adaptations associated with the collection of moisture/water

Observations reveal that the chimpanzees of Tongo have developed a number of behavioural adaptations for obtaining water. These behaviours were linked to the scarcity of water and adapted to obtaining moisture from places generally less accessible to the chimpanzees. The chimpanzees were frequently observed chewing on the wood of fallen trees, especially when the wood was rotten and saturated with rainwater. The chimpanzees would chew the wood and afterwards spit it in dry wadges. A fallen tree was frequently surrounded by many dry chimpanzee wadges, indicating that many animals had chewed on the wood.

Table 3.1. *Observations of chimpanzees drinking with tools or with hands*

Observation	Age/sex category	Party size and composition	Use of tool	Method used
23 November 1988	Adult male	5 adult males; 1 subadult male	Tool	Moss sponge between index and forefinger, arrived in tree with sponge already prepared
23 November 1988	Juvenile male	5 adult males; 1 subadult male	Tool	Moss sponge with whole hand
13 March 1989	Adult male	3 adult males; 1 subadult male	Hand	Finger licking
13 March 1989	Adult male	3 adult males; 1 subadult male	Tool	Moss sponge in fingers (not clear)
13 March 1989	Subadult male	4 males	Tool	Moss sponge
11 September 1989	Adult male	6 males	Tool	Unclear but probably moss sponge
18 September 1989	Adult male	6 mixed party: 5 males, 1 female	Tool	Moss sponge between index and forefinger, collected some 20 m before reaching drinking tree
20 November	Subadult male	8 mixed party: 5 adult males, 1 female + infant, 2 subadult male	Tool	Moss sponge, unclear which fingers
20 November	Adult male	Same as above	Tool	Sponge, probably moss

SPONGES

During the 28 months in which the chimpanzees were monitored systematically in Tongo, nine observations were made of chimpanzees drinking water collected in the branches of trees. On eight occasions, when the water was not accessible directly, the animals prepared tools, such as sponges made from moss. In one instance an individual used his hand to scoop out the water (Table 3.1).

The chimpanzees obtained the water from the hollows in trees or the forks between branches where rainwater collected. When wide enough, the chimpanzees could simply drink the water straight from the branches. Usually, however, these pools were deep or narrow, allowing only a hand or fingers to be inserted. Although some chimpanzees would insert their hand and then lick their fingers, the Tongo chimpanzees generally used moss sponges to collect the water from these trees. The chimpanzees collected the moss off the bark of the trees, loosely rolled it into a bundle, generally not bigger than a few centimetres wide, and inserted it into the hollow of the branches. The sponge would then be extracted and sucked of its moisture. The chimpanzees repeatedly re-inserted the sponge to drink.

Frequently these 'sponges' were prepared well before the chimpanzee arrived at the tree. This suggests that they knew which trees had water, but also that they visited these trees with the express intention of drinking and prepared their sponge in anticipation.

Observations of chimpanzees drinking with the use of sponges were common. During the rainy season, when most of the little pools of water between the branches were full, this could be seen on an almost daily basis by the project staff. From the observations of the habituation teams and other staff, all age categories of chimpanzees utilised this method and it was not restricted to a few individuals. Occasionally individuals were also observed drinking directly from pools of water collected at the base of trees or in wide forks.

ROOT/TUBER COLLECTION

During the prolonged dry season, from May to August, the forest became dry and brittle, and there was very little saturated fallen wood to chew on or water between the branches. Although the region is never extremely dry, with a minimum monthly rainfall of between 80 mm and 100 mm, this is the season when the chimpanzees were observed searching for moisture, climbing the Rugomba, Katwa, and Kanyanbundu hills more frequently than at any other time. There were few fruiting trees on these hills and little that would attract the

Table 3.2. *Observations of chimpanzees digging and eating tubers*

Observation	Party size and composition	Individuals digging and eating tubers	Individuals sharing tubers	Total duration of observation
9 August 1988	6, all adult males	All observed	None	57 minutes
25 July 1989	6, mixed party; 3 adult males; 2 subadult males; 1 non-tumescent female	All observed	None	73 minutes
14 July 1989	6, mixed party; 4 adult males;1 subadult male; 1 non-tumescent female	Only three adult males	None	44 minutes
19 May 1989	10, mixed party: 6 adult males; 3 subadult males; 1 nulliparous female	All observed	One adult male begging from another adult male, not sure if received any	210 minutes
13 July 1989	3, all adult males	All observed	None	51 minutes
18 September 1989	6, mixed party: 3 adult males; 2 subadult males; 1 tumescent female	All observed	One adult male received entire tuber from another adult male who was holding two	39 minutes
21 September 1989	5, mixed party: 2 adult males; 1 subadult male; 2 tumescent adult females	All observed	None	Approximately 55 minutes
Average				76 minutes

chimpanzees during the rest of the year. During the dry season, however, they would climb the hills and collect tubers found deep beneath the sandy soil. Seven episodes of this behaviour were observed in 3 years.

During these dry periods, the chimpanzees climbed the hills as a group and dug in the soil, at times inserting their entire arms into the holes (Table 3.2). Either they knew the exact location of the plants, or they recognized the bare lianas climbing into the canopy. It generally took several (up to 15) minutes to dig up the tubers and the animals would spend long periods chewing on the tubers (Figure 3.3), sucking all the moisture out of them, and then spitting out the wadges of dry, white pulp. No nutritional analysis has been done to determine if they were eaten purely for their moisture content.

Although the tuber species collected by the chimpanzees of Tongo has not been identified for certain, I believe it is a species from the family Ranunculaceae. Five species of Ranunculaceae have been identified in the vegetation of the High and Low Lava Plains of North Kivu (Lebrun 1960). These are:

Clematis simensis
Clematis hirsuta
Ranunculus multifidus
Ranunculus bequaerti
Thalictrum rhynchocarpum.

Both *Clematis* species are adapted to dry conditions, and the tubers are believed to be one of these two species (Dr Terèse Hart, personal communication).

The tubers grow primarily in the sandy soil of these hills. The liana has only a thin spiralling stem, with few leaves on it at the lower strata, and it rapidly climbs to the upper branches of the trees. Beneath the soil, to a depth of approximately 50 cm, the tubers grow to a diameter of 20 cm or more. The tubers have a very thin, dark cortex, a thin, reddish inner layer, and then a uniform, white core. The tubers are saturated with moisture, similar to that of a potato. When tasted, they have no discernable flavour. One medium-sized tuber, 25 cm in diameter and 70 cm in circumference, weighed 5 kg.

The chimpanzees were never observed collecting the

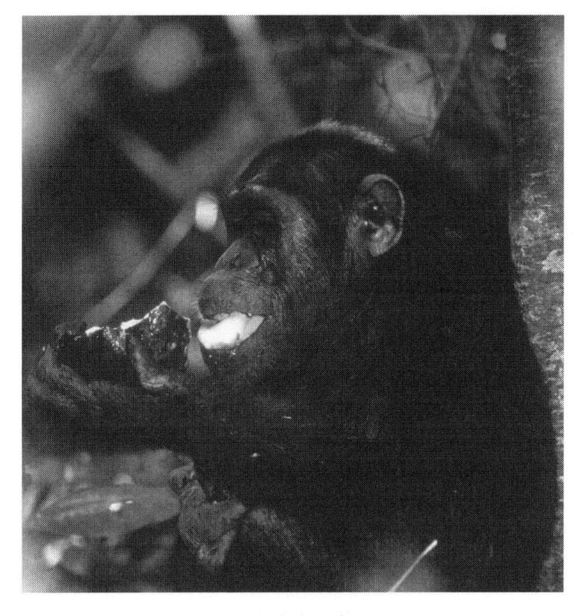

Fig. 3.3. Tongo chimpanzee eating a tuber.

tubers on their own, but always in a group. The digging of the tubers was usually accompanied by a great deal of excitement and social interactions, similar to when chimpanzees enter a valued fruit tree. Occasionally, begging by younger or less dominant individuals was observed, although juvenile animals were seen to successfully dig up smaller tubers. Each individual would eventually have a tuber or piece of it, and sit and chew the pulp. The animals sometimes walked around with armloads of tubers and sat apart protecting their bounty.

On three occasions, the chimpanzees carried their tubers for many hours after they had descended the hill and moved to other areas. A young adolescent male was observed carrying his large tuber (diameter approximately 20 cm, weight approximately 3–4 kg) on the following day. The actual consumption of the tuber was not observed, but he carried the heavy root over a distance of at least 1.5 km.

DISCUSSION

A number of different individual chimpanzees, ranging from juvenile males to adult males, were observed using a moss sponge to collect water for drinking. This behaviour can be considered habitual, if not customary, for the Tongo chimpanzees, based on the definition provided by Whiten *et al.* (1999). It can therefore be considered a unique cultural adaptation discovered and socially transmitted by individuals within the Tongo population.

It appears that the chimpanzees in Tongo learned to obtain the moisture collected inside the tubers during periods of water scarcity and even occasionally 'carried' the water away, stored in the tuber. As mentioned earlier, although the forest at Tongo is in a high rainfall zone, it is relatively dry due to the particular characteristics of the lava soil upon which it grows. In many respects, the Tongo forest provides a unique transition between rainforest and woodland, and where the chimpanzees have adapted to this condition of water scarcity.

Tuber collection only occurred during the dry season, when other sources of water were no longer available inside the forest. To understand fully the significance of these observations and to confirm that the Tongo chimpanzees have culturally adapted to water scarcity using these tubers, additional and regular observations of tuber collection are needed. The plants need to be identified with certainty and analysed for nutritional content. If they truly only contain significant water content and have little other nutritive value, it will be possible to conclude that the chimpanzees of Tongo have adapted to water scarcity by exploiting these tubers. It is also possible, of course, that the animals chew the tubers for both nutrition and moisture. We require a fuller understanding of when and under what circumstances tubers are being exploited, and by which individuals. In addition, it will be necessary to determine whether these tubers are present in other areas, in order to determine if the tuber-collection behaviour is a unique cultural adaptation, or a purely ecological one. The chimpanzees of Tongo are unique in their use of the tubers if similar, water-laden tubers are present (but not used) at other sites (i.e., the behaviour is due to differential cultural learning rather than availability).

The adaptation of the Tongo chimpanzees to eating tubers has also not been reported for any other populations of chimpanzees. Wrangham & Peterson (1996) speculate that this particular eating behaviour may explain how prehistoric apes made the transition from rainforest to drier, more woodland habitats. Hunt & McGrew (Chapter 2) also suggest that studying the overlap in habitat types utilised by chimpanzees and Mio-Pliocene hominids may help us understand the evolutionary ecology of hominids.

It may be some time before more information about tuber eating among Tongo chimps can be collected. Given the current political and security situation in the region, limited conservation efforts continue to be directed specifically at Tongo due to the recognised conservation and potential economic value of the site. Security conditions are slowly improving and short visits are again possible. Although long-term research may not be feasible at the moment, the guards at Tongo are trained and experienced observers, and it should be possible to collect basic but reliable data through them. Research and conservation efforts in Tongo will help improve the protection of the animals in general, as well as safeguard the unique culture of this community of chimpanzees.

REFERENCES

Bermejo, M., Illera, G. & Sabater Pi, J. (1989). New observations on the tool-behavior of the chimpanzees from Mt. Assirik (Senegal, West Africa). *Primates*, 30, 65–73.

Boesch, C. (1991). Teaching in chimpanzees. *Animal Behaviour*, 41, 530–2.

Collins, D. A. & McGrew, W. C. (1988). Habitats of 3 groups of chimpanzees (*Pan troglodytes*) in western Tanzania compared. *Journal of Human Evolution*, 17, 553–74.

Goodall, J. (1986). *The Chimpanzees of Gombe: Patterns of Behavior*. Cambridge, MA: Harvard University Press.

Hunt, K. D., Cleminson, A. J. M., Latham, J. Weiss, R. I. & Grimmond, S. (1999). A partly habituated community of dry-habitat chimpanzees in the Semliki Valley Wildlife Reserve, Uganda. *American Journal of Physical Anthropology*, Suppl. 28, 157.

ICCN. (1990). *Données Météorologiques Annuels*. Institut Congolais pour la Conservation de la Nature.

Lebrun, J. (1960). *Exploration du Parc National Albert (Mission J. Lebrun 1937–1938): Études sur la Flore et la Végétation des champs de lave au Nord du Lac Kivu (Congo Belge)*. Brussels: Institut des Parcs Nationaux du Congo Belge.

McGrew, W.C. (1983). Animal foods in the diets of wild chimpanzees: Why cross-cultural variation? *Journal of Ethology*, 1, 46–61.

McGrew, W. C., Baldwin, P. J. & Tutin, C. E. G. (1981). Chimpanzees in a hot, dry and open habitat: Mt. Assirik, Senegal, West Africa. *Journal of Human Evolution*, 10, 227–44.

National Research Council (1981). *Techniques for the Study of Primate Population Ecology*, Committee on Nonhuman Primates.

Sugiyama, Y. & Koman, J. (1992). The flora of Bossou: its utilization by chimpanzees and humans. *African Study Monographs*, 13, 127–69.

Tazieff, H. (1951). *Cratères en feu*. Paris: B. Arthaud.

Tutin, C. E. G. & Fernandez, M. (1993). Composition of the diet of chimpanzees and comparisons with that of sympatric gorillas in the Lopé Reserve, Gabon. *American Journal of Primatology*, 30, 195–211.

Whiten, A., Goodall, J., McGrew, W. C., Nishida, T., Reynolds, V., Sugiyama, Y., Tutin, C. E. G., Wrangham, R. W. & Boesch, C. (1999). Cultures in chimpanzees. *Nature*, 399, 682–5.

Wrangham, R. & Peterson, D. (1996). *Demonic Males: Apes and the Origins of Human Violence*. New York: Houghton Mifflin Co.

4 • Bonobos of the Lukuru Wildlife Research Project

JO A. MYERS THOMPSON

INTRODUCTION

Beginning in the early 1990s, intraspecies behavioural variations across a number of chimpanzee (*Pan troglodytes*) populations became the focus of increasing attention, and interspecies behavioural variations between chimpanzees and bonobos (*Pan paniscus*) have been the popular mainstay of bonobo recognition (variations such as female social role, use of social sex, etc.). Interspecies differences (larger groups, less aggression, and morphology) have largely been attributed to ecological variation across species ranges, bonobos having been considered to be isolated in an arboreal existence (MacKinnon 1978; Susman 1979) and restricted to a lowland, dense forest ecology. Behavioural diversity in wild chimpanzees first became evident by comparing tool use in the context of foraging as observed across populations inhabiting broad ecological conditions (McGrew 1992). The study of intraspecies behavioural diversity of bonobos is in its infancy, so ecological differences must be considered. When examining behavioural variation between bonobo study sites, White (1992) considered food provisioning at Wamba and the temporal effects of habituation at Lomako to explain variation.

In the relatively few years of bonobo field research, only two populations have been widely documented: Wamba (00°10′N, 22°30′E) and Lomako (00°50′N, 21°05′E). These two populations are both located within uniformly hot, wet, low-lying and flat topographies sheltered by closed-canopy, moist, evergreen lowland forest vegetation. Previous attempts to consider bonobo behavioural diversity (White 1992) have largely been restricted by the disproportionately small number of sites where bonobos have been studied, the inability of field researchers to conduct continuous field studies over the course of years, the unavailability of findings based on the comparative study of multiple bonobo communities, the relatively proximate geographic location of these two bonobo sites, and the potentially cognate nature of the subjects at these two sites where their mutual habitat corridor is not broken by a major river system or geographic barrier. The apparent absence of recognizable tool use by wild bonobos (Savage-Rumbaugh *et al.* 1989; McGrew 1992; except reports of the use of leafy twigs for rain cover by Kano 1982, and nest use as 'information centers' by Fruth & Hohmann 1996) and the fixed ecological patterns across these known sites (Kano 1992; White 1992; Malenky & Wrangham 1994) has led us to look for a broader sample to consider in comparison. Because tool use during food acquisition has not yet been observed in bonobos (except indirectly by Badrian *et al.* 1981) and food resources have been considered spatially and temporally more abundant for bonobos (Malenky & Stiles 1991; Malenky & Wrangham 1994; Malenky *et al.* 1994), distinct variations in foraging strategy across the Wamba and Lomako populations have not yet become apparent. By considering any or all of the issues presented, we may begin to illustrate aspects of bonobo intraspecies behavioural diversity.

Preliminary field data from Lukuru (03°45′S, 21°21′E) may provide information that we can use to begin to study the dimensions of intraspecific variations in foraging strategies and in social behaviours. Unlike the other study sites, bonobos at Lukuru live in dry forest/savanna mosaic habitat. The hilly terrain of the area consists of an irregular dry forest and grassland mosaic habitat, increasing in elevation out of the southern periphery of the topographic Congo Basin. Lukuru is divided by the navigable Lukenie River, which obstructs genetic flow between the southernmost bonobo population (including the Bososandja bonobo community) and the core habitat that includes the Lomako and Wamba populations. To determine whether bonobo foraging is potentially more varied than previously recognized, we must briefly describe the types of habitat they use, substrates where food is gathered, types of foods eaten, and techniques used.

METHODS

To address intraspecies behavioural diversity in the bonobo we must first determine variation in ecological conditions.

Since Lomako and Wamba are proximate, most comparisons presented here are based on findings from Lukuru (Myers Thompson 1997), the southernmost bonobo study site, related to Lomako (Badrian & Malenky 1984; Susman *et al.* 1985; Malenky 1990; Thompson-Handler 1990; Fruth 1995), the most northern bonobo study site. Standardized weather data were systematically collected as daily values of minimum/maximum temperature and rainfall amounts for 16 continuous months (June 1994 through September 1995) at Lukuru and intermittently for 13 years (August 1981 through May 1994) at Lomako. Validity of the 16-month, short-term extreme fluctuations in precipitation (582 mm in November 1994) during data collection and long-term continuous records for 22 years from proximate meteorological stations (1941–63 for Ilebo and Lodja) were considered to represent Lukuru.

Climatic constraints on species distribution are expressed through the development of vegetation. To determine the distribution and proportion of major gross vegetation types, a grid of 1280 blocks was superimposed over the topographic sample area (71.67 km^2) map such that each grid cell represented approximately 0.0559 km^2 of surface area at Lukuru (Myers Thompson 1997). Modified from White (1983) and Richards (1952), vegetation types were categorized as climax forest, secondary forest, riparian slope/swamp forest, and grassland. The broad presence of each vegetation type was visually assessed from the ground based on the physiognomy or external features of the plant community. The numbers of blocks containing each gross vegetation type were then totalled. Where grid blocks contained more than one vegetation type, the block was counted as the predominant vegetation type represented within its boundaries.

Foraging hypothesis posits that temporal and spatial distribution and abundance of food resources influence group size. The availability of regular and widespread food resources in the bonobo habitat of Lomako and Wamba has been used to explain party size where party size of bonobos is on average higher than that of chimpanzees. The increased size of feeding patches due to simultaneously/concurrently available Terrestrial Herbaceous Vegetation (THV) and fleshy fruits of secondary tree species promotes reduced feeding competition. Vegetation resources, particularly plants that yield foods, may affect foraging strategies through party size (defined as associations, subunits or temporary clusters of individuals who aggregate for daily foraging, typified by mother–offspring, including adult sons). In order to estimate party size, one encounter was recorded whenever contact was made with bonobos on any single day within a radius area of 1 km from initial sighting at Lukuru (Myers Thompson 1997). A group included parties of individuals or solitaries travelling in the same direction that could maintain auditory or vocal contact and occupied a similar geographic area. The number of independent individuals seen and estimated (as detected by distinct vocalization or movement in the foliage) to be present was recorded.

Foraging stategy is ultimately determined by food preference. Preliminary identification of bonobo food choices was based on direct observations of feeding, collection of discarded portions of foods during feeding bouts, and location of feeding remains. Plant food specimens were identified at the Belgian National Botanical Garden (Jardin Botanique National de Belgique). Encounters were biased as the study progressed since we searched more frequently in the area of secondary forest where the bonobos were more tolerant of our presence.

Estimates of the presence, abundance, and distribution of ground-level plant foods consumed by the study group were considered to further illuminate variation across the bonobo habitat. Estimates were derived from the record of rooted stems of known ground-level plant foods identified by the Belgian National Botanical Garden. Using the random plot site method with the consecutively numbered blocks demarked for assessing vegetation types, the density of rooted stems from ten species of identified plants were counted (Myers Thompson 1997). The centre of each block was translated into a geographic position and determined as a 4.52-m^2 circular plot on the ground. Stem density was recorded during the periods of February through May and July through September 1995. During this study, 240 plots were examined and a total of 479 stems were counted within those plots (Myers Thompson 1997). Nine species in the study were stems of food plants consumed by the study group as identified from direct observation of feeding or food remains. One species, locally called *Nsusa* (*Aframomum* sp.), grows abundantly in the grassland areas and is very similar to the forest relative *Aframomum subsericeum*. Bonobos were not known to feed on the fruit of this plant during the course of this study, although it is particularly palatable (personal experience) and considered to be a potential food source. Findings were then used in comparison with data from the Lomako site.

RESULTS

Data presented in Table 4.1 illustrate the variation and environmental breadth of bonobo distribution. Bonobos inhabit environments with a perennially wet climate throughout the

Table 4.1. *Environmental variables across bonobo sites*

Variable	Lukuru[a]	Lomako[b]	Lomako[c]
Latitude	03°45′S	00°50′N	00°50′N
Longitude	021°21′E	021°05′E	021°05E
Elevation	564 m	390 m	up to 400 m
Average daily max.	35.5 °C	28.8 °C	ann. avg. temp. reported[d]
Average daily min.	19.5 °C	21.7 °C	ann. avg. temp. reported[d]
Average annual precipitation	1666/1778 mm[e]	1843.6 mm	above 2000 mm
Predominant vegetation	Dry forest and grassland	Moist and swamp forest	Moist and swamp forest
Time period measured	1941–1963 June 1994–Sept 1995[f]	August 1981–May 1982[f]	October 1990–June 1991 March 1992–July 1992 February 1993–July 1993 April 1994–May 1994[g]

Notes:
[a] Myers Thompson 1997; [b] Thompson-Handler (1990); [c] Fruth (1995); [d] Average temperature was reported by Fruth (1995) as 25 °C; [e] Average annual precipitation for the proximate area of Lukuru is based on long-term (1941–1963) assessment comparison of proximate sites at 1666 mm for Ilebo and 1778 mm for Lodja. For the time period measured (June 1994–September 1995), the Lukuru weather station recorded 2009.5 mm of precipitation; [f] Time period data reflect only months when variables were measured during the entire period; [g] Time period data reflect dates when researchers were at the site and do not indicate if complete months were recorded.

year and extend their distribution into areas with a more variable climate marked by a distinct dry season (May–August). It is worth pointing out that the vegetation development of Lukuru is a mosaic of dry forest and savanna grassland, but physiologically it is not considered open woodlands. This is unique in the comparison of bonobo study sites. Bonobos inhabiting a mosaic landscape may serve as a model when considering the archaic environment where the common ape/human ancestor (*Australopithecus ramidus*) evolved (Thompson 2001).

Table 4.2 compares percentage of different types of vegetation cover between Lomako and Lukuru. Although percentages of climax forest are similar across sites and it is the dominant cover, other vegetation types are variable. Lukuru consists of predominantly mature, undisturbed dry forest cover, although fields and clearings of secondary forest, narrow strips of riparian forest on seasonally moist soil shouldering shallow streams, and large expanses of perennially dry grassland also occur. Lomako (Fruth, August 1996 poster presentation at the International Primatological Society/American Society of Primatology Joint Congress held in Madison, Wisconsin) has extensive riparian forest and Lukuru is unique for dry grassland. The grassland microenvironments of Lukuru consist of continuous, rolling or level substrate of either short or tall grasses.

At Lukuru, bonobos were located by walking slowly and quietly along a network of paths where they had most recently been sighted or most frequently been encountered over time. When encountered, the Lukuru bonobos always departed on the ground, travelled on the ground, and regularly fed on the ground. Some sampling bias may have occurred due to unobstructed observations of individuals on the ground, which were of better quality. Lukuru bonobos were encountered in climax forest 38% of the time, secondary forest 56.2% of the time, grassland 4.1%, and riparian forest 1.7% (Myers Thompson 1997) of the time. At Lukuru, where bonobos were not habituated, the mean party size was 6.4 ($n = 121$; range = 1–22; SD = 4.63) independent individuals observed with reliability. All parties with more than eight members were encountered either within secondary vegetation where feeding patches included simultaneously abundant fruits, stem pith, leaf shoots and flowers, or at the perennial pools of Bososandja Refuge (Myers Thompson 2001). This number may actually be an underrepresentation because counts were conservatively

Table 4.2. *Percentage of different vegetation types across sites*

Vegetation type	Lukuru[a]	Lomako[b]
Climax forest	64.6	64
Secondary forest	11.3	4
Riparian forest	3.9	32
Grassland	20.2	NI[c]

Notes:

[a] Myers Thompson 1997; [b] Fruth, August 1996 poster presentation at the International Primatological Society/American Society of Primatology Joint Congress held in Madison, Wisconsin; [c] NI, not identified.

made and visibility was often obstructed. Otherwise, foraging parties did not change size when moving from forest to grassland. The average party size for Lomako bonobos ranged from 4.8 (Malenky & Wrangham 1994) to 9.69 (White 1988) and may be a result of the effects of progressing habituation. The average party size for Wamba bonobos has been reported as ranging from four (Myers Thompson 1997, calculated from data presented by Furuichi 1989) to 22.7 (Idani 1991) and in some instances may have been affected by provisioning. The mode (most common) party size at Lomako was four and at Lukuru three.

Feeding behaviour

During the preliminary investigation at Lukuru, 38 plant and one invertebrate food were recorded (Myers Thompson 1997). Of the 38 recognized plant foods, 22 were identified to species level and six to genus level (Table 4.3 lists the 28 identified plants). The remaining recorded plants are known only by the local name. The majority of the identified food plants (53.6%) were forest tree fruits. However, this does not indicate percentage of exploitation. Table 4.3 compares the presence or absence of plant food species across the three sites.

On five occasions, Lukuru bonobos were observed to move through the grassland and consume the fruits of *Annona senegalensis*, *Anisophyllea quangensis* and *Landolphia lanceolata* (Myers Thompson 1997). Except for *L. lanceolata* being found at Lomako (Hohmann, personal communication, 2001), these foods are not identified at any other bonobo study site. When the bonobos fed in the grassland they remained quiet, making no vocalizations even when ripe fruits were found.

Table 4.3. *Plant food species present at Lukuru, compared to other sites*

Present at Lukuru[a]	Present at Wamba[b]	Present at Lomako[c]
Aframomum subsericeum		
Anisophyllea quangensis		
Annona senegalensis		
Anonidium mannii	+	+
Anubias sp.		
Caloncoba welwitschii	+	+
Cola diversifolia		
Dacryodes edulis	+	+
Dictyophleba lucida		
Ficus sp.	+	+
Gambeya lacourtianum	+	+
Haumania liebrechtsiana	+	+
Icacina guessfeldtii		
Laccosperma sp.		
Landolphia lanceolata		+
Landolphia owariensis	+	
Megaphrynium macrostachyum	+	+
Musanga cecropioides	+	+
Myrianthus arboreus	+	
Ongokea gore	+	+
Plagiostyles africana	+	
Platycerium sp.		
Pycnanthus angolensis	+	+
Raphia sp.		+
Sarcophrynium prionogonium	+	+
Synsepalum stipulatum	+	
Treculia africana	+	+
Uapaca sp.	+	+

Notes:

[a] Myers Thompson 1997; [b] Kano & Mulavwa 1984, Kano 1992; [c] Badrian & Malenky 1984, Malenky 1990, Dupain & Van Krunkelsven 1996 unpublished report, and Gottfried Hohmann, personal communication 2000.

+ = Present.

Another aspect of bonobo food resources that has been of interest to scientists is the significance of the reliance on THV as a food source, and its continuous and widespread availability for bonobos of Wamba and Lomako. Chimpanzees are known to rely on the pith of herbaceous ground plants when fruit is seasonally scarce. Thus, the drier habitat of Lukuru posed a question about the availability of THV. A

Table 4.4. *Comparison of stem densities (per m²) counted by vegetation type*

Vegetation type	Lukuru[a]	Lomako[b]
Climax forest	1.71	2.02
SD (CV)	0.88 (51.72)	0.65 (32.20)
Secondary forest	3.24	NI
SD (CV)	2.97 (91.64)	
Riparian forest	1.82	NI
SD (CV)	1.53 (83.96)	
Grassland	0.75	NI
SD (CV)	0.98 (131.0)	
Mean	2.00	2.02
	1.96 (98.05)	0.65 (32.20)

Notes:
[a] Myers Thompson 1997; [b] Malenky 1990.
NI, not identified.

favoured type of herb, *Haumania liebrechtsiana* can be observed growing in different forms. In a closed forest canopy, it grows at ground level as individually detached stem patches, while it may dominate light gap clearings and open forest in dense thickets of vine towers climbing up to the level of the canopy. The vine towers produce preferred feeding patches through increased density of new leaves (unfurled) and shoots (new stems) at the terminal end of stems, and through greater biomass of immature pith correlated with larger average basal stem diameter, which creates local variation in density and food productivity due to interannual variation associated with rainfall amounts. *H. liebrechtsiana* towers sometimes occur in clearings associated with tree falls and along road edges intermingled with superabundant fruit patches associated with secondary trees and colonies of *Megaphrynium macrostachyum*. Abundant resources, found particularly in the concurrently occurring fruits and nonreproductive plant parts of secondary vegetation, allow bonobos to aggregate in larger groups with less interference from feeding competition.

Malenky (1990) counted rooted stems of *Haumania liebrechtsiana* and *Palisota* spp. between October 1984 through July 1985 and October 1985 through July 1986, which resulted in a measure of stem density at Lomako of 2.02 stems m^{-2} (Table 4.4). When comparing stem density from Lomako (climax forest only; Malenky 1990) and Lukuru (including all vegetation types; Myers Thompson 1997) no statistical difference was found ($t = 0.027$, $p < 0.90$) Further, when comparing stem density at Lomako (Malenky 1990) with climax forest only from this study, there is still no statistical difference ($t = 0.91$, $p < 0.90$). It is important to note that the stem density statistics from Lomako shown in Table 4.4 account only for two species, while stem densities from Lukuru account for five species within the climax forest statistics and for ten species across all vegetation types. This finding suggests that available ground-level foods consumed by bonobos may be more abundant at the Lomako site.

Malenky (1990) further recorded the abundance (based on visual assessment according to ratings of abundant, common, sparse or absent) of seven species of plant consumed by Lomako bonobos within plots where the species was observed. As recorded in Table 4.5, at both study sites (Lomako and Lukuru) *H. liebrechtsiana* was present within plots more often than other recorded species. However, in the drier Lukuru habitat, *H. liebrechtsiana* was present in only 57.14% of the climax forest plots, while it was present in 96.4% of the plots in the moister Lomako habitat. This difference may be a result of annual or seasonal variation between the two studies, so conclusions must be cautious.

When calculating all stems counted across vegetation types (Myers Thompson 1997), plant stem density was ranked highest in secondary forest (3.24 m^{-2}) and ranked lowest within grassland vegetation (0.75 m^{-2}). The mean density of stems across all vegetation types was 2.0 m^{-2}. Using Kruskal–Wallis analysis to test stem densities across the four vegetation types resulted in significant difference ($H = 45.63$, $p < 0.001$ two-tailed). Only between climax forest and riparian forest, and between secondary forest and riparian forest is there no statistical difference in stem densities of the known plant species.

The distribution of plant species varied across vegetation types with the tendency for rooted plant species to be more clumped in the grassland vegetation (coefficient of variation, $CV = 131.0$) and more evenly distributed within the climax forest ($CV = 51.72$) (Myers Thompson, 1997). The smaller coefficient of variation indicates the greater uniformity of individual species stem distribution. Therefore, the species within the grasslands are more concentrated, yet sparse, while species within the climax forest are more widespread (SD = 0.88), and stems within the secondary forest are densely concentrated at particular sites (SD = 2.97) (Myers Thompson 1997).

Within Lukuru, intercommunity encounters have only been observed around a series of perennial pools in the Bososandja Refuge, where bonobos regularly gather to feed on sub-aquatic vegetation, sometimes in large parties of 18–25 individuals. In these instances, the defence of this preferred resource is played out through avoidance. As a

Table 4.5. *Comparison of frequency of occurrence of terrestrial herbaceous vegetation (THV) plants in plots of climax forest at Lukuru and Lomako*

Species	Lukuru[a] (126) percentage (no.) of plots	Lomako[b] (112) percentage (no.) of plots
Haumania liebrechtsiana	57.14 (72)	96.4 (108)
Palisota spp.	NI	83.9 (94)
Trachyphryium braunianum	NI	25.9 (29)
Sarcophrynium schweinfurthii	NI	24.1 (27)
Sarcophrynium prionogonium	12.70 (16)	NI
Renealmia africana	NI	4.5 (5)
Megaphrynium macrostachyum	19.05 (24)	0
Aframomum sp.	0	0
Aframomum subsericeum	5.56 (7)	NI
Costus afer	NI	0
Anisophyllea quangensis	0	NI
Landolphia lanceolata	0	NI
Annona senegalensis	0	NI
Anubias sp.	0	NI
Icacina guessfeldtii	3.97 (5)	NI

Notes:
[a] Myers Thompson 1997; [b] Malenky 1990.
NI, not identified; *n*, total number of plots.

larger group approaches the pool they can be heard to vocally announce their approach more than 15 minutes before arrival. This broadcast provides the smaller group an opportunity to depart. However, on two occasions the smaller group remained along the shore of the pool and quietly observed from the security of a vegetative cover as the larger group entered the arena and fed in the pool.

Bipedalism

Observation of bonobos in and around the perennial pools provides confirmation that bonobos are not afraid to enter waist-deep water. It is worth comparing the behaviour of bonobos in these pools to the behaviour of chimpanzees at Bossou. Matsuzawa (2000) reported that chimpanzees (*Pan troglodytes verus*) at Bossou had invented a tool to 'sweep' for algae (similar to that eaten by bonobos at the Bososandja pools of Lukuru), but the chimpanzees worked from the safety of the shore, thereby avoiding the water. As an aquatic foraging strategy, the adult bonobos of Lukuru do not hesitate to wade waist-deep into the pools to gather algae and subaquatic vegetation. For these bonobos, inventing a specialized tool for food acquisition is unnecessary. Susman (1984) reported indirect evidence of terrestrial bipedal locomotion along shallow streambeds, where footprints (without associated knuckle prints) indicated bonobos' transition to a bipedal posture while travelling along and crossing streams.

The perennial pools, with substrate comprised of thick organic matter, offer opportunities to observe terrestrial bipedal locomotion. Morphological differences between chimpanzees and bonobos, including more centrally positioned *foramen magnum* (Shea 1984), longer thigh bones (Zihlman 1984), lower intermembral index (Coolidge & Shea 1982; Napier & Napier 1985), heavier lower limb muscle (Zihlman 1984), longer feet (Fleagle 1988) and differential distribution of body weight (Zihlman 1984) have lead to the claim that bonobos demonstrate a greater predisposition for bipedal locomotion (Susman 1987). However, few behavioural observations have been presented across bonobo field studies that confirm or dispute the assertion that bonobos exhibit more habitual use of bipedal locomotion. Susman (1984) and Doran (1993) examined locomotion at Lomako. Susman (1984) found terrestrial bipedal locomotion when bonobos walked short distances dragging branches or food during display behaviour. Doran (1993) concluded that bonobos are more arboreal than chimpanzees (*Pan troglodytes verus*), although caution was urged due to the wide difference in habituation of the study subjects. In addition, arboreal bipedalism inversely decreased as habituation increased. Further, Doran reported intraspecies variation between Lomako and Wamba. Well-habituated bonobos at Wamba travel exclusively on the ground (Doran 1993; personal observation and personal communication from S. Kuroda to D. M. Doran) and exhibit terrestrial bipedality when carrying provisioned foods. More recently, Videan (2000) and Videan & McGrew (2001) examined chimpanzees in comparison with bonobos in order to determine differences in rates of bipedality. Videan reported that, in the captive environment, no interspecies differences were found in frequency rates of bipedalism, although functionally, bonobos exhibited bipedalism more often when carrying or when in a state of vigilance, whereas chimpanzees did so more often during display behaviour.

Terrestrial bipedalism may be observed when bonobos of Lukuru walk upright through open short grass plains, along secondary vegetation and grasses of roadsides, and especially in the pools. Except for observations at the pools, the effect of observer presence on locomotion is immediately evident. Upon detection, bonobos drop to a quadrupedal posture and flee on the ground. However, at the pools bonobos display more boldness while observers remain behind a constructed observation blind. Bipedalism is used when bonobos step into and wade through the water. On any single day of observation at the Bososandja pools of Lukuru, if any bonobos were observed striding two or more consecutive steps bipedally, a count of one was recorded to estimate frequency of bipedalism. In all instances, bonobos using the pools were matures (adult females carrying infants, however, did not enter the water). Immatures remained on the shore, playing only in the shallow waters that could be reached with an extended hand, sometimes supported by holding onto vegetation to lean out over the water. There was no difference between sexes in aquatic foraging.

When they were observed feeding on the ripe fruits of *Annona senegalensis*, which has the growth form of a shrub, bonobos would stand beside the plant and pick the small fruits either with their hands or lips. (Chimpanzees have also been reported to consume the fruits of *A. senegalensis.*) On one occasion (20% of grassland encounters), when bonobos were seen moving between shrubs, they did so in short spurts (more than two steps) of bipedal locomotion. Fruits of *Landolphia lanceolata* were found when the bonobos moved quadrupedally through tall grasses. Walking slowly along worn trails, parting areas of tall grasses within their reach, the bonobos periodically swept the backs of their hand/wrist through the grasses, thus uncovering clumps of fruits. The fruit was carried and eaten as the bonobo progressed through the grassland habitat. On several occasions, evidence of areas of pressed or flattened tall grasses indicated that the bonobo party had been sitting for extended periods in the open grasslands. No feeding remains were located nor food patches associated within or near these resting sites. These incidents occurred during the rainy season in the morning hours when temperatures were cool and may have provided opportunities for the bonobos to warm themselves in the morning sun.

Although drawn from small sample sizes, the incidence of terrestrial bipedal locomotion observed at Lukuru indicates that the transition to bipedalism may have been prompted by exploitation of a wider range of ecological opportunities. These compelling initial findings suggest that bipedalism developed in a habitat where selective pressures emphasized the variety of adaptations to a more complex environment. Bonobos foraging in the pools exhibited terrestrial (aquatic) bipedal locomotion in 24.14% of encounters. Susman (1984) reported arboreal bipedal locomotion in 6% of his observations. During Susman's study, terrestrial locomotion was observed only a fraction of the time, due to the low degree of habituation. However, the propensity for bonobos to transition into bipedal locomotion during aquatic foraging as determined by this initial study, requires further examination and more extensive observation time. Also, caution must be exercised when drawing any inferences from these findings.

DISCUSSION AND CONCLUSIONS

Prior to the Lukuru study, scientists (Kano 1992) considered bonobo distribution to be confined within an area where limited monthly fluctuations in temperature varied between means of 20 °C and 30 °C. However, bonobos at the Lukuru site experience broader ecological parameters, challenging previous descriptions of bonobo ecology.

Except for Lukuru, all bonobo field study sites have been located in areas of swamp vegetation associated with major waterways (Lomako – Lomako River, Wamba – Luo River, Lilungu – Tshuapa River, Lake Tumba – Lake Tumba, and Yalosidi – Njali River). The Lukuru area includes two major navigable waterways, the Lukenie and Sankuru Rivers, but does not have the extensive associated swamp vegetation shouldering these rivers, due to the higher elevation rising out of the Congo Basin onto the Southern Highland Plateau. So far, there is no evidence to indicate that the Lukuru bonobos inhabit terrain in proximity to these two rivers. Bonobo habitat seems to differ from chimpanzee habitat in the absence of an open woodland zone where small trees are scattered across the terrain. The peripheral habitat of bonobos is distinguished by an abrupt southern vegetation transition into grassland, without the presence of acacia or baobab trees. Observations from Lukuru are unique in that bonobos access bodies of perennially pooled fresh surface water, while at other bonobo study sites bonobos access moving water in streams.

Contrary to expectations of larger party size associated with more closed habitat, interspecies habitat variation does not seem to have a marked affect on bonobo party size or aggregation across sites.

Although many food types are available across the geographic distribution of bonobos, regional differences in food choices can be identified. For example, bonobos at Lukuru were observed feeding on the leaves of *Platycerium* sp., an epiphyte. No other field site observations have identified epiphytes as a food source for bonobos. It is not clear whether this species is available at the other locations. Further, it is not clear whether different choices in plant foods are cultural or due to possible differences in dry versus moist forest composition. Cooperative research is required to determine the frequency of species representation and the presence or absence of species across bonobo sites. Some bonobo behavioural variation across sites may result from the differences in availability of cultivated food species, higher percentage of secondary plant forms (such as at Wamba and Lukuru) and the opportunity for more varied choices. Further investigation should take into account that, although some secondary plants species are available at Lomako, bonobos at Wamba and Lukuru may have access to a wider variety of secondary plant foods and rely on them more consistently.

In conclusion, bonobos, who do not compete with chimpanzees for food resources, have not needed to develop different foraging strategies; therefore, parallel behaviours would be expected between these allopatric sister-species, especially where broader environmental circumstances are similar. There are a number of ecological differences across these two species, but clearly foraging is more varied than previously recognized. These field data from Lukuru challenge descriptions of the bonobo as an exclusively lowland, moist forest living ape. We find that bonobos habitually occupy and utilize hilly, mosaic terrain where the topography climbs out of the Central Congo Basin toward the Southern Highland Plateau.

Frequently, when considering behavioural diversity, we conclude differential opportunities for innovation and acquisition of new behaviours. Perhaps, as may be the case for the bonobo, differential strategies are a result of the forfeit of relic behaviours that become unprofitable as the habitat changes. Although it is clear that bonobos require access to forest, findings from this study conclude that they also inhabit a drier and more open habitat, extending our knowledge of their foraging strategies and adaptability. Thus, we find that bonobos are more similar to chimpanzees than previously considered.

REFERENCES

Badrian, N. L. & Malenky, R. K. (1984). Feeding ecology of *Pan paniscus* in the Lomako Forest, Zaire. In *The Pygmy Chimpanzee: Evolutionary Biology and Behavior*, ed. R. L. Susman, pp. 275–99. New York: Plenum Press.

Badrian, N., Badrian, A. & Susman, R. L. (1981). Preliminary observations on the feeding behavior of *Pan paniscus* in the Lomako forest of central Zaire. *Primates*, 22(2),173–81.

Coolidge, H. J. Jr & Shea, B. T. (1982). External body dimensions of *Pan paniscus* and *Pan troglodytes* chimpanzees. *Primates*, 23(2), 245–51.

Doran, D. M. (1993). Comparative locomotor behavior of chimpanzees and bonobos: The influence of morphology on locomotion. *American Journal of Physical Anthropology*, 91, 83–98.

Fleagle, J. G. (1988). *Primate Adaptation and Evolution*. San Diego: Academic Press, Inc.

Fruth, B. (1995). Nests and nest groups in wild bonobos (*Pan paniscus*): Ecological and behavioral correlates. PhD dissertation. Ludwig-Maximilians Universitat München and Max-Planck-Institut für Verhaltensphysiologie Seewiesen.

Fruth, B. & Hohmann, G. (1996). Nest building behavior in the great apes: the great leap forward? In *Great Ape Societies*, ed. W. C. McGrew, L. F. Marchant and T. Nishida, pp. 225–40. Cambridge: Cambridge University Press.

Furuichi, T. (1989). Social interactions and the life history of female *Pan paniscus* in Wamba, Zaire. *International Journal of Primatology*, 10(3), 173–97.

Idani, G. (1991). Social relationships between immigrant and resident bonobo (*Pan paniscus*) females at Wamba. *Folia Primatologica*, 57, 83–95.

Kano, T. (1982). The use of the leafy twigs for rain cover by the pygmy chimpanzees of Wamba. *Primates*, 23(3), 453–7.

Kano, T. (1992). *The Last Ape: Pygmy Chimpanzee Behavior and Ecology*. Stanford, CA: Stanford University Press. (English translation; first published in 1986 in Japanese.)

Kano, T. & Mulavwa, M. (1984). Feeding ecology of the pygmy chimpanzees (*Pan paniscus*) of Wamba. In *The Pygmy Chimpanzee: Evolutionary Biology and Behaviour*, ed. R.L. Susman, pp. 233–74. New York: Plenum Press.

MacKinnon, J. (1978). *The Ape Within Us*. New York: Holt Rinehart and Winston.

Malenky, R. K. (1990). Ecological factors affecting food choice and social organization in *Pan paniscus*. PhD dissertation. State University of New York at Stony Brook.

Malenky, R. K. & Stiles, E. W. (1991). Distribution of terrestrial herbaceous vegetation and its consumption by *Pan paniscus* in the Lomako Forest, Zaire. *American Journal of Primatology*, 23, 153–69.

Malenky, R. K. & Wrangham, R. W. (1994). A quantitative comparison of terrestrial herbaceous food consumption by *Pan paniscus* in the Lomako Forest, Zaire, and *Pan troglodytes* in the Kibale Forest, Uganda. *American Journal of Primatology*, 32, 1–12.

Malenky, R. K., Kuroda, S., Vineberg, E. O. & Wrangham, R. W. (1994). The significance of terrestrial herbaceous foods for bonobos, chimpanzees, and gorillas. In *Chimpanzee Cultures*, ed. R. W. Wrangham, W. C. McGrew, F. B. M. de Waal & P.G. Heltne, pp. 59–75. Cambridge, MA: Harvard University Press.

Matsuzawa, T. (2000). Development of tool use in chimpanzees at Bossou. Paper presented at the conference on *Behavioural Diversity in Chimpanzees and Bonobos, 11–17 June, Seeon, Germany.*

McGrew, W. C. (1992). *Chimpanzee Material Culture: Implications for Human Evolution*. Cambridge: Cambridge University Press.

Myers Thompson, J. A. (1997). The History, Taxonomy and Ecology of the Bonobo (*Pan paniscus*, Schwarz 1929) with a first description of a wild population living in a forest/savanna mosaic habitat. PhD dissertation. University of Oxford, UK.

Myers Thompson, J. (2001). The status of bonobos within their southern-most geographic range. In *All Apes Great and Small Volume 1: Chimpanzees, Bonobos and Gorillas*, ed. B. M. F. Galdikas, N. Briggs, L. K. Sheeran, G. L. Shapiro & J. Goodall, New York: Kluwer Academic Press.

Napier, J. R. & Napier, P. H. (1985). *The Natural History of the Primates*. Cambridge: Cambridge University Press.

Richards, P. W. (1952). *The Tropical Rain Forest: An ecological study*. Cambridge: Cambridge University Press.

Savage-Rumbaugh, S., Romski, M. A., Hopkins, W. D. & Sevcik, R. A. (1989). Symbol acquisition and use by *Pan troglodytes*, *Pan paniscus*, *Homo sapiens*. In *Understanding Chimpanzees*, ed. P. G. Heltne & L. A. Marquardt, pp. 266–95. Cambridge, MA: Harvard University Press.

Shea, B. T. (1984). An allometric perspective on the morphological and evolutionary relationships between pygmy (*Pan paniscus*) and common (*Pan troglodytes*) chimpanzees. In *The Pygmy Chimpanzee: Evolutionary Biology and Behavior*, ed. R.L. Susman (ed.), pp. 89–130. New York: Plenum Press.

Susman, R. L. (1979). Comparative and functional morphology of hominoid fingers. *American Journal of Physical Anthropology*, 50, 215–36.

Susman, R. L. (1984). The locomotor behavior of *Pan paniscus* in the Lomako Forest. In *The Pygmy Chimpanzee: Evolutionary Biology and Behavior*, ed. R.L. Susman, pp. 369–93. New York: Plenum Press.

Susman, R. L. (1987). Pygmy chimpanzees and common chimpanzees: Models for the behavioral ecology of the earliest hominids. In *The Evolution of Human Behavior: Primate Models*, ed. W.G. Kinzey, pp. 72–86. Albany, NY: State University of New York Press.

Susman, R., Badrian, N. L., Badrian, A. J. & Handler, N. T. (1985). Positional behavior and feeding ecology of the pygmy chimpanzee (*Pan paniscus*): first year results of the Lomako Forest Pygmy Chimpanzee Project. *National Geographic Society Research Reports*, 20, 725–39.

Thompson, J. A. M. (2001). A model of the biogeographical journey from *Proto-pan* to *Pan paniscus*: the Westside story. In *Evolutionary Neighbors: Fossils and DNA*, ed. T. Matsuzawa & J. Yamagiwa. Kyoto: Kyoto University Press.

Thompson-Handler, N. E. (1990). The pygmy chimpanzee: Sociosexual behavior, reproductive biology and life history patterns. PhD dissertation. Yale University.

Videan, E. N. (2000). Bipedality in bonobo (*Pan paniscus*) and chimpanzee (*Pan troglodytes*): Implications for the evolution of bipedalism in hominids. Master's thesis. Miami University.

Videan, E. N. & McGrew, W. C. (2001). Are bonobos (*Pan paniscus*) really more bipedal than chimpanzees (*Pan troglodytes*)? *American Journal of Primatology*, 54, 233-9.

White, F. (1983). *The Vegetation of Africa: A descriptive memoir to accompany the UNESCO/AETFAT/UNSO Vegetation Map of Africa*. Natural Resources Research XX, UNESCO publication.

White, F. J. (1988). Party composition and dynamics in *Pan paniscus*. *International Journal of Primatology*, 9(3), 179–93.

White, F. J. (1992). Pygmy chimpanzee social organisation: variation with party size and between study sites. *American Journal of Primatology*, 26, 203–14.

Zihlman, A. L. (1984). Body build and tissue composition in *Pan paniscus* and *Pan troglodytes*, with comparisons to other Hominoids. In *The Pygmy Chimpanzee: Evolutionary Biology and Behavior*, ed. R.L. Susman, pp. 179–200. New York: Plenum Press.

5 • Grooming-hand-clasp in Mahale M Group chimpanzees: implications for culture in social behaviours

MICHIO NAKAMURA

INTRODUCTION

Culture in social behaviours

Humans (*Homo sapiens*) have various types of behaviours that vary across populations, generations or ethnic groups, and which we label 'culture'. Among these behaviours, those used in social interactions are quite familiar to us. For example, even the inter-individual distances in particular social situations differ between different ethnic groups (Hall 1966), and the ways of greeting often vary between different countries (e.g. Collet 1993). Handshaking is common among westerners (Kendon & Ferber 1973), whereas Japanese usually do not shake hands but bow to each other (Nomura 1994), and Tongwe people in Tanzania traditionally clap their hands to each other (Itani *et al.* 1973). Such behaviours are often so deeply embedded in our everyday lives that we are not aware that they are culturally rooted.

The most interesting characteristic of such behavioural diversity in the social domain may be that some of the differences seem almost irrelevant to the functions of these behaviours. Also, the rationale for a particular variation in a behaviour versus another variation in that behaviour does not significantly help us to understand the differences. Rather, we perform these behaviours in the ways we do 'because the person in the front performs like this', or 'because people say it is the proper etiquette to follow'. In other words, these cultural behaviours often only have significance in interactions among individuals. Many of us are ready to admit that our human cultures have such a characteristic. However, is this characteristic unique to humans? What about our closest relative species, the chimpanzees (*Pan troglodytes*)? In this chapter, I focus on those aspects of chimpanzee cultures that are possibly shaped in and/or by their everyday interactions.

It is increasingly clear from long-term studies of chimpanzees that they too have many behaviours that vary between populations (see reviews by Nishida 1987; McGrew 1992, 1998; Thompson 1994; Tomasello 1994; Wrangham *et al.* 1994; Boesch 1996; Boesch & Tomasello 1998; Whiten *et al.* 1999). Among such diverged behaviours, the ones that most attract the attention of researchers are related to material culture, especially various techniques of tool-use. Since Goodall (1964) first reported tool-use in wild chimpanzees, several authors have presented lists of tools from various populations (e.g. van Lawick-Goodall 1973; Sugiyama & Koman 1979; Nishida & Hiraiwa 1982; Goodall 1986; Boesch & Boesch 1990; McGrew 1992, 1994; Boesch 1995; 1996; Matsuzawa & Yamakoshi 1996; Sugiyama 1997). By contrast still less is known about cultures in the social domains. Itani (1991) worried that the concept of 'culture' proposed by Imanishi (1952) might have become limited to 'the knowledge and the techniques for subsistence', and that 'the domains of social interactions or social organizations have been seldom stated in terms of culture'.

Of course, some social behaviours are also listed in the studies of chimpanzee cultures. Drawing from recent studies that included a wide range of behaviours (Boesch & Tomasello 1998; Whiten *et al.* 1999; Nishida *et al.* 1999), Table 5.1 summarizes the possible cultural behaviours in social domains. I limited the list to behaviours that differ qualitatively among populations, thus I did not include differences such as the different frequency structure of pant hoots (Mitani *et al.* 1992) or different proportions of mutual grooming (Boesch & Boesch-Achermann 2000). Nishida *et al.* (1999) categorized some behaviours as possibly cultural, but they did not fully confirm whether these are really absent in different populations, nor whether ecological explanations are possible (T. Nishida personal communication). However, it is notable that in their ethogram there are many behaviours that the former studies did not even discuss. For example, social scratch, or scratching another individual in the context of social grooming, is revealed to be absent in Gombe and other long-term studied populations (Nakamura *et al.* 2000), with the exception of the Ngogo community in Kibale, where D. P. Watts (personal communication) observed a different type of scratch (much shorter stroke). Inconsistencies in the descriptions of behaviour by

Table 5.1. *Summary of possible cultural behaviours in social domains from recent studies*

Behaviour	Use of object	Sites								Boesche & Tomasello 1998	Whiten *et al.* 1999	Nishida *et al.* 1999	Descriptions
		Bs	Ta	Go	Mm	Mk	Kk	Bd	Other sites				
Aimed-throw	rock etc.	+	+	+	+	[7]−	+	+		○[a]	○	×	Goodall, 1964
Branch haul[1]	branch	+	+	−	−		−			○	×[d]		Sugiyama & Koman, 1979
Branch-slap	branch	+	+	−	+	−	−	+			○		V. Reynolds, pers. com.[14]
Bump	—				+							○	Nishida *et al.*, 1999
Grooming-hand-clasp	—	−	+	−	+	+	+	−	Kn[9], Kz[10], Lp[11]	○[b]	○[b]	○	McGrew & Tutin, 1978
(ghc palm-to-palm)	—				−	+	+[8]						McGrew *et al.*, 2001
Knuckle knock	tree trunk etc.	+	+	+	+	+	−	−		○	○	○[g]	Boesch, 1995
Leaf groom[2]	leaf	−	−	+	+	+	+	+		○	○[e]	○[e]	van Lawick-Goodall, 1973
Leaf clipping with fingers	leaf	−	+	−	+	+	+	+			○		Whiten *et al.*, 1999[14]
Leaf clipping with mouth	leaf	+	+	[6]−	+	+	+	+	Nd[12]	○[c]	○	○[c]	Nishida, 1980
Leaf-strip	leaf	+	−	+	+	−	+	−			○		R. W. Wrangham, pers. com.[14]
Lift rock[3]	rock				+							○	Nishida *et al.*, 1999
Play start[4]	branch	+	+	+	+	+	+	+		○	×		Goodall, 1986
Rain dance	branch etc.[5]	−	+	+	+	+	+	+			○	×	Goodall, 1989
Shrub bend	shrub	+	−	−	+	−	−	+			○	○[h]	Nishida, 1997
Slap wall	wall				+							○	Nishida *et al.*, 1999
Snub	—				+							○	Nishida *et al.*, 1999
Social scratch	—	−	−	−	+	+	−	−	Kn[13]			○[i]	Nakamura *et al.*, 2000
Stem pull-through	shrub	+	−	+	+	−	+	−			○	×	Nishida, 1997
Stick club	stick	+	+	+	+	−	+	−		○	○[f]	×	Kortlandt & Kooij, 1963
Throw dry leaves	leaf				+							○	Nishida *et al.*, 1999
Throw splash[3]	rock				+							○	Nishida *et al.*, 1999

Notes:

+, present; −, absent; ○, cited as culture; ×, cited but not as culture. Sites: Bs, Bossou (Guinea); Ta, Taï (Ivory Coast); Go, Gombe (Tanzania); Mm, Mahale M (Tanzania); Mk, Mahale K (Tanzania); Kk, Kibale Kanyawara (Uganda); Kn, Kibale Ngogo (Uganda); Bd, Budongo (Uganda); Kz, Kalinzu (Uganda); Lp, Lopé (Gabon); Nd, Ndoki (Congo).

1: Boesch & Tomasello (1998) listed this as 'communicative behaviour' but, according to the original description by Sugiyama & Koman (1979), the branch haul ('branch hook' in Whiten *et al.* 1999) at Bossou seems to be neither communicative nor social. It is a behaviour whose purpose is to haul a remote branch with a stick. The communicative one may correspond to 'drag branch' of Whiten *et al.* (1999) (A. Whiten, personal communication). 2: Boesch & Tomasello (1998) also listed this as 'communicative behaviour' but its communicative function is not clear. According to Boesch (1995), it is a technique to squash ectoparasites at Gombe. 3: 'Lift rock' and 'throw splash' are two parts of one display: males in Mahale first lift a rock and then throw it into water to make a splash. 4: McGrew (1992) and Boesch & Tomasello (1998) put this in their lists of cultures (absent at Bossou) but Whiten *et al.* (1999) did not (also present at Bossou). 5: Use of an object is not an essential component of rain dance but it usually accompanies it. 6: It was also twice observed in Gombe (Nishida, 1987). 7: Also present at Mahale K group? (T. Nishida, personal communication). 8: M. Nakamura, personal observation. 9: J. C. Mitani & D. P. Watts, personal communication. 10: C. Hashimoto & T. Furuichi, personal communication. 11: C. E. G. Tutin, personal communication. 12: Kuroda, 1998. 13: D. P. Watts, personal communication. 14: Descriptions of these behaviours are cited on the WorldWide Web (http://chimp.st-and.ac.uk/cultures).

The following key is used for the names of behaviours in the original articles: [a], missile throw; [b], hand clasp; [c], clip leaf; [d], branch hook; [e], groom leaf; [f], club; [g], rap, thump; [h], bend shrub; [i], scratch socially.

Fig. 5.1. Grooming-hand-clasp (*a*) and branch clasp grooming (*b*). Photographs show chimpanzees from Mahale M Group (*a*), and chimpanzees from Gombe (*b*) (courtesy of McGrew & Tutin).

different authors may be a sign that studies in this domain are still in the early stages of development. More detailed descriptions of possible cultural behaviours from studied populations would stimulate researchers from other sites to determine whether or not equivalent behaviours are actually absent from the communities they are investigating. Such investigations would, I believe, add a number of behaviours to the list presented in Table 5.1.

Among 21 behaviours in Table 5.1, 17 are related to material objects such as leaves, branches, rocks or vegetation. In these behaviours, an individual's performance is directed to an object, and its outcome is used socially. For example, in leaf clipping in Mahale (Nishida 1980), an individual clips a leaf or leaves and its outcome, that is a conspicuous sound, is used to attract attention. It is also notable that many of these behaviours (namely, branch-slap, knuckle knock, leaf clipping, leaf-strip, shrub-bend, and stem pull-through) are used in the context of courtship to attract the mating partner (leaf clipping is also known to be used in different contexts at Bossou (Sugiyama 1981) and at Taï (Boesch 1995)). In courtship displays, such cultural signals are usually accompanied with universal signals, that is erected penis and tumid sexual swelling; therefore, it is less likely that the intended partner misunderstands the arbitrary signals produced by such behaviours. What a performer has to learn is the relationship between the manipulation of objects and its outcome, for example the clipping of a leaf always produces a sound, and it is sufficient to attract the attention of the prospective mate. Thus, in this sense, most of the cultural behaviours that occur in a social context, and that have been reported so far, can also be classified as a version of material cultures. Actually, some authors included some of the behaviours, such as leaf clipping or aimed-throw, in the list of tool-use (e.g. McGrew 1992).

GROOMING-HAND-CLASP (GHC)

Grooming-hand-clasp (McGrew & Tutin 1978) is one of the four 'pure' social behaviours in the list. The behavioural

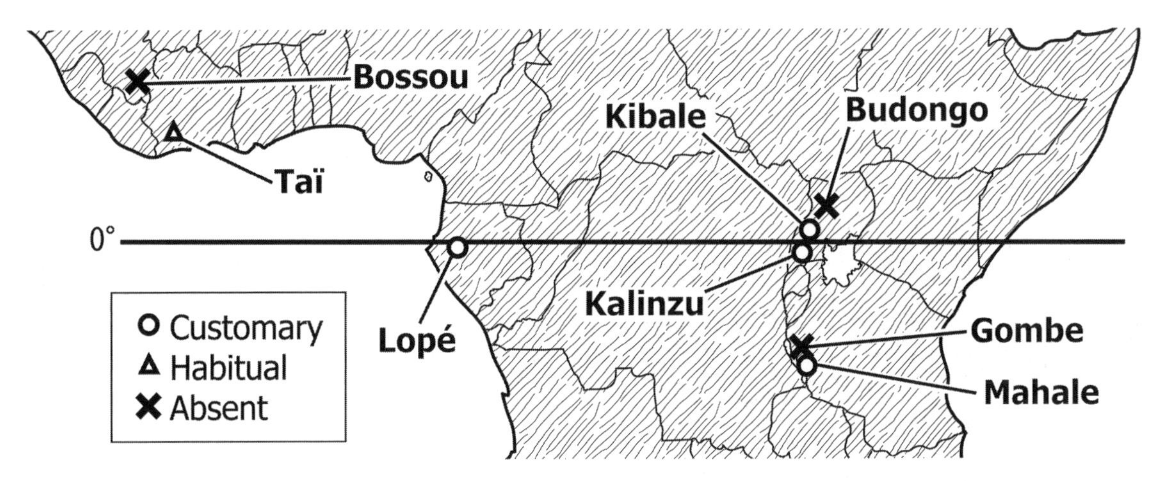

Fig. 5.2. Distribution of grooming-hand-clasp across chimpanzee populations.

pattern of grooming-hand-clasp (Figure 5.1(*a*)) is very symmetrical in that two chimpanzees sit face-to-face, clasp corresponding hands overhead, and simultaneously groom each other's underarm regions (elbow to waist) with their free hands. This is basically a variation of mutual grooming, although some individuals, in rare cases, do not groom the partner while being groomed (Nishida 1994). At the onset of the behaviour, one individual raises his/her hand, and the other individual responds immediately. However, in many cases observers cannot tell which of the two is the initiator because they raise their hands almost simultaneously. McGrew *et al.* (2001) reported that there was an inter-group difference between the Mahale K group and M group in the styles of grooming-hand-clasp: the K group members truly clasped their hands with palmar contact (palm-to-palm pattern), whereas members of the M group usually only crossed their wrists.

The grooming-hand-clasp is observed in some populations but not in others (Figure 5.2). It was customarily observed in the Mahale K group (McGrew & Tutin 1978; Nishida 1988) and M group (McGrew *et al.* 2001; this study), Kanyawara (Ghiglieri 1984) and Ngogo (J. C. Mitani & D.P. Watts, personal communication) groups in Kibale, Lopé (C. E. G. Tutin, personal communication), and in the Kalinzu (C. Hashimoto & T. Furuich, personal communication). It also appeared in a captive colony of Yerkes Regional Primate Research Center (de Waal & Seres 1997). Only two adult males habitually perform this pattern in Taï (C. Boesch, personal communication), and it has never been observed in Gombe, Bossou or Budongo. We do not have further details about this behaviour from living wild populations. T. Furuichi (personal communication) once observed that bonobos (*Pan paniscus*) in Wamba performed grooming-hand-clasp, and Kano (1998) reported that they sometimes hold each other's hand overhead while grooming. However, Kano (1998) also stated that mutual grooming is rare in bonobos and when it occurs it takes place as a slight overlap when the two participants alter the roles of groomer and groomee. This is consistent with what other bonobo researchers say (Wamba: S. Kuroda, personal communication; Lomako: B. Fruth, personal communication). It may be that bonobos sometimes clasp hands overhead, but grooming itself rarely becomes mutual. This is different from chimpanzees, whose grooming in hand-clasp posture is almost always mutual.

BRANCH-CLASP-GROOMING (BCG)

Goodall (1965) reported that Gombe chimpanzees performed a grooming pattern quite similar to the grooming-hand-clasp, called branch-clasp grooming by Whiten *et al.* (1999). Branch-clasp differs from hand-clasp only in that chimpanzees do not clasp their hands but instead grasp a branch that is overhead (Figure 5.1(*b*)). This pattern has been observed in all long-term studied chimpanzee populations including Gombe, Bossou and Budongo, where grooming-hand-clasp is absent (Whiten *et al.* 1999), and is also common in captive chimpanzees (de Waal & Seres 1997). Therefore, it can be considered a universal behavioural pattern in chimpanzees, and some authors (McGrew & Tutin 1978; Goodall 1986; de Waal & Seres 1997) suspect that this behaviour may be the origin of the grooming-hand-clasp. There is no information on this behaviour for bonobos.

Despite the common existence of the branch-clasp and

Table 5.2. *Frequency of grooming-hand clasp (GHC)*

Data set	Site	Year of study	Number of GHC[a]	Observation hours	No. GHC per observation h	No. GHC per grooming h	Source
M94	Mahale M	1994	65	451.0	0.14	—	This study
M96	Mahale M	1996–97	95	480.0	0.20	0.69	This study
K	Mahale K	1975	14	33.5	0.42	—	McGrew & Tutin, 1978
Y92	Yerkes	1992–93	22	654.0	0.03	—	de Waal & Seres, 1997
Y94	Yerkes	1994–96	67	412.5	0.16	—	de Waal & Seres, 1997

Note:

[a] Total numbers of grooming-hand-clasps in different studies.

its similarity to the hand-clasp, no one has ever compared these two behaviours directly. By comparing these two similar patterns in Mahale, I would like to present the possibility that chimpanzees also arbitrarily shape their everyday social interactions. First, I report qualitative and empirical data on the grooming-hand-clasp from the Mahale M group and make some comparisons with the behaviour of Mahale K group and the Yerkes captive group. I then make some comparisons with branch-clasp grooming to investigate how chimpanzees differentiate these two patterns.

METHODS

I collected two data sets on M group chimpanzees in the Mahale Mountains, Tanzania, from different periods. For detailed information about Mahale, see Nishida (1990). The main data set (referred to as M96 hereafter) was collected from July 1996 to May 1997, following ten males and ten females of various ages (Table 5.3, see below). I continuously recorded all of the grooming behaviour that occurred around the focal individual throughout the day. Grooming behaviours, including grooming-hand-clasp and branch-clasp grooming, were recorded not only for focal individuals but also for non-focal individuals within the grooming cluster in which focal individuals were engaged. Observation time totalled 480 hours, and the accumulated individual grooming duration (including non-focal) was 137 hours. The other data set (referred to as M94) was collected from June to November 1994, during which I studied nine adolescent males for 451 hours. This observation would have been biased toward males, since adolescent males tend to be in the proximity of adult males (Hayaki 1988; Pusey 1990). Therefore, this data set is used only for supplementary purposes.

For comparative purposes, I used published data sets from Mahale K group (referred to as K) (McGrew & Tutin 1978) and from Yerkes (de Waal & Seres 1997). De Waal & Seres divided their data into two periods because after the first observation of the grooming-hand-clasp in 1992, its frequency dramatically increased in 1994. Thus, the first period in Yerkes will be referred to as Y92 and the latter as Y94.

RESULTS

Frequency

Table 5.2 compares the overall frequency of grooming-hand-clasp among the four data sets. In M96, I observed 0.20 grooming-hand-clasps per observation hour (95/480 h), which is once per 5.1 hours. It occurred 0.69 times per grooming hour, or once per 1.4 hours. In M94, the frequency was 0.14 times per hour (65/451 h), less than M96 but about the same frequency. In Y92, it occurred 0.03 times per hour (22/654 h). The frequency was low, probably because propagation of the behaviour was still in the beginning stages. It increased to 0.16 times per hour in Y94 (67/412.5 h), almost equal to the two M group data sets. In K, it was observed 0.42 times per hour (14/33.5 h), which is more than twice the others.

Grooming-hand-clasp of 20 focal individuals in M96 (Table 5.3) averaged 0.09 ± 0.15 (SD) times per focal hour and 0.44 ± 0.67 times per grooming hour. The most frequent performer was the oldest male, Kalunde, who performed it 0.51 times per focal hour and 1.58 times per grooming hour. Some individuals never performed it when they were focal individuals.

Table 5.3. *Frequency of grooming-hand-clasp by focal targets of M96*

Names of focal targets	Abbreviates	Age–sex class[a]	Hours of focal follows	Number of GHC during follows	GHC/ focal follow	Grooming hours during follows	GHC/ grooming	GHC not during follows
Nsaba	NS	adult male (alpha)	15.67	2	0.13	3.61	0.55	12
Kalunde	DE	adult male (beta)	29.55	15	0.51	9.47	1.58	5
Fanana	FN	adult male (gamma)	30.85	0	0.00	4.80	0.00	1
Dogura	DG	adult male	31.10	12	0.39	5.55	2.16	7
Hanby	HB	adult male	20.74	0	0.00	3.05	0.00	0
Bonobo	BB	adult male	13.72	0	0.00	0.34	0.00	0
Alofu	AL	adolescent male	25.50	0	0.00	0.38	0.00	2
Carter	CT	adolescent male	24.63	0	0.00	0.71	0.00	1
Sinsi	SS	adolescent male	13.82	0	0.00	1.00	0.00	0
Darwin	DW	juvenile male	33.19	1	0.03	2.75	0.36	0
Ikocha	IK	adult female (lactating)	26.18	0	0.00	4.15	0.00	0
Fatuma	FT	adult female (lactating)	22.80	0	0.00	4.30	0.00	3
Pinky	PI	adult female (lactating)	26.84	0	0.00	2.93	0.00	10
Gwekulo	GW	adult female (cycling)	25.84	5	0.19	5.48	0.91	16
Christina	XT	adult female (lactating)	23.22	4	0.17	3.25	1.23	4
Nkombo	NK	adult female (cycling)	28.03	2	0.07	3.30	0.61	0
Abi	AB	adult female (cycling)	23.75	6	0.25	4.22	1.42	3
Serena	SE	adolescent female (cycling)	21.30	0	0.00	1.88	0.00	0
Maggy	MG	adolescent female	22.40	0	0.00	0.60	0.00	0
Ai	AI	juvenile female	20.95	0	0.00	1.78	0.00	0

Notes:

[a] Definition of age–sex classes basically follows Haraiwa-Hasegawa *et al.* (1984). However, note that, during this study, a 15-year-old male Dogura was fourth-ranking (third-ranking in 1997), overtaking some older males, and was the most important coalition partner for the beta male. Therefore, that age was used as adult in this study.

Duration

Mean duration of grooming-hand-clasp was 21.7±19.8 seconds in K, 36.9±20.5 seconds in M96, 52.2±41.9 seconds in Y92, and 99.3±66.9 seconds in Y94. Both Yerkes periods showed longer duration than K and M96. The longest in Y94 was 296 seconds, while the longest was 94 seconds in M96 and only 60 seconds in K, both less than the average of Y94. In M96, mean duration between adult males was 59.1±18.6 seconds ($n=8$), while those between adult females was 36.5±19.3 seconds ($n=30$) and those between sexes was 33.2±17.8 ($n=35$). The difference was significant (Kruskal–Wallis test, $H=9.95$, $p<0.01$).

Performers

Total numbers of performers were 20 in M96, 24 in M94, and 9 in K (Table 5.4).

The three data sets showed similar tendencies: most adult males performed the behaviour, while about half of the adult females did. Numbers of different dyadic combinations (36, 28 and 14, respectively for M96, M94 and K) were relatively smaller than expected from the number of individuals. The youngest individual that performed this behaviour in M96 was a 7-year-old orphan female (Pippi), who performed it with her most frequent associate adult female (Gwekulo); no juveniles or infants performed it in K or

Table 5.4. *Proportion of performers of grooming-hand-clasp*

	K		M94		M96	
	+	−	+	−	+	−
Adult males	5	0	8	1	5	2
Adult females	3	6	12	12	10	8
Adolescent males	0	0	3	6	2	3
Adolescent females	1	3	1	7	0	5
Juveniles and infants	0	10	0	22	3	15
Total	9	19	24	48	20	33

Note:
+ Number of individuals who performed GHC.
− Number of individuals who did not perform GHC.

M94. However, at different periods a 5.7-year-old female (Penelope) (Nishida 1988) and a 5.3-year-old male (Christmas) (M. Nakamura, personal observation) each performed it with their mother. Two of the adult males who did not perform this in M96 were both young adults whose behaviours in general were still like those of adolescent males. Although the number of grooming-hand-clasps per grooming hour is positively correlated with age (Spearman's rank correlation, $r_s=0.51, p<0.001, n=53$), there is no correlation when we see only adults ($r_s=0.04, p=0.84, n=25$), showing that after maturity is reached the frequency seems to depend more on individuals than on age. No significant sex differences in the frequencies were exhibited either among any of the combined ages (Mann–Whitney U-test, $U=306.5$, $p=0.77$, n[male]=20, n[female]=32) or among adults ($U=58.0$, $p=0.75$, n[male]=7, n[female]=18).

Figure 5.3 shows the percentages of the combinations of individuals that performed the grooming-hand-clasp in K and M96. Unlike the frequency of performers among age and sex classes, the combinations may be seen as showing quite divergent tendencies. Although grooming-hand-clasps between females were never observed in K, they accounted for 36.8% (35/95) in M96. The proportion of combinations including non-adult individuals was small in both data sets. Grooming-hand-clasps between adolescent males were rare, and juveniles always performed it with adults, never with each other.

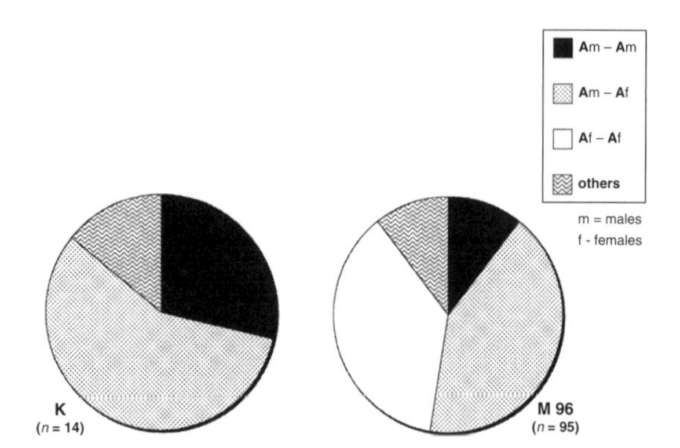

Fig. 5.3. Proportion of combinations of the grooming-hand-clasp in K and M96. Am = adult males; Af = adult females.

Incomplete grooming-hand-clasp

There were 23 cases of incomplete grooming-hand-clasp in M96 (Table 5.5). These incompletions occurred most frequently (12/23) when one chimpanzee raised a hand that was not clasped by the partner. In seven cases, both of the participants raised hands but failed to clasp them. The other four occurred when two participants clasped their hands but one did not groom the other. In the first type, only four of 12 cases were observed among GHC dyads that performed at least one grooming-hand-clasp during the study period. This was significantly less than the other two types, in which nine of 11 cases were GHC dyads (Fisher's exact probability test, $p<0.05$). The first type can be viewed as one party ignoring the other since they were not GHC dyads, and the latter two types can be considered real mistakes (e.g. failure to synchronize timing) between GHC dyads who often performed the behaviour.

Comparison of grooming-hand-clasp and branch-clasp

The two cases below may indicate that chimpanzees differentiate branch-clasp grooming and grooming-hand-clasp on the basis of the partner.

Case 1

The adult female Opal raised her hand in front of the third ranking adult male Fanana, probably in order to solicit grooming-hand-clasp. Fanana did not accept her solicitation but instead clasped the overhead branch. Then Opal also took the same branch, and the two started to perform branch-clasp grooming.

Table 5.5. *Incomplete grooming-hand-clasp in M96*

Types of incompletion	GHC dyads[a]	non-GHC dyads[a]	Total
Clasped hands but one did not groom	3	1	4
Both raised hands but failed to clasp	6	1	7
Only one raised the hand, therefore failed to clasp	4	8	12
Total	13	10	23

Note:

[a] Dyads that showed grooming-hand-clasp at least once are defined at GHC dyads. Those that never performed this behaviour are non-GHC dyads.

Case 2

The alpha male Nsaba and the young adult female Totzy groomed each other in branch-clasp grooming. After a while, Totzy stopped grooming and moved aside. Then Masudi, a low ranking adult male, arrived. He sat in the place where Totzy had been sitting a few moments before. The two males began a grooming-hand-clasp, despite the same branch being available.

The first case may be interpreted as Fanana not wanting to perform grooming-hand-clasp with Opal, but unable to ignore her entirely, offering branch-clasp grooming instead. Both Fanana and Opal were observed performing grooming-hand-clasps during the study period, but never with each other. The second case may indicate that Nsaba distinguished between branch-clasp and hand-clasp despite the availability of a branch for branch-clasp. Totzy also knew the GHC behaviour because she was observed performing it with others during the study period. Though there are other possible explanations, these cases suggest that chimpanzees somehow differentiate these two behaviours on a dyadic basis.

In Figure 5.4 (*a*), the number of grooming-hand-clasp events is plotted against the number of branch-clasp grooming events on an individual basis ($n=34$). Those who performed more branch-clasp grooming also performed more grooming-hand-clasp (Spearman's rank correlation, $r_s=0.58$, $p<0.01$, $n=34$). This positive correlation is produced perhaps because both of the numbers are correlated with total grooming duration. It is clear, at least, that no individual specialized in either branch-clasp or hand-clasp. Notably, with one exception, individuals that performed grooming-hand-clasp also performed branch-clasp grooming. However, when dyads ($n=68$) are plotted (Figure 5.4(*b*)), the dyads with more branch-clasp grooming showed less grooming-hand-clasp, and the dyads with more hand-clasp showed less branch-clasp.

It is difficult, however, to decide how they differentiate these two patterns. Some male pairs are hand-clasp pairs and some are branch-clasp pairs. That also goes for female pairs. Neither the duration of total grooming between the dyads (Spearman, $r_s=0.26$, $p=0.12$, $n=36$) nor the reciprocity of grooming among the dyads (Spearman, $r_s=0.21$, $p=0.23$, $n=36$) was correlated with the number of grooming-hand-clasps.

DISCUSSION

Grooming-hand-clasp in the sphere of cultures in social behaviours

Among the list of candidates of cultures, the grooming-hand-clasp was one of the rare examples of culture that is performed in an exclusively social manner, that is without the use of any material objects. The infrequent occurrence of these behaviours in the list could be because they actually are rare: perhaps chimpanzees learn how to manipulate objects (e.g. tool-use) more easily because the consequences of such behaviours are more constant and predictable than the consequences of social behaviours. Also, although idiosyncratic object manipulations are sometimes observed, such as tool-use to clear a blocked nasal passage (Nishida & Nakamura 1993), or using a kerosene can in the charging display (Goodall 1986), this kind of idiosyncratic behaviour seems to be rare in a 'purely' social domain. This may be because when these kinds of idiosyncratic behaviours do occur in this domain, they may be repeatedly corrected by the individual who directly receives and reacts to the behaviour; this may make it more difficult for such novel behaviours to survive.

However, I assume that these kinds of behaviours are rare in part because of the human bias toward tool-use-like behaviours. 'Pure' social behaviours are not as conspicuous as tool-use, and tend to be classified into some particular behavioural categories, so they are rarely described in detail

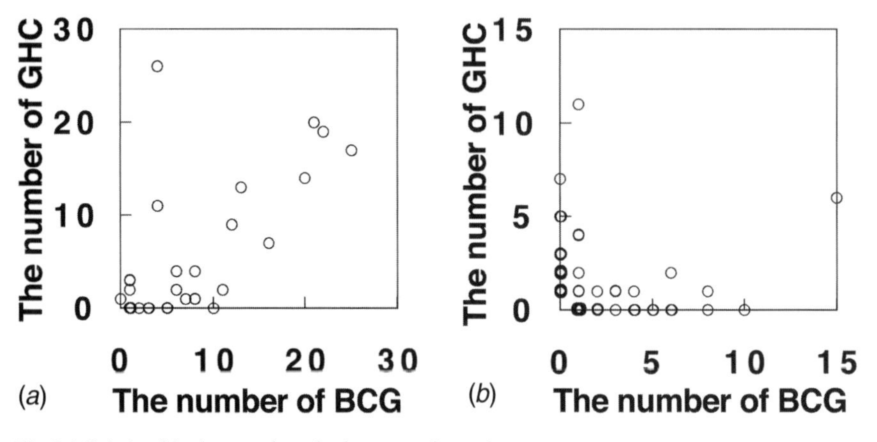

Fig. 5.4. Relationships between branch-clasp grooming and grooming-hand-clasp on an individual basis (*a*) ($n = 34$) and on a dyadic basis (*b*) ($n = 68$).

on their own merits. For example, social scratch in Mahale had been included in the larger category of 'social grooming'. Without a description of the subtle behavioural elements within the larger category, it would be difficult to detect locality-specific patterns. Also, as major topics of research, behaviours without concrete functions are sometimes difficult to study. Most instances of tool-use have concrete and apparent functions, while it is sometimes difficult to determine the direct functions of behaviours used in social contexts. It is relatively easy to suggest the function of courtship displays because the outcome of the displays usually leads to mating. This may be one of the reasons that courtship displays often appear in the list of cultures. However, it is rather difficult to determine the function of the grooming-hand-clasp. One of the possible functions may be cleaning of the underarm region. Tomasello & Call (1997) proposed this as a function because they suspected that individuals could easily learn it with only local enhancement if they remembered that there were 'tasty parasites' in the underarm region. However, it does not seem to be an efficient way for cleaning the underarm, since chimpanzees in this position can use only one hand for grooming, and the inspection of hair is also limited in this fixed posture. In order to serve an hygienic function, it would be more efficient to groom alternately using both hands. Also, given the ubiquity of branch-clasp grooming, chimpanzees would not need to invent the hand-clasp for the sake of cleaning their underarm regions, provided that one-handed grooming served the same purpose as two-handed grooming.

Performers of grooming-hand-clasp

The grooming-hand-clasp was performed mostly among adults. This is not simply because the behaviour is difficult to perform for younger individuals. Adolescent males often try to perform it with adult males but are usually ignored completely. Youngsters seem to know how to perform the behaviour but do not engage in it because others do not easily accept them as partners. It is also not simply because of the physical differences between juveniles and adults, because juveniles never performed it with similar-sized juveniles. Thus, it is possible that there is some kind of consensus that this is a behaviour of adults.

Only half of the adult females performed grooming-hand-clasp in Mahale. The absence of GHC in some females may indicate that this custom does not exist in their particular natal group. However, it is not easy to decide whether such females completely lack the habit or only that their frequencies are too low. Although one focal adult female, Ikocha, never performed this behaviour in M96, she once performed it with her adolescent son, Iwan, in M94. More observations of these females that do not display hand-clasp would be needed in order to determine whether they completely lack this habit.

Differences between populations

Mahale K group showed a higher frequency of the behaviour than did M group and Yerkes colony. This result might be derived from different group compositions or observation methods. For example, the Yerkes colony included only one adult male. This could have lowered the frequency since adult males are usually the most active performers of the grooming-hand-clasp. It may be that the data on K group were collected when relatively more individuals gathered together in the provisioning site (T. Nishida, personal communication). This in turn may have caused the higher frequency of K group given that chimpanzees are likely to be active in grooming when they assemble together.

There were also differences in duration; the Yerkes group showed longer durations than the two Mahale groups. This is not easily explained by differences in observation methods. It is possible that in performing the behaviour in everyday interactions, slight differences can appear even in the same behavioural pattern.

Why are branch-clasp grooming and grooming-hand-clasp differentiated?

The results suggest that the M group chimpanzees use the two types of mutual grooming differently on a dyadic basis. This implies at least the following two possibilities. One is that there is some critical functional difference derived from the different forms of these two behaviours. Perhaps branch-clasp is better when branches are available because they can rest their hands on the branch, and hand-clasp may have some value when there is no branch because they can still mutually groom underarm without the help of branches. However, this does not explain why some dyads tend to perform hand-clasp more often than other dyads. It also does not explain why hand-clasp is necessary only in some populations such as Mahale, but not in others, such as Gombe. It is not surprising that chimpanzees perform grooming-hand-clasp when there are no branches available because they cannot clasp a non-existing branch. However, there were some cases in which they hand-clasped even when branches were available. Indeed, one individual performed both hand-clasp and branch-clasp in the very same place with different partners, as shown in Case 2. These examples imply that chimpanzees do not simply differentiate these two patterns with respect to the availability of branches. There is also no evidence that branches are less frequently available in Mahale than in Gombe, Bossou or Budongo. It seems difficult to assume that certain functional differences are required for only particular dyads.

The second possibility is that there are no critical differences between these two behavioural patterns, but after chimpanzees acquired the hand-clasp custom they added a new social meaning to the behaviour. Thus, these two grooming patterns are now different for them. It seems relatively easy for chimpanzees to invent the form of hand-clasp, given that they already groom each other's underarm mutually with the help of branches and that they often raise their hands even without the help of branches. Probably, physical adaptation for unimanual arm-hanging (Hunt 1991) makes it possible for chimpanzees to raise hands easily. Accordingly, what they have to do in hand-clasp is simply to hold or to touch the other's hand in the air.

Unfortunately, it is difficult to clarify how chimpanzees differentiate grooming-hand-clasp and branch-clasp grooming. However, I would like to propose the possibility that, at least in the adult male dyads, social relationships seem more stable in hand-clasp pairs than branch-clasp pairs. For example, the alpha male Nsaba and the old beta male Kalunde were a hand-clasp pair while Nsaba and the young gamma male Fanana were a branch-clasp pair, perhaps because old Kalunde was no longer a threat for Nsaba, but the relationship with young Fanana was still unstable.

After all, adaptive functions alone can hardly explain why grooming-hand-clasp occurs only in some locations and why it is used differently from branch-clasp grooming at Mahale. I would argue that, no matter what the actual function is, the most important determinant for an individual in performing such a behaviour is 'because the partner performs the behaviour', and this is the most interesting feature of culture in the social domain. Of course, we cannot prove that chimpanzees really determine their behaviours in this way. However, we humans often behave because of the performance of the other person in a face-to-face situation. For example, we shake hands not because of the function of the behaviour itself, such as showing friendship, equality, and so on. These functions can also be expressed with other behaviours, so the particular form the handshaking takes has no direct or apparent functional connection. In this sense, we perform this behaviour arbitrarily, and Boesch (1995) notes that such arbitrariness can also be found in chimpanzees.

In cultural behaviours used for subsistence, such as most tool-uses, an individual's motivation to perform the behaviour may be directly related to the function of the behaviour. For example, motivation for nut cracking is to crack open a nut, which is the very function of the behaviour. It also seems plausible that individuals can perform such behaviours on their own once they acquire the behaviour, although the acquisition itself is affected by the way other individuals perform the behaviour (frequency, techniques, etc.). In contrast, the performance of cultural behaviours in social domains always depends on other individuals. The behavioural elements in leaf-clipping and social scratch, clipping and scratching, respectively, do not seem complex or novel.

Only the context and the partner are important here. What appears to be a social custom in chimpanzees may be some kind of mutual anticipation of other individuals performing in particular ways that are shaped in everyday face-to-face interactions. However, in chimpanzees, such anticipation seems limited to the individuals who are interacting at the moment because there is no case of sanction by a third party for not performing a behaviour like grooming-hand-clasp. When the anticipation for such a behaviour is extended to a generalized 'public' and *not* performing as others do is seen as an aberration, it can be considered institutionalized.

ACKNOWLEDGEMENTS

This chapter was originally prepared for the conference 'Behavioural Diversity in Chimpanzees and Bonobos' held 12–17 June, 2000 in Seeon, Germany. I express my thanks to: the Tanzania Commission for Science and Technology, the Tanzania Wildlife Research Institute, Tanzania National Parks, Mahale Mountains Wildlife Research Centre, and Mahale Mountains National Park for permission to do the research; K. Kawanaka, S. Uehara, W.C. McGrew, L.F. Marchant, H. Sasaki, R. Kitopeni, M.B. Kasagula, M. Matumula and K. Athumani for cooperation in the field; F. Fukuda, T. Nemoto, and H. Kayumbo for help in various ways in Tanzania; T. Nishida, M.A. Huffman, N. Itoh and the three editors of this book, C. Boesch, L.F. Marchant, and G. Hohmann for comments on an earlier version of the manuscript. The study was supported by grants from the Japanese Ministry of Education, Science, Sports, and Culture (No. 07041138 to T. Nishida and No. 12740482 to M. Nakamura).

REFERENCES

Boesch, C. (1995). Innovation in wild chimpanzees (*Pan troglodytes*). *International Journal of Primatology*, 16, 1–16.

Boesch, C. (1996). Three approaches for assessing chimpanzee culture, In *Reaching into Thought*, ed. A. E. Russon, K. A. Bard & S. T. Parker, pp. 404–29, Cambridge: Cambridge University Press.

Boesch, C. & Boesch, H. (1990). Tool use and tool making in wild chimpanzees. *Folia Primatologica*, 54, 86–99.

Boesch, C. & Boesch-Achermann, H. (2000). *The Chimpanzees of the Taï Forest*, Oxford: Oxford University Press.

Boesch, C. & Tomasello, M. (1998). Chimpanzee and human cultures. *Current Anthropology*, 39, 591–614.

Collett, P. (1993). *Foreign Bodies*, London: Simon & Schuster Ltd.

de Waal, F. B. M. & Seres, M. (1997). Propagation of handclasp grooming among captive chimpanzees. *American Journal of Primatology*, 43, 339–46.

Ghiglieri, M. P. (1984). *The Chimpanzees of Kibale Forest*, New York: Colombia University Press.

Goodall, J. (1964). Tool-using and aimed throwing in a community of free-living chimpanzees. *Nature*, 28, 1264–6.

Goodall, J. (1965). Chimpanzees of the Gombe Stream Reserve. In *Primate Behavior*, ed. I. de Vore, pp. 423–73, New York: Holt, Rinehart and Winston.

Goodall, J. (1986). *The Chimpanzees of Gombe*, Cambridge, MA: Harvard University Press.

Goodall, J. (1989). Glossary of chimpanzee behaviors. Tanaka, M. & Matsuzawa, T. (Japanese translation, 1992) *Primate Research*, 8, 123–52.

Hall, E. T. (1966). *The Hidden Dimension*, New York: Doubleday & Co., Inc.

Hayaki, H. (1988). Association partners of young chimpanzees in the Mahale Mountains National Park, Tanzania. *Primates*, 29, 147–61.

Hiraiwa-Hasegawa, M., Hasegawa, T. & Nishida, T. (1984). Demographic study of a large-sized unit-group of chimpanzees in the Mahale Mountains, Tanzania: a preliminary report. *Primates*, 25, 401–13.

Hunt, K. D. (1991). Positional behavior in the Hominoidea. *International Journal of Primatology*, 12, 95–118.

Imanishi, K. (1952). Evolution of humanities, In *Man*, ed. K. Imanishi, pp. 36–94, Tokyo: Mainichi-Shinbunsha, in Japanese.

Itani, J. (1991). The concept of culture: subsequence of identification theory, In *Cultural Histories of Monkeys and Apes*, ed. T. Nishida, K. Izawa & T. Kano, pp. 271–7, Tokyo: Heibonsha, in Japanese.

Itani, J., Nishida, T. & Kakeya, M. (1973). *The Eastern Hinterland of Lake Tanganyika*, Tokyo: Chikuma-Shobou, in Japanese.

Kano, T. (1998). A preliminary glossary of bonobo behaviors at Wamba. In *Comparative Study of the Behavior of the Genus* Pan *by Compiling Video Ethogram*, ed. T. Nishida, pp. 39–81, Kyoto: Nisshindo, in Japanese.

Kendon, A. & Ferber, A. (1973). A description of some human greetings. In *Comparative Ecology and Behavior of Primates*, ed. R. P. Michael & J. H. Crook, London: Academic Press.

Korlandt, A. & Kooij, M. (1963). Protohominid behaviour in primates. *Symposia of the Zoological Society of London*, 10, 61–88.

Kuroda, S. (1998). Preliminary ethogram of Tschego chimpanzees. In *Comparative Study of the Behavior of the Genus* Pan *by Compiling Video Ethogram*, ed. T. Nishida, pp. 39–81, Kyoto: Nisshindo, in Japanese.

Matsuzawa, T. & Yamakoshi, G. (1996). Comparison of chimpanzee material culture between Bossou and Nimba, West Africa, In *Reaching into Thought*, ed. A. E. Russon, K. A. Bard & S. T. Parker, pp. 211–32, Cambridge: Cambridge University Press.

McGrew, W. C. (1992). *Chimpanzee Material Culture*, Cambridge: Cambridge University Press.

McGrew, W. C. (1994). Tools compared: the material of culture. In *Chimpanzee Cultures*, ed. R. W. Wrangham, W. C. McGrew, F. B. M. de Waal & P .G. Heltne, pp. 25–40, Cambridge, MA: Harvard University Press.

McGrew, W. C. (1998). Culture in nonhuman primates? *Annual Review of Anthropology*, 27, 301–28.

McGrew, W. C., Marchant, L. F., Scott, S. E. & Tutin, C. E. G. (2001). Inter-group differences in a social custom of wild chimpanzees: the grooming hand-clasp of the Mahale Mountains, Tanzania. *Current Anthropology*, 42, 148–53.

McGrew, W. C. & Tutin, C. E. G. (1978). Evidence for a social custom in wild chimpanzees? *Man*, 13, 234–51.

Mitani, J. C., Hasegawa, T., Grous-Louis, J., Marler, P. & Byrne, R. (1992). Dialects in wild chimpanzees? *American Journal of Primatology*, 27, 233–43.

Nakamura, M., McGrew, W. C., Marchant, L. F. & Nishida, T. (2000). Social scratch: another custom in wild chimpanzees? *Primates*, 41, 237–48.

Nishida, T. (1980). The leaf-clipping display: a newly-discovered expressive gesture in wild chimpanzees. *Journal of Human Evolution*, 9, 117–28.

Nishida, T. (1987). Local traditions and cultural transmission. In *Primate Societies*, ed. B. B. Smuts, D. L. Cheney, R. M. Seyfarth, R. W. Wrangham & T. T. Struhsaker, pp. 462–74, Chicago: University of Chicago Press.

Nishida, T. (1988). Development of social grooming between mother and offspring in wild chimpanzees. *Folia Primatologica*, 50, 109–23.

Nishida, T. (1990). *The Chimpanzees of the Mahale Mountains*, Tokyo: University of Tokyo Press.

Nishida, T. (1994). Review of recent findings on Mahale chimpanzees: implication and future research directions. In *Chimpanzee Cultures*, ed. R. W. Wrangham, W. C. McGrew, F. B. M. de Waal & P .G. Heltne, pp. 373–96, Cambridge, MA: Harvard University Press.

Nishida, T. & Hiraiwa, M. (1982). Natural history of tool-using behavior by wild chimpanzees in feeding upon wood-boring ants. *Journal of Human Evolution*, 11, 73–99.

Nishida, T. & Nakamura, M. (1993). Chimpanzee tool use to clear a blocked nasal passage. *Folia Primatologica*, 61, 218–20.

Nishida, T., Kano, T., Goodall, J., McGrew, W. C. & Nakamura, M. (1999). Ethogram and ethnography of Mahale chimpanzees. *Anthropological Science*, 107, 141–88.

Nomura, M. (1994). *Reading Body-Languages*, Tokyo: Heibonsha Library, in Japanese.

Pusey, A. E. (1990). Behavioral changes at adolescence in chimpanzees. *Behaviour*, 115, 203–46.

Sugiyama, Y. (1981). Observations on the population dynamics and behavior of wild chimpanzees at Bossou, Guinea, in 1979–1980. *Primates*, 22, 435–44.

Sugiyama, Y. (1997). Social tradition and the use of tool-composites by wild chimpanzees. *Evolutionary Anthropology*, 6, 23–7.

Sugiyama, Y. & Koman, J. (1979). Tool-using and -making behavior in wild chimpanzees at Bossou, Guinea. *Primates*, 20, 513–24.

Thompson, J. A. M. (1994). Cultural diversity in the behavior of *Pan*. In *Hominid Culture in Primate Perspective*, ed. D. Quiatt & J. Itani, pp. 95–115, Colorado: University Press of Colorado.

Tomasello, M. (1994). The question of chimpanzee culture. In *Chimpanzee Cultures*, ed. R. W. Wrangham, W. C. McGrew, F. B. M. de Waal & P .G. Heltne, pp 301–17, Cambridge, MA: Harvard University Press.

Tomasello, M. & Call, J. (1997). *Primate Cognition*, Oxford: Oxford University Press.

van Lawick-Goodall, J. (1973). Cultural elements in a chimpanzee community. In *Precultural Primate Behavior*, ed. E. W. Menzel, pp. 144–84, Basel: Karger.

Whiten, A., Goodall, J., McGrew, W. C., Nishida, T., Reynolds, V., Sugiyama, Y., Tutin, C. E. G., Wrangham, R. W. & Boesch, C. (1999). Cultures in chimpanzees. *Nature*, 399, 682–5.

Wrangham, R. W., de Waal, F. B. M. & McGrew, W. C. (1994). The challenge of behavioral diversity. In *Chimpanzee Cultures*, ed. R. W. Wrangham, W. C. McGrew, F. B. M. de Waal & P .G. Heltne, pp. 1–18, Cambridge, MA: Harvard University Press.

Part II
Social relations

Introduction

VERNON REYNOLDS

This section of *Behavioural Diversity in Chimpanzees and Bonobos* addresses two topics: (1) the influence of ecological and social factors on party size, party composition and inter-community relations, and (2) male relationships. These two topics are related to one another in a number of ways. Ecological factors (mainly food abundance and dispersion) underlie the fission–fusion social organisation found in chimpanzees and bonobos. In forming and reforming parties, chimpanzees and bonobos are influenced by the presence of estrous females, as well as by food. Male relationships within and between communities are mainly shaped by the intensity of intragroup rivalries and intercommunity territoriality.

A number of hypotheses have been advanced relating ecological variables to social ones. Fission–fusion social organisation itself has been seen as a response to food abundance and dispersion, with animals seeking to adjust competition for food to the changing pattern of food available. At the same time, individuals are engaged in competition for dominance and for mating partners and, as shown in the following chapters, these factors have a determining effect on the size and composition of parties within the fission–fusion system.

Anderson *et al.* (Chapter 6) set out to test factors influencing fission–fusion grouping in chimpanzees with data from their research in the Taï Forest, Ivory Coast. They note a number of factors that influence party size: food patch size, habitat-wide food abundance and dispersion, presence of estrous females, location within the range (core versus periphery), and activity (hunting, foraging, travelling, nesting).

The results indicate that party size is indeed increased by the presence of estrous females, and to a greater extent than it is influenced by food abundance. The authors find an interaction between estrus and food abundance, leading them to conclude that 'chimpanzees appear to sacrifice efficient foraging in order to obtain the social, and/or reproductive benefits of associating with estrous females'. While party size is therefore mainly driven by presence/absence of estrous females, they note that both the food supply and the number of females in estrus vary seasonally, and they end with the question of whether ecological constraints underlie variation in estrous cycles.

Mitani *et al.* (Chapter 7) address a similar set of issues based on their work on the Ngogo community in Kibale Forest, Uganda. Starting from a theoretical basis similar to that of Anderson *et al.*, they set out to test the effects of food supply and the presence of estrous females on party size, as well as looking at the gregariousness of different age–sex groups. They find both food abundance and presence of estrous females to be positively correlated with party size, both factors together explaining 87% of party size variance. Using partial correlation they also show that food abundance and presence of estrous females have independent effects.

Males are more gregarious than females in Ngogo chimpanzees, as has been found at other chimpanzee study sites, but, interestingly, the gregariousness of anestrous and lactating females is enhanced by the presence of estrous females in the group. Mitani *et al.* suggest various fitness gains for such females: assessment of male competitive ability, protection from males, and parasitising the food-finding abilities of males. This appears to be a fruitful area for future research. Another finding is that adult and adolescent males are, contrary to expectations, less gregarious when estrous females are present. The authors suggest this may be due to the large number of males in the Ngogo community (24 adult males out of 140 individuals in the whole community), which results in both a high level of male–male competition for females and the occurrence of coalitionary mate-guarding at Ngogo.

The third chapter with the theme of ecology and social organisation is that of Hohmann & Fruth (Chapter 10), this time on bonobos. They studied the Eyengo community of bonobos at Lomako, in Congo. They start by setting out the commonly accepted contrast between bonobos and chimpanzees: bonobos are more egalitarian than chimpanzees, females are more dominant, parties are larger and more female biased, and intercommunity relations are peaceful. They relate this to one current theory that holds that feeding competition between bonobos may be more relaxed than between chimpanzees.

Hohmann & Fruth found that bisexual parties were most common (69% of all parties), followed by all-female parties (21%), all-male parties (8%) and solitaries (4%). Female–female association was the most frequent type. The number of males in parties increased with the number of estrous females. Party size was not correlated with the amount of fruit available. Two adult males joined the Eyengo community in September 1997 and one of them actually transferred; they suggest that his may not be an isolated case since both males and females disappear from time to time and possibly the males have transferred. All of this indicates that male bonobos are less aggressive than male chimpanzees, and the authors suggest that this may be a result of an increased level of female choice in bonobos. Overall, Hohmann & Fruth find both similarities and differences between bonobos and chimpanzees, and their chapter emphasises some of the similarities. They conclude that as more data from more populations become available, 'the existing gap between the two *Pan* species may close'.

The second topic discussed in this section is male–male relationships. Muller (Chapter 8) studied agonistic relationships in the chimpanzees of Kanyawara community, Kibale Forest, Uganda. He found large sex differences in the amount of aggression, males being the more aggressive sex, and concludes that the higher costs of aggression for males are counterbalanced by the increased mating success accruing to dominant, aggressive males. His comparison of the contexts of aggression in males and females showed that males were mostly aggressive in the context of dominance rivalry, with 49% of male aggression directed at other males. However, males also directed their aggression at parous females, especially when they were in estrus, suggesting a degree of intimidation in male–female sexual interactions, although aggression against females occurred mainly in the context of feeding competition. The alpha male was the most frequent aggressor. Frequency of attacks correlated with the number of males in the party and the number of females in estrus rather than with party size. All these facts build up a picture of males striving for dominance, making and breaking alliances, engaging in political manoeuvring, and doing everything in their power to achieve a degree of dominance that will lead to mating opportunities and reproductive success.

Muller reports on two new cases of intercommunity aggression in the Kanyawara community. Fierce intercommunity hostility has now come to be recognised as a standard feature of intercommunity relationships in chimpanzees. He notes that such intercommunity violence is not found at Taï, and finds a possible answer in the fact that due to greater food availability at Taï, few individuals forage alone, and the most common situation in which intercommunity killings occur is when a number of males from one community discover a single individual from another.

Information about the males of the Sonso community in the Budongo Forest, Uganda, is provided by Newton-Fisher (Chapter 9). He explores the theme (also mentioned by Muller) of the fickleness of males in their political alliances with one another. His study showed that there was little long-term loyalty in the relations of adult males. Rather, their relations seem to be based on current need and current perception of the balance of power. This leads to the formation of alliances between adult males in order to strengthen their own power base, followed by the break-up of these alliances and the formation of new ones when circumstances make that more expedient. Males express solidarity by grooming each other, and Newton-Fisher found that males spent more time grooming than females. Most grooming went 'up the hierarchy' indicating that males were often currying favour from the more dominant males in the community. He tested whether or not this tendency to groom up the hierarchy affected association patterns and thus party composition, and also whether an accurate picture of male–male relationships could be drawn on the basis of association, proximity and grooming.

The results of this investigation brought to light four status groups: low (individuals of low status), middle, high, and alpha–beta alliance partners. Grouped in this way, it was clear that males preferred to groom with males of higher status than themselves. Grooming partners were not the same as association partners. Besides 'upwards' grooming, Newton-Fisher also found that high status males groomed more frequently than others, this being interpreted as a 'social tactic' on their part.

Overall, the chapters in this section give us a good insight into chimpanzee and bonobo society. They show the overriding roles of food abundance, availability, and distribution in determining the fission–fusion society and to some extent the size of parties. Within these parties, social factors predominate. The presence of estrous females increases party size over and above the effect of food availability. It also increases male rivalry and male aggression. Moving to the intercommunity level, in chimpanzees, but not in bonobos, we find a very striking and rather unexpected level of violence in some, but not all, communities, and it may be that where competition for resources is great, there is a tendency for intercommunity relations to become violent, with parties of adult males seeking opportunities to inflict damage on their neighbours, and (as the case of Gombe showed) to eventually take over their territories. The authors in this section point to a number of areas where more research is needed, in particular a point-by-point comparison of chimpanzees and bonobos that should help shed light on the ecological forces shaping the social organisation and behaviour patterns we see today in these two closely related species.

6 • Factors influencing fission–fusion grouping in chimpanzees in the Taï National Park, Côte d'Ivoire

DEAN P. ANDERSON, ERIK V. NORDHEIM,
CHRISTOPHE BOESCH & TIMOTHY C. MOERMOND

INTRODUCTION

The relationship between primate social group size and ecological factors has been a focal point of numerous theoretical and empirical research efforts (van Schaik & van Hooff 1983; Terborgh & Janson 1986; Chapman *et al.* 1995; Janson & Goldsmith 1995; Chapman & Chapman 2000). An increase in group size may confer benefits of increased protection from predation or infanticide, and/or increased effectiveness in intergroup competition (Wrangham 1980; van Schaik & van Hooff 1983; Sterck *et al.* 1997; Boesch & Boesch-Achermann 2000). Furthermore, individuals may prefer to associate with others to build coalitions (Boesch & Boesch-Achermann 2000), or to facilitate the socialization of offspring (Williams *et al.*, Chapter 14). However, the benefits of sociality come at the cost of an increase in intragroup foraging competition and an increase in traveling between food patches (Chapman *et al.* 1995; Janson & Goldsmith 1995). Consequently, the ecological constraints model predicts that the size of primate social groups should have an upper limit that is determined by the abundance and distribution of food (Chapman & Chapman 2000).

Chimpanzees and other primates that live in fission–fusion social organizations have been the focus of efforts to understand better how group size is constrained by ecological factors. In these societies, individuals form parties that change in size and composition many times a day (Goodall 1986; Boesch 1996). Substantial empirical evidence suggests that party sizes are correlated with patch size (Symington 1988; Strier 1989; de Moraes & de Carvalho 1998; Newton-Fisher *et al.* 2000), and that seasonal variations in party sizes are related to seasonal fluctuations in the habitat-wide abundance and distribution of food (Chapman *et al.* 1995). However, recent evidence from several sites suggests that estrous females have an equal if not greater effect on chimpanzee fission–fusion dynamics (Matsumoto-Oda *et al.* 1998; Boesch & Boesch-Achermann 2000; Wrangham 2000). In this chapter, we seek to expand our understanding of the factors that influence group size by examining how party sizes in chimpanzees are influenced by the presence of estrous females as well as by food availability. In our analysis, we account for additional factors that may affect party sizes, such as location within the home range, behavioral activities, and time of day.

Chimpanzee females are most sexually receptive when they exhibit a maximal anogenital swelling, which has a duration of approximately 10–12 days (Goodall 1986; Wallis 1997). Large party sizes could be formed by males seeking estrous females, or by estrous females seeking males for copulation. Males and estrous females tend to be more gregarious than anestrous females (Pepper *et al.* 1999; Williams *et al.*, Chapter 14), and parties with more estrous females contain more males (Matsumoto-Oda *et al.* 1998; Wallis & Reynolds 1999). Additionally, non-estrous females may seek companionship with estrous females, an action shown to stimulate the resumption of postpartum cycles (Wallis 1992; Wallis *et al.* 1995), and in adolescent females, initiate the first full anogenital swelling (Wallis 1994). Interestingly, seasonal peaks in the number of estrous swellings in chimpanzees have been suggested as coinciding with seasonal peaks in food availability (Riss & Busse 1977; Goodall 1986; Wallis *et al.* 1995; Boesch 1996; Matsumoto-Oda *et al.* 1998). Consequently, the positive relationship between food availability and party size demonstrated in previous studies may have been, at least in part, influenced by estrous females.

Chimpanzees may exhibit different party sizes depending on their location within the home range. In order to secure long-term access to food resources and reproductive opportunities, chimpanzees exhibit aggressive territorial behavior in the peripheral areas of their range (Nishida *et al.* 1985; Goodall 1986; Wrangham 1999; Boesch & Boesch-Achermann 2000; Muller, Chapter 8). In addition, when the chimpanzees are in the peripheral area, they have the added pressure of having to avoid aggressive conspecifics from neighboring communities. Females with infants are particularly at risk of losing their offspring to infanticide near the boundary of their range (Watts & Mitani

2000). Consequently, the influences on chimpanzee gregariousness of food availability and the presence of estrous females may be different in the core area than in the periphery.

Party sizes in chimpanzees vary with different behavioral activities (Boesch 1996). In the flexible fission–fusion society, individuals may choose to travel in a direction that is different from other members of the social group. For example, after depleting a food patch, individuals in a party may choose to travel alone to another food patch, or to travel at different speeds (Wrangham 2000; Williams *et al.*, Chapter 14). Therefore, foraging parties are expected to be larger than traveling parties. Similarly, hunting success increases with the number of chimpanzees present in a party (Stanford *et al.* 1994; Mitani & Watts 1999), suggesting that hunting parties should be relatively large. Following a successful hunt, party sizes are expected to increase even more as individuals congregate to gain access to meat (Boesch 1996). Consequently, party sizes may be influenced by activity patterns independent of plant food availability, and/or the number of estrous females.

Lastly, the chimpanzees may exhibit different party sizes at different times of the day. Activity patterns and party sizes are expected to vary substantially over the course of the day, with the exception of early morning and late afternoon hours. It is at these times that the chimpanzees are waking and leaving their night nest, or making their night nest and falling asleep. If chimpanzees tend to congregate in the evening to nest in proximity with each other, as has been observed in bonobos (Fruth & Hohmann 1994), then we would expect party sizes to be relatively large at the beginning and end of the day. If chimpanzees do not congregate in the late afternoon, then we would expect party sizes to vary as they do at all other times of the day.

This study was conducted from February 1997 through July 1998 in the Taï National Park, Côte d'Ivoire. We measured the abundance and distribution of food over time at the scale of the habitat as well as within cells of a 500-m × 500-m grid cell system. As we tracked the chimpanzees between cells of varying levels of food availability, the number of estrous females, activities, and time of day all varied. This formed the basis of our natural experiment in which we aimed to determine the relative predictive capabilities of these factors on party size. Based on relative use of different areas of the home range, we identified core and peripheral regions, and compared party sizes in the two areas. Our findings have implications for current hypotheses on the factors that affect social organization in primates.

METHODS

Study site

The study site is located on the western border of the Taï National Park in the Côte d'Ivoire (5°52′N, 7°20′W). The Taï forest is the largest remaining tract (4500 km^2) of undisturbed lowland rain forest in West Africa (Boesch & Boesch-Achermann 2000). The climate is characterized by two rainy seasons (March–June and September–October) and two dry seasons (July–August and November–February), with approximately 1800 mm rainfall per year (Boesch & Boesch 1983). The mean monthly temperature varies between 24 °C and 28 °C. The chimpanzee community we studied was fully habituated to observers (Boesch & Boesch 1983). During the period of this study there was a monthly mean of 31 individuals in the community, including three adult males, eleven adult females, three subadults, and several juveniles and infants. One adult male and one juvenile died, and three infants were born over the 18 months of the study. To organize our vegetation sampling, and to coordinate our food data with behavioral data, we used a 500-m × 500-m grid cell system that had been previously superimposed over the known range of the chimpanzees (± 17.75 km^2).

Behavioral sampling

Instantaneous focal animal samples (Altmann 1974) were collected every 15 minutes on single animals for approximately 8 hours each day. Behavioral sampling was conducted between the 4th and the 26th day of each month, with a monthly average of 148 hours (SD ± 33 hours) of observation. Behavioral sampling was centered around the middle of each month so as to coincide better with tree phenology observations, which were also conducted in the middle of each month (see below). Focal animals were changed daily (when possible) and all adult chimpanzees were followed as focal subjects. A systematic plan for selecting focal animals was established in which attempts were made to have an even distribution of observations for all potential focal animals.

Behaviors were classified into one of the following categories: foraging, traveling, resting, hunting, or meat-eating. For each observation in which the focal animal was foraging, the type of food eaten (fruit, leaves, pith, insects, mushrooms, or meat) and the species were recorded. Hunting and meat-eating were expected to have a strong influence on grouping dynamics (Boesch & Boesch-Achermann 2000),

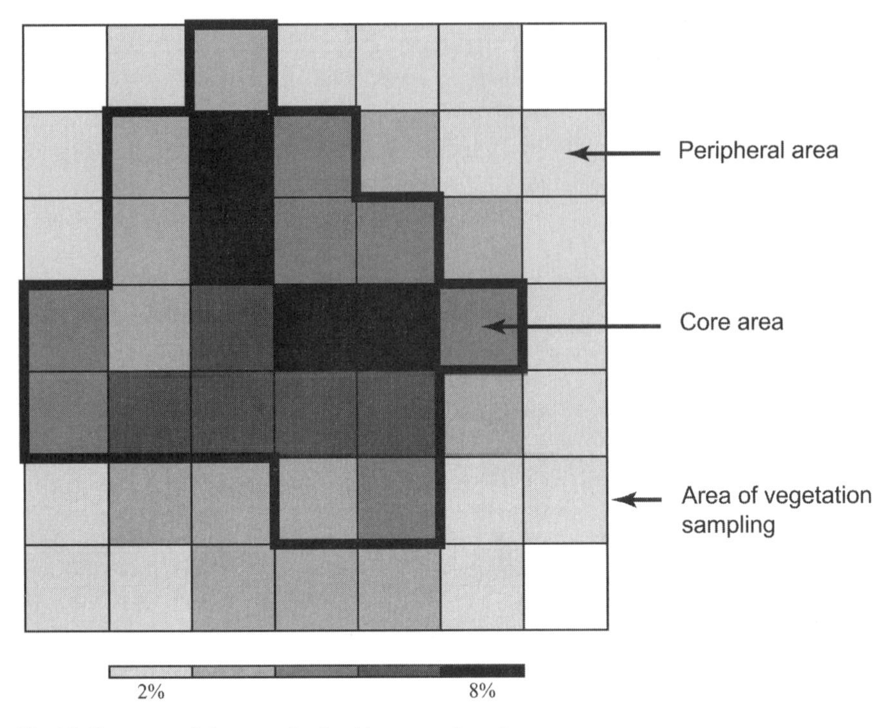

Fig. 6.1. Percentage of time spent by the chimpanzees in each 500-m × 500-m grid cell. Darker shaded cells are more frequently used than the lighter ones. Vegetation sampling was conducted in the entire area represented, except the three corner cells that are white.

but these behaviors were, in fact, relatively rare. Consequently, a behavioral observation was classified as hunting or meat-eating if any animal in the party was either hunting or eating meat, even if the focal animal was not directly involved.

Immediately following each 15-minute instantaneous focal animal sample, the number of individuals present in the party was counted. Party size was defined as the number of individuals present that travel and feed independently, that is all individuals, excluding juveniles and infants. Because we were analyzing how the number of estrous females influences party sizes, we subtracted the number of estrous females in the party from the total number of independent individuals present. Had this adjustment not been performed, there would be some artifactual confounding between the dependent variable (party size) and the independent variable (estrous females). The party size was scored as a zero if the focal animal was in estrus and if she was the only animal present in a party at a given time.

During the course of the study, the number of estrous females found in a party varied from zero to three. Every 15 minutes we noted the location of the party on the 500-m × 500-m grid cell system. In order to determine the effect of time of day on party sizes, behavioral observations were grouped into 2-hour blocks, beginning at 6:00 a.m. and ending at 6:00 p.m.

All behavioral sampling was conducted by two trained field assistants: Grégoire Nohon and Honora Kpazahi. Interobserver reliability (IOR) tests were conducted for data collection on party size and activity categories (Martin & Bateson 1986). The Pearson correlation coefficient for party size ($n = 102$) IOR was 0.97, and the index of concordance ($n = 102$) was 0.90 for activity categories, indicating high reliability.

We defined the size of the chimpanzees' home range as the area covered by all of the 500-m × 500-m grid cells visited by the chimpanzees during the course of the study. Although we followed the chimpanzees to the extreme boundaries of their range, our analysis was restricted to the central 12.25 km^2 on which we collected data on the vegetation (Figure 6.1). We refer to this area as the study area. The core area of the home range was defined as the area of the grid cells in which the chimpanzees spent 75% of their total time. The peripheral area was defined as the area outside the core area, but still within the study area. In this study, 20%

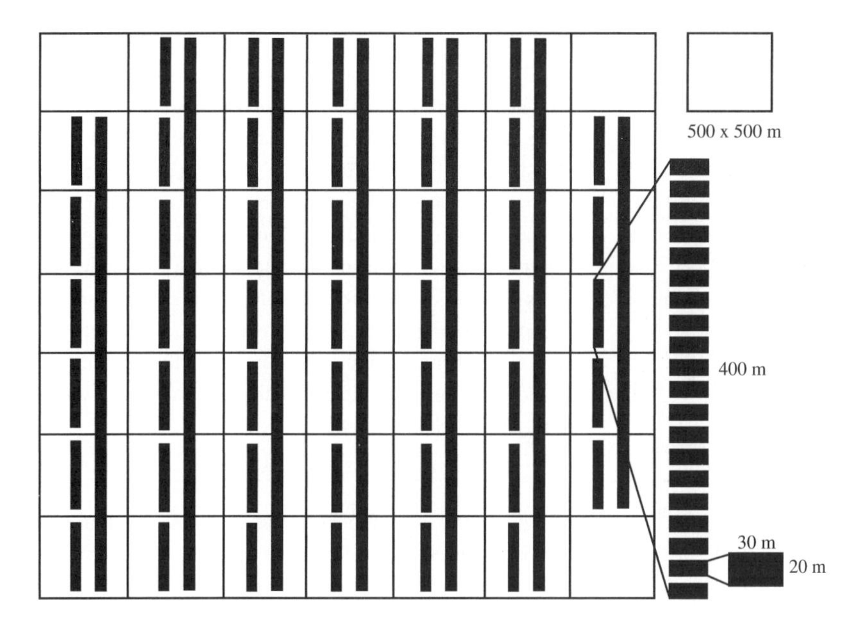

Fig. 6.2. Layout of transects within the 500-m × 500-m grid cell system. Seven long belt transects oriented in the north–south direction within the columns of grid cell system. Additional belt transects of 400 m × 30 m were placed within the grid cells. All transects were divided into contiguous quadrats of 20 m × 30 m.

of the total number of behavioral observations occurred in the peripheral area. The remaining 5% occurred outside the study area.

Food resource sampling

There are many potential variables of food availability that may influence party sizes in chimpanzees. In this study, we measured the abundance and aggregation of food at two scales; within each 500-m × 500-m grid cell, and for the entire study area. Quantification of these 'variables of food availability' necessitated extensive vegetation sampling that included the measurement of fruit and leaf phenology, and density and basal area of food trees. We placed phenology observation routes (sensu Malenky *et al.* 1993) along access trails that covered all areas of the range, excluding the extreme edges. We marked 840 individual trees, representing 45 species. These species were previously known to be food sources for the chimpanzees (Boesch, unpublished data). We returned to each tree monthly and noted the presence or absence of fruit, and new or mature leaves. Our phenology observations were conducted over a 7-day period in the middle of each month.

We placed seven long, parallel belt transects of 3400 m × 30 m in the north–south direction of the columns of the grid cell system (Figure 6.2). In addition, short belt transects of 400 m x 30 m were put in each of the 500-m × 500-m grid cells. All transects were subdivided into contiguous quadrats of 20 m × 30 m. We measured the density and basal area of all individual trees with a diameter at breast height (dbh) of equal to or more than 20 cm in every quadrat along these transects.

QUANTIFICATION OF FOOD ABUNDANCE

To calculate food abundance we had to identify 'important' tree species to include in the analysis. The definition of 'important' is unavoidably arbitrary. Therefore, we conducted a series of preliminary analyses in which we calculated a food abundance index (see below) using several different definitions of important species, and analyzed the relationship between chimpanzee party sizes and each of the different food abundance indices. We examined the following list of 'important' species criteria: (1) all species consumed by chimpanzees in a given month; (2) all species that made up at least 5% of the total diet for a given month; and (3) all species that made up at least 10% of the total diet for a given month. In each of the preceding definitions, a particular species could be important in one month, but not important in a subsequent month. All approaches gave qualitatively similar results, but the 5% rule resulted in slightly lower p-values than the other two rules. We selected the 5% rule for all of our analyses because it was the middle approach and resulted in a slightly better relationship with party size. The range in the number of species in each month that met the 5% rule was two to eight species. The monthly percentage of total diet covered by these species ranged from 51% to 86%, and the mean was 66%.

Using the data of the phenology, density, and basal area (as measured by dbh) of important tree species, we calculated a food abundance index for the study area. The food abundance index for the entire study area, A_m, for month m is defined as follows:

$$A_m = \sum_{k=1}^{n} D_k B_k P_{km} \qquad (6.1)$$

where D_k (stems per hectare) is the density of species k across the study area; B_k is the mean basal area (meters squared) of species k across the study area; and P_{km} is the percentage of observed trees of species k across the study area that present important food items in month m. We obtained 18 different food abundance values for the study site, one for each month of the study.

Food abundance values for each grid cell were obtained using the data about the density and mean basal area from the short belt transect, and the central 400-m portion of the long belt transect found in each grid cell (Figure 6.2). The food abundance index A_{mc} for month m in cell c is defined as follows:

$$A_{mc} = \sum_{k=1}^{n} D_{kc} B_{kc} P_{km} \qquad (6.2)$$

where D_{kc} is the density of species k in cell c; B_{kc} is the mean basal area of species k in cell c; and P_{km} is the percentage of observed trees of species k across the study area that provide 'important food items' in month m.

QUANTIFICATION OF THE SPATIAL DISTRIBUTION OF FOOD TREES

We employed Moran's I spatial autocorrelation statistic (Sokal & Oden, 1978) to quantify the intensity of aggregation of food, for each month of the study. We quantified this variable of food distribution at the same two scales described above; across the entire study area and within each 500-m × 500-m grid cell. In order to calculate the Moran's I, we first obtained a food abundance index (A_{mq}) for each 20-m × 30-m quadrat q in month m:

$$A_{mq} = \sum_{k=1}^{n} D_{kq} B_{kg} P_{km} \qquad (6.3)$$

where D_{kq} is the density of species k in quadrat q; B_{kq} is the mean basal area of species k in quadrat q; and P_{km} is the percentage of observed trees of species k across the study area that provide important food items in month m. Moran's I was calculated on the food abundance indices of the quadrats.

The Moran's I statistic evaluates the correlation of food index values between pairs of quadrats separated by a specified distance. The value of Moran's I ranges from −1 to +1, where the expected value in the absence of significant spatial autocorrelation is around zero (Cliff & Ord 1981). If there is high positive correlation of food indices between quadrats at a given distance class, there is high autocorrelation and the Moran's I will approach +1. If the food trees in a grid cell have regular distribution, the Moran's I will approach –1. For adjacent quadrats, (20 meter distance), Moran's I values were calculated as follows:

$$I = [n\Sigma\Sigma w_{ij}(y_i - \bar{y})(y_j - \bar{y})]/[W\Sigma(y_i - \bar{y})^2] \qquad (6.4)$$

where n equals the number of quadrats across the transects; y's are the index values of adjacent quadrats; ($\bar{y}$) is the mean index value over the study area; w_{ij}s take the value of 1 when the pair (i, j) pertains to adjacent quadrats, and otherwise 0; W is the number of pairs used in computing the coefficient (Sokal & Oden 1978).

Because our transects were only in the north–south direction, our analysis of aggregation is unidirectional. Consequently, we were unable to test for isotropy (spatial pattern is equal in all directions). However, as we limited the calculation of Moran's I to adjacent quadrats, and given the mild topographic relief in Taï, we did not expect anisotropy to be a problem.

Data analysis

We conducted a series of exploratory regression analyses to determine which factors influence party size of chimpanzees. A square root transformation of party size was used in all of our analyses in order to stabilize the variance. We used a mixed regression model in the Proc Mixed procedure of SAS 8 (Littell *et al.* 1996). The independent variables – abundance and aggregation (Moran's I) of food within each grid cell, abundance and aggregation of food across the study area, estrous females, time of day, and activity – were entered into the model as fixed effects. Time of day and activity were used as indicator variables. The experimental unit of the measures of food abundance and aggregation within the grid cells were the grid cells by month combinations since each grid cell had a unique value for each of the 18 months of the study. The experimental unit of estrous females and activity was the 15-minute sampling interval, and for time of day the 2-hour time block. The values for estrous females and activity could potentially have changed every 15 minutes, while time of day changed in 2-hour blocks, as previously described. Each of these experimental units contributed a random error term to the model.

The unbalanced nature of the data required careful modeling of the covariance structure so as to ensure valid conclusions of the tests. The grid cell by month error (mon–cell) was quantified by a unique number assigned to each combination of month and grid cell. There were 18 months in the study and 46 grid cells, resulting in 828 values for mon–cell. Each mon–cell value was associated with unique measures of food abundance and aggregation. We allowed the variance of the random mon–cell term to differ for each value of the number of estrous females. Thus, mon–cell was nested within the variable estrous females. This nesting was done because the number of estrous females had a substantial impact on the relationship between food abundance within a grid cell and party size. The variables of food abundance and aggregation within grid cells were tested against this mon–cell error term. The variables of food abundance and aggregation across the study area were tested against month (as a random effect). In order to account for the data imbalance, standard errors and approximate degrees of freedom were obtained using Satterthwaite's approximation method (Littell *et al.* 1996).

Because our data were collected over time and space, we needed to control for both temporal and spatial autocorrelation of errors. We used a repeated measures analysis with a 'spatial power' structure in order to model the covariance between behavioral observations taken at 15-minute intervals (Littell *et al.* 1996). Observations on different days were considered independent of each other. Similarly, we included a 'spatial power' adjustment component into our mixed model to correct for spatially correlated errors in adjacent grid cells in the respective months (Littell *et al.* 1996).

It was impossible to run the regression model with the entire data set at one time due to the complex temporal and spatial covariance structure and computer limitations. Consequently, we adopted the following approach. First, using the entire data set, we ran the model with the temporal autocorrelation adjustment component and not the spatial component. After finding the best model, we checked the residuals for spatial autocorrelation and found autocorrelated errors in four of the 18 months. We then incorporated the spatial adjustment component into the model and ran the regression on 3–4 months of data at a time. After running all of the months of data using both the temporal and spatial correction components, we found that the factors that had a significant impact on party size were the same as those in the model that incorporated only the temporal autocorrelation adjustment component. Because the results remained consistent across trials, we concluded that the results were robust. Results of regression analyses presented here are from the regression model that included the full 18 months of data, but only the temporal autocorrelation correction component. However, we emphasize that the same general biological conclusions were obtained from all of the analyses we considered.

Table 6.1. *Results of regression analysis when data was restricted to the core area. Party size was regressed against the number of estrous females, behavioral activities, time of day, and variables of food availability at the scale of the grid cell and the study area. Results are from type III tests of fixed effects using the likelihood ratio method of Proc Mixed procedure in SAS*

Source	df-error	F-value	p-value
Estrous Females	564	210.53	<0.0001
Activity	6423	23.36	<0.0001
Time of day	4366	3.46	<0.0012
Food abundance–grid cell	187	2.54	0.1114
Food abundance–grid cell[a] estrous	343	3.40	0.0344
Food abundance–study area	14.5	1.32	ns
Aggregation of food–study area	13.4	0.85	ns
Aggregation of food–grid cell	178	1.02	ns

Notes:

[a] indicates an interaction effect between food abundance and the number of estrous females; ns, not significant.

RESULTS

Over the course of the 18 months of this study the chimpanzees spent time in a total of 71 different grid cells; however, 75% of this time was spent in 21 centrally located cells (Figure 6.2). Thus, they ranged over an area of approximately 17.75 km^2, and had a core area of 5.25 km^2. The food abundance index per grid cell within the core area was not significantly different from that in the peripheral area (t-test, df error = 44, t-value = 0.74, $p = 0.47$).

Results of the regression of party size against the potential influencing factors indicated that party size was affected by the number of estrous females, the particular activity of the focal animal, and the time of day (Table 6.1). Of all the

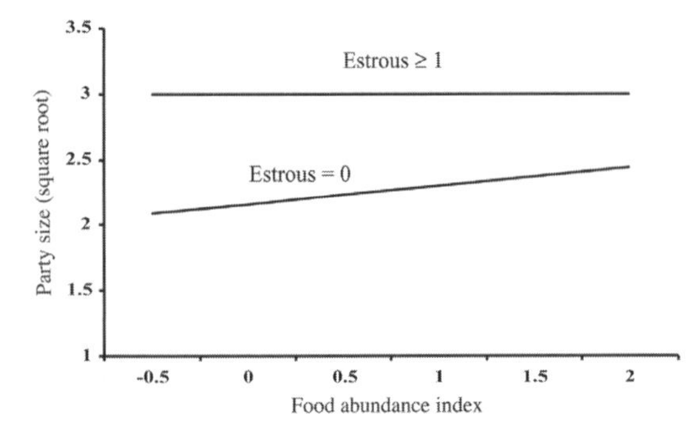

Fig. 6.3. Regression lines of party size on grid-cell food abundance. Results demonstrate the relative effects of food abundance in grid cells on party size when there were no estrous females present, and when there were one or more estrous females present. Parties with no estrous females were small relative to parties with one or more estrous females. However, parties with no estrous females showed a significant relationship with food abundance ($f=4.54$, $p=0.0341$). The food abundance index was standardized so that the mean value equals zero.

variables of food availability, only the food abundance within grid cells was found to be important in this relationship. Food abundance in grid cells as a main effect (i.e. by itself) was nonsignificant ($p=0.1114$), but the interaction between food abundance and the number of estrous females was significant ($p=0.0344$). A closer examination of this interaction revealed that when there were no estrous females in the party, the abundance of food in grid cells had a positive effect on party size (Figure 6.3). However, if there were one or more estrous females present, gregariousness was no longer influenced by the availability of food. All other potential interactions between independent variables were explored and found to be nonsignificant.

We examined in more detail the relationship between the number of estrous females present in a party and the party size. The mean party size showed a positive relationship with the number of estrous females present in a focal party (Figure 6.4). The results of pair-wise tests of the least square means indicate that the mean party sizes were significantly different between focal parties with 0 estrous females present and all other parties. Similarly, the mean with one (1) estrous female was significantly different from all others. However, the mean party size for two and three estrous females did not differ. (Note that this analysis considers the number of estrous females only, and ignores the other independent variables.)

The relationship between activity of the focal animal and party size is illustrated in Figure 6.5 (as above, this analysis does not consider the other independent variables). Pair-wise tests of the least square means of the party size for the respective behavioral categories indicate that the largest party sizes were observed when meat-eating was occurring within the party. The smallest party sizes occurred when the focal animal was traveling. Party sizes during meat-eating and traveling were significantly different from all other behaviors. Group hunting was significantly different from resting, but not different from foraging.

Figure 6.6 demonstrates the significant relationship between the 2-hour time blocks and party size. Results of the pair-wise tests of time blocks indicate that party sizes were largest in the early morning and in the late afternoon. Parties were significantly larger before 10:00 a.m. and after 4:00 p.m. than they were in the middle of the day.

Although party size was not influenced by an interaction effect between activity and time of day, the percentage of time spent foraging, resting, and hunting varied consistently with time of day. An analysis of variance indicates that the percentage of time spent foraging in each two-hour time block was not equal (df-error $=102$, $f=4.11$, $p=0.002$). The chimpanzees tended to spend more time foraging from 6:00 a.m. to 8:00 a.m. and from 4:00 p.m. to 6:00 p.m. than during the other times of the day (Figure 6.7(*a*)). The chimpanzees tended to rest more in the middle part of the day (df-error $=102$, $f=7.73$, $p=0.0001$); nearly the opposite pattern of the time spent foraging (Figure 6.7(*b*)). The chimpanzees showed two peaks in the diurnal hunting frequency; between 10:00 a.m. and 12:00 p.m., and between 2:00 p.m. and 4:00 p.m. (df-error $=102$, $f=2.13$, $p=0.067$) (Figure 6.7(*c*)). Finally, the percentage of time spent traveling and eating meat did not vary over the course of the day.

In a separate regression analysis, we examined how the independent variables influence party size when the data set was restricted to data collected on behavior and vegetation in the peripheral area. All variables were nonsignificant predictors of party size, except for the number of estrous females (df $=1859$, F-value $=77.2$, $p<0.0001$). The number of estrous females was a strong predictor of party size regardless of location.

The monthly food abundance index for the entire study area has been shown elsewhere to influence party sizes in primates with fission–fusion societies (Chapman *et al.* 1995). However, our results indicate that this index did not make a significant contribution to the model when estrous females, grid-cell food index, activity, and time of day were all in the model.

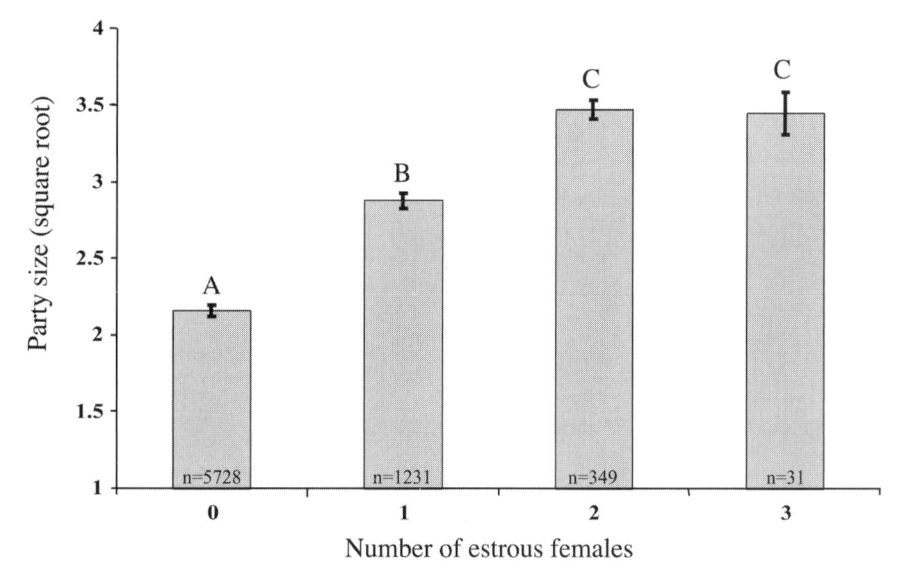

Fig. 6.4. Least square means and standard errors of the square root of party size plotted against the number of estrous females. The sample size of each of the respective estrous females categories is indicated in each of the columns. Estrous female categories that have the same letter (A,B,C) above the error bars were not significantly different from each other at the level $\alpha = 0.05$. All significant differences have a p-value <0.001.

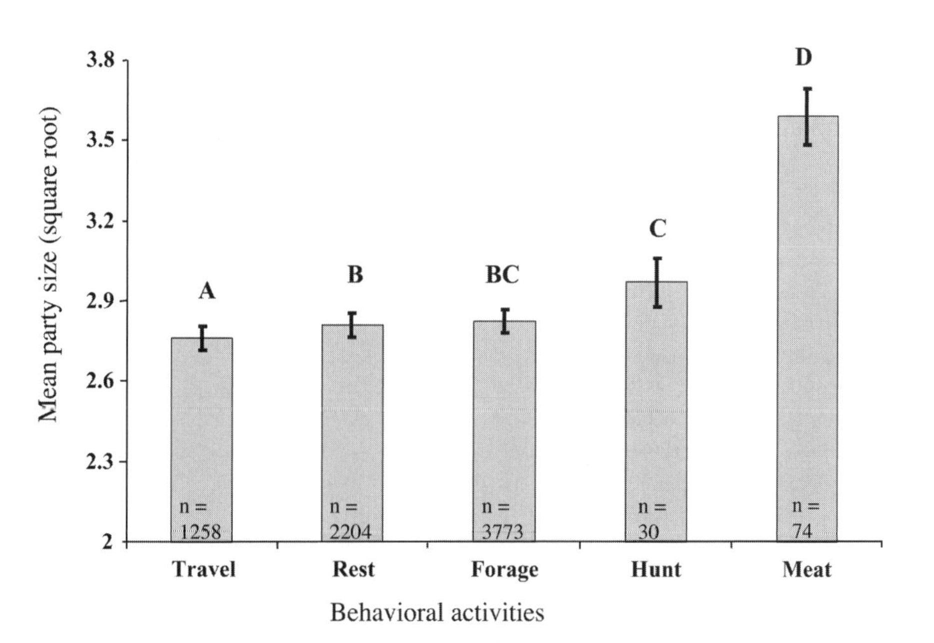

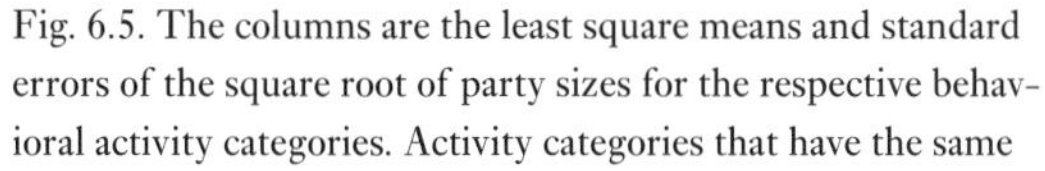

Fig. 6.5. The columns are the least square means and standard errors of the square root of party sizes for the respective behavioral activity categories. Activity categories that have the same letter (A,B,C) above the error bars were not significantly different from each other. All significant differences have a p-value <0.05.

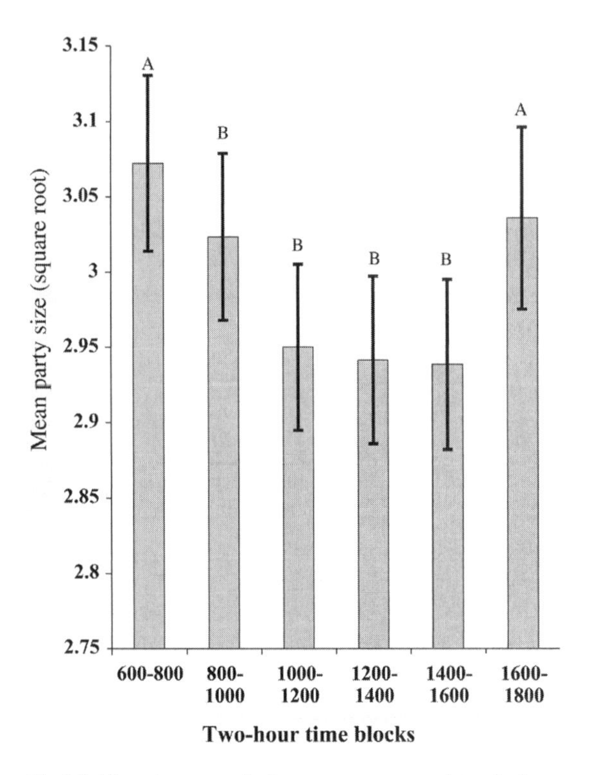

Fig. 6.6. The columns are the least square means and standard errors of the square root of party sizes for each 2-hour time block. Time blocks that have the same letter (A or B) above the error bars were not significantly different from each other at the level $\alpha = 0.05$.

We conducted a regression analysis to determine how party sizes within the core area were affected by the food abundance for the study area if all other factors were removed from the model. A positive trend was found, but the relationship was not significant (df = 15.6, *F*-value = 3.86, *p*-value = 0.0675).

In one of our analyses, we showed that the grid-cell food abundance index was a significant predictor of party size only as an interaction term with estrous females. When party size was regressed against the grid-cell food abundance index, with all other variables removed, the grid-cell index showed a significant relationship (df = 232, *F*-value = 8.57, *p*-value = 0.0038).

DISCUSSION

Results presented here demonstrate that food availability had a significant effect on party size, but its influence was limited to when the chimpanzees were in the core area and no estrous females were present. Estrous females had a

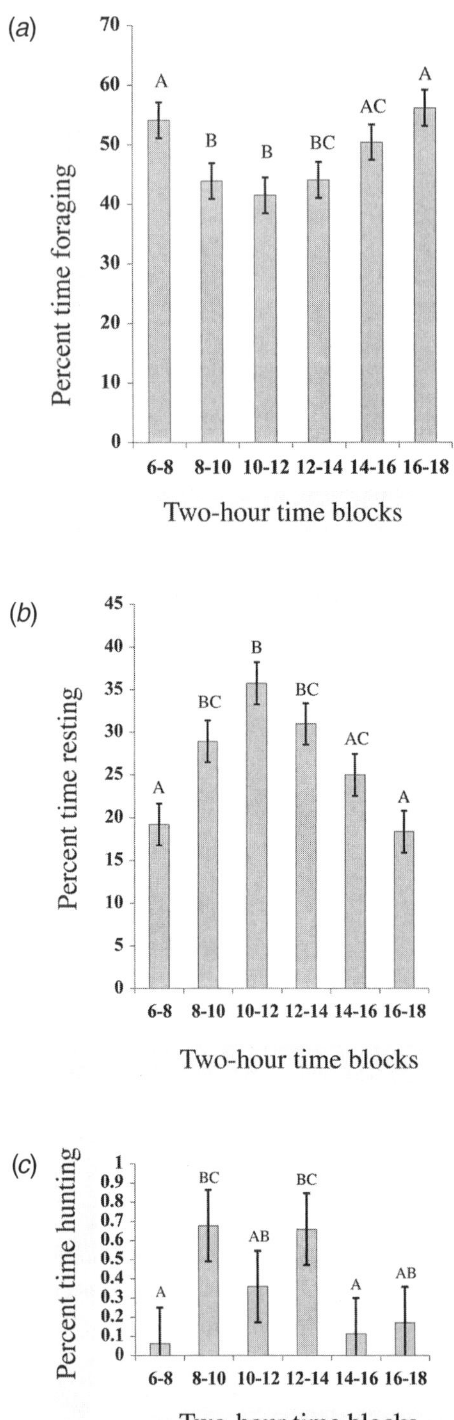

Fig. 6.7. Percentage of time spent foraging (*a*), resting (*b*), and hunting (*c*) at each of the 2-hour time blocks. Time blocks that have the same letter (A,B,C) above the error bars were not significantly different from each other at the level $\alpha = 0.05$.

strong impact on party sizes regardless of activity, time of day, or location within the home range. When estrous females were present, parties tended to remain large regardless of food availability. Associating with estrous females is expected to increase opportunities for copulation for males, and may stimulate anestrous females to come into estrous (Wallis 1992, 1994, 1995). Consequently, estrous females create a social condition within the chimpanzee community that does not follow the prediction of the ecological constraints model that group sizes are constrained by food availability (Chapman & Chapman 2000).

While the grid-cell food index had a positive relationship with party size when no estrous females were present, the aggregation of food (Moran's I) did not. When food trees are highly aggregated, foraging efficiency should be high since travel time between patches is expected to be low (Chapman *et al.* 1995; Wrangham 2000). Consequently, we had anticipated large party sizes when food trees were highly aggregated due to reduced foraging competition, but this pattern was not observed.

Our results indicate that grouping dynamics changed depending on the location within the home range. The chimpanzees spent 75% of their time in an area (core area) that made up less than one-third of their total home range. Food productivity was not meaningfully greater in the core area than in the periphery, so it was clearly not food that was attracting them to this area. The observation that the chimpanzees spent most of their time in this area is most likely the result of how this community and the neighboring communities have distributed their territories across the landscape. The reduced effect of food on foraging parties in the periphery may be explained by the added challenges of defending their territory and/or avoiding aggressive conspecifics from neighboring communities (Nishida *et al.* 1985; Goodall 1986; Boesch & Boesch-Achermann 2000; Watts & Mitani 2000; Muller, Chapter 8). In contrast, party size was affected by the presence of estrous females in both the core area and in the periphery.

In addition, we suggest that it is important to control for the effect of behavioral activities when examining party sizes in chimpanzees. Seasonal variation in activity budgets (Doran 1997) may influence the temporal changes in party sizes. Consistent with the findings of Boesch (1996), results presented here indicate that parties were largest when the chimpanzees were eating meat, and smallest when they were traveling. Consequently, relatively large shifts in the time spent either meat-eating or traveling are likely to show corresponding shifts in party sizes. Meat-eating, in particular, appeared to have a large seasonal effect on party size. Chimpanzees at Taï consistently hunt more frequently during the months of September and October than during other months (Boesch & Boesch-Achermann 2000). It has been hypothesized that the hunting season is related to the onset of the rainy season, reduced food availability, or the birthing season of the main prey species, red colobus monkey (Boesch & Boesch-Achermann 2000). Alternatively, the seasonal hunting pattern may be influenced by a seasonal peak in the number of estrous females, which results in larger party sizes. This is consistent with the finding that hunting success is correlated with party size (Stanford *et al.* 1994; Mitani & Watts 1999). More research is necessary to elucidate the relationship between hunting, food availability, prey vulnerability, estrous cycles, and party sizes.

Results presented here demonstrate that time of day had an important impact on party size, and that its effect was independent of activity. Parties were largest late in the afternoon and early in the morning (Figure 6.6), which coincided with the making of nests in the evening, and the departure from the nests in the morning. The chimpanzees also spent a greater percentage of time foraging in the early morning and late afternoon than they did at other times of the day (Figure 6.7(*a*)). However, foraging parties were not larger than resting parties (Figure 6.5), and resting occurred more frequently in the middle part of the day (Figure 6.7(*b*)). Hunting parties tended to be larger than both resting and foraging parties (Figure 6.5), and hunting frequencies were high from 12:00 p.m. to 4:00 p.m. Hunting events are relatively rare, and are unlikely to influence the effect that time of day has on party size. Lastly, the time spent meat-eating and traveling did not differ over the course of a day, and these two activities usually had the largest and smallest party sizes, respectively.

These results suggest that the chimpanzees tend to congregate in the late afternoon and forage intensively until they make their night nests. At dawn, they once again forage in relatively large groups, and by 10:00 a.m. individuals have usually fissioned into smaller parties and travel in independent directions. This finding is consistent with the observation that night parties in bonobos are larger than daytime parties (Fruth & Hohmann 1994).

In conclusion, the seasonal variation in chimpanzee party sizes has previously been attributed to changes in food availability (Wrangham 1977; Chapman *et al.* 1995). Our results indicate that grouping dynamics are mainly driven by estrous females. We showed that food availability had some predictive influence only when no estrous females were present and the chimpanzees were in the core area. However, the underlying importance of food availability may be understated by these conclusions in that the number of estrous females present in a given month appears to be influenced by food variables (Anderson 2001). Given that the number of estrous females varies seasonally (Wallis 1995; Matsumoto-Oda *et al.* 1998; Boesch & Boesch-Achermann 2000; Wallis, Chapter 13), future research should be focused not only on the ecological constraints of group size but also on the ecological constraints of the estrous cycle and how estrous females affect social dynamics of primate groups.

ACKNOWLEDGEMENTS

Funding for this research was provided by a graduate fellowship from the Max Planck Society, the L.S.B. Leakey Foundation, the National Science Foundation (Timothy Moermond, P.I.), and the USAID – Program for Science and Technology (Timothy Moermond, P.I.). We are grateful to the ministries of Eaux et Forêts and La Recherche Scientifique for granting permission to conduct this study. The Projet Autonome pour la Conservation du Parc National de Taï, and the Centre de Recherche Ecologique were valuable collaborators in this research. We thank all of the assistants of the Taï Chimpanzee Project, especially those involved with the behavioral and vegetation data collection: Honora Kpazahi, Grégoire Nohon, Glebeo Pierre Polé, Bally Wabo Albert, Tahou Mompeho Jonas, and Dji Troh Camille. Susanne Anderson assisted with data collection, data entry, and provided invaluable support in the field. Thomas Gordon provided computer assistance in the field. Téré Henri helped with the identification of plant species. Jakob and Marie Zinsstag of the Centre Suisse pour la Recherche Scientifique (CSRS) provided logistic support in the Ivory Coast. Peter Crump provided assistance with syntax of the SAS programming language. Helpful comments on the manuscript were provided by Erica Cochrane, Gottfried Hohmann, Su Liying, Karen B. Strier, Thomas J. Givnish, Monica G. Turner, Susanne Rust, and Mariano Sironi.

REFERENCES

Altmann, J. (1974). Observational study of behavior: Sampling methods. *Behaviour*, 49, 227–67.

Anderson, D. P. (2001). Tree Phenology and Distribution, and their Relation to Chimpanzee Social Ecology in the Taï National Park, Côte d'Ivoire. PhD thesis, University of Wisconsin-Madison.

Boesch, C. (1996). Social grouping in Taï chimpanzees. In *Great Ape Societies*, ed. W. C. McGrew, L. F. Marchant & T. Nishida, pp. 101–13. Cambridge: Cambridge University Press.

Boesch, C. & Boesch, H. (1983). Optimization of nut-cracking with natural hammers by wild chimpanzees. *Behaviour*, 83, 265–86.

Boesch, C. & Boesch-Achermann, H. (2000). *The Chimpanzees of the Taï Forest: Behavioural Ecology and Evolution*. Oxford: Oxford University Press.

Chapman, C. A. & Chapman, L. J. (2000). Determinants of group size in primates: the importance of travel costs. In *On the Move: How and Why Animals Travel in Groups*, ed. S. Boinski & P. A. Garber, pp. 24–42. Chicago: University of Chicago Press.

Chapman, C. A., Wrangham, R. W. & Chapman, L. J. (1995). Ecological constraints on group size: an analysis of spider monkey and chimpanzee subgroups. *Behavioral Ecology and Sociobiology*, 36, 59–70.

Cliff, A. D. & Ord, J. K. (1981). *Spatial Processes: Models and Applications*. London: Pion Limited.

de Moraes, P. L. R. & de Carvalho, O. Jr (1998). Population variation in patch and party size in Muriquis (*Brachyteles arachnoides*). *International Journal of Primatology*, 19, 325–37.

Doran, D. (1997). Influence of seasonality on activity patterns, feeding behavior, ranging, and grouping patterns in Taï chimpanzees. *International Journal of Primatology*, 18, 183–206.

Fruth, B. & Hohmann, G. (1994). Comparative analyses of nest building behavior in bonobos (*Pan paniscus*) and chimpanzees (*Pan troglodytes*). In *Chimpanzee Cultures*, ed. R.W. Wrangham, W.C. McGrew, F. de Waal & P. Heltne, pp. 109–28. Cambridge, MA: Harvard University Press.

Goodall, J. (1986). *The Chimpanzees of Gombe: Patterns of Behavior*. Cambridge, MA: Belknap Press of Harvard University Press.

Janson, C. H. & Goldsmith, M. L. (1995). Predicting group size in primates: foraging costs and predation risks. *Behavioural Ecology*, 6, 326–36.

Littell, R. C., Milliken, G. A., Stroup, W. W. & Wolfinger, R. D. (1996). *SAS System for Mixed Models*. Cary: SAS Institute.

Malenky, R. K., Wrangham, R. W., Chapman, C. A. & Vineberg, E. O. (1993). Measuring chimpanzee food abundance. *Tropics*, 2, 231–44.

Martin, P. & Bateson, P. (1986). *Measuring Behavior: an Introductory Guide*. Cambridge: Cambridge University Press.

Matsumoto-Oda, A., Hosaka, K., Huffman, M. A. & Kawanaka, K. (1998). Factors affecting party size in chimpanzees of the Mahale Mountains. *International Journal of Primatology*, 19, 999–1011.

Mitani, J. C. & Watts, D. P. (1999). Demographic influences on the hunting behavior of chimpanzees. *American Journal of Physical Anthropology*, 109, 439–54.

Newton-Fisher, N. E., Reynolds, V. & Plumtre, A. J. (2000). Food supply and chimpanzee (*Pan troglodytes schweinfurthii*) party size in the Budongo Forest Reserve, Uganda. *International Journal of Primatology*, 21, 613–28.

Nishida, T., Hiraiwa-Hasegawa, M., Hasegawa, T. & Takahata, Y. (1985). Group extinction and female transfer in wild chimpanzees in the Mahale National Park, Tanzania. *Zeitschrift für Tierpsychologie*, 67, 284–301.

Pepper, J. W., Mitani, J. C. & Watts, D. P. (1999). General gregariousness and specific social preferences among wild chimpanzees. *International Journal of Primatology*, 20, 613–32.

Riss, D. & Busse, C. (1977). Fifty day observation of a free-ranging adult male chimpanzee. *Folia Primatologica*, 28, 283–97.

Sokal, R. R. & Oden, N. L. (1978). Spatial autocorrelation in biology 1. Methodology. *Biological Journal of the Linnean Society*, 10, 199–228.

Stanford, C., Wallis, J., Mpongo, E. & Goodall, J. (1994). Hunting decisions in wild chimpanzees. *Behavior*, 131, 1–18.

Sterck, E. H. M., Watts, D. P. & van Schaik, C. P. (1997). The evolution of female social relationships in nonhuman primates. *Behavioral Ecology and Sociobiology*, 41, 291–309.

Strier, K. B. (1989). Effects of patch size on feeding associations in muriquis (*Brachyteles arachnoides*). *Folia Primatologica*, 52, 70–7.

Symington, M. (1988). Food competition and foraging party size in the black spider monkey (*Ateles paniscus chamek*). *Behavior*, 105, 117–34.

Terborgh, J. W. & Janson, C. H. (1986). The socioecology of primate groups. *Annual Review of Ecology and Systematics*, 17, 111–35.

van Schaik, C. P. & van Hooff, J. A. R. A. M. (1983). On the ultimate causes of primate social systems. *Behaviour*, 85, 91–117.

Wallis, J. (1992). Socioenvironmental effects on timing of first postpartum cycles in chimpanzees. In *Topics in Primatology, Human Origins*, ed. T. Nishida, W. C. McGrew, P. Marler, M. Pickford & F. deWaal, pp. 119–30. Tokyo: University of Tokyo Press.

Wallis, J. (1994). Socioenvironmental effects on first full anogenital swellings in adolescent female chimpanzees. In *Current Primatology: Social Development, Learning, and Behavior*, ed. J. J. Roeder, B. Thierry, J. R. Anderson & N. Herrenschmidt, pp. 25–32. Strasbourg: Université Louis Pasteur.

Wallis, J. (1995). Seasonal influence on reproduction in chimpanzees of Gombe National Park. *International Journal of Primatology*, 16, 435–51.

Wallis, J. (1997). A survey of reproductive parameters in the free-ranging chimpanzees of Gombe National Park. *Journal of Reproduction and Fertility*, 109, 297–307.

Wallis, J. & Reynolds, V. (1999). Seasonal aspects of sociosexual behavior in two chimpanzee populations: a comparison of Gombe (Tanzania) and Budondo (Uganda). *American Journal of Primatology*, 49, 111 (abstract).

Wallis, J., Mbago, F., Mpongo, E. & Chepstow-Lusty, A. (1995). Sex differences and seasonal effects in the diet of chimpanzees. *American Journal of Primatology*, 36, 162 (abstract).

Watts, D. P. & Mitani, J. C. (2000). Infanticide and cannibalism by male chimpanzees at Ngogo, Kibale National Park, Uganda. *Primates*, 41, 357–65.

Wrangham, R. W. (1977). Feeding behavior of chimpanzees in Gombe National Park, Tanzania. In *Primate Ecology*, ed. T. H. Clutton-Brock, pp. 503–38. London: Academic Press.

Wrangham, R. W. (1980). An ecological model of female-bonded primate groups. *Behaviour*, 75, 262–300.

Wrangham, R. W. (1999). The evolution of coalitionary killing: the imbalance-of-power hypothesis. *Yearbook of Physical Anthropology*, 42, 1–30.

Wrangham, R. W. (2000). Why are male chimpanzees more gregarious than mothers? A scramble competition hypothesis. In *Male Primates*, ed. P. M. Kappeler, pp. 248–58, Cambridge: Cambridge University Press.

7 • Ecological and social correlates of chimpanzee party size and composition

JOHN C. MITANI, DAVID P. WATTS &
JEREMIAH S. LWANGA

INTRODUCTION

Why primates live in social groups and what factors account for variation in group size and composition have been two central questions in the study of primate behavioral ecology (Alexander 1974; Altmann 1974; Wrangham 1980; van Schaik 1983; Rodman 1988; Isbell 1994; Janson & Goldsmith 1995). Theory suggests that the relative costs and benefits of grouping will influence variations in group size and composition. Several factors, such as feeding competition, predation risk, and competition for mates, affect these costs and benefits for group members, but not necessarily equally (ibid.). For example, food generally limits female reproduction in most mammals (Trivers 1972; Bradbury & Vehrencamp 1977; Emlen & Oring 1977; Wrangham 1980; Clutton-Brock 1989), and feeding competition consequently affects females to a greater extent than males. Alternatively, females are the limiting source for reproduction by males (ibid.), and the availability of mates accordingly influences male behavior more than that of females.

The fission–fusion social system of chimpanzees provides a model system for investigating sources of variation in group size. Wild chimpanzees live in large, fluid unit-groups or communities, whose members form temporary parties that vary in size and composition (Nishida 1968; Sugiyama 1968; Halperin 1979; Boesch 1996). In keeping with predictions stemming from current theory, the availability of both food and estrous females have been implicated as important determinants of chimpanzee party size (Riss & Busse 1977; Wrangham & Smuts 1980; Ghiglieri 1984; Goodall 1986; Isabirye-Basuta 1988; Sakura 1994; Stanford *et al.* 1994; Chapman *et al.* 1995; Boesch 1996; Doran 1997; Matsumoto-Oda *et al.* 1998; Boesch & Boesch 2000; Wrangham 2000). Chimpanzees are highly frugivorous, and the size and spatial distribution of trees bearing ripe fruit varies considerably over time. Chimpanzees may reduce feeding competition by adjusting party size to the availability of food. Slow reproduction makes fertile females a scarce resource for males, and males are likely to aggregate around them regardless of associated ecological costs.

Despite an impressive body of data bearing on the problem of variation in chimpanzee grouping patterns, previous research has tended to dichotomize the issue by focusing exclusively on the effects of either food or females (e.g. Isabirye-Basuta 1988; Stanford *et al.* 1994; Chapman *et al.* 1995; Doran 1997). Few studies have examined the joint effects of these two variables on party size (but see Wrangham 2000). In addition, most prior studies provide only indirect measures of food availability, which renders their results problematic and difficult to interpret. For example, several researchers have employed rainfall as a proxy variable for food availability and have shown decrements in party size during dry seasons, the presumed period of food scarcity (Wrangham 1977; Boesch 1996; Doran 1997; Matsumoto-Oda *et al.* 1998). Direct measures of food availability have only rarely been collected, however (e.g. Wrangham *et al.* 1992; Chapman *et al.* 1995; Wrangham 2000) and, as a result, the relationship between food availability and rainfall remains largely inferential.

In this chapter, we present new observations of chimpanzees at the Ngogo study site in the Kibale National Park, Uganda. We combine phenological observations of fruit trees with data on tree species density and size to provide direct estimates of the amount of food available to chimpanzees. We utilize these estimates and the observed numbers of estrous females in parties to assess the effects that food and estrous females have on party size. We also examine the extent to which each of these variables affects the general levels of gregariousness shown by members of different age–sex classes. While several of our results conform to the predictions derived from conventional theory and observations from previous empirical research, other findings are not easily interpreted. We attempt to resolve these paradoxical observations within the context of the unique demographic structure of the Ngogo chimpanzee community.

METHODS

Study site and animals

We conducted observations of chimpanzees at the Ngogo study site in the Kibale National Park, Uganda. Data presented here were collected during two periods spanning 16 months from January to June 1998 and from October 1998 to August 1999. Ngogo has been the site of previous research on chimpanzees, first by Michael Ghiglieri (1984) between 1976 and 1981 and later by us (Watts 1998, 2000a,b; Mitani & Watts 1999, 2001; Mitani *et al.* 1999, 2000; Pepper *et al.* 1999; Watts & Mitani 2000, 2001, 2002). We have maintained observations of the Ngogo chimpanzees continuously since June 1995. Ngogo lies at an interface between lowland and montane rain forest and is covered primarily with moist, evergreen forest interspersed between large blocks of *Pennisetum purpureum* grassland. Ghiglieri (1984), Butynski (1990), and Struhsaker (1997) provide more detailed descriptions of the Ngogo study area.

The Ngogo chimpanzees have been of special interest, given their unusually large community size. Over 140 individuals reside within the Ngogo chimpanzee community. At the time of observations reported here, the community included 24 adult males, 15 adolescent males, and approximately 47 adult females, 9 adolescent females, 15 juveniles, and 34 infants.

Food availability and rainfall

We monitored the phenological stages of fruit trees composing the top 20 species in the diet of the Ngogo chimpanzees throughout the 16-month study period (Table 7.1). Preliminary analyses indicate that these 20 species form over 75% of the diet of the Ngogo chimpanzees (Watts, Mitani & Lwanga, unpublished data). We noted the presence or absence of ripe fruit for 20 trees of each species each month. Phenological observations were made during the first 5 days in each calendar month. We selected tree species to monitor on the basis of previous observations of the feeding behavior of chimpanzees both at Ngogo and at the nearby Kanyawara study site (Ghiglieri 1984; Wrangham *et al.* 1992; Wrangham, personal communication; Watts, Mitani & Lwanga, unpublished data). Individual trees selected for sampling were randomly scattered across the Ngogo chimpanzee community range. We used phenological data along with the sizes of each tree species and their densities

Table 7.1. *Fruit trees utilized by the Ngogo chimpanzees.*

Species	Density (trees ha^{-1})	Size (mean ± SD cm DBH)
Aningeria altissima	0.01	185 ± 58
Celtis durandii	39.70	40 ± 17
Chrysophyllum albidum	39.54	56 ± 25
Cordia millenii	0.46	110 ± 49
Ficus brachylepis	1.37	184 ± 128 (n = 26)
Ficus cyathistupula	0.15	65 ± 42 (n = 22)
Ficus dawei	0.15	207 ± 110
Ficus exasperata	0.01	135 ± 37
Ficus mucuso	0.15	245 ± 77
Ficus natalensis	0.76	150 ± 106 (n = 26)
Mimusops bagshawei	4.11	78 ± 36
Monodora myristica	4.56	66 ± 19
Morus lactea	0.30	84 ± 35
Pseudospondias microcarpa	2.59	121 ± 72
Pterygota mildbraedii	13.84	139 ± 72
Teclea nobilis	2.28	24 ± 9
Treculia africana	0.46	83 ± 25 (n = 23)
Uvariopsis congensis	64.64	19 ± 4
Warburgia ugandensis	0.76	99 ± 55 (n = 28)
Zanha golungensis	0.61	82 ± 27

Notes:
Densities and average tree sizes are shown for the top 20 tree species fed upon at Ngogo. Densities of each tree species were computed based on counts from 263 5-m × 50-m plots. Thirty trees of each species were measured to calculate mean DBH tree sizes, except where noted in parentheses.

recorded in 263 5-m × 50-m plots to compute an index of food availability each month:

$$\sum_{i=1}^{20} p_i \cdot d_i \cdot s_i \qquad (7.1)$$

where p_i is the percentage of the ith tree species possessing ripe fruit, d_i is the density of the ith tree species (trees per hectare), and s_i is the mean size of the ith species in centimeters DBH (diameter at breast height).

We used DBH as an estimate of tree size. Previous researchers (Leighton & Leighton 1982; Strier 1989; Chapman *et al.* 1992) have shown that tree DBH is a good proxy variable for the overall size of fruit crops in trees used by chimpanzees and other primates. For 15 tree species, we measured DBH of the 20 trees included in our phenological samples plus an additional ten randomly selected individuals to calculate average tree sizes. We could not locate a sufficient number for five of the tree species, and in these cases, DBH samples ranged from 22 to 28 individuals (Table 7.1). The 263 vegetation plots were a subsample of 392 plots originally established by Tom Struhsaker in 1975, and they covered the range occupied by the Ngogo chimpanzees.

We collected daily records of rainfall during the 16-month study period. Observations were made each morning at our base camp, with total amounts of rainfall measured to the nearest 0.1 mm.

Party sizes and compositions

We recorded party sizes and compositions during daily observations of chimpanzees. We defined parties as all individuals present upon first contact (cf. Tutin *et al.* 1983; Pepper *et al.* 1999). When in the field together, Mitani and Watts frequently split up to cover a larger portion of the Ngogo community range; sightings of multiple parties on a single day typically occurred when these two observers were watching animals in different locations and were separated by hundreds of meters. To ensure statistical independence of these data, we generally limited observations to one party per day per observer. Records of multiple parties on a single day did not include observations of individuals sighted previously in the day. We used von Neumann's test for serial independence to examine whether successive observations of party sizes were autocorrelated (Sokal & Rohlf 1995). Observations made during each of the 16-month sample periods showed no signs of interdependence ($p > 0.05$ for all 16 tests).

For each party sighted, we tallied the number of individuals in four age–sex classes as defined by previous researchers (Goodall 1986; Nishida *et al.* 1990): adult males, adult females, adolescent males, and adolescent females. We subdivided adult females into three classes: estrous, anestrous and lactating, and anestrous and nonlactating (cf. Matsumoto-Oda 1999). To maintain consistency with previous research (e.g. Wrangham 2000), we excluded infants and dependent juveniles from our counts of party size.

General gregariousness

To analyze chimpanzee sociability, we used an index of class association that reflects the social environment individuals experience. Specifically, we employed the mean number of associates experienced by members of each age–sex class to investigate their levels of overall gregariousness (Pepper *et al.* 1999). Numerically this 'general gregariousness' index is:

$$\frac{\sum_{i=1}^{n} a_i \cdot (s_i - 1)}{\sum_{i=1}^{n} a_i} \qquad (7.2)$$

where a_i is the number of occurrences of members of class a in group i, s_i is the size of group i.

We implemented a group randomization technique (Smolker *et al.* 1992; Pepper *et al.* 1999) to assess statistically the degree to which overall levels of gregariousness deviated from chance expectations. Here we created a matrix of party compositions by listing the age–sex class of each individual present. In this observation matrix each row corresponds to an observed group, each column represents an age–sex class, and the value of each cell represents the number of individuals of a given class observed in a given group. In the group randomization procedure all observations are removed from this matrix, then each is randomly reassigned to one of the cells, subject to the constraint that row and column totals are preserved. The resulting matrix contains the same number of groups of each size as the original data, and the same number of observations of each age–sex class, but with the patterns of association among age–sex classes randomized. The distribution of group sizes and the representation of each class are thus controlled when generating a null hypothesis of unbiased association. The randomization process is repeated several times, with the general gregariousness index computed for each class each time. This generates a frequency distribution for the expected gregariousness index for each class under the assumption of unbiased interactions. We used the mean of 10000 iterations as the expected value of the gregariousness index for each age–sex class. To quantify bias in association patterns, we employed a bias quotient ($O/E - 1$), where O and E represent the observed and expected gregariousness index for a given class, respectively. A positive bias quotient indicates that a given class was more gregarious than expected, that is more gregarious than the population as a whole, while a negative value indicates low gregariousness. As a statistical test of the null hypotheses that an observed gregariousness index could have arisen through unbiased interactions, we used the proportion of randomized indices that were at least as extreme as the observed index. This proportion represented a one-tailed *P*-value, which we doubled for use in a two-tailed statistical test.

Our analyses of general gregariousness involve multiple

comparisons between age–sex classes. To correct for the increased probability of committing a Type I error when making these multiple comparisons, we adjusted our criteria of significance downward using a sequential Bonferroni technique (Holm 1979). For k multiple tests, our adjusted alpha levels were set at:

$$P_i \leq a / (1 + k - i) \quad (7.3)$$

where $a = 0.05$ and is the overall experimentwise error rate and i is the ith sequential test.

Statistical analyses

We examined the effects of rainfall on food availability and chimpanzee party size by regressing the 16 summed monthly rainfall scores against measures of mean monthly food availability and average chimpanzee party size. We conducted a multiple regression analysis to determine the simultaneous effects of the availability of food and estrous females on observed chimpanzee party sizes. Here we regressed the average party size observed each month against our composite measure of monthly food availability and the mean number of estrous females sighted in each party per month. Raw scores satisfied the assumptions of the analysis of variance, and we conducted all analyses using the original data. We carried out partial correlation analyses to investigate the relationships between each of the two independent variables and party size after controlling for the effects of the other. For these and subsequent analyses, we constructed monthly scores of rainfall and party sizes and compositions centered around the times we recorded phenological observations. For example, our observations of party sizes and compositions for March 1999 included sightings of chimpanzees from 19 February to 18 March 1999, instead of those made during the calendar month of March.

We completed three additional analyses to examine how the presence of food and estrous females affected chimpanzee sociability. First, we calculated general gregariousness indices for different age–sex classes during periods of food abundance ($n = 60$ parties). We operationally defined 'food abundant' times as those months that showed food availability scores one standard deviation above the 16-month average. Three months, June, November, and December 1998, met this criterion and were subsequently included in this analysis (Figure 7.3(*b*), see below). To factor out the effect that the presence of estrous females may have had on this analysis, we excluded parties containing these individuals. We compared results of this analysis to a control period comprising 12 'food-normal' months whose food availability scores fell within plus or minus one standard deviation of the 16-month average. Excluded from this analysis were the three 'food-abundant' months noted above as well as a single 'food-poor' month, March 1998, whose food availability score fell greater than one standard deviation below the 16-month mean. To eliminate effects due to the presence of estrous females, we excluded from these analyses all parties in which they were present. We conducted a third analysis to determine the effect of reproductive females on chimpanzee sociability. Here we constructed general gregariousness indices for each chimpanzee age–sex class when estrous females were present and compared these to similar indices derived from parties that lacked estrous females. To exclude the potential effects of food availability, we restricted parties to those observed during the 12 'food-normal' months.

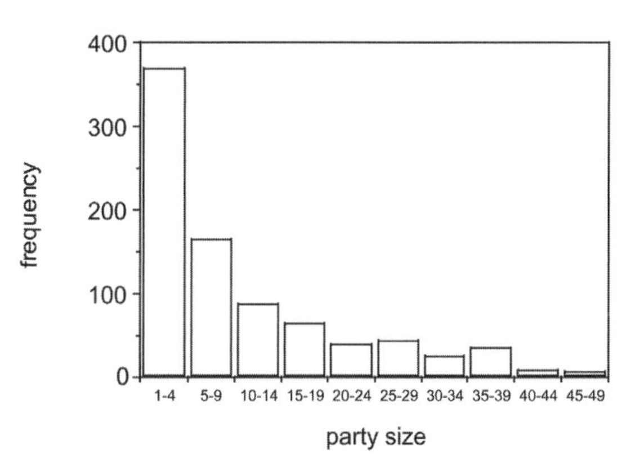

Fig. 7.1. Frequency distribution of chimpanzee party sizes at Ngogo ($n = 827$ parties).

RESULTS

Party size

Chimpanzee parties comprised ten individuals on average during the 16-month study period ($\bar{X} = 10.27$, $n = 827$ parties). Party size varied substantially (range $= 1 - 47$; SD $= 10.24$), however, with considerable skew toward the right (Figure 7.1). Median party size was consequently smaller at six individuals.

Rainfall and food availability

Long-term field observations indicate that the Kibale National Park experiences two dry and two wet seasons each

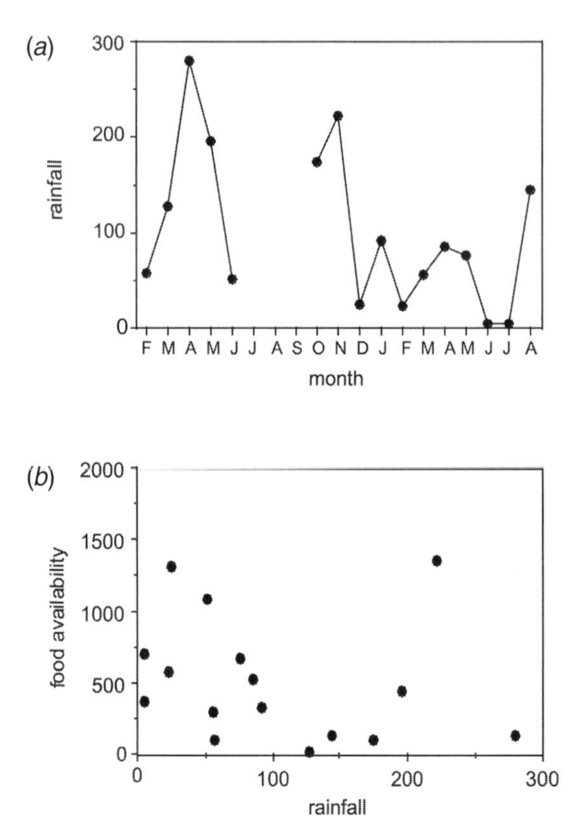

Fig. 7.2. Seasonal variation in rainfall and its effects at Ngogo. (*a*) Monthly variation in rainfall. (*b*) Relationship between monthly food availability and rainfall.

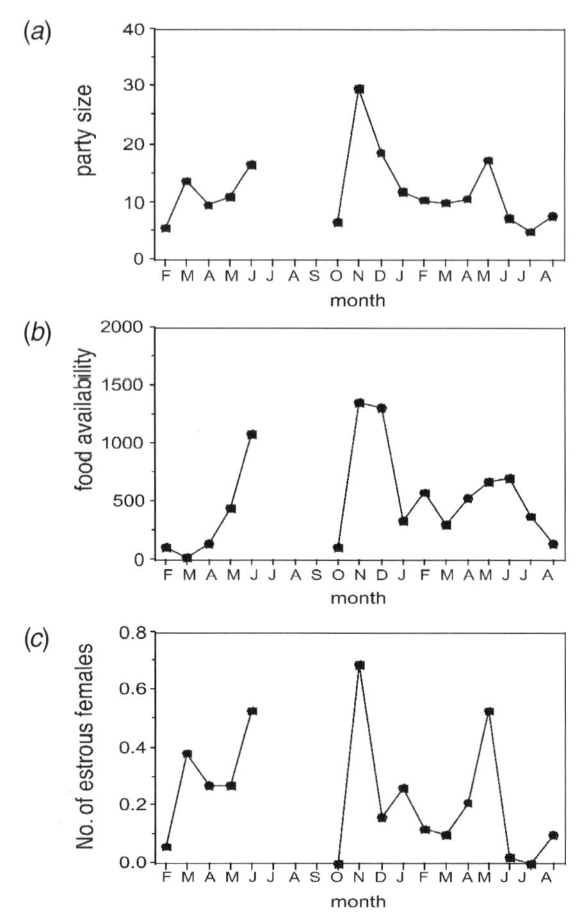

Fig. 7.3. Temporal variations in (*a*) party size, (*b*) food availability, and (c) presence of estrous females per party. Observations were made over a 16 month period between 1998 and 1999.

year. Minor and major dry seasons typically occur near the start and middle of each calendar year, with wetter periods falling outside these times (Struhsaker 1997). There was substantial monthly variation in rainfall during the 16-month study period (Figure 7.2(*a*)). Seasonal peaks in rainfall did not correspond precisely with those reported previously (cf. Struhsaker 1997), perhaps due to the lingering effects of El Niño. Food availability was not correlated with rainfall ($F_{1,14}=0.61, p>0.40$). Dry months did not correspond to times of poor fruit availability (Figure 7.2(*b*)).

Correlates of party size

Chimpanzee party sizes showed pronounced fluctuations over time (Figure 7.3(*a*)). Food availability and the number of estrous females per party also varied markedly over time (Figures 7.3(*b*)(*c*)). Food availability and the number of estrous females were both positively correlated with monthly party sizes (food: $t=4.13$, females: $t=5.79$, $p<0.001$ for both comparisons; Figures 7.4(*a*)(*b*)). Together the two independent variables explained 87% of the variation in monthly party sizes ($f_{2,13}=50.56$, $p<0.001$). In contrast to these strong relationships, there was no association between chimpanzee party size and rainfall ($f_{1,14}=0.85, p>0.35$).

If food availability is correlated with the number of estrous females observed in parties, then the documented relationships between each of the two independent variables and party size may be spurious, with either relationship due to a hidden and uncontrolled third variable. Additional analyses do not support this suggestion. Although there was a significant association between the number of estrous females and food availability ($f_{1,14}=4.90, r^2=0.21, p<0.05$), both independent variables were also positively correlated

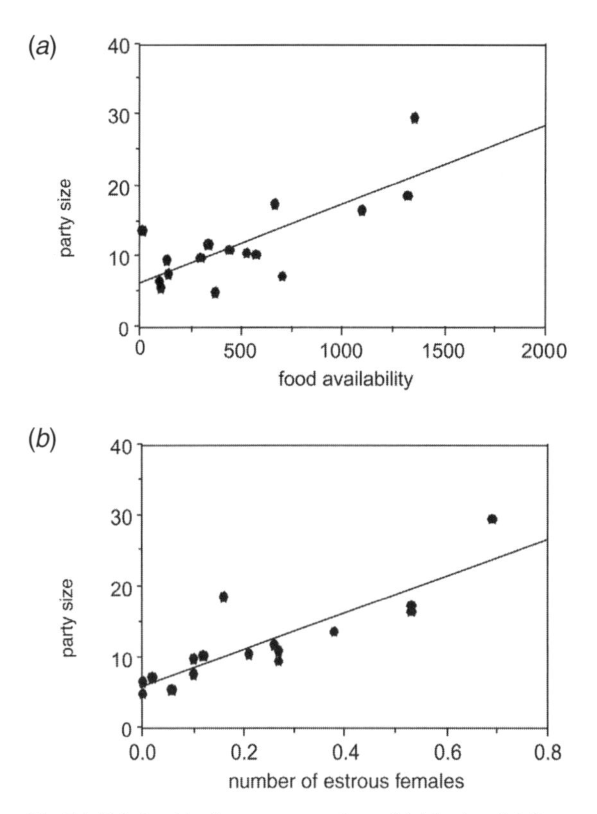

Fig. 7.4. Relationships between party size and (*a*) food availability and (*b*) number of estrous females per party. Data points represent mean monthly values for each variable.

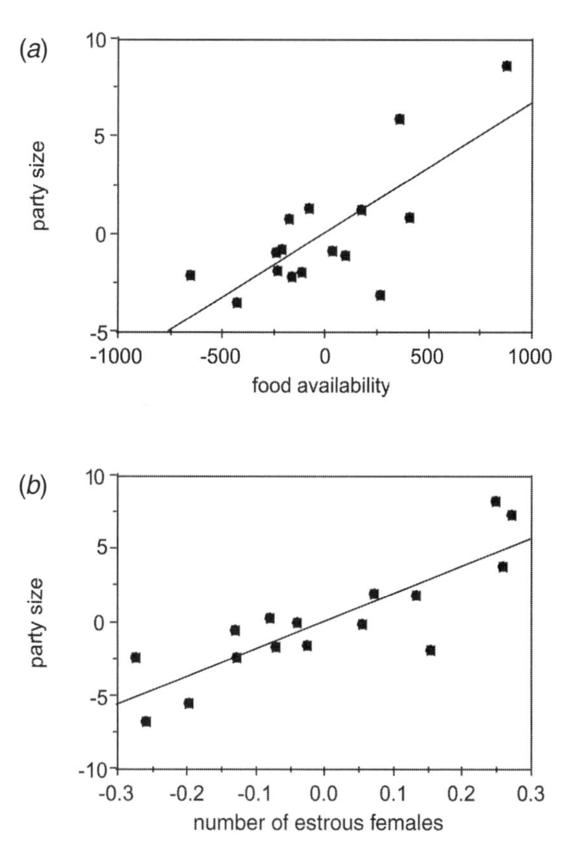

Fig. 7.5. Partial correlations between party size and (*a*) food availability after controlling for the number of estrous females and (*b*) the number of estrous females per party after controlling for food availability.

with party size after controlling for the effects of the other through partial correlation (food: $r = 0.75$, females: $r = 0.85$, $p < 0.001$ for both comparisons; Figures 7.5(*a*)(*b*)).

General gregariousness

In keeping with the findings of previous research both at Ngogo and elsewhere, male chimpanzees were more gregarious than females during control periods (Figure 7.6(*a*)). Only three of five age–sex classes, however, showed significant deviations from chance expectation in their measures of general gregariousness. Adolescent males were significantly more sociable and anestrous and lactating females were less gregarious than expected.

The presence of abundant food altered the sociability of three of five age–sex classes. Anestrous and lactating females became more sociable, while adolescent males became less gregarious, during periods of food abundance (Figure 7.6(*b*)). The presence of estrous females had equally striking effects on the sociability of chimpanzees of differing ages and sexes. The presence of estrous females did not increase adult or adolescent male sociability, nor did it increase that of other estrous females. In contrast, the presence of estrous females served to make anestrous or lactating females more gregarious than they were at other times (Figure 7.6(*c*)).

DISCUSSION

Previous field studies of chimpanzees consistently reveal temporal variations in party size (Wrangham 1977; Isabirye-Basuta 1988; Sakura 1994; Stanford *et al.* 1994; Chapman *et al.* 1995; Boesch 1996; Doran 1997; Matsumoto-Oda *et al.* 1998). The availability of food has been repeatedly implicated as a major determinant of this variability (ibid.). Several prior researchers have assumed that plant productivity is directly

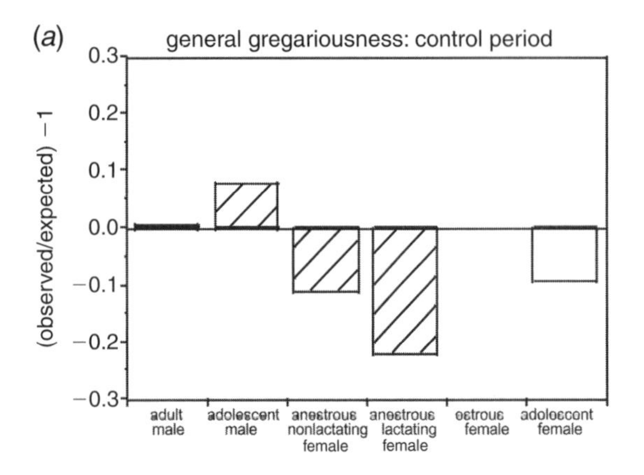

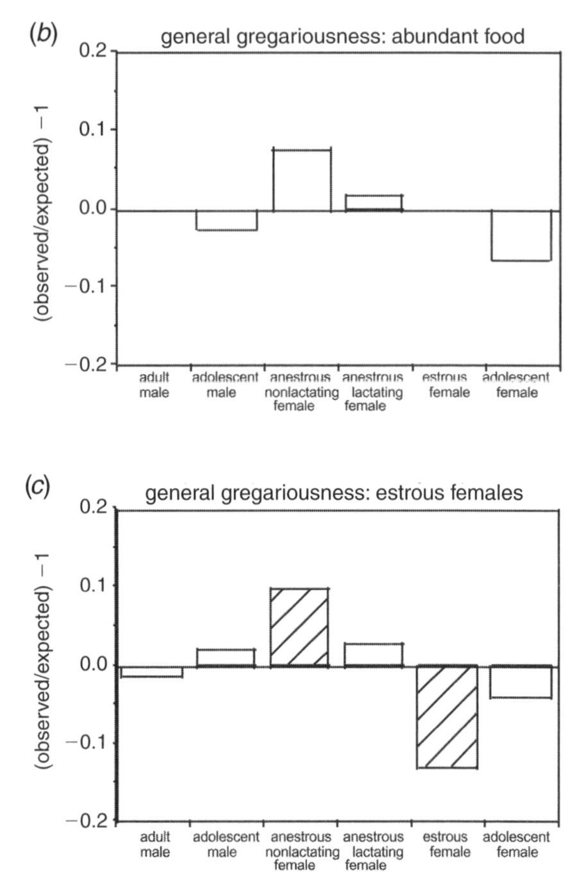

Fig. 7.6. General gregariousness indices for chimpanzees of different age–sex classes. (*a*) Control period. Control periods included parties without estrous females when food was neither abundant nor scarce. (*b*) Periods with abundant food. Food abundant times were operationally defined as those months that exceeded one standard deviation above the 16-month average. (*c*) Parties including estrous females. Ratios of observed to expected numbers of associates minus 1 are shown. Ratios greater than 0 indicate association with more individuals than expected, while values less than 0 indicate fewer associates than expected. Hatched bars indicate those age–sex classes that had significantly more or significantly fewer associates than expected by chance. See text for further explanation.

related to rainfall, and have employed rainfall as a proxy variable for food availability (Wrangham 1977; Boesch 1996; Doran 1997; Matsumoto-Oda *et al.* 1998). Negative relationships between party size and rainfall have been used as prima facie evidence that chimpanzee party size decreases as food abundance falls. Although dry seasons may appear to correspond to times of food scarcity (e.g. Taï: Doran 1997), our results indicate that rainfall is not associated with direct measures of food availability at Ngogo, and thus cannot be used to predict variations in chimpanzee party size at this site.

The presence of estrous females has also been hypothesized to play a significant role in determining the sizes of parties formed by chimpanzees (Riss & Buss 1977; Goodall 1986; Sakura 1994; Stanford *et al.* 1994; Boesch 1996; Matsumoto-Oda *et al.* 1998). Female chimpanzees reproduce only once every 5–6 years (Tutin & McGinnis 1981; Nishida *et al.* 1990; Boesch & Boesch 2000). Consequently, reproductively active females are an extremely limited and scarce resource for male chimpanzee reproduction and are a primary focus of male attention. Observations at Ngogo are consistent with this expectation: the presence of estrous females is a significant predictor of overall party size.

In sum, the analyses presented here are consistent with prior claims regarding the important effects that food and females have on chimpanzee social organization (Wrangham 1977; Isabirye-Basuta 1988; Sakura 1994; Stanford *et al.* 1994; Chapman *et al.* 1995; Boesch 1996; Doran 1997; Matsumoto-Oda *et al.* 1998; Boesch & Boesch 2000). Together, both of these variables combine to explain nearly all of the variance in observed party sizes at Ngogo. Recent studies concerning the determinants of chimpanzee party size have tended to emphasize the importance of one of these variables to the exclusion of the other (e.g. Stanford *et al.* 1994; Chapman *et al.* 1995; Doran 1997). Similar debates have ensued in studies of other primates as well, most notably orangutans (Sugardjito *et al.* 1987; te Boekhorst *et al.* 1990; Mitani *et al.* 1991; Utami *et al.* 1997). Our analyses underscore that the availability of food and estrous females do not necessarily work in opposition, but can operate together to produce the patterns of social organization observed in nature.

While our results regarding the ecological and social determinants of chimpanzee party sizes are clear, the effects that food and estrous females have on the sociability of different age–sex classes are less easily explained. The finding that food has a greater impact on female sociability than that of adult males fits well with conventional theory (Trivers 1972; Bradbury & Vehrencamp 1977; Emlen & Oring 1977; Wrangham 1980; Clutton-Brock 1989). Why greater food availability should lead to the observed decrement in gregariousness by adolescent males, however, is unclear. Equally unexpected was the finding that adult and adolescent males became less sociable in the presence of estrous females compared with other times. The general attraction that estrous females had on anestrous and lactating females was also not predicted on the basis of standard theory.

At first blush, the fact that adolescent males become less social with decreasing ecological costs of grouping seems paradoxical. Partial explanation for this phenomenon might lie in the unusual social pressures brought to bear upon adolescent males. Adolescence is a turbulent time for male chimpanzees (Pusey 1990). Shortly after weaning, males break their primary social ties with their mothers and attempt to integrate themselves into the world of adult males. During this period, most adolescent males are peripheralized socially, while receiving high amounts of aggression from adult males. These social costs are probably exacerbated in large parties that contain several adult males. Adolescent males at Ngogo might choose to lower these costs by simply opting out of the large groups that form when food abundance is high. Testing this hypothesis will require additional observations of how individual adolescent males respond to changing social situations.

Why males, in general, become less gregarious in the presence of estrous females is even more puzzling than the pattern of adolescent male sociality displayed in the face of lowered feeding costs. If females synchronize their estrous cycles and avoid each other, as observed (Figure 7.6(*c*)), then the reduced sociability of males may simply result from the latter mapping onto the movements of the former, who spread themselves out. Alternatively, it may be important to consider the extraordinary demographic structure of the Ngogo chimpanzee community. With 24 adult males and 15 adolescent males, the Ngogo group dwarfs all other chimpanzee communities that have been described thus far in the wild (summary in Wrangham 2000). Males within this community experience high levels of intrasexual competition, leading them to adopt unusual mating tactics such as coalitionary mate-guarding (Watts 1998). Nowhere will these high levels of mating competition be more evident than in the presence of estrous females, and reduced participation by males in groups may be a logical result. The preceding observations suggest that there will be significant interindividual variation in male participation in parties that contain reproductively active females. At Ngogo, male rank correlates positively with mating success (Watts, unpublished data), which leads to the specific prediction that high-ranking males will be over-represented in parties containing estrous females. Investigating this possibility and whether differential participation translates into direct fitness gains will be two important foci for future research.

Finally, why are anestrous and lactating females attracted to parties containing estrous females? Elsewhere, we have shown that after controlling for their overall levels of general sociability, females as a class associate with each other more than expected by chance (Pepper *et al.* 1999). Recently, Matsumoto-Oda and colleagues (1998) have suggested three possible fitness gains derived by females who participate in parties containing estrous females. First, these parties might create forums within which females can assess male competitive ability. Second, females might benefit by gaining protection against males, particularly those living outside their communities. Third, females could gain through parasitizing the food-finding abilities of males in these parties. Few data exist to evaluate these hypotheses at Ngogo or elsewhere, and the fitness benefits as well as costs derived by female chimpanzees remain critical and unexplored areas for further inquiry.

ACKNOWLEDGEMENTS

Our research was sponsored by the Makerere University, the Ugandan National Parks, and the Ugandan Wildlife Authority. We thank J. Kasenene, G. Isabirye-Basuta, and the staff of the Makerere University Biological Field Station for providing logistical assistance. We are grateful to J. Pepper and the editors, C. Boesch, G. Hohmann, and L. Marchant, for comments on the manuscript and to C. Businge, A. Magoba, G. Mbabazi, G. Mutabazi, L. Ndangizi, A. Tumusiime, and T. Windfelder for field assistance. We acknowledge the support of the following organizations for providing funds enabling us to participate in the Behavioral

Diversity of Chimpanzees symposium: The Wenner-Gren Foundation, Deutsche Forschungsgemeinschaft, Bayerisches Staatsministerium für Wissenschaft, Forschung und Kunst, and the Max-Planck-Gesellschaft. Our field research was supported by a National Science Foundation Presidential Faculty Fellowship Award SBR-9253590 and L.S.B. Leakey and Wenner-Gren grants to JCM, and by L.S.B. Leakey Foundation and National Geographic Society grants to DPW.

REFERENCES

Alexander, R. (1974). The evolution of social behavior. *Annual Review of Ecology and Systematics*, 5, 325–83.

Altmann, S. (1974). Baboons, space, time, and energy. *American Zoologist*, 14, 221–48.

Boesch, C. (1996). Social grouping in Taï chimpanzees. In *Great Ape Societies*, ed. W. McGrew, L. Marchant & T. Nishida, pp. 101–13. Cambridge: Cambridge University Press.

Boesch, C. & Boesch, H. (2000). *The Chimpanzees of the Taï Forest*. Oxford: Oxford University Press.

Bradbury, J. & Vehrencamp, S. (1977). Social organization and foraging in emballonurid bats. III. Mating systems. *Behavioral Ecology and Sociobiology*, 2, 1–17.

Butynski, T. (1990). Comparative ecology of blue monkeys (*Cercopithecus mitis*) in high- and low-density subpopulations. *Ecological Monographs*, 60, 1–26.

Chapman, C., Chapman, L., Wrangham, R., Hunt, K., Gebo, D. & Gardner, L. (1992). Estimators of fruit abundance in tropical trees. *Biotropica*, 24, 527–31.

Chapman, C., Wrangham, R. & Chapman, L. (1995). Ecological constraints on group size: an analysis of spider monkey and chimpanzee subgroups. *Behavioral Ecology and Sociobiology*, 36, 59–70.

Clutton-Brock, T. (1989). Mammalian mating systems. *Proceedings of the Zoological Society of London*, 236, 339–372.

Doran, D. (1997). Influence of seasonality on activity patterns, feeding behavior, ranging, and grouping patterns in *Taï* chimpanzees. *International Journal of Primatology*, 18, 183–206.

Emlen, S. & Oring, S. (1977). Ecology, sexual selection and the evolution of mating systems. *Science*, 197, 215–23.

Ghiglieri, M. (1984). *The Chimpanzees of the Kibale Forest*. New York: Columbia University Press.

Goodall, J. (1986). *The Chimpanzees of Gombe*. Cambridge: Belknap Press of Harvard University Press.

Halperin, S. (1979). Temporary association patterns in free ranging chimpanzees: An assessment of individual grouping preferences. In *The Great Apes*, ed. D. Hamburg & E. McCown, pp. 491–9. Menlo Park: Benjamin/Cummings.

Holm, S. (1979). A simple sequentially rejective multiple test procedure. *Scandanavian Journal of Statistics*, 6, 65–70.

Isabirye-Basuta, G. (1988). Food competition among individuals in a free-ranging chimpanzee community in Kibale Forest, Uganda. *Behaviour*, 105, 135–47.

Isbell, L. (1994). Predation on primates: ecological patterns and evolutionary consequences. *Evolutionary Anthropology*, 3, 61–71.

Janson, C. & Goldsmith, M. (1995). Predicting group size in primates: foraging costs and predation risks. *Behavioral Ecology*, 6, 326–36.

Leighton, M. & Leighton, D. (1982). The relationship of size and feeding aggregate to size of food patch: Howler monkeys (*Alouatta palliata*) feeding in *Trichilia cipo* fruit trees on Barro Colorado Island. *Biotropica*, 14, 81–90.

Matsumoto-Oda, A. (1999). Mahale chimpanzees: grouping patterns and cycling females. *American Journal of Primatology*, 47, 197–207.

Matsumoto-Oda, A., Hosaka, K., Huffman, M. & Kawanaka, K. (1998). Factors affecting party size in chimpanzees of the Mahale Mountains. *International Journal of Primatology*, 19, 999-1011.

Mitani J. & Watts, D. (1999). Demographic influences on the hunting behavior of chimpanzees. *American Journal of Physical Anthropology*, 109, 439-54.

Mitani, J. & Watts, D. (2001). Why do chimpanzees hunt and share meat? *Animal Behaviour*, 61, 915–24.

Mitani, J., Grether, G., Rodman, P. & Priatna, D. (1991). Associations among wild orang-utans: sociality, passive aggregations, or chance? *Animal Behaviour*, 42, 33–46.

Mitani J., Hunley, K. & Murdoch M.E. (1999). Geographic variation in the calls of wild chimpanzees: a reassessment. *American Journal of Primatology*, 47, 133–51.

Mitani J., Merriwether, D.A. & Zhang, C. (2000). Male affiliation, cooperation, and kinship in wild chimpanzees. *Animal Behaviour*, 59, 885-93.

Nishida, T. (1968). The social group of wild chimpanzees in the Mahali Mountains. *Primates*, 9, 167–224.

Nishida, T., Takasaki, H. & Takahata, Y. (1990). Demography and reproductive profiles. In *The Chimpanzees of the Mahale Mountains*, ed. T. Nishida, pp. 63–97. Tokyo: University of Tokyo Press.

Pepper, J., Mitani, J. & Watts, D. (1999). General gregariousness and specific social preferences among wild chimpanzees. *International Journal of Primatology*, 20, 613–32.

Pusey, A. (1990). Behavioral changes at adolescence in chimpanzees. *Behaviour*, 115, 203–46.

Riss, D. & Busse, C. (1977). Fifty-day observation of a free-ranging adult male chimpanzee. *Folia Primatologica*, 28, 283–97.

Rodman, P. (1988). Resources and group sizes of primates. In *The Ecology of Social Behavior*, ed. C. Slobodchikoff, pp. 83–108. New York: Academic Press.

Sakura, O. (1994). Factors affecting party size and composition of chimpanzees (*Pan troglodytes verus*) at Bossou, Guinea. *International Journal of Primatology*, 15, 167–83.

Smolker, R., Richards, A., Connor, R. & Pepper, J. (1992). Sex differences in patterns of association among Indian Ocean bottlenose dolphins. *Behaviour*, 123, 38–69.

Sokal, R. & Rohlf, F. J. (1995). *Biometry*. 3rd edition. New York: W.H. Freeman.

Stanford, C., Wallis, J., Mpongo, E. & Goodall, J. (1994). Hunting decisions in wild chimpanzees. *Behaviour*, 131, 1–18.

Strier, K. (1989). Effects of patch size on feeding associations in muriquis (*Brachyteles arachnoides*). *Folia Primatologica*, 52, 70–7.

Struhsaker, T. (1997). *Ecology of an African Rain Forest*. Gainesville: University Press of Florida.

Sugardjito, J., te Boekhorst, I. & van Hooff, J. (1987). Ecological constraints on the grouping of wild orangutans (*Pongo pygmaeus*) in the Gunung Leuser National Park, Sumatra. *International Journal of Primatology*, 8, 17–42.

Sugiyama, Y. (1968). Social organization of chimpanzees in the Budongo Forest, Uganda. *Primates*, 9, 225–58.

te Boekhorst, I., Schurmann, C., Sugardjito, J. (1990). Residential status and seasonal movements of wild orangutans in the Gunung Leuser Reserve (Sumatra, Indonesia). *Animal Behavior*, 39, 1098–109.

Trivers, R. (1972). Parental investment and sexual selection. In *Sexual Selection and the Descent of Man*, ed. B. Campbell, pp. 136–79. Chicago: Aldine.

Tutin, C. & McGinnis, P. (1981). Chimpanzee reproduction in the wild. In *Reproductive Biology of the Great Apes*, ed. C. Graham, pp. 239–64. New York: Academic Press.

Tutin, C., McGrew, W. & Baldwin, P. (1983). Social organization of savanna-dwelling chimpanzees, *Pan troglodytes verus*, at Mt. Assirik, Senegal. *Primates*, 24, 154–73.

Utami, S., Wich, S., Sterck, E. & van Hooff, J. (1997). Food competition between wild orangutans in large fig trees. *International Journal of Primatology*, 18, 909–27.

van Schaik, C. (1983). Why are diurnal primates living in groups? *Behaviour*, 87, 120–44.

Watts, D. (1998). Coalitionary mate-guarding by male chimpanzees at Ngogo, Kibale National Park, Uganda. *Behavioral Ecology and Sociobiology*, 44, 43–55.

Watts, D. (2000a). Grooming between male chimpanzees at Ngogo, Kibale National Park, Uganda. I. Partner number and diversity and reciprocity. *International Journal of Primatology*, 21, 189–210.

Watts, D. (2000b). Grooming between male chimpanzees at Ngogo, Kibale National Park, Uganda. II. Male rank and possible competition for partners. *International Journal of Primatology*, 21, 211–38.

Watts, D. & Mitani, J. (2000). Infanticide and cannibalism by male chimpanzees at Ngogo, Kibale National Park, Uganda. *Primates*, 41, 357–65.

Watts, D. & Mitani, J. (2001). Boundary patrols and intergroup encounters in wild chimpanzees. *Behaviour*, 138, 299–327.

Watts, D. & Mitani, J. (2002). Hunting behavior of chimpanzees at Ngogo, Kibale National Park, Uganda. *International Journal of Primatology*, 23, 1–28.

Wrangham, R. (1977). Feeding behaviour of chimpanzees in Gombe National Park, Tanzania. In *Primate Ecology*, ed. T. Clutton-Brock, pp. 503–38. London: Academic Press.

Wrangham, R. (1980). An ecological model of female-bonded primate groups. *Behaviour*, 75, 262–300.

Wrangham, R. (2000). Why are male chimpanzees more gregarious than mothers? A scramble competition hypothesis. In *Primate Males*, ed. P. Kappeler, pp. 248–58, Cambridge: Cambridge University Press.

Wrangham, R. & Smuts, B. (1980). Sex differences in the behavioral ecology of chimpanzees in the Gombe National Park, Tanzania. *Journal of Reproduction and Fertility*, 28, 13–31.

Wrangham, R., Clark, A. & Isabirye-Basuta, G. (1992). Female social relationships and social organization of Kibale Forest chimpanzees. In *Topics in Primatology*, Volume 1, *Human Origins*, ed. T. Nishida, W. McGrew, P. Marler, M. Pickford & F. deWaal, pp. 81–98. Tokyo: Tokyo University Press.

8 • Agonistic relations among Kanyawara chimpanzees

MARTIN N. MULLER

INTRODUCTION

Although wild chimpanzees may spend hours resting and grooming peacefully in mixed social groups, and affiliative interactions among them may be more common than agonistic ones, intraspecific aggression is none the less a frequent occurrence in chimpanzee society. Both males and females exhibit an array of aggressive behaviors (from mild threats to lethal attacks) in a variety of contexts (from infant protection to sexual competition) against a range of competitors (from extracommunity males to newly immigrated females). Aggression, or merely the threat of aggression, can have a profound impact on individual patterns of ranging and association. For example, female immigrants at Gombe tend to settle in peripheral areas of the range and away from dominant females, where the risk of infanticide may be lower (Williams *et al.*, Chapter 14).

The most dramatic examples of chimpanzee aggression come from observations of intercommunity encounters (e.g. Goodall *et al.* 1979). Male chimpanzees are philopatric, and they aggressively defend their community range against incursions from neighboring males (Nishida 1979; Goodall 1986; Watts & Mitani 2001). In the course of such defense, they sometimes cooperate to inflict lethal wounds on vulnerable strangers (reviewed in Wrangham 1999). Lethal coalitionary aggression is rare among mammals, having previously been documented as a major source of adult mortality only in wolves (Mech *et al.* 1998) and humans (van der Dennen 1995). In male chimpanzees, it may be part of a larger strategy to reduce the coalitionary strength of neighboring groups and to expand territorially at their expense (Wrangham 1999).

Within communities, male chimpanzees compete aggressively both for status within a linear dominance hierarchy and for access to estrous females. Coalitions can play an important role in both of these contexts, as males may cooperate either to challenge rivals (Nishida 1983; Nishida & Hosaka 1996) or to maintain exclusive access to an estrous female (Watts 1998). Intracommunity aggression is normally less brutal than that between communities, perhaps in part because relatedness among males is higher within a community (Morin *et al.* 1994). However, dominance struggles are sometimes marked by intense dyadic aggression and potentially lethal wounding (e.g. Goodall 1992; Nishida *et al.* 1995; Fawcett & Muhumuza 2000).

Male chimpanzees appear to incur large costs as a result of competition for dominance. These include not only the risk of severe injury in escalated fighting, but the energetic demands of agonistic display. The presumption that such costs must be offset by considerable reproductive benefits has previously been supported by behavioral data indicating that high rank is associated with increased mating success (e.g. Nishida 1983). More recently, genetic tests of paternity have allowed for direct measures of reproductive success, and corroboration of the behavioral evidence (Constable *et al.* 2001; for bonobos see Gerloff *et al.* 1999).

The potential influence of dominance rank on female reproductive success is not as well understood. Dominance relationships among female chimpanzees are less conspicuous than those of males, such that observers often find it difficult to rank them (Bygott 1974). On the one hand this is not surprising, as the limiting factor on female ape reproduction is normally considered to be food, and competition for food among female chimpanzees generally takes the form of scramble, rather than contest, competition (Wrangham 2000). On the other hand, the lack of overt competition among female chimpanzees is puzzling since recent evidence suggests that female dominance rank does influence factors such as infant survivorship and interbirth interval, probably through access to food (Pusey *et al.* 1997).

This chapter describes patterns of between- and within-group aggression among chimpanzees in Kibale National Park, Uganda. There are four main objectives. First, sex differences in rates, contexts, and targets of aggression are described, and accounted for in terms of reproductive strategies. Second, the relationship between aggression and dominance is explored, and the possible costs and benefits of high rank in this species are considered. Third, new incidents of intercommunity aggression, including a lethal coalitionary attack, are reported, and evaluated in light of the 'imbalance-of-power' hypothesis (Manson & Wrangham 1991). Finally, these data are compared with observations from other long-term study sites in Tanzania (Bygott 1974; Goodall 1986; Nishida 1989; Nishida & Hosaka 1996) and Ivory Coast (Boesch & Boesch-Achermann 2000) in an attempt to identify the broader patterns underlying behavioral variation.

METHODS

The Kanyawara community in Kibale National Park has been studied continuously by Richard Wrangham and colleagues since 1987. Struhsaker (1997) provides a detailed description of the research site. At the beginning of this study, the community consisted of 50 chimpanzees, including 11 adult males, 15 adult females, 1 subadult male, 2 nulliparous females, 8 juveniles, and 13 infants. During the course of the study, three infants were born, and five individuals died.

The data in this study were collected by me from January through December, 1998 as part of a larger project investigating the hormonal correlates of dominance and aggression in chimpanzees. With the help of long-term field assistants, chimpanzees were followed, whenever possible, from the time that they woke in the morning, until the time that they constructed their night nests. If observers lost track of a chimpanzee party, a new one was located by either listening for long-distance vocalizations, or waiting near a fruiting tree. All-male and bisexual parties were followed preferentially, in order to facilitate data collection on male aggression.

Forty-minute group focals were used to generate rates of aggression for individual chimpanzees. This was equivalent to all-occurrence sampling (Altmann 1974), which was possible because the boisterous nature of male chimpanzee agonism renders it highly conspicuous to observers. When observation conditions did not allow focal data to be collected, ad libitum observations of aggression and submission were made. These were combined with focal data to rank the adult males in a linear dominance hierarchy (see below).

Behavioral categories followed those of Bygott (1979) and Goodall (1986); these are summarized in Nishida *et al.* (1999). *Charging displays* involved exaggerated locomotion, piloerection, slapping, stamping, branch swaying and throwing; they were classified as either *vocal* or *nonvocal*, depending on the presence of the pant-hoot call. *Chases* were recorded when an individual pursued a fleeing conspecific, who was generally screaming. *Attacks* were recorded for all contact aggression. This included hits, kicks, or slaps delivered in passing (Goodall's level 1), more extended episodes of pounding, dragging, and biting (Goodall's level 2), and incidents lasting more than 30 seconds or leading to serious injury (Goodall's level 3). In behavioral analyses, charging displays were categorized as *low-level aggression*, while chases and attacks were classified as *high-level aggression*. Whenever both members of a dyad could be observed for 10 minutes following an aggressive interaction, affiliative contact (e.g. grooming, embracing, kissing) between the pair during that period was recorded as a *reconciliation* (de Waal 1993).

Behavioral contexts of aggression were recorded in the following categories: *reunion* (incidents occurring within 5 minutes of a reunion), *social excitement* (incidents occurring upon hearing distant chimpanzee calls, or arrival at a fruiting tree), *sexual competition*, *meat competition*, *plant-food competition*, *protection* (of offspring), and *no obvious context*. For further description of these contexts see Bygott (1974) and Goodall (1986).

Dominance ranks are commonly assigned to male chimpanzees based on the distribution of pant-grunt vocalizations (e.g. Bygott 1979). Pant-grunt orientation is highly directional, and reliably correlates with several other measures of dominance (Bygott 1974). In the current study, however, observed pant-grunts were insufficient to distinguish male rank beyond the basic categories of *alpha*, *high*, *medium*, and *low* (cf. Bygott 1974).

In order to acquire enhanced resolution on relationships among the adult males, I assigned dominance ranks based on the outcome of decided agonistic bouts. To do this, I employed a probabilistic model (Jameson *et al.* 1999) that takes into account the number of opponents that an individual has successfully defeated, and the relative success of those opponents in their own agonistic encounters. This method has two major advantages over the standard techniques that provide ordinal dominance rankings. First, it can be used to predict dominance relationships for dyads that have not yet been observed encountering each other. Second, it provides information about how dominant individuals are over others. Specifically, scale values indicate the probabilities associated with pairs of ranks (males in the current study received scale values falling between −1 and 1). This probabilistic method is similar to those used for ranking chess players. Further discussion of the method,

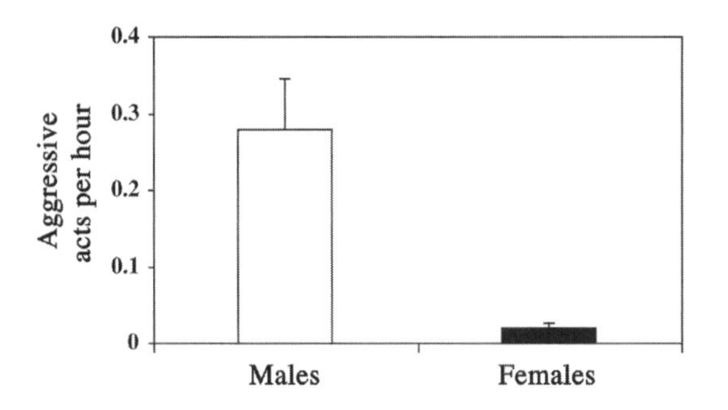

Fig. 8.1. Mean rates of aggression (+SE) by male and female chimpanzees in Kanyawara, 1998. Aggression includes charging displays, chases, and attacks. Data are from 11 males and 10 females with at least 25 observation hours (male median = 144.7, female median = 69.7). Follows of lone targets have been excluded. Rates of male aggression are significantly higher than those of females (Rust–Fligner rank-based test, $p < 0.00001$).

and the formulae used to calculate the rank scores are available in Jameson *et al.* (1999) and de Vries & Appleby (2000).

Statistical procedures

I used the Rust–Fligner rank-based test for comparisons between independent groups (Rust & Fligner 1984). When applied to two groups, this is equivalent to the Mann–Whitney–Wilcoxon test. However, the Rust–Fligner procedure eliminates the assumption of equal variances that can result in poor power and wide confidence intervals in the Mann–Whitney–Wilcoxon test (reviewed in Wilcox 1996). The Storer–Kim procedure was used for comparisons between two independent binomials (Storer & Kim 1990). Reported correlations employ Pearson's product moment correlation coefficient (r).

RESULTS

Sex differences in aggression

Male chimpanzees at Kanyawara exhibited much more frequent and intense aggression than did females. Figure 8.1 presents mean rates of aggression (charging displays, chases, and attacks) for both sexes. Though variability exists within each sex, adult males were aggressive approximately 14 times more often than adult females (males: mean = 0.28 acts per observation hour, $n = 11$; females: mean = 0.02 acts per observation hour, $n = 10$). This figure probably overestimates female aggression because it excludes observations made when focal individuals were traveling alone (or solely with dependent offspring), and female chimpanzees are more likely to be solitary than males (Nishida 1979; Goodall 1986; Wrangham 2000). Table 8.1 shows attack rates for males and females at both Kanyawara and Gombe. The same pattern of male-biased aggression is evident.

Aggression by female chimpanzees tended to take place in different contexts from aggression by males. Table 8.2 illustrates this sex difference with data from three long-term study sites; it only includes episodes in which a context could be clearly determined. Fifty-two percent of male aggression at Kanyawara and 14% at Gombe took place in indeterminate contexts. Most of these episodes appeared to be related to male dominance rivalry (Goodall 1986; personal observation). When contexts of male aggression could be determined, they were similar at Gombe and Kanyawara. A large proportion of male agonism took place either within 5 minutes of a reunion (34% at Kanyawara, 38% at Gombe) or during sexual competition (36% at Kanyawara, 17% at Gombe). Much aggression at Gombe also took place during competition for meat (17%); this figure was lower at Kanyawara (4%), where little hunting took place during 1998.

Contexts of female aggression were less ambiguous than those of male aggression; 6.3% of observations at Gombe, 11% at Kanyawara, and 28.6% at Mahale could not be assigned to a distinct context. The most frequent contexts of female aggression were competition for plant food (80% at Kanyawara, 38% at Gombe, 35% at Mahale) and protection of offspring (12% at Kanyawara, 38% at Gombe, 35% at Mahale). If interspecific interactions are considered at Kanyawara, an even higher percentage of female aggression took place there during feeding competition; female chimpanzees occasionally chased blue monkeys or mangabeys from preferred feeding sites in fruiting trees. These episodes were excluded from the present data set, but the aggression appeared identical to that directed toward conspecific feeding competitors.

Males and females tended to direct aggression at different targets. Table 8.3 shows the distribution of victims at two long-term study sites. At Kanyawara, most aggression by adult males (49.2%) was directed at other adult males; somewhat less (24.6%) was directed toward parous females. Few parous females were cycling during this study, but preliminary data from one popular female, Lia, suggest that parous females receive more aggression when they exhibit sexual swellings than at other times. During 33 hours of

Table 8.1. *Frequency of attacks by males and females at two long-term study sites*

Community	Year	Sex	Focal hours	Focal IDs	Total attacks	Observation hours per attack
Kanyawara	1998	Male	1428.3	11	45	31.7
Gombe	1978	Male	1570	7	23	68.3
Gombe	1976	Male	829	6	16	51.8
Gombe	1970	Male	—	—	—	33
Kanyawara	1998	Female	679.8	10	5	136
Gombe	1978	Female	1647	7	15	109.8
Gombe	1976	Female	897	6	9	99.7

Note:
Kanyawara data include adults with more than 25 observation hours. Gombe data from 1978 and 1976 include only adults with more than 100 observation hours (Goodall 1986). Gombe data from 1970 are from Bygott (1974). Follows of lone targets (including females with dependent offspring) have been excluded.

Table 8.2. *Contexts of aggression in three long-term study sites*

	Kanyawara 1998		Gombe 1978		Mahale
Context	Males	Females	Males	Females	Females
Reunion	34	0	38	5	5
Sexual competition	36	8	17	2	7.5
Plant-food competition	15	80	7	38	35
Meat competition	4	0	17	5	0
Social excitement	8	0	13	1	2.5
Protection	1	12	5	38	35
Miscellaneous	2	0	3	11	15
Total observations	212	25	274	149	40

Note:
Numbers refer to the percentage of aggressive acts that take place in each context. Cases where the context was not clear to observers have been omitted (Gombe: 14% of male observations, 6.3% of female observations; Kanyawara: 52% of male observations, 11% of female observations: Mahale: 28.6% of female observations). Gombe data are from Goodall (1986). Only female data are available from Mahale; these were collected over three field seasons in the early 1980s by Nishida (1989).

focal observation with no sexual swelling, Lia received aggression from adult males every 8.3 hours; during 46 hours of observation with a partial or maximal swelling, Lia was the victim of male aggression every 3.8 hours. This increased rate of received aggression was partially due to the fact that more adult males were present when Lia was swollen. However, males also became more aggressive when traveling in parties with estrous females (see below). Adult females were not observed to exhibit any aggression towards other adult females; more than 70% of their aggression was directed at juvenile and subadult females.

At Gombe, adult males directed slightly more of their

Table 8.3. *Victims of directed aggression in two long-term study sites*

	Aggressor							
	Kanyawara 1998				Gombe 1978			
Victim	AM	(%)	AF	(%)	AM	(%)	AF	(%)
Adult males	120	49.2	3	12.5	109	34	7	4
Adult females	60	24.6	0	0	154	48	53	33
Subadult males	15	6.1	1	4.2	22	7	54	34
Subadult females	18	7.4	5	20.8	13	4	33	21
Juvenile males	8	3.3	0	0	12	4	8	5
Juvenile females	22	9	12	50	9	3	4	3
Infants	1	0.4	0	0	0	0	0	0
Other	0	0	3	12.5	0	0	0	0

Note:
Columns show both the total number and percentage of agonistic acts perpetrated by adult males (AM) and adult (parous) females (AF) against each age/sex class on the left.

aggression at adult females (48%) than adult males (34%) (Goodall 1986). Adult females at Gombe also appeared to be involved in aggressive interactions with other adult females more frequently than at Kanyawara; 33% of observed aggression by adult females at Gombe was directed at other adult females (Goodall 1986).

Dominance relationships

Most pant-grunts by adult males (64%) were directed toward the alpha male ($n=89$). The two lowest-ranking males produced 39% of male pant-grunts. Consistent with data from Mahale (Nishida & Hosaka 1996), male dominance rank was positively and significantly correlated with the total number of pant-grunts received ($n=11$, $r=0.66$, $p=0.03$), and the total number of agonistic initiations ($n=11$, $r=0.69$, $p=0.02$) (both uncorrected for observation hours).

During the study period, the alpha-male, Imoso, was just beginning his tenure at the top. For 3 years, starting in mid-1994, Big Brown had been the alpha-male at Kanyawara, and his ally, Tofu, had held the beta position (Kibale Chimpanzee Project, long-term data). In mid-1997 Imoso, together with his ally Johnny, began to challenge these top-ranking males. By the end of that year they were clearly established as the new alpha and beta males. Tofu fell to third, and Big Brown to fourth place in the hierarchy.

Although the observed dominance hierarchy was of relatively recent origin, it appeared to be quite stable during the study period; no reversals were observed in male–male pant-grunt interactions. Only three reversals were recorded in 107 decided agonistic bouts.

It was not possible to rank the 15 adult females in a linear hierarchy because dominance interactions between them were extremely rare. In 680 hours of focal observation, not one aggressive interaction was recorded between parous females. Also, although parous females frequently pant-grunted to adult males, they were rarely observed pant-grunting to other parous females.

Patterns of male aggression

The most frequent form of male aggression was the charging display. Display rates differed among males, but the alpha male, Imoso, was clearly the most aggressive member of the community. His display rate (0.69 per observation hour) was 4.5 times the male average (0.15 per observation hour), and more than twice that of the next highest male (Tofu: third-ranking male, 0.29 per observation hour). Dominance rank and frequency of display were significantly correlated across the 11 adult males ($r=0.75$, $p=0.008$, Figure 8.2).

Consistent with reports from Gombe (Bygott 1974; Goodall 1986), vocal displays were rarely directed toward

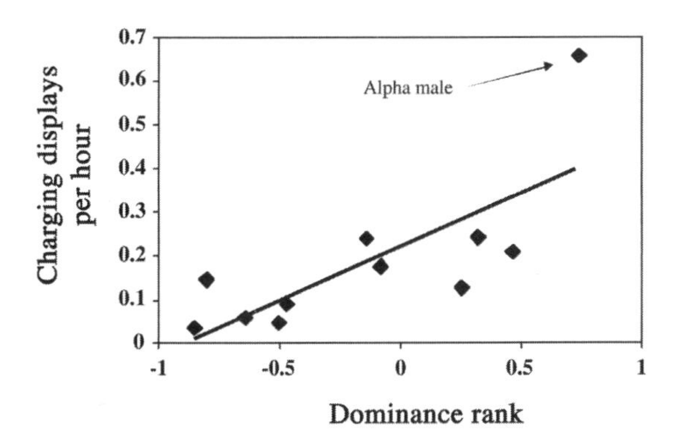

Fig. 8.2. Dominance rank and display rate in Kanyawara males (1998). The alpha male displayed much more frequently than any other chimpanzee in the community. Across 11 adult males, there was a significant correlation between rank and display rate ($r = 0.75$, $p = 0.008$). Display rates are from 1429 hours of focal observation (male median = 144.7 hours).

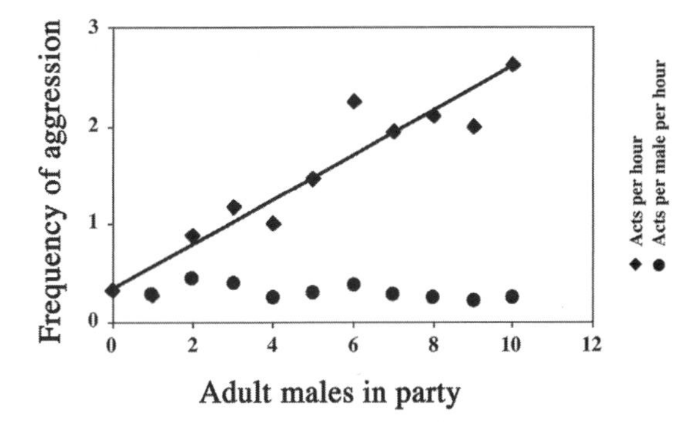

Fig. 8.3. Frequency of aggression in relation to the number of adult males in a party. The number of aggressive acts per hour is highly correlated with the number of adult males present ($r = 0.95$, $p < 0.0001$). However, males do not become more aggressive with increasing party size (i.e. there is no significant relationship between party size and the number of aggressive acts per male per hour). Parties containing maximally swollen females have been omitted. Data are from Kanyawara, 1998.

specific individuals (only 6.8% of 148 vocal displays). Instead, they appeared to be directed either toward the group in general, or toward distant chimpanzees. Nonvocal displays, on the other hand, were frequently directed toward specific targets within a party (53.2% of 186 displays: Storer and Kim procedure, $p < 0.000001$).

When only high-level agonistic acts (chases and attacks) are considered, the alpha-male was still the most frequent aggressor in the community. Furthermore, rank and high-level aggression were positively and significantly correlated across the adult males ($n = 11$, $r = 0.71$, $p = 0.014$). Table 8.1 shows mean attack rates for males at both Kanyawara and Gombe. Male attack rates from Kanyawara in 1998 are comparable to those reported from Gombe in 1970; they are higher than those reported from Gombe in 1976 and 1978.

The frequency of aggression in a given party at Kanyawara was positively and significantly correlated with the number of adult males in that party ($r = 0.95$, $p < 0.0001$, Figure 8.3). However, the rate of aggression per male did not increase with party size. Party composition, in contrast, had a substantial effect on rates of male aggression. When maximally swollen, parous females were present in a party, both the rate and intensity of male aggression increased. The average daily rate of aggression in parties containing eight to nine adult males and no estrous females was 1.68 incidents per hour. These incidents tended to be mostly low-level aggression (70% charging displays only, 30% chases and attacks). The average daily rate of aggression in parties containing eight to nine adult males and a parous female in her periovulatory period was significantly higher, at 3.2 incidents per hour (Rust–Fligner rank-based test: $Z = 4.01$, $p < 0.05$) of mostly high-level aggression (59% chases and attacks).

Table 8.4 shows rates of reconciliation at Kanyawara together with those reported from captivity. Adult males at Kanyawara rarely reconciled with each other following a conflict (less than 9% of interactions). Across all age and sex classes, only 10% of conflicts were reconciled within 10 minutes. This is considerably lower than the rate of reconciliation reported from captivity (de Waal 1993). Similarly low rates of reconciliation reported from Taï (Wittig 2000) suggest that chimpanzees in the wild are less motivated to reconcile conflicts, perhaps because in the short-term they can always avoid particular conspecifics by fissioning from a party.

Intercommunity aggression

Intercommunity relations among male chimpanzees are predictably hostile. Border patrols, aggressive territorial defense, and border avoidance by lone individuals have been reported from all the long-term study communities except Bossou, which has no neighbors (reviewed in Wrangham 1999). Coalitionary intergroup attacks, sometimes lethal, appear to be a regular feature of chimpanzee society. Two such attacks occurred at Kanyawara during this study, one of which resulted in a fatality.

Table 8.4. *Reconciled conflicts in wild and captive chimpanzees*

Study Population		Dyads observed	Number reconciled	Percentage reconciled
Kanyawara 1998	All dyads	174	18	10.3
	Male–male	91	8	8.8
	Male–female	62	8	12.9
	Female–female	8	1	12.5
	Other	13	1	8.3
Captive chimpanzees	Arnhem 1975–76	150	52	34.7
	Arnhem 1976 (outdoors)	200	59	29.5
	Arnhem 1980 (outdoors)	395	105	26.6

Note:
Reconciliation is defined as any affiliative contact between individuals within ten minutes of their aggressive interaction. 'Other' includes interactions involving at least one juvenile. Interactions with juveniles have been excluded from the other categories. Captive data are from de Waal (1993).

On 12 August, 1998, five adult males from the Kanyawara community traveled to the southern boundary of their territory, where they encountered a nulliparous female and a subadult male from the neighboring community. The Kanyawara males attacked the pair, chasing them into a low tree. Both leaped approximately 7 m to the ground, and the subadult male fled to the south. Three Kanyawara males surrounded the female, striking her repeatedly. She was able to escape, however, and ran quickly to the southwest. Four adult males from the southern community arrived within 2 minutes of the female's departure, presumably attracted by the screaming. They displayed and exchanged aggressive vocalizations with the Kanyawara males, who subsequently retreated further than 1 km into their own territory.

In a second incident on 25 August, 1998, a group of ten adult males from the Kanyawara community patrolled the northern border of their territory. Observers lost the group at 17:05, but located them nearby the following morning with a dead male (aged 20–25 years) from the neighboring community. He had been killed the previous evening.

The site and the body exhibited signs of a protracted and vicious assault. The dead male lay at the base of a steep slope in a 7-m × 12-m patch of trampled vegetation (Figure 8.4). His arms and legs were extended, suggesting that he had been immobilized by some individuals while others attacked (as occurs in both captive chimpanzees and bonobos: Amy Parish, personal communication). The pattern of injuries was consistent with this scenario, as more than 30 puncture wounds and lacerations, ranging from 0.5 cm to 12 cm in diameter, were distributed across the male's face and the ventral surface of his arms, legs, and abdomen. The dorsal surface was undamaged.

Compound fractures in four of the left ribs attested to severe blows delivered by fists or feet. The testicles had been ripped from the body, and were recovered nearby. Five fingernails and one toenail had been torn from the digits with significant portions of flesh attached. The immediate cause of death appeared to be massive trauma to the throat, including a severed trachea (Figure 8.4).

The Kanyawara males were found with the dead stranger on the morning of 26th August. Several of the males were beating on the corpse when we arrived. As they had also been seen near the attack site on the evening of 25th August, when the male was killed, it seems likely that they were responsible. At approximately 10:45, at least three males from the neighboring community arrived near the attack site, and the two groups exchanged aggressive vocalizations. At 11:45 the Kanyawara males retreated to the south.

DISCUSSION AND CONCLUSIONS

Female–female competition

Consistent with reports from Gombe (Goodall 1986) and Mahale (Nishida 1989), female chimpanzees at Kanyawara were aggressive primarily in the context of feeding compe-

Fig. 8.4. Dead male chimpanzee in Kibale National Park. This stranger was apparently the victim of lethal coalitionary aggression when he encountered a group of ten males from Kanyawara patrolling the boundary of their territory. (*a*) The arms and legs of the victim were extended, and wounds were restricted to the ventrum, suggesting that some males immobilized him while others attacked. (*b*) The immediate cause of death appeared to be massive trauma to the throat, including a severed trachea. (Photos by Martin Muller.)

tition. Such aggression was rarely severe; it consisted primarily of mild threats or brief charging displays used to supplant feeding competitors. The prolonged attacks characteristic of male dominance rivalry were seen on only three occasions; these were all directed at one recently immigrated female. Parous females were rarely aggressive or submissive towards each other.

Nishida (1989) suggests that parous females may direct aggression primarily at newly immigrated females because such immigrants represent potential resource competitors. After dominance relationships are established between females, they remain stable because the costs of escalated aggression, which include potential danger to offspring, outweigh the benefits of increased dominance standing. Since parous females are already settled in their core areas, they 'have no pressing reason to strive for higher rank' (Nishida 1989, p. 86).

Given that core areas may differ in both size and quality, however, and that core areas frequently overlap (e.g. Pusey *et al.* 1997), the rarity of aggressive interactions between adult females remains puzzling. Data from Gombe suggest that high dominance rank accords female chimpanzees significant reproductive advantage. For example, high-ranking females there appear to maintain access to higher-quality core areas (Pusey *et al.* 1997; Williams 1999). These females live longer than low-ranking females, and enjoy shorter inter-birth intervals and higher offspring survival. They also produce daughters who reach sexual maturity earlier than those of low-ranking mothers, presumably because of improved nutrition (Pusey *et al.* 1997). With so much at stake, one might expect female chimpanzees to show more overt competition over dominance than they appear to.

One possible explanation for the low rates of aggressive competition observed among females at Kanyawara is that competition for space is not as pronounced there as it is at Gombe; thus, the benefits of high rank to Kanyawara females are correspondingly less. This hypothesis has not yet been tested directly, but several observations suggest that female competition at Gombe is particularly intense. First, young females at Gombe exhibit a relatively low rate of transfer (Williams *et al.*, Chapter 14). A female who stays in her natal community presumably bears increased costs associated with inbreeding, but may benefit from associating with a high-ranking mother, for example by settling in her core area. Second, both infanticide and attempted infanticide by high-ranking females against low-ranking mothers appear to be more common at Gombe than at other sites (Pusey *et al.* 1997; Clark Arcadi & Wrangham 1999). Third, aggressive interactions between parous females appear to be more common at Gombe than at Kanyawara (Table 8.3). In the present study, no aggressive interactions were observed between adult females at Kanyawara in more than 680 hours of focal observation. At Gombe in 1978, adult females targeted other adult females approximately 33% of the times that they were aggressive ($n = 159$ aggressive acts; Goodall

1986). It is difficult to say whether this reflects a real difference, as Goodall does not provide overall rates of female aggression. Long-term data on female dominance rank and reproductive success at Kanyawara will eventually help to clarify this issue.

Male–male competition

Male–male competition in chimpanzees takes two general forms, both of which were evident in this study. In the long-term, male chimpanzees cooperate to defend a territory against neighboring males. This cooperation includes border patrols and coalitionary attacks on vulnerable rivals, and is discussed below. In the short-term, males within a community compete aggressively (e.g. Watts 1998) and through sperm competition (Hasegawa & Hiraiwa-Hasegawa 1990) for access to estrous females.

Male chimpanzees are sometimes able to avoid short-term mating competition by escorting an estrous female to a peripheral part of the range in an exclusive consortship (Goodall 1986). Genetic data from Gombe indicate that this is sometimes a successful reproductive strategy, particularly for low-ranking males (Constable *et al.* 2001). Of fourteen infants genotyped in that community, three were sired in the consortship context, and two more were probably sired in the consortship context, all by low- and middle-ranking males.

Behavioral data from three long-term study sites (Gombe, Mahale, and Taï), however, indicate that most chimpanzee conceptions occur in the context of multi-male parties (75–94%: Hasegawa & Hiraiwa-Hasegawa 1990; Wallis 1997; Boesch & Boesch-Achermann 2000). Although copulations sometimes occur in such parties with little male aggression or coercion (Tutin 1979), this is generally limited to matings with nulliparous females, or parous females in the mid-follicular phase. Once parous females reach their periovulatory period, they become much more attractive to males, and coercion and aggression increase markedly (reviewed in Muller & Wrangham 2001). In the current study, males at Kanyawara showed both increased frequency and intensity of aggression in the presence of periovulatory females.

Among male chimpanzees, success in short-term mating competition is closely predicted by dominance rank. Because low-ranking males may have access to parous females in the mid-follicular phase, and nulliparous females throughout the cycle, overall copulation rates may or may not correlate with dominance rank. However, alpha and high-ranking males consistently exhibit the highest copulation rates with parous females in the periovulatory period (reviewed in Muller & Wrangham 2001).

Copulation frequency is not necessarily an accurate proxy for reproductive success. However, preliminary genetic data from Gombe support the idea that high rank is reproductively advantageous. Of nine infants sired there in the group-mating context, five were sired by the alpha male, two by males who subsequently became alpha, one by a high-ranking male, and one by a middle-ranking male (Constable *et al.* 2001; for similar results from bonobos see Gerloff *et al.* 1999).

The positive correlation between dominance rank and aggression reported here is consistent with observations from other long-term study sites. At Gombe in 1970, Bygott (1974) found significant correlations between male dominance rank and rates of both charging display and attack. At Taï in 1993, the two highest-ranking males performed 80% of all agonistic displays (Boesch & Boesch-Achermann 2000). At Mahale in 1992, dominance rank was strongly correlated with total number of agonistic initiations (Nishida & Hosaka 1996). Finally, long-term data from Kanyawara indicate that two previous alpha males there also had the highest rates of display during their tenures at the top (Wrangham, unpublished data).

Among primates generally, however, high levels of aggression are not consistently associated with high dominance rank. Olive baboons (*Papio anubis*) are a well-studied example. Sapolsky (1982) found that among male baboons in Masai Mara, dominance rank in copulatory success was related to success in a number of competitive behaviors. During periods of dominance stability, however, frequency of aggression was not associated with high rank. In fact, the initiators of aggressive interactions lost those interactions 80% of the time when the hierarchy was stable. Only during periods of dominance instability do dominant male baboons consistently show the highest rates of aggression (reviewed in Sapolsky 1993). Sapolsky characterizes a baboon hierarchy as unstable when the overall rate of reversals in approach–avoidance interactions exceeds 10%. Such periods of instability are relatively rare in the wild, but may follow shifts in troop membership (e.g. Alberts *et al.* 1992).

The chimpanzee data, however, suggest that high rates of aggression are characteristic of high-ranking chimpanzee males, even in stable hierarchies. The associations between rank and aggression reviewed above all took place during periods of relative dominance stability. The rate of pant-grunt reversals among adult males at Gombe during 1970 was only 0.5% (2 of 442 interactions: Bygott 1974, table 5.7). At Mahale in 1992 the rate was 1.1% (3 of 268 interactions: Nishida & Hosaka 1996, Table 9.3). At Taï in 1993, the rate was 0.9% (1 of 109 interactions, Boesch & Boesch-Achermann 2000, Table 6.3). In the present study, no reversals were recorded in 89 pant-grunt interactions between adult males.

The overall rate of reversals for decided agonistic bouts between adult males in this study was only 2.8%. A comparable figure is not available from Gombe during 1970, but the overall rate of reversals for attacks there was only 3.3% (calculated from Bygott 1974, Table 5.4). At Taï in 1993, the overall rate of reversals for agonistic interactions among adult males was, at 12.9%, slightly higher (calculated from Boesch & Boesch-Achermann 2000, Table 6.3).

Although the alpha male at Kanyawara had just started his tenure during the present study, at other sites high rates of aggression were exhibited by males who had been dominant for some time. During Bygott's (1974) Gombe study, the alpha male, Mike, was in the 6th year of his tenure. At Taï in 1993, Fitz was in the 3rd year of a 4-year tenure (Boesch & Boesch-Achermann 2000). At Mahale, Ntologi reacquired his alpha status approximately 5 months prior to Nishida's 1992 observations; he had been the alpha male in 1991, but was briefly ousted from the position by a coalition of two other males (Nishida & Hosaka 1996).

Thus, male chimpanzees living in what are normally considered to be stable hierarchies tend to exhibit patterns of aggression similar to those of male baboons living in extremely unstable hierarchies. What accounts for this element of perpetual instability in dominance relationships among chimpanzee males? One possibility is that the fission–fusion nature of chimpanzee social organization makes it relatively more difficult for high-ranking individuals to keep track of social relationships among other males in the group. Male baboons range in relatively stable groups, so their knowledge of the dominance hierarchy and their place in it should be fairly accurate. Chimpanzee males, in contrast, frequently leave groups and rejoin them hours, days, or even weeks later. Since a high-ranking male can never be certain what political maneuvering has occurred in his absence, it is necessary for him continually to reestablish his dominance when parties come together. This could help to explain the large proportion of aggression that takes place in the context of reunions.

Coalitions are important to chimpanzee males in both maintaining and improving their dominance standing (Nishida 1983; Goodall 1986; Nishida & Hosaka 1996). These coalitions can be very fluid, with males showing high degrees of 'allegiance fickleness' (i.e. a male may frequently turn against his former alliance partner if it is in his interest to do so) (Nishida 1983). The fluid nature of both male chimpanzee coalitions and party composition, then, is likely to explain some, but not all of the apparent instability in male chimpanzee hierarchies. For, despite their frequent use of coalitions, males can generally be ranked in a linear hierarchy (Boesch & Boesch-Achermann 2000), and although having a coalition partner may raise a male's confidence to the point that he challenges a higher-ranking individual, actual reversals in rank generally result from dyadic fights (Goodall 1986; Boesch & Boesch-Achermann 2000).

The possibility that dominance hierarchies among male chimpanzees are inherently unstable has important implications for the presumed costs and benefits of social dominance in this species. When baboon hierarchies are unstable, high rank is often associated with elevated glucocorticoid secretion, probably reflecting severe psychological stress (Sapolsky 1992). Chronic exposure to high levels of circulating glucocorticoids is associated with a suite of adverse physiological effects, including depressed immune function (e.g. Sapolsky 1993). Furthermore, among baboons the most aggressive males also tend to exhibit the highest levels of circulating testosterone (Sapolsky 1993). Additional costs associated with chronically high levels of circulating testosterone include increased metabolic rate and immunosuppression (reviewed in Wingfield *et al.* 1997). If the most aggressive, highest-ranking chimpanzee males are maintaining chronically elevated levels of cortisol and testosterone, then these would represent previously unknown costs to social dominance. Preliminary hormonal data from Kanyawara indicate that this is indeed the case; during the study period both testosterone and cortisol were positively and significantly correlated with rank across the adult males (Muller 2002).

Intercommunity aggression

Coalitionary killing has previously been reported from two long-term chimpanzee study sites in Tanzania. In Gombe, males from the Kasakela community made systematic incursions into the neighboring Kahama group's territory over a period of 3 years, eliminating their rivals one by one in a series of vicious gang attacks (Goodall *et al.* 1979; Goodall 1986). Subsequently, the Kasakela males appropriated much of the Kahama group's range. A similar group extinction was later documented in Mahale National Park (Nishida *et al.* 1985). Because chimpanzees at both of these sites were provisioned, it has been argued that lethal intergroup attacks are an adverse effect of artificial feeding, not a natural aspect of chimpanzee behavior (Power 1991).

The coalitionary attacks described above add to a growing body of evidence that lethal, intergroup aggression is a common strategy employed by male chimpanzees to reduce the coalitionary strength of their neighbors and expand their territories (Wrangham 1999). The evolutionary benefits of such expansions are clear. First, larger territories may include the ranges of more females. Second, females living in larger territories may have shorter interbirth intervals and improved offspring survival (Williams 1999). After the group extinctions at Gombe and Mahale, the aggressors appropriated both territory and females from their defeated neighbors (Goodall *et al.* 1979; Nishida *et al.* 1985; Goodall 1986).

Animals generally avoid escalated dyadic aggression because of the inherent risk of severe injury from other, similarly armed adults (e.g. Maynard Smith & Pryce 1973). Coalitionary aggression, however, may allow chimpanzees to mitigate these costs. At both Gombe and Kibale, cooperating males inflicted lethal injuries without sustaining appreciable wounds. Selection may therefore have favored a strategy by which male chimpanzees utilize lethal aggression against foreign males whenever there is an extreme imbalance of power (Manson & Wrangham 1991; Wrangham 1999). This idea is supported by the observations from Kanyawara, in which cooperating males attacked a subadult male and a nulliparous female, and appear to have attacked and killed a lone male, but in both cases retreated upon the arrival of additional strangers. This may also explain why killings have not yet been witnessed at Taï, where food availability is high, party size is consistently large, and individuals are rarely forced to forage alone (Boesch & Boesch-Achermann 2000).

ACKNOWLEDGEMENTS

I thank Christophe Boesch and Gottfried Hohmann for organizing the conference 'Behavioural Diversity in Chimpanzees and Bonobos,' held 11–17 June, 2000 in Seeon, Germany. I thank the Uganda Wildlife Authority and Makerere University Biological Field Station for sponsoring long-term research in Kibale. I am grateful to Richard Wrangham for the opportunity to do fieldwork at Kanyawara. For assistance in the field, I thank John Barwogeza, Christopher Katongole, Francis Mugurusi, Donor Muhangyi, Christopher Muruuli, Peter Tuhairwe, and Ross Wrangham. For technical assistance, I thank Randall Collura. For helpful comments on the manuscript, I thank Peter Gray, Sonya Kahlenberg, John Mitani, Craig Stanford, and the editors. Research was supported by NSF awards SBR-9729123 and SBR-9807448, and grants to Richard Wrangham and MNM from the L.S.B. Leakey Foundation.

REFERENCES

Alberts, S. C., Sapolsky R. M. & Altmann, J. (1992). Behavioral, endocrine, and immunological correlates of immigration by an aggressive male into a natural primate group. *Hormones and Behavior*, 26, 167–78.

Altmann, J. (1974). Observational study of behavior: sampling methods. *Behaviour*, 49, 227–67.

Boesch, C. & Boesch-Achermann, H. (2000). *The Chimpanzees of the Taï Forest: Behavioural Ecology and Evolution.* Oxford: Oxford University Press.

Bygott, J. D. (1974). Agonistic behaviour and dominance in wild chimpanzees. PhD thesis. University of Cambridge.

Bygott, J. D. (1979). Agonistic behaviour, dominance and social structure in wild chimpanzees of the Gombe National Park. In *The Great Apes*, ed. D. Hamburg & E. R. McCown, pp. 404–27. Menlo Park, CA: Benjamin Cummings.

Clark Arcadi, A. & Wrangham, R. W. (1999). Infanticide in chimpanzees: review of cases and a new within-group observation from the Kanyawara study group in Kibale National Park. *Primates*, 40, 337–51.

Constable, J. L., Ashley, M. V., Goodall, J. & Pusey, A. E. (2001). Noninvasive paternity assignment in Gombe chimpanzees. *Molecular Ecology*, 10, 1279–1300.

de Vries, H. & Appleby, M. C. (2000). Finding an appropriate order for a hierarchy: a comparison of the I&SI and the BBS methods. *Animal Behavior*, 59, 239–45.

de Waal, F. B. M. (1993). Reconciliation among primates: a review of empirical evidence and theoretical issues. In *Primate Social Conflict*, ed. W. A. Mason & S. Mendoza. Albany, NY: SUNY Press.

Fawcett, K. & Muhumuza, G. (2000). Death of a wild chimpanzee community member: possible outcome of intense sexual competition. *American Journal of Primatology*, 51, 243–7.

Gerloff, U., Hartung, B., Fruth, B., Hohmann, G. & Tautz, D.

(1999). Intracommunity relationships, dispersal pattern and paternity success in a wild living community of bonobos (*Pan paniscus*) determined from DNA analysis of faecal samples. *Proceedings of the Royal Society of London, Series B*, 266, 1189–95.

Goodall, J. (1986). *The Chimpanzees of Gombe: Patterns of Behavior.* Cambridge, MA: Harvard University Press.

Goodall, J. (1992). Unusual violence in the overthrow of an alpha male chimpanzee at Gombe. In *Topics in Primatology*, ed. T. Nishida, W. C. McGrew, P. Marler, M. Pickford & F. B. M. de Waal. Tokyo: University of Tokyo Press.

Goodall, J., Bandora, A., Bergmann, E., Busse, C., Matama, H., Mpongo, E., Pierce, A. & Riss, D. (1979). Inter-community interactions in the chimpanzee population of the Gombe National Park. In *The Great Apes*, ed. D. Hamburg & E. R. McCown, pp. 12–53. Menlo Park, CA: Benjamin Cummings.

Hasegawa, T. & Hiraiwa-Hasegawa, M. (1990). Sperm competition and mating behavior. In *The Chimpanzees of the Mahale Mountains: Sexual and Life History Strategies*, ed. T. Nishida, pp. 115–32. Tokyo: University of Tokyo Press.

Jameson, K. A., Appleby, M. C. & Freeman, L. C. (1999). Finding an appropriate order for a hierarchy based on probabilistic dominance. *Animal Behaviour*, 57, 991–8.

Manson, J. H. & Wrangham, R. W. (1991). Intergroup aggression in chimpanzees and humans. *Current Anthropology*, 32, 369–90.

Maynard Smith, J. & Price, G. R. (1973). The logic of animal conflict. *Nature*, 246, 15–18.

Mech, D. L., Adams, L. G., Meier, T. J., Burch, J. W. & Dale, B. W. (1998). *The Wolves of Denali.* Minneapolis: University of Minnesota Press.

Morin, P. A., Moore, J. J., Chakraborty, R., Jin, L., Goodall, J. & Woodruff, D. S. (1994). Kin selection, social structure, gene flow, and the evolution of chimpanzees. *Science*, 265, 1193–201.

Muller, M. N. (2002). Endocrine aspects of aggression and dominance in chimpanzees of the Kibale Forest. PhD dissertation. University of Southern California.

Muller, M. N. & Wrangham, R. W. (2001). The reproductive ecology of male hominoids. In *Reproductive Ecology and Human Evolution*, ed. P. T. Ellison, pp. 397–427. New York: Aldine de Gruyter.

Nishida, T. (1979). The social structure of chimpanzees of the Mahale Mountains. In *The Great Apes*, ed. D. Hamburg & E. R. McCown. Menlo Park, CA: Benjamin Cummings.

Nishida, T. (1983). Alpha status and agonistic alliance in wild chimpanzees *(Pan troglodytes schweinfurthii)*. *Primates*, 24, 318–36.

Nishida, T. (1989). Social interactions between resident and immigrant female chimpanzees. In *Understanding Chimpanzees*, ed. P. G. Heltne & L. A. Marquardt, pp. 68–89. Cambridge, MA: Harvard University Press.

Nishida, T. & Hosaka, K. (1996). Coalition strategies among adult male chimpanzees of the Mahale Mountains, Tanzania. In *Great Ape Societies*, ed. W. C. McGrew, L. F. Marchant & T. Nishida, pp.114–34. Cambridge: Cambridge University Press.

Nishida, T., Hiraiwa-Hasegawa, M., Hasegawa, T. & Takahata, Y. (1985). Group extinction and female transfer in wild chimpanzees in the Mahale National Park, Tanzania. *Zeitschrift für Tierpsychologie*, 67, 284–301.

Nishida, T., Hosaka, K., Nakamura, M. & Hamia, M. (1995). A within-group gang attack on a young adult male chimpanzee: ostracism of an ill-mannered member? *Primates*, 36, 207–11.

Nishida, T., Kano, T., Goodall, J., McGrew, W. C. & Nakamura, M. (1999). Ethogram and ethnography of Mahale chimpanzees. *Anthropological Science*, 107, 141–88.

Power, M. (1991). *The Egalitarians – Human and Chimpanzee: An Anthropological View of Social Organization.* Cambridge: Cambridge University Press.

Pusey, A., Williams, J. & Goodall, J. (1997). The influence of dominance rank on the reproductive success of female chimpanzees. *Science*, 277, 828–30.

Rust, S. W. & Fligner, M. A. (1984). A modification of the Kruskal–Wallis statistic for the generalized Behrens–Fisher problem. *Communications in Statistics – Theory and Methods*, 13, 2013–27.

Sapolsky, R. M. (1982). The endocrine stress response and social status in the wild baboon. *Hormones and Behavior*, 16, 279–92.

Sapolsky, R. M. (1992). Cortisol concentrations and the social significance of rank instability among wild baboons. *Psychoneuroendocrinology*, 17, 701–9.

Sapolsky, R. M. (1993). Endocrinology alfresco: psychoendocrine studies of wild baboons. *Recent Progress in Hormone Research*, 48, 437–68.

Storer, B. E. & Kim, C. (1990). Exact properties of some exact test statistics for comparing two binomial proportions. *Journal of the American Statistical Association*, 85, 146–55.

Struhsaker, T. T. (1997). *Ecology of an African Rain Forest.* Gainesville: University Press of Florida.

Tutin, C. E. G. (1979). Mating patterns and reproductive strategies in a community of wild chimpanzees. *Behavioral Ecology and Sociobiology*, 6, 39–48.

van der Dennen, J. (1995). *The Origin of War: The Evolution of a Male-Coalitional Reproductive Strategy.* Groningen: Origin Press.

Wallis, J. (1997). A survey of reproductive parameters in the free-ranging chimpanzees of Gombe National Park. *Journal of Reproduction and Fertility*, 109, 297–307.

Watts, D. P. (1998). Coalitionary mate guarding by male chimpanzees at Ngogo, Kibale National Park, Uganda. *Behavioral Ecology and Sociobiology*, 44, 43–55.

Watts, D. P. & Mitani, J. C. (2001). Boundary patrols and intergroup encounters in wild chimpanzees. *Behaviour*, 138, 299–327.

Wilcox, R. R. (1996). *Statistics for the Social Sciences.* New York: Academic Press.

Williams, J. (1999). Female strategies and the reason for territoriality in chimpanzees: lessons from three decades of research at Gombe. PhD Thesis. University of Minnesota.

Wingfield, J. C., Jacobs, J. & Hilgarth, N. (1997). Ecological constraints and the evolution of hormone–behavior interrelationships. *Annals of the New York Academy of Sciences*, 807, 22–41.

Wittig, R. (2000). Conflict management in wild chimpanzees. Paper presented at the conference on *Behavioural diversity in chimpanzees and bonobos, 11–17 June, Seeon, Germany.*

Wrangham, R. W. (1999). The evolution of coalitionary killing: the imbalance-of-power hypothesis. *Yearbook of Physical Anthropology*, 42, 1–30.

Wrangham, R. W. (2000). Why are male chimpanzees more gregarious than mothers? A scramble competition hypothesis. In *Primate Males*, ed. P. M. Kappeler, pp. 248–58. Cambridge: Cambridge University Press

9 • Relationships of male chimpanzees in the Budongo Forest, Uganda

NICHOLAS E. NEWTON-FISHER

INTRODUCTION

Relationships between individuals are essentially the product of individual efforts to cope with environmentally imposed selection pressures in a social context. They emerge from repeated interactions between individuals, providing the social structure within which the individuals operate (Hinde 1976). In chimpanzees (*Pan troglodytes*), this structure is especially fluid (Goodall 1965; Reynolds & Reynolds 1965; Nishida 1979). The relationships that an adult male chimpanzee has with other males within his community appear to be both key components in the struggle for high social status within a community, and crucial for successful inter-community territorial encounters (Bygott 1979; Wrangham & Smuts 1980; Wrangham 1986; Nishida & Hosaka 1996; Boesch & Boesch-Achermann 2000). High social status is thought to increase the ability of males to monopolise females and so achieve higher mating success (Sugiyama & Koman 1979; Tutin 1979; Hasegawa & Hiraiwa-Hasegawa 1983). In social situations where a single male may not be able to monopolise access, a pair of males may be able to do so, providing a direct mating benefit to the cooperating males (Goldberg & Wrangham 1997; Watts 1998). Relationships may also provide proximate benefits beyond support in agonistic confrontations, which may, in turn, lead to higher social status, and serve to reduce both social tension and individual stress (Kawanaka 1990; de Waal 1996).

Despite the value of cooperative relationships to male chimpanzees, they appear to show little long-term loyalty to one another and can be extremely fickle in their allegiances (Nishida 1983; Uehara *et al.* 1994; Nishida & Hosaka 1996). Resident for their entire lives within their natal community (unless their mother transfers while they are still young and dependent upon her) adult males will share a long history of interactions and, ultimately, of unreliable relationships. Social status is relative, and there can be only one alpha male. In a fission–fusion society in which the composition of any one grouping is ephemeral and unpredictable (Chapman *et al.* 1993), the shifting sands of male relationships are likely to produce social tension with males vying with one another to build and maintain supportive relationships with individuals they cannot trust because they themselves are pursuing their own selfish interests.

Each relationship that a male has with the other males of his community may at any one time be affiliative, neutral, or antagonistic, and this may change repeatedly. Keeping a relationship affiliative, and thus supportive in some sense, means investing time and energy in that relationship. Grooming of one individual by another is an obvious indication of a willingness to make such an investment, as both time and energy are limited. When there are many possible partners and little time, individuals may be forced to invest disproportionately, focusing on the particular relationships that are currently or potentially more important (Dunbar 1988; Watts 2000b). The pattern of this investment is likely to be demonstrated by the interactions between individuals and will reflect the current, and possibly influence the future, state of relationships.

Male chimpanzees typically spend more time grooming than do females, and more time grooming each other (Wrangham 1986; Nishida & Hiraiwa-Hasegawa 1987). If particular males are desirable and the number of supportive relationships any individual can form is limited, there may be competition for these partners. Males holding high social status may be attractive grooming partners for other male chimpanzees, as high status individuals may be more able to provide support during agonistic encounters. In support of this idea, grooming between males appears (sometimes, but not always) to be directed 'up the hierarchy' (Simpson 1973; Takahata 1990; Hemelrijk & Ek 1991; Nishida & Hosaka 1996; Watts 2000b).

Where grooming does not conform to this pattern, it may be that other characteristics are more important in determining which individuals will be most desired (e.g. age: Bygott 1979), or that high status males need to develop relationships that buffer them from challenges to their status (Nishida & Hosaka 1996). If grooming is a behaviour that is used tactically – with flexibility to increase the future success of the actor most effectively – as it has been described (de Waal 1982; Nishida 1983; Koyama & Dunbar 1996; Nishida & Hosaka 1996), then the way it is distributed may appear more idiosyncratic and not be directly influenced by existing positions in a status hierarchy. Any effect of status may only be apparent when many males are present in a community (cf. Sambrook *et al.* 1995; Watts 2000b): with only a few males, each may be a valuable ally irrespective of his status.

For direct interactions, such as grooming and those that lead to establishment of social status, to take place, individuals must first associate with each other. Under a fission–fusion social system, male chimpanzees may have considerable freedom to alter these association patterns. If chimpanzee parties – the fluid subgroups that make up the social environment – are not passive aggregations of individuals drawn independently to the same resource patches, but the result of active behaviour aimed at altering the make-up of that social environment, then their composition will be a compromise between the optimal mix for each individual (Newton-Fisher 1999). Males may, for example, limit their association with individuals who might interfere with their grooming, or tend to associate with individuals with whom they could engage in grooming (Watts 2000b).

If this is the case, then association patterns themselves may well be an expression of male relationships, and a similar logic applies to the spatial positioning of individuals within parties. Although close proximity is an obvious prerequisite to grooming it may also allow for future possibilities of grooming, perhaps by simply providing an easy choice of grooming partner, or possibly by attempting to exclude others from approaching: a 'zone of control' approach, as seems to work for males who are mate-guarding females (personal observation). Greater distances between individuals, if a regular occurrence, may indicate a degree of avoidance between males who find themselves in the same party. An individual consistently on the periphery of parties may be suffering from some form of social exclusion.

This chapter examines the expression of male relationships for chimpanzees from the Sonso community in the Budongo Forest, Uganda. Data are drawn from the first behavioural study of chimpanzees from this community (Newton-Fisher 1997), which was the first in Budongo since the 1960s (Reynolds & Reynolds 1965; Sugiyama 1968; Suzuki 1971). In particular, this chapter addresses the extent to which observed patterns of association and proximity are indicative of male relationships, and the extent to which the expression of male relationships is governed by social status. The study took place during a period of social instability, with a change in alpha male, and this chapter examines changes in association, proximity, and grooming and their relationship to concurrent changes in social status.

METHODS

Study site

Male chimpanzees of the Sonso community in the Budongo Forest were studied intensively between August 1994 and December 1995. The forest, covering some 428 km^2 in western Uganda, is classified as moist, semi-deciduous, tropical forest and has a history of selective logging (Eggeling 1947; Synnott 1985; Plumptre 1996). The Sonso region (1°44′N, 31°33′E) lies close to the centre of the forest. Members of the study community were individually recognised, named and assigned a two-letter identification code. During this study period, the community contained 12 adult males (DN, VN, BK, MG, KK, MA, BY, MU, TK, CH, JM, NJ) and three adolescent males (ZT, AY, ZF). In addition, there were 14 adult females, and with births and immigrations, the total community size increased to 46 by the end of the study.

Data collection

Focal animal and scan sampling techniques (Altmann 1974) were used to systematically record behaviour and interactions of the 15 subjects between October 1994 and December 1995. During this period, individual males were sufficiently habituated to human observation so that they could be followed at close quarters on the ground, although they were noticeably nervous when unfamiliar observers attempted to follow them.

Scan sampling recorded the identity of each chimpanzee present. These data were used to determine patterns of association. A total of 5117 scan samples, collected every 15 minutes during systematic observation, were collected on parties containing at least one of the adult or adolescent males.

Individual males were subjects of 30-minute focal samples, during which a continuous, timed, record of the behaviour of both the focal and of his nearest neighbour was made. Behaviours relevant to these analyses were: grooming, pant-grunt vocalisations and agonistic acts (detailed below). The 30-minute sampling duration was chosen during initial observations as the maximum length of time for which subjects could be kept under continuous observation with intense data collection. Focal subjects were rotated according to a randomised list, and a minimum time interval of 15 minutes separated consecutive focal samples. Sampling of individual focal subjects was distributed across daylight hours. A total of 1023 30-minute focal samples were collected.

At the start of each focal sample, an instantaneous scan sample of the relative locations of all individuals within a 10-m radius of the focal male was taken. These were used to determine measures of within-party proximity. All proximity scans were separated by at least 45 minutes. The 10-m limit was imposed by habitat-related visibility constraints. A record of the movements of individuals within this area was maintained during the focal sample.

Data analysis

Association was defined as membership of the same party. To associate, or to be an associate, was to be in the same party. Any pair of subjects, whether or not in association, was regarded as a specific dyad. The tendency of males to associate was calculated using a standard twice-weight association index, which was then expressed as a Z-score. Details of the analysis technique and method used to reduce dependency between consecutive samples are presented elsewhere (Newton-Fisher 1999).

To measure positioning of individuals within parties, three Indices, expressed as Z-scores, were derived from proximity scans, after discarding records where no male was within 10 m of the focal. These Indices were (1) the frequency of each dyad existing as 'focal–nearest neighbour'; (2) this frequency weighted by the reciprocal of the distance (estimated by eye to the nearest 0.5 metre) separating the individuals; and (3) the frequency with which an individual was nearest, or second-nearest, neighbour to the focal. In generating Index 2, physical contact was arbitrarily assigned a distance of 0.5 m. Visual estimation of distance was as accurate as using a range finder ($\leq$25 m: $t_s=1.08$, df$=19$, $p=0.29$; $\leq$10 m: $t_s=0.35$, df$=6$, $p=0.74$). Using reciprocal distance attached greater importance to close proximity. As with association, a standard index formed the basis of these measures:

$$P_{AB}=(N_{AB}+N_{BA}) / (F_A+F_B) \quad (9.1)$$

[N_{AB}, number of observations with B as nearest neighbour to A; N_{BA}, number of observations with A as nearest neighbour to B; F_A, as number of focal samples of A; F_B, number of focal samples of B.]

Frequencies of grooming were extracted from the focal animal samples, recording which individual was performing the grooming, and which was being groomed. Breaks in grooming in excess of 1 minute were used to mark the end of individual bouts. Frequencies of grooming were corrected for different levels of association by dividing grooming frequency by the frequency of association for each particular dyad. To examine how dyads differed in their grooming, these frequencies were converted into Z-scores, which give a measure of grooming relative to the average level across all male dyads.

A cardinal index of social status was constructed by combining observations of pant-grunt vocalisations and agonistic acts. These were primarily from focal samples, but included relevant ad libitum observations. Five behaviour patterns were lumped together as 'agonistic': threats, displacements/supplants, displays, attacks and chases. Frequencies of interactions were corrected for different observation times of individuals, and ratios based on win/loss (agonistic) and received/given (pant-grunts) criteria were calculated (Fournier & Festa-Bianchet 1995). For each individual these were summed, such that pant-grunts received were combined with agonistic acts performed, and that sum transformed using natural logarithms of square-rooted data. This transformation normalised the data, and the resultant measure was a cardinal index of social status (Newton-Fisher 1997). The broad applicability of this method has been confirmed with data from the Mahale M-group (unpublished results). Individuals who cooperated repeatedly in agonistic interactions were defined as alliance partners. Only one such partnership could be clearly identified: DN & VN, who performed joint displays. A second partnership, between MG & BY, was thought by observers to exist but could not be unambiguously confirmed.

These indices, constructed from data collected throughout the 15-month study period, were used to investigate whether association, proximity and grooming could be regarded as expressions of the relationships between males, and what produced the variation in these behaviour patterns between individual males. Grooming was examined both at the group level and at the dyadic level using observations of grooming-partner choice. These were recorded during focal sampling. The choosing individual was the male that moved into close proximity to a second (the chosen individual); whichever male started to groom first was regarded as initiating the grooming bout.

Table 9.1. *Social status of males in the Sonso community*

	Full study				
Identity	Ordinal rank	Status level	Status index	Jan–Jun 1995 Status index	Jul–Dec 1995 Status index
DN	1	Alpha	2.64	2.15	2.80
VN	2	Beta	2.50	2.15	2.62
BK	3	High	1.12	0.33	1.27
MG	4	High	0.94	1.13	0.87
MA	5	High	0.66	0.62	0.38
CH	6	High	0.43	0.34	0.23
JM	7	High	0.27	−0.32	0.33
BY	8	Mid	−0.18	0.22	−1.28
KK	9	Mid	−0.27	−0.16	−0.36
NJ	10	Mid	−0.55	−0.92	−1.07
MU	11	Mid	−0.65	−1.23	−0.11
ZF	12	Low	−1.20	—	−1.25
TK	12	Low	−1.20	−0.92	−1.18
ZT	14	Low	−1.41	−1.41	−1.02
AY	15	Low	−1.47	−1.92	−1.56

Note:
Indices of social status are given for the full 15-month study period, and for each of the two 6-month blocks in 1995.

The analyses focused particularly on investigating the degree to which social status influenced variation in the way male relationships were expressed. Since this study took place while the alpha male was being replaced, data collected over the 12 months of 1995 were partitioned into two 6-month time blocks, which were then compared to investigate the impact of changing social status on the expression of male relationships.

Parametric statistics were used when the data appeared not to violate underlying assumptions; where these were violated non-parametric methods were used. Extensive use was made of matrix correlation procedures (Hemelrijk 1990) and Mantel regressions (Smouse *et al.* 1986) to avoid problems stemming from possible dependencies in the data. These tests determine significance by generating repeated permutations of the data matrices to produce a distribution of the test statistic against which its probability can be assessed (Adams & Anthony 1996). Matrix correlations are expressed as Kendall's τ values to make the strength of the correlation intuitively obvious (Dietz 1983; Hemelrijk 1990).

RESULTS

Status

In common with male chimpanzees elsewhere, the Budongo males showed clear dominance components in their relationships. While each male could be assigned a unique social status, these clustered into at least four distinct groups: low status (four males), mid-status (four males), high status (five males) and an alpha/beta alliance partnership (Table 9.1).

Patterns of association

For all adult and adolescent males, the average (median) number of associates was either seven or eight. With the exception of the old and disabled TK, all of the adult males spent similar amounts of time associating with other males, and were with at least one other adult male for over 90% of the time (mean percentage time as only male in party = $7.89 \pm 4.44\%$).

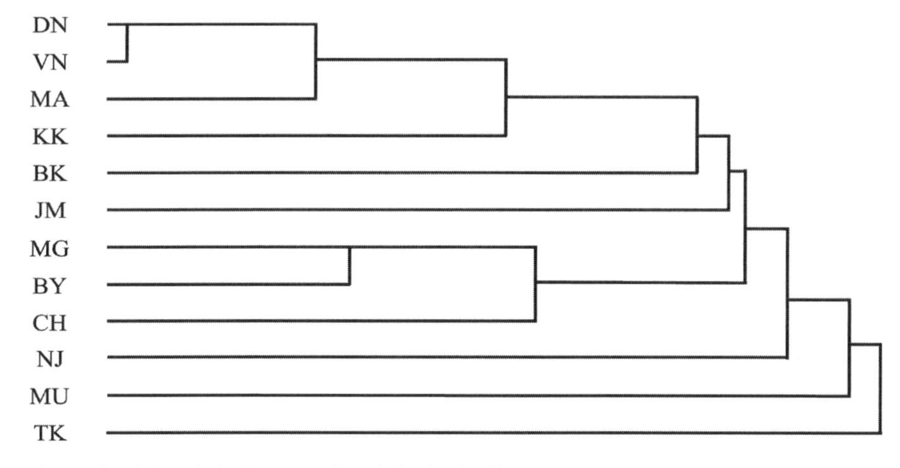

Fig. 9.1. Spatial proximity. Patterns of proximity for the 12 adult males in the study community. Dendrogram produced from UPMGA cluster analysis of weighted proximity indices.

WHAT INFLUENCED ASSOCIATION?

There was a large degree of variation in the tendency males had to associate with others; some dyads were more frequently together than others. Association indices for all 105 dyads varied from −0.88 to +3.87 (Z-scores, 15 individuals, calculated over 15 months). The presence of swollen 'oestrus' females increased association between males (Kruskal–Wallis $H=492.11$, df=4, $p<0.0001$), but this effect was asymptotic, with the average number of males remaining stable at around four while numbers of 'oestrus' females increased to two and above. Association patterns of male chimpanzees were apparently dominated by preferential association with particular other males. Males preferred to associate in small parties (Newton-Fisher 1999), and showed stronger association with higher status males ($K_r=609$, $\tau=0.45$, $p<0.01$).

Proximity

The average (median) distance between nearest adult males within a party was 1 m (inter-quartile range 0.5–3 m, $n=72$ dyads).

WAS PROXIMITY AN INCIDENTAL RESULT OF ASSOCIATION?

The frequency with which males were nearest neighbours varied between dyads (−1.83 to +3.72; Figure 9.1). Much of this variation was due to variation in association (Mantel regression: $Z=105$, $p<0.0001$; $r^2=0.65$). As might be expected, the proximity index based on nearest and second nearest neighbours was even more the product of association ($Z=110$, $p<0.0001$; $r^2=0.72$.). However, some dyads were frequently nearest neighbours within a party, but rarely close to each other, while others were infrequently nearest neighbours but, when they were, the distance between them was short. Known (DN & VN) and presumed (MG & BY) alliance partners, for example, had high positive values for proximity, greater than would be predicted on the basis of their positive association tendencies.

Variation in association explained far less of the variance in proximity when this was weighted by the reciprocal distance separating nearest neighbours (Mantel regression: $Z=82$, $p<0.05$; $r^2=0.40$), supporting the intuitive idea that distance between nearest neighbours was an important measure, and that spatial positioning within parties was an expression of male relationships beyond that demonstrated by association.

WHAT WAS THE EFFECT OF SOCIAL STATUS?

Relationships between high status males (excluding the alpha and beta males) influenced their proximity. Dyads consisting of two of these males had significantly greater values for weighted proximity than dyads containing only one, or none, of these males (Table 9.2: $F_{2,63}=6.86$, $p<0.01$; post hoc Scheffé test). This difference did not exist when the (unweighted) frequency of proximity was used ($F_{2,63}=0.83$, ns). While many of these dyads had close to average levels of association and proximity, when they were nearest neighbours, inter-individual distances were particularly short.

Male chimpanzees showed no apparent preference for individuals of similar status as nearest neighbours, instead demonstrating a preference for nearest neighbours who

Table 9.2. *Proximity of high status males*

	Mean residual	
Number of males	Unweighted	Weighted
None	-0.085 ± 0.51	-0.304 ± 0.67
One	-0.024 ± 0.60	-0.076 ± 0.72
Two	0.183 ± 0.69	0.626 ± 0.76

Note:
Mean residuals (± standard deviation) from regressions of unweighted and weighted frequencies of dyadic proximity against dyadic association, grouped by the number of high status (excluding alpha and beta) males comprising the dyad.

Table 9.3. *Status and nearest neighbour preference*

	Proximity measured by		
	Nearest male		Nearest/
Hypothesis Preference for:	Frequency	Weighted frequency	second nearest frequency
Same status level	0.06	0.09	0.03
Other status level	−0.04	−0.09	−0.01
Higher status level	0.16[b]	0.27[a]	0.23[a]
Lower status level	−0.17	−0.19	−0.25[a]

Notes:
Results of K_r tests (matrix correlations) are presented as τ values.
[a] Indicates significant results at $\alpha = 0.05$; [b] indicates an apparent trend (i.e. just not significant at $\alpha = 0.05$).

were of higher status and an avoidance of being one of the nearest two males to lower status males (Table 9.3).

Grooming frequencies

WHAT WAS THE EXTENT OF INDIVIDUAL VARIATION?

The frequency at which males groomed one another varied between dyads, even once grooming frequencies were corrected for differing levels of association. Some males groomed more frequently than others, showing marked preferences for particular partners (Figure 9.2). Whether or not males groomed often was indicated by the deviation in grooming frequency from an average level across all males, and variation in this measure within an individual across potential partners provided an indication of preference. The maximum (most positive) deviation for individual males varied from −0.33 to 3.25. The lowest of these (−0.33) was for the late adolescent male ZT, who showed extremely low levels of grooming other males. This may have been related to his age, but may also have been symptomatic of a problem which culminated in his death at the hands of other males in the community in 1998 (Fawcett & Muhumuza 2000).

WAS THIS VARIATION DUE TO ASSOCIATION PATTERNS?

Association patterns appeared to have little impact on grooming frequencies at a group level (Mantel regression: $Z = 108.14$, $p < 0.01$, $r^2 = 0.07$), although individual males groomed other males in a way that was related to their level of dyadic association ($K_r = 466$, $\tau = 0.37$, $n = 15$, $p < 0.01$) and received grooming similarly ($K_r = 323$, $\tau = 0.27$, $n = 15$, $p < 0.01$). Dyads with higher than average levels of association had higher than average levels of grooming. This pattern was apparent despite correcting the grooming frequencies for differing degrees of association. However, the seven dyads with grooming frequencies far greater than the average (Z-score >1.96) were not the same pairs of males who showed similarly high levels of association (six dyads), and while 18 dyads (grooming) and 15 dyads (association) showed frequencies greater than one standard deviation above average, only eight dyads showed these levels of both grooming and association.

WAS THIS VARIATION DUE TO PROXIMITY PATTERNS?

As might be expected, there were strong correlations between grooming and proximity. Males groomed individuals with whom they were in frequent proximity (frequency: $K_r = 220$, $\tau = 0.34$, $n = 12$, $p < 0.001$; weighted frequency: $K_r = 363$, $\tau = 0.56$, $n = 12$, $p < 0.001$), and received grooming similarly (frequency: $K_r = 170$, $\tau = 0.27$, $n = 12$, $p < 0.01$; weighted frequency: $K_r = 275$, $\tau = 0.43$, $n = 12$, $p < 0.001$). As grooming requires close proximity, investigating whether males simply groomed those that were nearby requires looking directly at the choice of grooming partners (below).

WAS THIS VARIATION INFLUENCED BY SOCIAL STATUS?

Higher status males appeared to groom more frequently ($K_r = 533$, $\tau = 0.43$, $n = 15$, $p < 0.001$) and also to be groomed

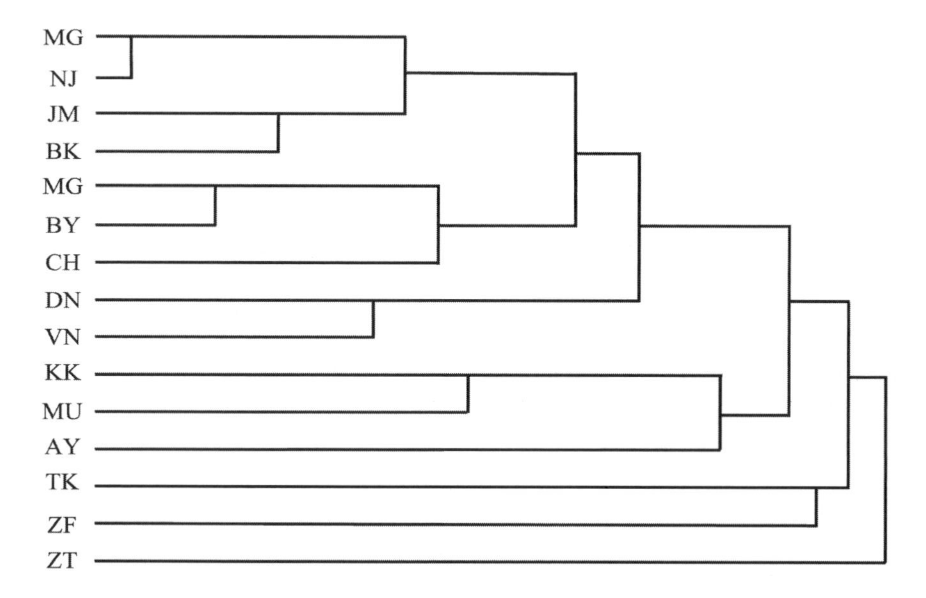

Fig. 9.2. Grooming. Patterns of grooming for the 15 adult and adolescent males in the study community. Dendrogram produced from UPMGA cluster analysis of grooming indices combining grooming performed and received.

more frequently, although this was a weaker effect ($K_r=251$, $\tau=0.21$, $n=15$, $p<0.05$). The relationship between social status and grooming was curvilinear (grooming: $r^2=0.51$, $F_{2,12}=8.15$, $p<0.01$; groomed: $r^2=0.71$, $F_{2,12}=18.23$, $p<0.001$), with the high, but not highest, status males performing most of the grooming. This pattern probably accounted for the reciprocation of grooming, both at the 'relative' level (a rank order distribution of grooming: $K_r=694$, $\tau=0.61$, $n=15$, $p<0.001$) and at the 'absolute' level (a distribution of grooming in proportion to that received: Mantel $Z=39544$, $n=15$, $p<0.001$).

Choice of grooming partners

Males were observed selecting a grooming partner from two or more potential partners on 81 occasions. In 41 cases the choice was between two males, in 40 cases between three to six males. In 74 cases (91.4%) the choosing individual was responsible for initiating grooming.

DID PROXIMITY AFFECT GROOMING CHOICE?

The distance between individuals appeared to have only minimal impact on the choice of grooming partner. In 31 cases (38%) at least one rejected potential partner was as close to the choosing male as was the selected partner. In the remaining 50 cases, the nearest individual was chosen in 24 cases, a more distant individual in 26 cases. The average (median) distance to the chosen individual was less than to rejected potential partners (Mann–Whitney $U=5235.5$, $n_{selected}=81$, $n_{rejected}=155$, $p<0.05$), but when selecting one of only two possible partners there were no significant differences in the distance between the choosing individual and the selected and rejected partners (Wilcoxon sign ranks test: $z=-0.99$, ns).

WERE HIGHER STATUS MALES MORE ATTRACTIVE GROOMING PARTNERS?

The average (median) status of chosen individuals was higher than those rejected (Mann–Whitney $U=464$, $n=36,36$, $p<0.05$). The number of potential grooming partners was correlated with the status of the choosing individual ($r_s=0.25$, $n=79$, $p<0.05$). This presumably was the result of the preference for higher status individuals as proximity partners, giving higher status males a slightly wider range of potential grooming partners. High status individuals were also approached more rapidly than low status individuals ($r=-0.026$, $n=60$, $p<0.05$). The response of the groomed individual (mutual grooming, reversal of grooming, terminating grooming bout) was not related to either the status difference between the individuals (Kruskal–Wallis: $H=5.6$, df$=3$, ns) or the status of the selecting individual ($H=1.52$, df$=3$, ns), although there was a trend for higher status individuals to terminate without returning the

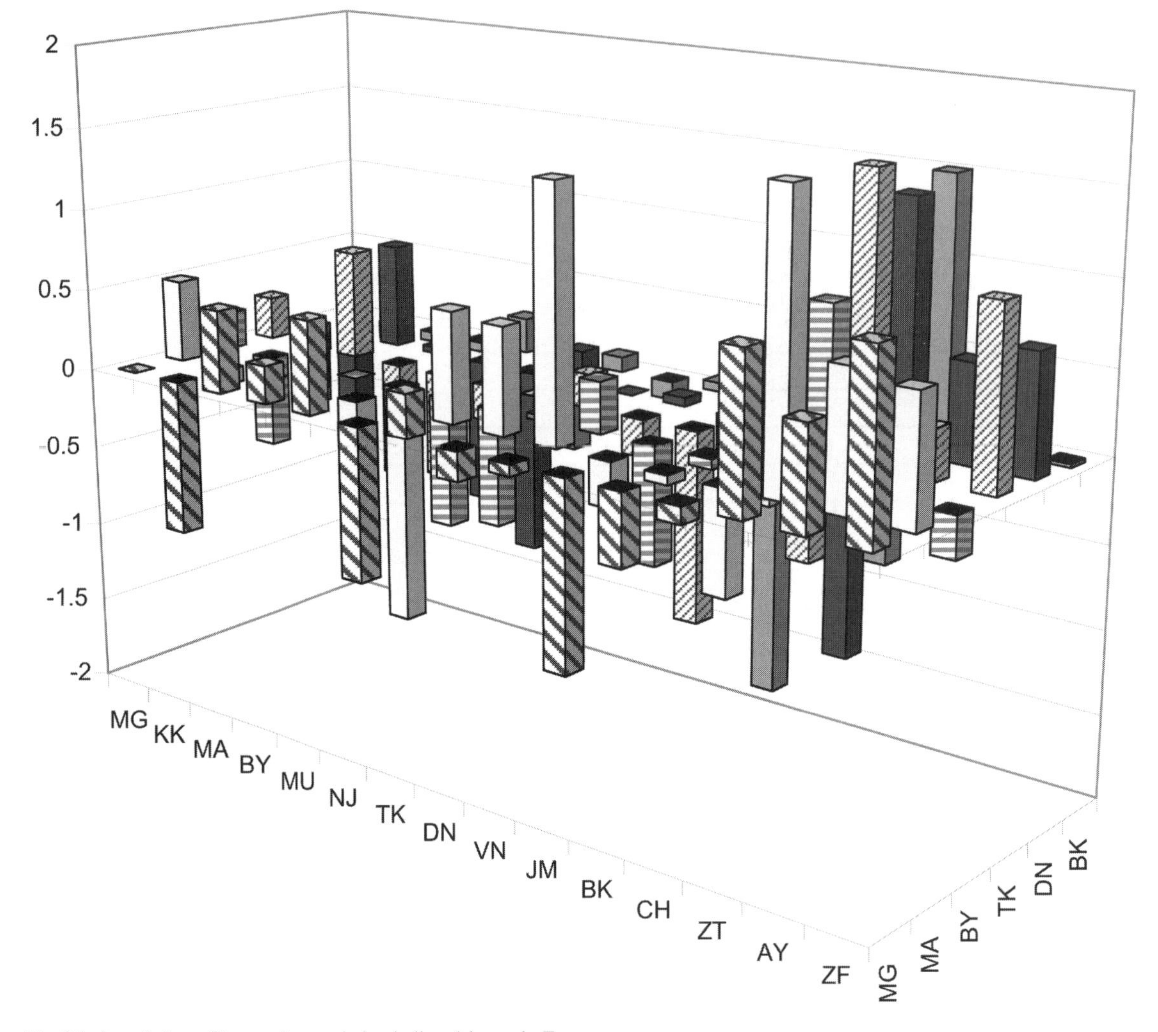

Fig. 9.3. Associations. Changes in association indices (change in Z scores) from the first 6 months in 1995 to the second 6 months. Association changes shown for six males (MG, MA, BY, TK, DN, BK) discussed in the text.

grooming, and for lower status individuals to return grooming (H=7.21, df=3, p=0.065).

WHAT DETERMINED GROOMING CHOICE?

Association patterns appeared to influence the choice of grooming partner. For the choices between two possible partners, dyadic association was the only significant predictor of choice in a stepwise logistic regression (χ^2=7.98, df=1, $p<0.01$) that included relative measures of distance, status, dyadic association and dyadic proximity and absolute measures of status. Of the two possible options, males chose to groom the individual with whom they more commonly associated.

Relationship dynamics

DID CHANGING SOCIAL STATUS EXERT AN EFFECT ON EXPRESSION OF MALE RELATIONSHIPS?

Relationships between adult males changed over the course of the study. Social status increased for some males and decreased for others. Patterns of association, proximity and grooming all changed. An increase in grooming was associated with an increase in proximity (K_r=222, τ=0.34, $p<0.001$), but changing patterns of association were not linked to changes in grooming (K_r=−31, ns) and proximity

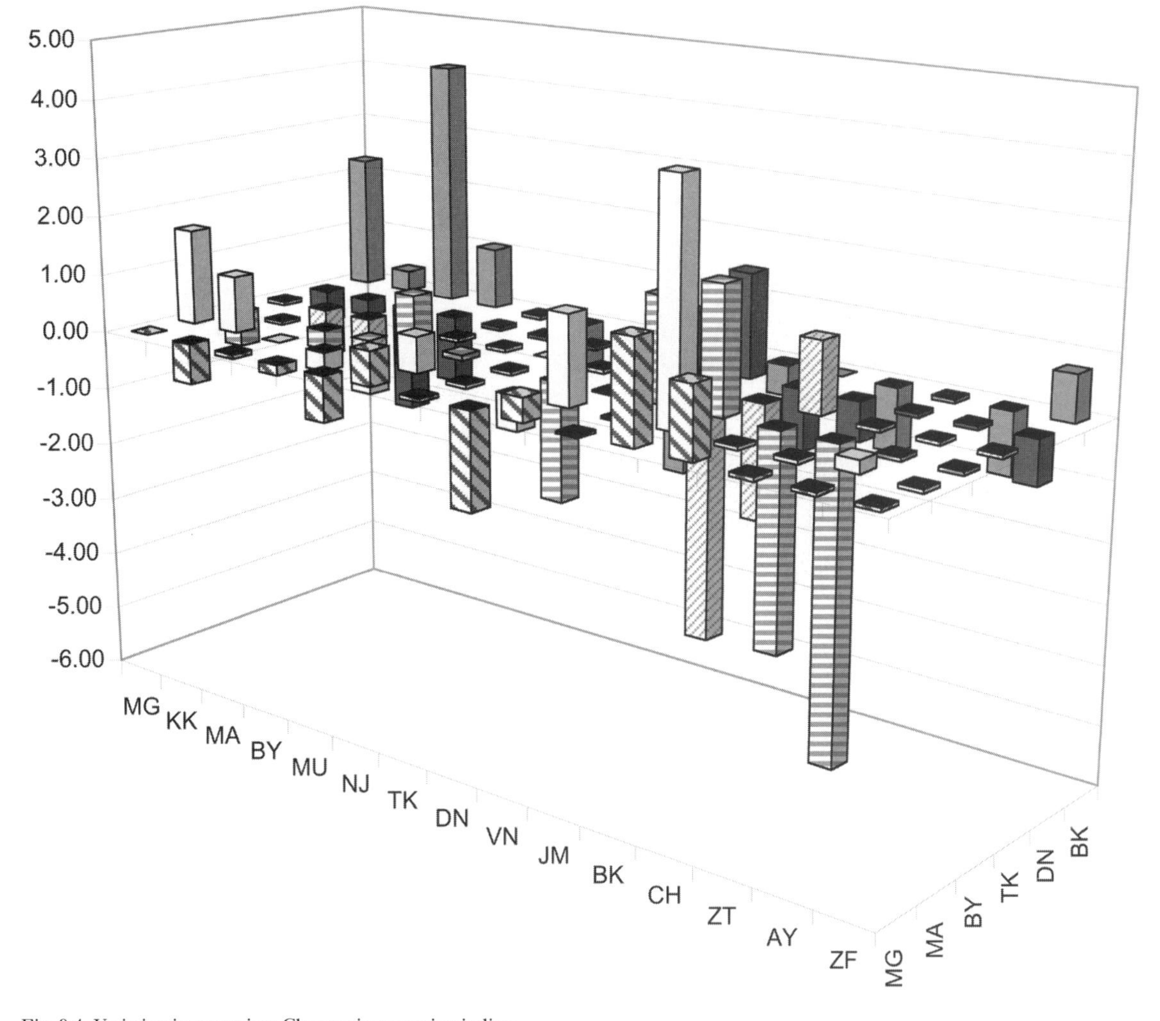

Fig. 9.4. Variation in grooming. Changes in grooming indices (change in Z scores) from the first 6 months in 1995 to the second 6 months. Changes in the amount of grooming performed shown for six males (MG, MA, BY, TK, DN, BK) discussed in the text.

(K_r=96, ns). Interestingly, the changes in association (Figure 9.3), proximity and grooming (Figure 9.4) were not related to changes in social status, whether males became more or less similar in status from the first 6-month period to the second, at least when looked at across all subjects (association: $K_r = -27$, ns; proximity: $K_r = -78$, ns; grooming: $K_r = -11$, ns).

WHAT CHANGES WERE SHOWN?

The deposed alpha male (MG) showed a reduction in association with seven of the twelve adult males, increasing association only with MA, MU, BY, and TK (an old, asocial male). MG and BY had absolutely high levels of association (Z-scores: Jan–Jun: +1.83, Jul–Dec: +2.06) and grooming (MG grooming BY: Jan–Jun: +1.71, Jul–Dec: +1.54; BY grooming MG: Jan–Jun: +2.60, Jul–Dec: +2.00), and both fell in status through 1995. BY provided coalitional support for MG and they may have been allies, although this was unconfirmed.

MG changed grooming patterns during 1995, particularly showing more frequent grooming of BK who was rapidly rising in status (MG grooming BK: Jan–Jun: −0.66, Jul–Dec: +1.01). MG received much of his grooming from BK, and also from MA who was one of the most prolific in his grooming of other males. MG showed slightly reduced levels of association (Jan–Jun: +1.89, Jul–Dec: +1.71), and was involved in fewer grooming interactions, with the new alpha male (DN) in the second half of 1995 (MG grooming DN: Jan–Jun: +1.74, Jul–Dec: +0.11; DN grooming MG: Jan–Jun: +0.87, Jul–Dec: +0.32). BY and BK also decreased the grooming they gave the new alpha male. BK changing grooming patterns, including a decrease in grooming DN (Jan–Jun: +1.82, Jul–Dec: −0.43), may have been part of a strategy to increase in social status. Over the course of 1995, BK moved from somewhere in the middle of the high status males (ranked 6 out of 15 males) to an unambiguous gamma male (3rd ranked) position. He used agonistic acts more frequently than all except the alpha and beta males, and was only seen pant-grunting to DN, never to the beta male VN.

The new alpha male maintained high levels of association with five of the adult males, and showed levels of association less than average only with TK and the adolescent males. The only large changes were an increase in association with MA (Jan–Jun: +2.40, Jul–Dec: +3.03) and a decrease in association with CH (Jan–Jun: +0.75, Jul–Dec: −0.53). The frequency with which DN groomed other males went down or remained unchanged, except for an increase in grooming his alliance partner VN (Jan–Jun: +0.69, Jul–Dec: +1.60) and the infrequent associate JM (Jan–Jun: −0.66, Jul–Dec: +1.06). These males both increased the frequency with which they groomed the alpha male (VN: Jan–Jun: +0.35, Jul–Dec: +2.06; JM: Jan–Jun: −0.66, Jul–Dec: +1.51).

DISCUSSION

In the general character of their interactions, as well as in their relationships, males from the Sonso community showed great similarities with male chimpanzees in other communities (Goodall 1986; Nishida & Hosaka 1996; Goldberg & Wrangham 1997; Mitani *et al.* 2000; Boesch & Boesch-Achermann 2000).

In the Sonso community, males with higher social status were preferred as associates and nearest neighbours, but only to a limited extent as grooming partners. However, higher status may have led to a wider choice of potential grooming partners and possibly allowed males to receive grooming without being obliged to reciprocate grooming within a particular bout: they still received grooming despite an apparent trend towards terminating bouts without reciprocation. Across bouts, however, grooming was strongly reciprocal. Males in this community distributed their grooming in proportion to the amount of grooming received. Such absolute reciprocation in chimpanzee grooming (in contrast to a simple rank-order response – relative reciprocation) was first demonstrated in captivity (Hemelrijk 1990) and subsequently in the wild for three communities (Newton-Fisher 1997; Watts 2000a). This reciprocation suggests that chimpanzees attach great importance to the allocation of grooming effort, which supports the idea that grooming is used as a social tactic (de Waal 1982; Nishida & Hosaka 1996), as does the existence of marked preferences for certain grooming partners.

Frequent associates were often, but not always, the preferred grooming partners for individual males, and the stronger the association between males, the greater the preference and the smaller the size of the parties in which they would come together. This would make sense if grooming were related to the current and future state of relationships between males. Smaller parties would provide an environment with fewer disturbances, reducing the probability that grooming would be disrupted by other males. Such separating interventions (de Waal 1982), whereby a male scatters a group of individuals grooming or resting in close proximity, have been observed in the Sonso community (unpublished data).

The use of separating interventions to scatter individuals in close proximity, but not otherwise interacting, suggests that proximity alone may be a component of male relationships and neither simply a precursor to grooming nor a by-product of association. Some dyads, particularly of higher status males, were found in close proximity when nearest neighbours within a party although their frequencies of association and proximity were close to average. These males may have been those most likely to need or exploit supportive relationships: above them in the status hierarchy was the well-established alliance between alpha and beta males, below them the young and old, unlikely to challenge for status. Any fickleness in relationships would most likely be occurring amongst these higher status males. Indeed it was these males who were responsible for most of the observed grooming, which is consistent with the idea that they are attempting to modify or maintain their relationships with each other. This contrasts with the established alpha–beta alliance. Although each was the other's most frequent grooming partner, neither showed particularly high levels of grooming.

Changes in social status did not appear to account for changes in the observed patterns of association, proximity, and grooming during the study. This appears to conflict with the observations that social status had an important influence on these behavior patterns. This conflict might be resolved by considering the time-scale of the study, and the changes in social status themselves. The change in alpha male was already underway at the beginning of 1995, with the alliance between DN & VN apparently well established. DN was not clearly alpha male until the second half of the year, and may have continued to strengthen his position thereafter; he remains alpha male in 2001 (V. Reynolds, personal communication).

The changes in association, proximity and grooming may have been the result of males opportunistically shifting the investment they made in each of their relationships during this period of social instability. Each male is likely to have attempted to strengthen supportive relationships while neglecting others. Most – if not all – of the males may have been attempting to do this with the same range of partners at the same time. Such opportunism, possibly compounded by individual personality (Murray, 1995) and relationship history, may have produced the apparently idiosyncratic variation showing no systematic relationship to status changes. A longer time frame, covering periods of stability and instability, may be required to examine properly the impact of social status on the dynamics of relationships.

These analyses support, or are least consistent with, a view of male chimpanzee sociality in which the observed grouping and interaction patterns are products of the relationships between individuals, with association and proximity being the direct, rather than incidental, result of male behaviour. The idiosyncratic variation across individuals and over time, the link between stronger association and smaller party size (Newton-Fisher 1999), and the relationships between association, proximity and grooming all support this view.

The availability of reproductive opportunities in the form of females showing sexual swellings is a factor influencing party size in Sonso, as in other communities (Gombe: Goodall 1986; Stanford *et al.* 1994, Taï: Boesch & Boesch-Achermann, 2000). These females are attractive to males, and their presence seems likely to increase not only party size but also the number of males in the party. However, the pattern seen in this study was more complex than this.

The presence of females with sexual swellings did increase the number of males but only on average from two males to four males, even if more than one swollen female was present. Sexual swellings last for around 9–12 days in chimpanzee females, with ovulation occurring in the last 2 or 3 days of maximum swelling (Tutin & McGinnis 1981; Goodall 1986; Nishida & Hiraiwa-Hasegawa 1987; Takahata *et al.* 1996; Boesch & Boesch-Achermann 2000). Parties, in contrast, can change in composition from moment to moment; average duration for unchanging party composition varies between communities, but is on the scale of minutes, not days (Boesch & Boesch-Achermann 2000, Table 5.1). For the Sonso community, average duration was 14 minutes (V. Reynolds, personal communication).

Male associations could, therefore, change repeatedly without compromising access to swollen females, particularly if sexual swellings were not entirely reliable as indicators of ovulation and males could rely on sperm competition. For most males monopolising access to swollen females may not be a feasible option. Only males of highest social status appear to be successful at monopolising access, even where mate-guarding is conducted cooperatively (Watts 1998). For the rest, their time may be more profitably spent cultivating relationships with an eye to the future, and grabbing copulations when they can.

If the presence of swollen females increases tension and aggression between males (de Waal 1986; Shefferly & Fritz 1992), then males may attempt to avoid such tension by altering their associations. For individual males, the size of the party may be less important than the number of other males, and the number of males may be less important than the identities of those other males. Associating with preferred individuals while avoiding others may provide a male with greater flexibility in forming, maintaining and modifying relationships, although this flexibility will inevitably be constrained by the association patterns of the other individuals in the community.

REFERENCES

Adams, D. C. & Anthony, C. D. (1996). Using randomization techniques to analyse behavioural data. *Animal Behaviour*, 51, 733–8.

Altmann, J. (1974). Observational study of behaviour: sampling methods. *Behaviour*, 49, 227–65.

Boesch, C. & Boesch-Achermann, H. (2000). *The Chimpanzees of the Taï Forest*. Oxford: Oxford University Press.

Bygott, J. D. (1979). Agonistic behavior, dominance, and social structure in wild chimpanzees of the Gombe National Park. In *The Great Apes*, ed. D. A. Hamburg & E. R. McCown, pp. 405–28. Menlo Park, CA: Benjamin/Cummings.

Chapman, C. A., White, F. J. & Wrangham, R. W. (1993). Defining subgroup size in fission–fusion societies. *Folia Primatologica*, 61, 31–4.

de Waal, F. B. M. (1982). *Chimpanzee Politics*. New York: Harper and Row.

de Waal, F. B. M. (1986). The brutal elimination of a rival among captive chimpanzees. *Ethology and Sociobiology*, 7, 237–51.

de Waal, F. B. M. (1996). Conflict as negotiation. In *Great Ape Societies*, ed. W. C. McGrew, L. Marchant & T. Nishida, pp. 159–72. Cambridge: Cambridge University Press.

Dietz, E. J. (1983). Permutation tests for association between two distance matrices. *Systematic Zoology*, 32, 21–6.

Dunbar, R. I. M. (1988). *Primate Social Systems*. London: Croom Helm.

Eggeling, W. J. (1947). Observations of the ecology of the Budongo Rain Forest, Uganda. *Journal of Ecology*, 34, 20–87.

Fawcett, K. & Muhumuza, G. (2000). Death of a wild chimpanzee community member: possible outcome of intense sexual competition. *American Journal of Primatology*, 51, 243–7.

Fournier, F. & Festa-Bianchet, M. (1995). Social dominance in adult female mountain goats. *Animal Behaviour*, 49, 1449–59.

Goldberg, T. & Wrangham, R.W. (1997). Genetic correlates of social behavior in wild chimpanzees: evidence from mitochondrial DNA. *Animal Behaviour*, 54, 559–70.

Goodall, J. (1965). Chimpanzees of the Gombe stream reserve. In *Primate Behavior*, ed. I. Devore, pp. 425–73. New York: Holt, Rinehart and Winston.

Goodall, J. (1986). *The Chimpanzees of Gombe*. Cambridge, MA: Belknap Press of Harvard University Press.

Hasegawa, T. & Hiraiwa-Hasegawa, M. (1983). Opportunistic and restrictive mating among wild chimpanzees in the Mahale Mountains, Tanzania. *Journal of Ethology*, 1, 75–85.

Hemelrijk, C. K. (1990). Models of, and tests for, reciprocity, unidirectionality and other social interaction patterns at a group level. *Animal Behaviour*, 39, 1013–29.

Hemelrijk, C. K. & Ek, A. (1991). Reciprocity and interchange of grooming and 'support' in captive chimpanzees. *Animal Behaviour*, 41, 923–35.

Hinde, R. A. (1976). Interactions, relationships and social structure. *Man*, 11, 1–17.

Kawanaka, K. (1990). Alpha males' interactions and social skills. In *Chimpanzees of the Mahale Mountains*, ed. T. Nishida, pp. 149–70. Tokyo: University of Tokyo Press.

Koyama, N. F. & Dunbar, R. I. M. (1996). Anticipation of conflict by chimpanzees. *Primates*, 37, 79–86.

Mitani, J. C., Merriwether, D. A. & Zhang, C. (2000). Male affiliation, cooperation and kinship in wild chimpanzees. *Animal Behaviour*, 59, 885–93.

Murray, L. E. (1995). Personality and individual differences in captive African apes. Unpublished PhD thesis, University of Cambridge.

Newton-Fisher, N. E. (1997). Tactical behaviour and decision making in wild chimpanzees. Unpublished PhD thesis, University of Cambridge.

Newton-Fisher, N. E. (1999). Association by male chimpanzees: a social tactic? *Behaviour*, 136, 705–30.

Nishida, T. (1979). The social structure of chimpanzees of the Mahale Mountains. In *The Great Apes*, ed. D. A. Hamburg & E. R. McCown, pp. 73–122. Menlo Park, CA: Benjamin/Cummings.

Nishida, T. (1983). Alpha status and agonistic alliance in wild chimpanzees. *Primates*, 24, 318–36.

Nishida, T. & Hiraiwa-Hasegawa, M. (1987). Chimpanzees and bonobos: cooperative relationships between males. In *Primate Societies*, ed. B. B. Smuts, D. L. Cheney, R. M. Seyfarth, R. W. Wrangham & T. T. Struhsaker, pp. 165–77. Chicago: University of Chicago Press.

Nishida, T. & Hosaka, K. (1996). Coalition strategies among adult male chimpanzees of the Mahale Mountains, Tanzania. In *Great Ape Societies*, ed. W. C. McGrew, L. Marchant & T. Nishida, pp. 114–34. Cambridge: Cambridge University Press.

Plumptre, A. J. (1996). Changes following 60 years of selective timber harvesting in the Budongo Forest Reserve, Uganda. *Forest Ecology and Management*, 89, 101–13.

Reynolds, V. & Reynolds, F. (1965). Chimpanzees of the Budongo Forest. In *Primate Behaviour*, ed. I. Devore, pp. 368–424. New York: Holt, Rinehart and Winston.

Sambrook, T. D., Whiten, A. & Strum, S. C. (1995). Priority of access and grooming patterns of females in a large and a small group of olive baboons. *Animal Behaviour*, 50, 1667–82.

Shefferly, N. & Fritz, P. (1992). Male chimpanzee behavior in relation to female ano-genital swelling. *American Journal of Primatology*, 26, 119–131.

Simpson, M. J. A. (1973). The social grooming of male chimpanzees. In *Comparative Ecology and Behaviour of Primates*, ed. R. P. Michael & J. H. Crook, pp. 411–506. London: Academic Press.

Smouse, P. E., Long, J. C. & Sokal, R. R. (1986). Multiple regression and correlation extensions of the Mantel test of matrix correspondence. *Systematic Zoology*, 35, 627–32.

Stanford, C. B., Wallis, J., Mpongo, E. & Goodall, J. (1994). Hunting decisions in wild chimpanzees. *Behaviour*, 131, 1–18.

Sugiyama, Y. (1968). Social organisation of chimpanzees in the Budongo Forest, Uganda. *Primates*, 10, 197–225.

Sugiyama, Y. & Koman, J. (1979). Social structure and dynamics of wild chimpanzees at Bossou, Guinea. *Primates*, 20, 323–39.

Suzuki, A. (1971). Carnivory and cannibalism observed among forest living chimpanzees. *Journal of the Anthropological Society of Japan*, 79, 30–48.

Synnott, T. J. (1985). *A Checklist of the Flora of the Budongo Forest Reserve, Uganda, with Notes on Ecology and Phenology*. Commonwealth Forest Institute Occasional Paper 27, Oxford Forestry Institute, Oxford.

Takahata, Y. (1990). Social relationships among male chimpanzees. In *The Chimpanzees of the Mahale Mountains*, ed. T. Nishida, pp. 133–48. Tokyo: University of Tokyo Press.

Takahata, Y., Ihobe, H. & Idani, G. (1996). Comparing copulations of chimpanzees and bonobos: do females exhibit proceptivity or receptivity? In *Great Ape Societies*, ed. W. C. McGrew, L. Marchant & T. Nishida, pp. 146–55. Cambridge: Cambridge University Press.

Tutin, C. E. G. (1979). Mating patterns and reproductive strategies in a community of wild chimpanzees (*Pan troglodytes schweinfurthii*). *Behavioral Ecology and Sociobiology*, 6, 29–38.

Tutin, C. E. G. & McGinnis, P. R. (1981). Chimpanzee reproduction in the wild. In *Reproductive Biology of the Great Apes*, ed. C. E. Graham, pp. 239–64. New York: Academic Press.

Uehara, S., Hiraiwa-Hasegawa, M., Hosaka, K. & Hamai, M. (1994). The fate of defeated alpha male chimpanzees in relation to their social networks. *Primates*, 35, 49–55.

Watts, D. (1998). Coalitionary mate guarding by male chimpanzees at Ngogo, Kibale National Park Uganda. *Behavioral Ecology and Sociobiology*, 44, 43–55.

Watts, D. (2000a). Grooming between male chimpanzees at Ngogo, Kibale National Park. I. Partner number and diversity and grooming reciprocity. *International Journal of Primatology*, 21, 189–210.

Watts, D. (2000b). Grooming between male chimpanzees at Ngogo, Kibale National Park. II. Influence of male rank and possible competition for partners. *International Journal of Primatology*, 21, 210–38.

Wrangham, R.W. (1986). Ecology and social relationships in two species of chimpanzee. In *Ecology and Social Evolution: Birds and Mammals*, ed. D. I. Rubenstein & R. W. Wrangham, pp. 352–78. Princeton, NY: Princeton University Press.

Wrangham, R.W. & Smuts, B.B. (1980). Sex difference in the behavioural ecology of chimpanzees in the Gombe National Park, Tanzania. *Journal of Reproduction and Fertility, Supplement*, 28, 13–31.

10 • Dynamics in social organization of bonobos (*Pan paniscus*)

GOTTFRIED HOHMANN & BARBARA FRUTH

INTRODUCTION

In the last 20 years, chimpanzee (*Pan troglodytes*) and bonobo (*Pan paniscus*) research has produced contrasting pictures of these two sister species. Chimpanzee society has been characterized as male dominated and structured by a linear hierarchy amongst males, with more egalitarian relations amongst females. Male dominance rank is often based on alliances with other males and exerted by intense aggression (Riss & Goodall 1977; Goodall 1986; McGrew 1996; Watts 1998). Parous females, except when they are in oestrus, tend to avoid travelling with males in order to prevent aggression and to improve their foraging efficiency (Williams *et al.*, Chapter 14; Wrangham, Chapter 15). Consequently, parties are relatively small and often male biased (Nishida 1979; Wrangham 1986; Wrangham *et al.* 1992; Boesch & Boesch-Achermann 2000). In comparison, bonobo society is characterized by egalitarian relations between the sexes (Furuichi 1997) and females may collaborate to defend food sources against males (Idani 1991; Parish 1994; Hohmann & Fruth 1996; Vervaecke *et al.* 2000). Males establish dominance relationships with each other but aggression amongst males and between the sexes is less intense than in chimpanzees, and conflicts are often settled in a non-agonistic way (Furuichi & Ihobe 1994; de Waal 1995). Compared to chimpanzees, bonobo parties are large and biased towards females. Recently, the behavioural contrasts between the two *Pan* species have been questioned (Stanford 1998) for various reasons. These include scarcity of information from wild bonobos, and the failure to compare data from wild chimpanzees with what is known from studies of captive bonobos. The goal of this chapter is to address these issues by providing new data on the social organization of wild bonobos.

Information on social organization and social relations of bonobos comes principally from two field sites, Lomako and Wamba, both located in the province of Equateur (Democratic Republic of Congo). Studies from both sites report that cohesion of community members is high and parties are relatively large. Parties usually contain mature individuals of both sexes with more females than males (Kano 1982; White 1988; Fruth 1995). Males tend to disperse when food is scarce, but females remain gregarious (White 1998). Although aggressive encounters between community members occur, social relations among males and among females are characterized by affiliative relationships, frequent non-agonistic displays of social status, and dominance relations that are enforced by agonistic displays (Furuichi & Ihobe 1994; Parish 1994; Furuichi *et al.* 1998; Hohmann & Fruth 2000). Although the bonobos at these two sites share a number of social features, several population-specific differences do exist. First, White (1992, 1996) noted site-specific differences in party size as well as affiliative and aggressive relations between the sexes. Second, unlike Lomako, where communities regularly split into smaller parties (Fruth 1995), community members at Wamba remain together most of the time (Kano 1992). Third, females at Lomako but not at Wamba hunt duikers (*Cephalophus*) and share the meat with other party members (Hohmann & Fruth 1993; Fruth & Hohmann, Chapter 17). Another difference might also exist concerning the adult sex ratio: in one community at Lomako the adult sex ratio was strongly female biased (Fruth 1995; Hohmann *et al.* 1999), while numbers of males and females in two communities at Wamba were almost the same (see Kano 1992, p. 80). However, whether this applies to other communities at the two sites is not clear. The precise reason for these differences in social organization and social relations between the two sites remain to be discovered, but the observed variation demonstrates flexibility on various levels.

While the intraspecific variation of bonobo sociality has gained little attention, differences between chimpanzees and bonobos have been emphasized in a number of studies (Wrangham 1986; de Waal 1988; Kano 1992). Several explanations for interspecific differences are based on presumed differences in the intensity of feeding competition that result from differences in the availability and distribution of food. One hypothesis related the relaxed feeding competition of bonobos to the consumption of herbaceous vegetation, abundantly available to bonobos (Badrian & Badrian 1984). An alternative hypothesis assumed that bonobos use food patches that are larger or more densely distributed than those in areas occupied by chimpanzees (White & Wrangham 1988). Another hypothesis suggested that the availability of high quality food is less affected by seasonal variation for bonobos (Malenky & Wrangham 1994). Initial comparison of selected data sets from both species supported all three hypotheses, but re-analysis of these data sets confirmed none of them (Chapman *et al.* 1994). In sum, the differences in grouping patterns between both *Pan* species remain to be explained.

However, before starting to tackle this question, one can go back one step and ask how different are the grouping patterns of the two *Pan* species? There are good reasons for posing this particular question: new data from long-term projects on both species are now available for further interspecific comparison. For example, information on bonobos living in a forest–savanna mosaic reveal new dimensions of the behavioural ecology of this species and challenge previous views about their limited social and ecological flexibility (Myers-Thompson, Chapter 4; Furuichi & Hashimoto, Chapter 11).

Differences in grouping patterns are the result of variation in social relations. Group size is determined by the distribution of food and the intensity of risks from predators or conspecifics. According to socio-ecological theory, relationships among females are set by ecological conditions, while those relations among males are related to mating opportunities (Emlen & Oring 1977). Elaborating on this perspective, Wrangham (1980) and van Schaik (1989) suggested that social relations among females are determined by the nature of resource competition: if resources are potentially monopolizable, then direct or 'contest' competition predominates, differences in social status affect resource accessibility, and behaviours signalling dominance are advantageous. In contrast, if access to resources is not limited, then indirect or 'scramble' competition predominates, differences in social status are less likely to affect accessibility, and signals of dominance become less beneficial. Additionally, female–female relationships vary along three interrelated social dimensions (Sterck *et al.* 1997): (1) transfer (female exogamy vs. female philopatry); (2) type of dominance relations (despotic vs. egalitarian); and (3) mode of expressing dominance (tolerance vs. intolerance).

Tests of the socio-ecological model have focused on published data from various primate species and, occasionally, other mammals and birds (van Schaik 1996; Sterck *et al.* 1997; Crook 1964; Ingold & Marbacher 1991; Wrangham *et al.* 1993). The model explains a large proportion of the variety in social relations within and between the sexes, but not all species follow its predictions. Bonobos appear to be one of these outliers. Their food patches vary in size, but are often large and occur at relatively high density. According to Wrangham's (1980) female-bonded model, this should promote philopatry, but female bonobos undergo natal transfer and probably even secondary transfer (Furuichi 1989). Van Schaik's (1989) ecological model can accommodate female transfer, but only if predation pressure is low. Considering the scarcity of information on mortality, the latter assumption remains speculative. Female transfer means that close genetic ties among resident females are absent (Hashimoto *et al.* 1996; Gerloff *et al.* 1999). Nevertheless, resident females are reported to establish affiliative relationships with each other (Idani 1991), contrary to the predictions of the socio-ecological model. Sterck *et al.* (1997) have suggested that alliances among unrelated females may be a strategy against infanticide. However, since information about the causes of infant mortality is almost non-existent, the function of female alliances remains ambiguous.

How consistent are the patterns of bonobo society, and what accounts for any variations? The aim of this study is to investigate variation in social dynamics; to assess how variation in food availability, the presence of oestrous females, and encounters between communities influence this variation; and to compare results of this study with information from other populations of *Pan*.

METHODS

Data collection

Records on party size and behavioural interactions were collected by event sampling (Altmann 1974) during six field seasons (1993–98), when identification of subjects was complete. Annual field seasons ranged between 2 and 7 months in duration. The study involved members of the Eyengo community inhabiting the eastern part of the Lomako study site (Badrian & Badrian 1984), as well as individually known and unknown members of neighbouring communities. The number of adult and adolescent female co-residents ranged from 11 to 14; the number of adult and adolescent male co-residents ranged from four to seven. More details on demographic changes of this community are presented elsewhere (Fruth 1995; Hohmann *et al.* 1999).

Party size and composition

The term party size refers to all adolescent and adult individuals travelling, resting or foraging in relative proximity. Infants and juveniles are not included. Age estimates of individuals born before the onset of this project (January 1991) were based on dentition, body size, testes size, swelling patterns and other morphological traits. Separation of immatures (infants, juveniles) from matures (adolescent, adult) follows the classification used by Hashimoto (1997). Counts of party members were made when individuals were relatively stationary (e.g. foraging in one food patch or in several small adjacent patches), when travel parties crossed gaps in the vegetation, and at nest sites. Within the same day, a new score for party size/composition was given when individuals joined/left the party under observation. Relative party size refers to the number of party members in relation to the number of community members.

Dyadic associations

To assess dyadic associations, a Twice-Weight association index (dyadic association index: DAI) (Cairns & Schwager 1987) was applied. Indices are functions of the time two individuals (A and B) attend the same party (T_t), the time A was travelling without B (T_a), and the time B was travelling without A (T_b): $\mathrm{DAI}=(T_t)/(T_t)+(T_a)+(T_b)$. Records represent events and do not consider the actual time two given individuals spend together. In species living in a fission–fusion society, observed associations may be random. Therefore, we used records of party attendance to calculate a second index that compares observed dyadic associations with random associations. This method was used before with a subset of the data presented below (Hohmann *et al.* 1999). Only one record per day was used and records collected within the same field season were lumped and analysed separately from records of other field seasons. The frequency of party attendance of a certain individual (p_i) was calculated by dividing the number of records when the individual was present (n_i) by the total number of records of party size (n). The expected value ($p_{ij\,\mathrm{exp}}$) that two given individuals (i and j) attended the same party was calculated by multiplication of the observed attendance of both individuals ($p_i \times p_j$). To decide whether the observed values of dyadic association deviated from the expected values, a randomisation test written by Lamprecht & Hofer using a RAN2 generator (Lamprecht 1985; Press *et al.* 1992) was applied. Values of random associations (RA) were computed by running 1000 random, uniformly distributed trials for each individual of a given dyad. The simulation showed how often an observed association (OA) was expected by chance. Using statistically significant deviations ($p<0.05$, two-tailed), dyads were assigned to one of three classes: close associates (OA>RA), random associates (OA=RA), or separatists (OA<RA). The latter class accounted for the smallest proportion (<5%) and will not be considered in this study.

Food availability

Assessments of the availability of food were available for 11 months. Data are based on monthly records of the number of trees with ripe fruit from 1400 trees of three line transects, and 100 additional food trees selected within the home range of the Eyengo community. The latter sub-sample included important food tree species that were either missing or underrepresented in the line-transect sample. All trees were tagged and most (95%) were taxonomically identified. In this study we consider only presence and absence of ripe fruit.

Number of oestrous females

Oestrous was defined as the period when genital swellings of females were maximally tumescent. Records of changing patterns of genital swellings were available from four field seasons (1995–98). Ratings considered two criteria: firmness and skin surface appearance. Four stages were distinguished: (1) tissue flaccid, swelling wobbly; skin dry with deep wrinkles; (2) tissue viscous, swelling wobbly, skin wrinkled; (3) tissue firm, swelling elastic, skin soft with reduced or no wrinkles; (4) tissue swollen, swelling sturdy during locomotion, skin shiny. Labial occlusion (Dahl 1986), moist skin, and vaginal secretion were used as additional characteristics to distinguish stage four from stage three.

Intercommunity encounters

Visual encounters between members of different communities were seen 23 times. An encounter was scored when parties from two communities were close enough that observers could see members of both. All but two encounters ended with spatial separation. In two cases, the two parties nested within 100 m of each other. Cases when single, non-resident individuals 'visited' the Eyengos for short periods but subsequently left, or those cases where individuals transferred into the community, are not included here. Aggressive interactions during encounters were recorded as events and divided by the number of mature individuals present at the site of the encounter. In five cases, encounters involved two very large parties and the total number of individuals had to be estimated. The number of aggressive events per individual present during the encounter was converted into hourly rates. This encounter rate of aggression was compared with rates of aggression by members of the Eyengo community during corresponding times one day following the encounter (matched sample).

Statistics

All statistical analyses are non-parametric and two-tailed using statistics packages provided by SPSS 8.0 for Windows and SsS[Rubisoft 1998]. Statistics are given as means ($\bar{x} \pm$ SD) or medians (M) and ranges. P-values are two-tailed.

RESULTS

Party size

Records from 827 parties collected during 24 cumulative months of field work gave an average of 4.85 ± 2.63 mature individuals (M = 4, range 1–16). Small parties (one to five members) accounted for the largest proportion (65%), followed by large parties (six to ten members, 32%) and very large parties (more than ten members, 3%). Party size varied within and between field seasons (Kruskal–Wallis test for all months, χ^2: 96.028, df = 29, $p<0.001$, Figure 10.1). Observed changes in party size may reflect changes in community size (Boesch 1996). Party size tended to increase when community size decreased, but this relationship was not significant (relative party size: $f_{1,4,5}=5.4$, $p=0.081$, $r^2=0.574$).

Party composition

Bonobos travelled alone, in mixed-sex parties, and in unisex parties. Proportions of different party types were as follows: bisexual parties (69%), all female (21%), all male (8%). Solitary individuals accounted for 36 (4%) of the records (females 17 times vs. males 19 times). Combining the data from the six field seasons, the median for the number of male party members was 1 (range: 0–6) and the median for females was 3 (range: 0–11). Comparison of the socionomic sex ratio of the community with the average socionomic sex

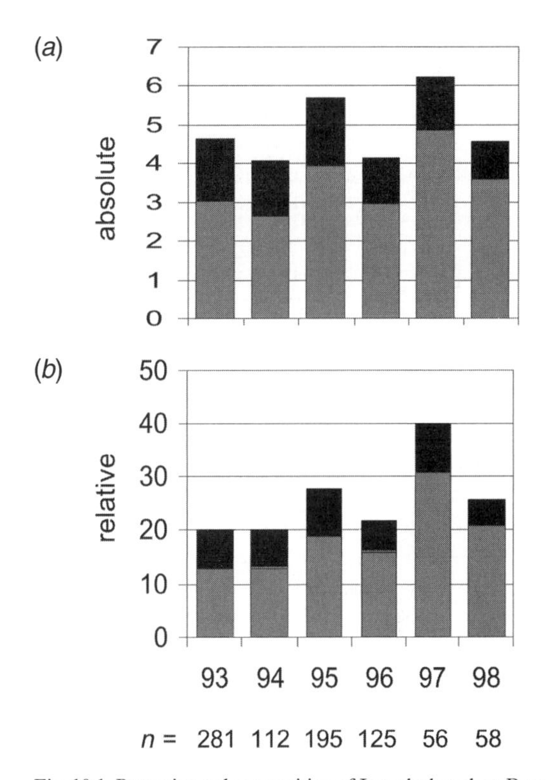

Fig. 10.1. Party size and composition of Lomako bonobos. Bars show average party size of mature bonobos as counted in the field (*a*) and in relation to the total number of the mature community members (*b*). The light part of the bar refers to females, the dark part to males. The *y*-axis gives the average number of mature male and female party members per field season (*a*) and the percentages that these parties represent in relation to community size (*b*). Figures on the *x*-axis indicate different field seasons by year (top) and the corresponding number of parties (bottom) used in this analysis.

ratio of parties shows relatively high correspondence during the first three field seasons and discrepancies during the last three field seasons (Figure 10.2).

Associations

In order to assess general patterns of spatial association, association indices were calculated for each dyad (average dyadic association). Average association indices were relatively consistent within field seasons but fluctuated between different field seasons (Table 10.1). Relatively strong dyadic associations between community members (association time >25%) occurred most frequently between females, followed

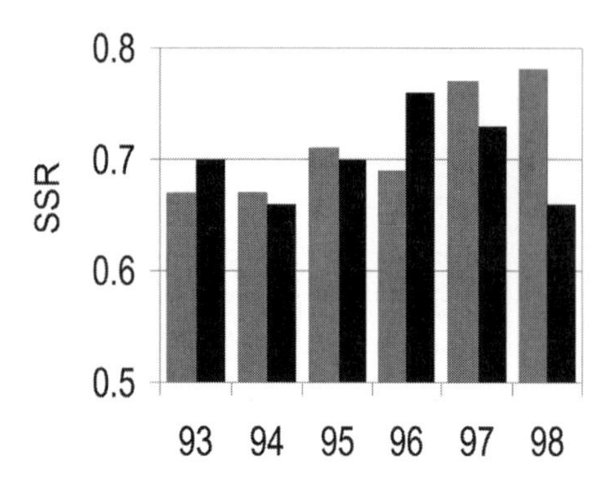

Fig. 10.2. Average socionomic sex ratio within parties (grey bars) in comparison to corresponding values of the entire community (black bars) across different field seasons. The *y*-axis shows the value of the socionomic sex ratio ($SSR = n_{\text{adult + adolescent females}} / n_{\text{adult + adolescent community members}}$) across different field seasons (*x*-axis).

Table 10.1. *Dyadic association indices of Lomako bonobos and four populations of chimpanzees, and variation of association indices of Lomako bonobos across six field seasons*

Population	M–M	M–F	F–F	Source
Gombe	0.24	0.07	0.05	Goodall 1986
Kibale	0.18	0.01	0.08	Wrangham *et al.* 1992
Taï	0.35	0.12	0.11	Boesch & Boesch-Achermann 2000
Lomako	0.18	0.22	0.25	This study
93	0.11	0.12	0.12	
94	0.13	0.13	0.11	
95	0.18	0.17	0.21	
96	0.18	0.21	0.19	
97	0.27	0.40	0.55	
98	0.23	0.31	0.32	

by mixed-sex dyads and male dyads (Table 10.1). In two field seasons (1997 and 1998), average values of dyadic association were higher than in the years before (Table 10.1).

Temporal associations of individuals living in a fission–fusion society (Kummer 1968) are likely to represent a compromise between costs and benefits of grouping (Wrangham *et al.* 1993; Chapman *et al.* 1994). Individual associations may be affected by various factors, including kinship (Wrangham 1980), age of infants (Golla *et al.* 1999; Williams *et al.*, Chapter 14), reproductive strategies (Smuts 1985), distribution of food resources (Rubenstein 1986), and dominance relations (East *et al.* 1993). Depending on the level of investigation, tests and corresponding predictions are likely to apply to certain factions rather than the entire community. For example, in species where males are philopatric and females disperse, investigations of the impact of kinship on dyadic associations will focus on males and ignore female–female associations. In this study, we wanted to test the strength of association between all dyads independent of the age, sex and reproductive status of individuals. Therefore, we compared observed values of dyadic association with values of random association (corrected dyadic association). Although the null hypothesis is admittedly arbitrary, the procedure was chosen as a first and general screening to identify dyads that travel together more often than would be expected by chance.

Close associations occurred most often between females ($\bar{x} \pm SD = 13.0 \pm 1.9$, $n = 6$ field seasons), followed by mixed-sex dyads ($\bar{x} \pm SD = 11.7 \pm 6.3$) and male–male dyads ($\bar{x} \pm SD = 2.0 \pm 1.1$). Considering the total number of dyads and the total number of close associations amongst two individuals as observed during this study, only males and females formed dyads of close association more often than would be expected (Wilcoxon sign test, $Z = -2.207$, $p = 0.027$). With mother–son dyads ($n = 2$) excluded, observed numbers of close associations between males and females did not deviate from expected values (Kolmogorov–Smirnov test, Z values: 0.805 for female–male, 1.025 for female–female, 0.782 for male–male, all p-values ns).

Causes for within-community variation of party size

FOOD

Phenology data from 11 months revealed that monthly party size was not correlated with the amount of ripe fruit available (Spearman rank correlation, $r_{\text{s trees with ripe fruit ha}^{-1}} = 0.469$, $r_{\text{s average party size}} = 1.0$, $p = 0.146$, Table 10.2). White (1998) used the number of different fruits eaten by bonobos as an indirect measure of fruit availability. Regressions of the number of different fruit species consumed by bonobos in each month against average party size, male party size, and female party size respectively, showed that only female party size increased moderately in the predicted way ($B = 0.05$, $SE = 0.022$, $\beta = 0.46$, $T = 2.317$, $p = 0.031$).

OESTROUS FEMALES

The number of females showing signs of maximum tumescence varied between zero and four per day ($M = 1.4$,

Table 10.2. *Number of trees with ripe fruit and average party size of the corresponding month*

Month	Trees with fruit ha^{-1}	Trees with ripe fruit ha^{-1}	Average party size
January	3.7	0.5	2.20
February	6.5	0.5	1.90
March	11.8	3.8	4.80
April	15.5	5.5	3.80
May	1.7	0.5	1.80
June	1.3	0.2	3.11
July	5.4	0.5	3.42
August	2.7	0.7	5.56
September	3.2	0.8	5.41
October	9.5	4.3	3.41
November	14.0	1.8	4.47

$n = 197$ days and 417 parties) and the number of male party members increased with the number of oestrous females (Spearman rank correlation, $r_{\text{s number of swollen females}} = 0.24$, $r_{\text{s male party size}} = 1.0$, $p = 0.048$, $n = 68$). Parties in which mating occurred were larger than average (M = 7 members, $n = 207$ parties versus M = 4 members, $n = 827$ parties) and the number of males but not females was significantly higher in mating parties (Mann–Whitney U test, $U = 12374.5$, $Z = -3.065$, $p = 0.002$, Figures 10.3(*a*) and (*b*)).

Mixed-sex close associations most frequently involved cycling females (Figure 10.4). As another measure of the impact of mating opportunities on male–female associations we used data from females giving birth during the course of this project ($n = 9$) and the number of unrelated males who were closely associated with these females, before and after birth, respectively. We found that females had more male close associates during the months before giving birth than after parturition (Wilcoxon sign test, $Z = -2.121$, $p = 0.034$, $n = 9$ females, Figure 10.5).

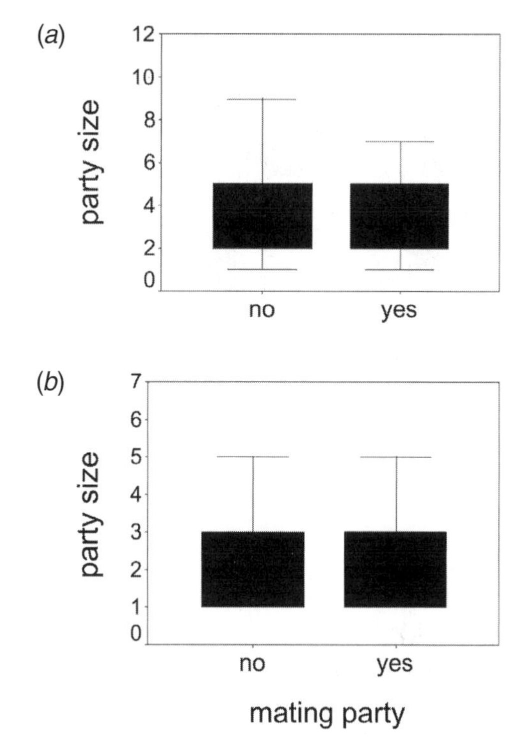

Fig. 10.3. Median number of females (*a*) and males (*b*) in parties without mating activity (left plot) and with mating activity (right plot). Horizontal bar in box indicates median with 75th percentile above and 25th percentile below. Bar outside box indicates range.

INTERCOMMUNITY ENCOUNTERS

Interactions between members of different communities included mating, social grooming, displays, and other agonistic behaviors. Aggressive exchanges that involved vocal and gestural signals and motor displays (e.g. branch dragging) occurred in 20 out of 23 (87%) encounters, and eight (35%) involved physical aggression between members of different communities. Compared with the matched samples collected during corresponding time periods on the day following a community encounter, the rate of agonistic behaviours was higher during encounters (Wilcoxon sign test, $Z = -2.803$, $p = 0.005$, $n = 20$). Party size was larger on the day following a community encounter (Wilcoxon sign test, $Z = -2.748$, $p = 0.006$, $n = 23$, Figure 10.6) due to a significant increase in male membership (Wilcoxon sign test, $Z = -2.909$, $p = 0.004$, $n = 23$). Female membership did not change in the same way (Wilcoxon sign test, $Z = -0.737$, $p = 0.461$, $n = 23$).

DISCUSSION

Compared to bonobos at Wamba, the size of parties at Lomako was small (see Table 10.3). Parties usually contained more females than males and the proportion of all-female parties was high, while that of all-male parties was low. More obvious than fluctuations in size was the change in party composition across field seasons: until 1996 the socionomic sex ratio of parties was relatively balanced and similar to or

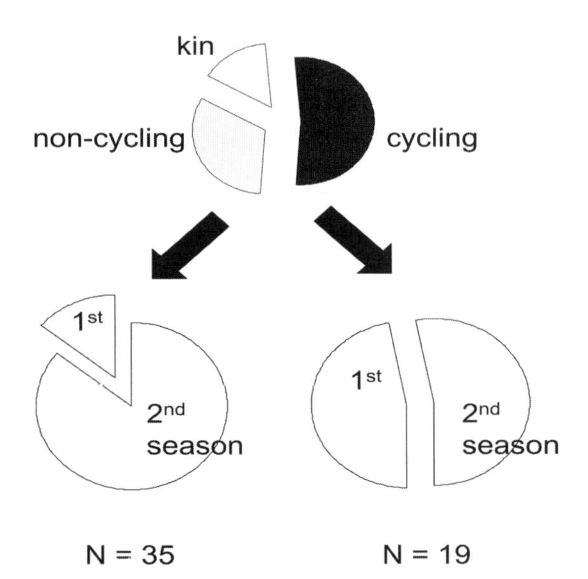

Fig. 10.4. Female partners in mixed-sex close associations. Within the category of closely associated male–female dyads (CA $_{\text{mixed sex}}$), cycling females (black section of graph) were more often involved than non-cycling females and kin. About 50% of close associations between cycling females and mature males exceeded one field season (below right), whereas the majority of close associations between non-cycling females with males lasted for two consecutive seasons or longer (below left).

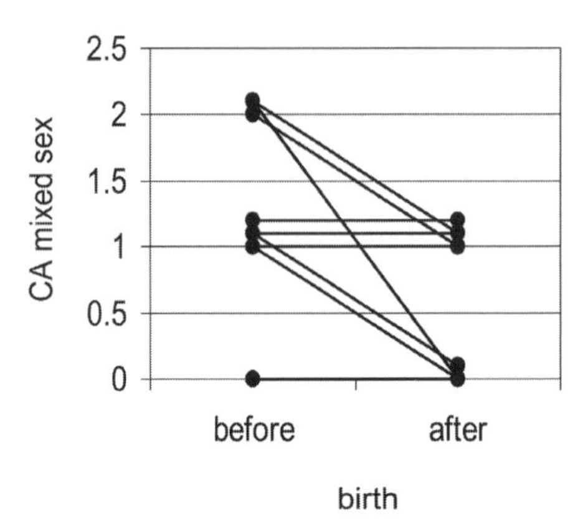

Fig. 10.5. Relation between the reproductive state of females and the number of male close associates. The *y*-axis shows the number of males closely associated to a given female ($n = 9$) before and after this female gave birth.

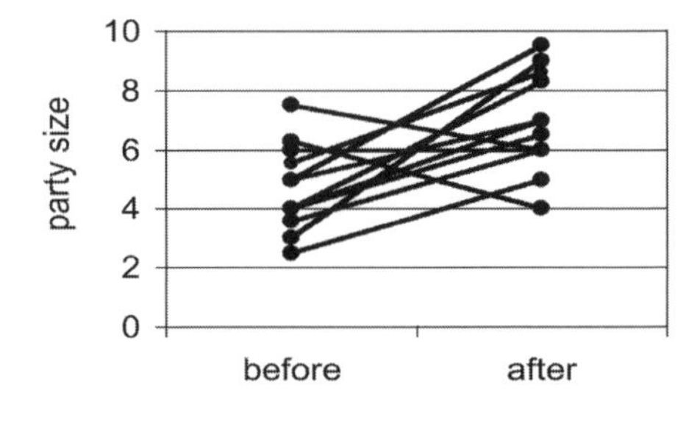

Fig. 10.6. Party size and intercommunity encounter. Size of nine parties that were involved in encounters with bisexual parties of another community. The *y*-axis gives the number of mature individuals the day before (left) and the day after (right) the intercommunity encounter.

below the socionomic sex ratio of the community. Later, socionomic sex ratio of parties rose and exceeded that of the community. This development coincided with demographic changes in the Eyengo community (Hohmann 2001, table 3), but may also reflect a decline in the tolerance of females for travelling together with males: in September 1997, two strange males were seen in the home range of the Eyengo community and, after 11 months, one of them was associated on a daily base with Eyengo females. During the first period of incursion, relations between residents and strangers were hostile and social tension was higher than usual. At the same time, some females appeared to avoid travelling in the mixed parties that included the two strangers. Judged by this coincidence, the shift of party composition may have been related to the increase of intraspecific aggression (Hohmann 2001). Nothing is known about the history of the two males who joined the Eyengo community. Judged by their body size, testes size and dentition, both males were adult, and it is almost certain that they had been associated with another community before their appearance here. Following conventional definitions of dispersal (e.g. Harcourt 1978), the behaviour of at least one of the two males qualifies as community transfer. No such case has ever been reported before. However, at both sites, Lomako and Wamba, males and females disappear from their community with similar frequencies (Thompson-Handler & Malenky 1993; Furuichi *et al.* 1998) and, with few exceptions, the fate of these individuals remained ambiguous. Although the evidence for occasional male transfer in *Pan paniscus* is still anecdotal, it raises the possibility that patterns of dispersal are more flexible. Male chimpanzees from the same community cooperate in mate guarding (Watts 1998), hunting (Boesch & Boesch 1989) and communal aggression (Wrangham 1999). This tendency of co-resident males to cooperate with each other

Table 10.3. *Average and relative party size of chimpanzees and bonobos*

Site	Average party size	Relative party size (%)	Reference
Bossou	4.0	20	Sakura 1994
Gombe	5.6	9	Goodall 1968
Kanyawara	5.6–6.1[a]	13	Wrangham *et al.* 1992, 1996
Ngogo	10.3	9	Mitani *et al.* Chapter 7
Mahale	6.1	—	Nishida 1979
Mahale	12.7[a]	30	Matsumoto-Oda *et al.* 1998
Taï	10.0	13	Boesch & Boesch-Achermann 2000
Wamba	16.9	29	Kuroda 1979
Wamba	14[a]	—	Kano 1992
Lomako	6.6[a]	—	White 1996
Lomako	4.9[a]	27	This study

Note:
Relative party size was calculated from the information on average party size and community size as given in the corresponding reference.
[a] Adult and adolescent only.

corresponds with xenophobia and intense physical aggression. Male bonobos from the same community may also engage in frequent affiliate interactions (Furuichi & Ihobe 1994), but agonistic aid during conflicts between members of different communities has never been reported and, compared with chimpanzees, aggression by bonobo males appears to be of low intensity. This brings up the question of why male bonobos are less aggressive than male chimpanzees. In a non-adaptive scenario, Parish (1994) proposed that intrasexual cooperation enables females to dominate males. Hemelrijk (2000) presented a model that showed that the more egalitarian dominance relations between the sexes could result from the strong cohesion of social groups. However, it is also possible that males may benefit from being less aggressive. In male sifakas, long-term reproductive success is thought to depend on '. . . competitive submission to females . . .' instead of aggressive male–male competition (Richard 1992, p. 403). In spotted hyenas, females are dominant over males (East *et al.* 1993). Since older males are dominant over younger males (but not females) and dominant males are more tolerated by females than subordinates, reproductive success of males is a function of tenure rather than status (East & Hofer 1991). Observations from Lomako on solicitation of copulation and mate guarding suggest that female choice plays a crucial role in mating activity (unpublished data), and future studies should address the relation between mating success, reproductive success and aggression of individual males.

Patterns of dyadic associations of male bonobos in this study did not differ from chimpanzees at Gombe and Kanyawara, but were lower than in Taï (Table 10.1). In some but not all field seasons, associations between females and between the sexes exceeded those of chimpanzees. Two seasons (1997 and 1998) had exceptionally high association rates between females as well as between the sexes. However, during the first two seasons, figures from Lomako are very similar to those from Taï and Kanyawara, respectively (Table 10.1). The striking variation in party size and composition over time at Lomako demonstrates considerable fluctuation in gregariousness and needs to be explained.

The frequencies of close associations between mixed-sex dyads and same-sex dyads observed within each field season corresponded with expected values. In a previous study, we found that mixed-sex dyads had higher association rates than other dyads (Hohmann *et al.* 1999), but when mother–son dyads are excluded (this study) this bias disappeared. Consistent with the previous study, females at Lomako do not form close associations with other females more often than do other dyads. The patterns of spatial association among Lomako bonobos correspond well with reports on dyadic relationships among bonobos at Wamba (Furuichi & Ihobe 1994). Reports on social grooming by Furuichi & Ihobe (1994) and by Hohmann *et al.* (1999) also suggest strong bonds between male bonobos (Wamba) and between the sexes (Wamba and Lomako). Similar relationships seem to be absent in East African chimpanzee populations (e.g. Kanyawara chimpanzees, Wrangham *et al.* 1992) but exist among chimpanzees of the Taï forest (Boesch & Boesch-Achermann 2000). Given the size and composition of parties, Lomako bonobos are within the range of variation of chimpanzees. However, differences remain on two levels: the proportion of female party members and the frequency of mixed parties.

Previous studies have specifically proposed that gregariousness is higher in bonobos than in chimpanzees because bonobos have access to large and superabundant food patches (White & Wrangham 1988; White & Lanjouw 1992). The finding that large parties have a more diverse daily diet than small ones may be due to a higher level of competition that forces some individuals to consume food items that would be neglected on other days. Variation of the number of trees with ripe fruit did not correlate with variation in party size. One could argue that our records from food transects covered a very brief period (11 months) and, therefore, the lack of correlation between party size and food availability does not exclude a potential causal relationship between these two factors. However, information from chimpanzees from Taï forest (Anderson *et al.*, Chapter 6) and from Ngogo (Mitani *et al.*, Chapter 7), respectively, provide alternative explanations for seasonal variation of party size, namely the presence/absence of oestrous females. Both studies provide evidence suggesting that, alone, the number of trees with fruit may not provide an accurate estimator of fruit abundance.

Several studies of chimpanzees have shown that party size was related to the presence and number of oestrous females (Matsumoto-Oda *et al.* 1998; Boesch & Boesch-Achermann 2000; Anderson *et al.*, Chapter 6; Mitani *et al.*, Chapter 7). In this study, party size appeared to be independent of the actual number of oestrous females, but the proportion of males per party increased when oestrous females were present. The reproductive stage of females also affected long-term associations between males and females. As shown above, the number of males closely associated with a particular female was higher when she had fertile cycles than when she was in an advanced stage of pregnancy or lactating. Associations between non-cycling females and males tended to last longer than associations between cycling females and males.

In chimpanzees, encounters with members of neighbouring communities are remarkably hostile and may have fatal consequences (Nishida & Hiraiwa-Hasegawa 1985; Wrangham 1999; Muller, Chapter 8). In bonobos, encounters between members of different communities are sometimes described as peaceful (Idani 1990; Kano 1992, p. 203), but aggressive displays also occur in this context (Kano 1992). Explanations for the intraspecific violence in chimpanzees have suggested links between hunting and eating the meat of other species on the one hand and intraspecific violence on the other (Eibl-Eibesfeldt 1975; van Hooff 1990). Chimpanzees sometimes do eat victims of infanticide, either alone or together with other residents (Boesch & Boesch-Achermann 2000; Watts & Mitani 2000). At Lomako, bonobos hunt duikers and share the meat; division and consumption of the prey may last several hours, participants show clear signs of arousal, and aggressive interactions are common (Hohmann & Fruth 1993, Fruth & Hohmann, Chapter 17). While collective hunting has not yet been seen, the intense arousal that accompanies meat sharing in bonobos may not be so different from that of chimpanzees.

At Lomako, encounters were characterized by high rates of aggression, and the number of males per party was higher on the day following an encounter than on the day when members of both communities met. Unlike chimpanzees, male bonobos at Lomako have never been seen to make border patrols; home ranges of neighbouring communities overlap extensively (Fruth 1995), and encounters with strangers can take place in the geographic centre of the community range (Hohmann 2001). Although the probability of an encounter is likely to increase with the distance from the centre of the home range, no part of the home range seems to be safe from incursion by strangers. Recent work at Lomako revealed substantial floristic heterogeneity throughout the range (Boubli *et al.* unpublished data). This raises the possibility that, at least seasonally, members of different communities may compete for the same resource and, by doing so, increase the likelihood of encounters.

Recently, Wrangham (1999) proposed a theoretical framework to explain variation of intraspecific violence: his hypothesis predicts that party size of bonobos from different communities should be more similar than in chimpanzees. Quantitative data on corresponding party sizes from different communities are not available, but repeated observations on encounters between parties of residents and two strangers (Hohmann 2001) suggest differences in group size: during encounters, the size of parties of residents ranged from five to twelve mature individuals. In this study, most encounters ended when intruders withdrew voluntarily or in response to charges from the Eyengos (ignoring the possible impact of human observers on non-habituated individuals). However, in three cases when all-female parties (two, three and five adult members) encountered mixed-sex parties from another community, residents showed signs of fear (screaming, hiding) and escaped by running away on the ground, remaining silent and highly vigilant for the rest of the day (personal observation). It should be noted that in all three cases, one or more resident females carried small infants. The increase of all-female parties at a time when two strange males appeared in the home range of the Eyengo community also indicates that females consider the presence of strangers as a potential risk.

During this study, for unknown reasons resident males were sometimes seen to charge females with small infants. Such male attacks provoked counter aggression by both females and males against the aggressor that were of remarkable intensity, and it is not clear whether or not these incidents were unsuccessful attempts of infanticide. The only case when the death of an infant was caused by another bonobo occurred when a newborn was kidnapped by a resident female. Consequently, any discussion on infanticide in wild bonobos remains speculative, but it seems premature to exclude the possibility that bonobos commit infanticide. Depending on the number, sex and, perhaps, identity of residents travelling together, encounters with members of another community may pose a high risk. Most encounters happened at large food patches with ripe fruit, either at single large trees (e.g. *Ficus*) or in areas with a high density of relatively small-sized food trees (e.g. *Polyalthia*). Large parties can deplete even large trees and another encounter at the same site is unlikely. In areas with a high density of food trees with ripe fruit, parties of different communities may meet several times. Under these conditions, the increase in the number of male party members may function to protect other community members from aggression by strangers, and may allow control of access to resources. Considering the hostile behaviour during intercommunity encounters and the increase in party size, we suggest here that bonobos may seek protection from other community members at times when conflicts are likely to occur. This idea was proposed by Wrangham (1986) but never tested with data from wild bonobos and still remains a challenge for future work.

Taken together, the results of this study suggest similarities between forest chimpanzees and bonobos with respect to party size and associations patterns. Evidence for the similarity in social cohesion in chimpanzees and bonobos comes from studies at Taï and applies to both associations within and between the sexes (Boesch & Boesch-Achermann 2000). This chimpanzee population lives in a dense forest habitat that is structurally and probably even floristically very similar to some of the forest ranges that are occupied by bonobos.

Female bonobos travel together more often than female chimpanzees, but measurements of dyadic associations do not support the idea of female bonding that exceeds that of some chimpanzee populations. Variation in size and composition of parties appeared to be related to the presence of oestrous females and the probability of encountering members of other communities. In certain patterns of social organization, obvious differences remain between bonobos from Lomako and Wamba on the one hand and chimpanzees living in more open habitats on the other. Although the current range of bonobo distribution is still unknown, it includes mosaics of forest and grassland in the south and swampy, lowland forests between the Lomami river and Lualaba river in the north-east (Hohmann & Eriksson 2001; Myers Thompson, Chapter 4). Given the wide range of environmental conditions in which bonobos live, it is our opinion that the behavioural diversity of bonobos resembles that of chimpanzees. As behavioural data from hitherto unknown populations become available, the existing gap between the two *Pan* species may close.

ACKNOWLEDGEMENTS

We thank the Ministère de l'Education Nationale, République Démocratique du Congo for permission to conduct fieldwork at Lomako. Financial support for fieldwork came from the Max-Planck-Society (MPG), Deutsche Forschungsgemeinschaft (DFG), and the University of Munich (LMU). The advice on statistics given by Jürg Lamprecht, Heribert Hofer and Redouan Bshary is gratefully acknowledged. For comments on earlier drafts we would like to thank David C. Watts, Ulrich Reichard, Linda F. Marchant and Christophe Boesch. The data shown above were presented at the conference on 'Behavioural Diversity in Chimpanzees and Bonobos' that was supported by the Max Planck Society, Wenner-Gren Foundation for Anthropological Research, Deutsche Forschungsgemeinschaft (DFG), and Bayerisches Staatsministerium für Wissenschaft, Forschung und Kunst.

REFERENCES

Altmann, J. (1974). Observational study of behavior: sampling methods. *Behaviour*, 49, 227–67.

Badrian, A. & Badrian, N. (1984). Social organization of *Pan paniscus* in the Lomako Forest, Zaire. In *The Pygmy Chimpanzee: Evolutionary Biology and Behavior*, ed. L. S. Susman, pp. 325–46. New York: Plenum Press.

Boesch, C. (1996). Social grouping in Taï chimpanzees. In *Great Ape Societies*, ed. W. C. McGrew, L. F. Marchant & T. Nishida, pp. 101–13. Cambridge: Cambridge University Press.

Boesch, C. & Boesch, H. (1989). Hunting behavior of wild chimpanzees in the Taï National Park. *American Journal of Physical Anthropology*, 78, 547–73.

Boesch, C. & Boesch-Achermann, H. (2000). The chimpanzees of the Taï Forest. *Behavioural Ecology and Evolution*. Oxford: Oxford University Press.

Cairns, S. J. & Schwager, S. J. (1987). A comparison of association indices. *Animal Behaviour*, 35, 1454–69.

Chapman, C. A., White F. J. & Wrangham R. W. (1994). Party size in chimpanzees and bonobos: a re-evaluation of theory based on two similarly forested sites. In *Chimpanzee Cultures*, ed. R. W. Wrangham, W. C. McGrew, F. B. M. de Waal & P. G. Heltne, pp. 41–57. Cambridge, MA: Harvard University Press.

Crook, J. H. (1964). The evolution of social organization and visual communication in weaverbirds (Ploceinae). *Behaviour*, 10, 1–178.

Dahl, J. F. (1986). Cyclic perineal swelling during the intermenstrual intervals of captive female pygmy chimpanzees (*Pan paniscus*). *Journal of Human Evolution*, 15, 369–85.

de Waal, F. B. M. (1988). The communicative repertoire of captive bonobos (*Pan paniscus*) compared to that of chimpanzees. *Behaviour*, 106, 183–251.

de Waal, F. B. M. (1995). Sex as an alternative to aggression in the bonobo. In *Sexual Nature, Sexual Culture*, ed. P. R. Abramson & S. D. Pinkerton, pp. 37–56. Chicago: University Press of Chicago.

East, M. L. & Hofer, H. (1991). Loud calling in a female-dominated mammalian society: II. Behavioral contexts and functions of whooping of spotted hyenas, *Crocuta crocuta*. *Animal Behaviour*, 42, 651–69.

East, M. L., Hofer, H. & Wickler, W. (1993). The erect 'penis' is a flag of submission in a female-dominated society: greetings in Serengeti spotted hyenas. *Behavioral Ecology and Sociobiology*, 33, 355–70.

Eibl-Eibesfeldt, I. (1975). *Krieg und Frieden aus der Sicht der Verhaltensforschung*. München: Piper Verlag.

Emlen, S. T. & Oring L. W. (1977). Ecology, sexual selection, and the evolution of mating systems. *Science*, 197, 215–23.

Fruth, B. (1995). *Nests and Nest Groups in Wild Bonobos (*Pan paniscus*): Ecological and Behavioural Correlates*. Aachen: Verlag Shaker.

Furuichi, T. (1989). Social interactions and the life history of female *Pan paniscus* in Wamba, Zaire. *International Journal of Primatology*, 10, 173–97.

Furuichi, T. (1997). Agonistic interactions and matrifocal dominance rank of wild bonobos (*Pan paniscus*) at Wamba. *International Journal of Primatology*, 18, 855–75.

Furuichi, T. & Ihobe, H. (1994). Variation in male relationships in bonobos and chimpanzees. *Behaviour*, 130, 212–28.

Furuichi, T., Idani, G., Ihobe, H., Kuroda, S., Kitamura, K., Mori, A., Enomoto, T., Okayasu, N., Hashimoto, C. & Kano, T. (1998). Population dynamics of wild bonobos (*Pan paniscus*) at Wamba. *International Journal of Primatology*, 19, 1029–43.

Gerloff, U., Hartung, B., Fruth, B., Hohmann, G. & Tautz, D. (1999). Intracommunity relationships, dispersal pattern and paternity success in a wild living community of bonobos (*Pan paniscus*) determined from DNA analysis of faecal samples. *Proceedings of the Royal Society of London, Series B*, 266, 1189–95.

Golla, W., Hofer, H. & East M. L. (1999). Within-litter sibling aggression in spotted hyenas: effect of maternal nursing, sex and age. *Animal Behaviour*, 58, 715–62.

Goodall, J. (1986). *The Chimpanzees of Gombe: Patterns of Behavior.* Cambridge, MA: Harvard University Press.

Harcourt, A. H. (1978). Strategies of emigration and transfer by primates, with particular reference to gorillas. *Zeitschrift für Tierpsychologie*, 48, 401–20.

Hashimoto, C. (1997). Context and development of sexual behavior of wild bonobos (*Pan paniscus*) at Wamba, Zaire. *International Journal of Primatology*, 18, 1–21.

Hashimoto, C., Furuichi, T. & Takenaka, O. (1996). Matrilineal kin relationship and social behavior of wild bonobos (*Pan paniscus*): sequencing the D-loop region of mitochondrial DNA. *Primates*, 37, 305–18.

Hemelrijk, C. K. (2000). Self-reinforcing dominance interactions between virtual males and females. Hypothesis generation for primate studies. *Adaptive Behavior*, 8, 13–26.

Hohmann, G. (2001). Association and social interactions between strangers and residents in bonobos (*Pan paniscus*). *Primates*, 42, 91–9.

Hohmann, G & Eriksson, J. (2001). Rapport de recherche d'une étude-pilot sur les bonobos (*Pan paniscus*) dans la région de Bolongo. Technical report to the Institut Congolais pour la Conservation de la Nature (ICCN), Kinshasa, RDC.

Hohmann, G. & Fruth, B. (1993). Field observations on meat sharing among bonobos, *Pan paniscus*. *Folia Primatol*ogica, 60, 225–9.

Hohmann, G. & Fruth, B. (1996). Food sharing and status in unprovisioned bonobos. In *Food and the Status Quest.*, ed. P. Wiessner & W. Schiefenhövel, pp. 47–67. Providence & Oxford: Berghahn.

Hohmann, G. & Fruth, B. (2000). Use and function of genital contacts among female bonobos. *Animal Behaviour*, 60, 107–20.

Hohmann, G., Gerloff, U., Tautz, D. & Fruth, B. (1999). Social bonds and genetic ties: kinship, association and affiliation in a community of bonobos (*Pan paniscus*). *Behaviour*, 136, 1219–35.

Idani, G. (1990). Relations between unit-groups of bonobos at Wamba, Zaire: encounters and temporary fusions. *African Study Monographs*, 11, 153–56.

Idani, G. (1991). Social relationships between immigrant and resident bonobo (*Pan paniscus*) females at Wamba. *Folia Primatologica*, 57, 83–95.

Ingold, P. & Marbacher, H. (1991). Dominance relationships and competition for resources among chamois, *Rupicapra rupicapra rupicapra* in female social groups. *Zeitschrift für Säugetierkunde*, 56, 88–93.

Kano, T. (1982). The social group of pygmy chimpanzees (*Pan paniscus*) of Wamba. *Primates*, 23, 171–88.

Kano, T. (1992). *The Last Ape: Pygmy Chimpanzee Behavior and Ecology*. Stanford, CA: Stanford University Press.

Kummer, H. (1968). *Social Organization of Hamadryas Baboons: a Field Study*. Biblioteca Primatologica, Vol. 6, Basel, New York: S. Karger.

Kuroda, S. (1979). Grouping of the pygmy chimpanzee. *Primates*, 20, 161–83.

Lamprecht, J. (1985). Distress call alternation in hand-reared goslings (*Anser indicus*): vocal co-operation between siblings? *Animal Behaviour*, 33, 839–48.

Malenky, R. K. & Wrangham, R. W. (1994). A quantitative comparison of terrestrial herbaceous food consumption by *Pan paniscus* in the Lomako Forest, Zaire, and *Pan troglodytes* in the Kibale Forest, Uganda. *American Journal of Primatology*, 32, 1–12.

Matsumoto-Oda, A., Hosaka, K., Huffman, M. & Kawanaka, K. (1998). Factors affecting party size in chimpanzees of Mahale mountains. *International Journal of Primatology*, 19, 999–1011.

McGrew, W. C. (1996). Dominance status, food sharing, and reproductive success in chimpanzees. In *Food and the Status Quest*, ed. P. Wiessner & W. Schiefenhövel, pp. 39–45. Providence & Oxford: Berghahn.

Nishida, T. (1979). The social structure of chimpanzees in the Mahale Mountains. In *The Great Apes*, ed. D. A. Hamburg & E. McCown, pp. 73–122. Menlo Park, CA: Benjamin Cummings.

Nishida, T. & Hiraiwa-Hasegawa, M. (1985). Responses to a stranger mother–son pair in the wild chimpanzee: a case report. *Primates*, 26, 1–13.

Parish, A. R. (1994). Female relationships in bonobos (*Pan paniscus*): evidence for bonding, cooperation, and female dominance in a male-philopatric species. *Human Nature*, 7, 61–96.

Press, W. H., Teukolsky, S. A., Vetterling, W. T. & Flannery, B. P. (1992). *Numerical Recipes in Fortran: The Art of Scientific Computing*, second edition. Cambridge: Cambridge University Press.

Richard, A. F. (1992). Aggressive competition between males, female-controlled polygyny and sexual monomorphism in a Malagasy lemur, *Propithecus verreauxi*. *Journal of Human Evolution*, 22, 395–406.

Riss, D. C. & Goodall, J. (1977). The recent rise to the alpha-rank in a population of freeliving chimpanzees. *Folia Primatologica*, 27, 134–51.

Rubenstein, D. I. (1986). Ecology and sociality in horses and zebras. In *Ecological Aspects of Social Evolution: Birds and Mammals*, ed. D. I. Rubenstein & R.W. Wrangham, pp. 282–302. Princeton, NJ: Princeton University Press.

Sakura, O. (1994). Factors affecting party size and composition in chimpanzees (*Pan troglodytes*) in Bossou, Guinea. *International Journal of Primatology*, 15, 167–83.

Smuts, B. B. (1985). *Sex and Friendship in Baboons*. New York: Aldine.

Stanford, C. B. (1998). The social behavior of chimpanzees and bonobos: empirical evidence and shifting assumptions. *Current Anthropology*, 39, 399–420.

Sterck, E. H. M., Watts, D. P. & van Schaik, C. P. (1997). The evolution of female social relationships in nonhuman primates. *Behavioral Ecology and Sociobiology*, 41, 291–309.

Thompson-Handler, N. & Malenky, R. (1993). Action Plan for *Pan paniscus*: Report on Free Ranging Populations and Proposals for their Preservation. Department of Anthropology, SUNY at Stony Brook, NY.

van Hooff, J.A.R.A.M. (1990). Intergroup competition and conflict in animals and man. In *Sociobiology and Conflict: Evolutionary Perspectives on Competition, Cooperation, Violence and Warfare*, pp. 23–54, London: Chapman and Hall.

van Schaik, C. P. (1989). The ecology of social relationships amongst female primates. In *Comparative Socioecology*, ed. V. Standen & R. A. Foley, pp. 195–218. Oxford: Blackwell.

van Schaik, C. P. (1996). Social evolution in primates: the role of ecological factors and male behavior. In *Evolution of Social Behavior Patterns in Primates and Man*, ed. W. G. Runciman & R. I. M. Maynard Smith, pp. 9–31. Oxford: Oxford University Press.

Vervaecke, H., de Vries, H. & Elsacker, L. (2000). Function and distribution of coalitions in captive bonobos (*Pan paniscus*). *Primates*, 41, 249–65.

Watts, D. P. (1998). Coalitionary mate guarding by male chimpanzees at Ngogo, Kibale National Park, Uganda. *Behavioral Ecology and Sociobiology*, 44, 43–55.

Watts, D. P. & Mitani, J. C. (2000). Infanticide and cannibalism by male chimpanzees at Ngogo, Kibale National Park, Uganda. *Primates*, 41, 357–64.

White, F. J. (1988). Party composition and dynamics in *Pan paniscus*. *International Journal of Primatology*, 9, 179–93.

White, F. J. (1992). Pygmy chimpanzee social organization: variation with party size and between study sites. *American Journal of Primatology*, 26, 203–14.

White, F. J. (1996). Comparative socioecology of *Pan paniscus*. In *Great Ape Societies*, ed. W. C. McGrew, T. Nishida & L. Marchant, pp. 29–41, Cambridge: Cambridge University Press.

White, F. J. (1998). Seasonality and socioecology: the importance of variation in fruit abundance to bonobo sociality. *International Journal of Primatology*, 19, 1013–27.

White, F. J. & Lanjouw, A. (1992). Feeding competition in Lomako bonobos: variation in social cohesion. In *Topics in Primatology, Vol. 1, Human Origins*, ed. T. Nishida, W. C. McGrew, P. Marler, M. Pickford & F. B. M. de Waal, pp. 67–79. Tokyo: University of Tokyo Press.

White, F. J. & Wrangham R. W. (1988). Feeding competition and patch size in the chimpanzee species *Pan paniscus* and *Pan troglodytes*. *Behaviour*, 105,148–63.

Wrangham, R. W. (1980). An ecological model of female-bonded primate groups. *Behaviour*, 75, 262–300.

Wrangham, R. W. (1986). Ecology and social relationships in two species of chimpanzees. In *Ecological Aspects of Social Evolution: Birds and Mammals*, ed. D. I. Rubenstein & R. W. Wrangham, pp. 352–78. Princeton: Princeton University Press.

Wrangham, R. W. (1999). The evolution of coalitionary killing: the imbalance-of-power hypothesis. *Yearbook of Physical Anthropology*, 42, 1–30.

Wrangham, R. W., Clark, A. P. & Isabirye-Basuta, G. (1992). Female social relationships and social organization of Kibale chimpanzees. In *Topics in Primatology, Vol. 1, Human Origins*, ed. T. Nishida, W. C. McGrew, P. Marler, M. Pickford & F. B. M. de Waal, pp. 81–98. Tokyo: University of Tokyo Press.

Wrangham, R. W., Gittleman, J. & Chapman, C.A. (1993). Constraints on group size in primates and carnivores: population density and day-range as assays of exploitation competition. *Behavioral Ecology and Sociobiology*, 32, 199–210.

Wrangham, R. W., Chapman, C., Clark-Arkadi, A. & Isabirye-Basuta, G. (1996). Social ecology of Kanyawara chimpanzees: implications for understanding the costs of great ape groups. In *Great Ape Societies*, ed. W. C. McGrew, L. F. Marchant & T. Nishida, pp. 45–57. Cambridge: Cambridge University Press.

Part III
Female strategies

Introduction. Female strategies: the females that did evolve[1]

MEREDITH F. SMALL

As a fresh anthropology graduate student in 1972, I was enthralled by the idea that you could observe nonhuman primates and gain clues about human behavior. Yet I was also confused by the limited information presented in the first primatology course I took. The portrait of females, for example, was rather simple. It appeared that monkey and ape females spent all their time making babies, having babies, and caring for them. If there was an evolutionary relationship between these animals and the human lineage, surely there was more to female behavior than this.

It is no coincidence, of course, that these thoughts occurred to me, a 25-year-old, independent woman, smack in the middle of the feminist revolution. So, after class one day, I approached the professor who was teaching the course and asked, 'What else do these females do besides have babies?' He looked at me thoughtfully and said, 'That's a good question. They must do something else. But I don't know what it is.' Surrounded by women who were occupied in many other ways besides mothering, and nursing ambitions for a career rather than nursing babies, I just knew that my nonhuman primate sisters had other things to do as well.

I went on to complete dissertation research at the University of California, Davis, on female macaques without infants. More significantly, this research was part of a much larger wave of research on female primate behavior. During the late 1970s and into the 1980s, animal behaviorists, both women and men, were starting to focus on female behavior. We soon found out that female nonhuman primates had sex lives, friends, social relationships, dominance hierarchies, could be nasty and competitive, make strategic decisions, and were often leaders. How they lived, mated, and made babies was highly influenced by their environment. Like males, these newly 'discovered' females were complex creatures with many alternative strategies for surviving and passing on genes.

This period, the renaissance of female behavior research, is symbolized best by Sarah Blaffer Hrdy's seminal book, *The Woman that Never Evolved* (1981), which opened our eyes to how inaccurately female primates had been portrayed up until that time. During those years, two journal articles also laid the groundwork for a theory of female behavior, Jeffrey Kurland's 1977 paper on the force of matrilines, and Richard Wrangham's 1980 paper on female-bonded primates. Then came a focus on female choice as a newly important issue of sexual selection. With hard work, a shift of focus, and new-found theory, females became respected, intriguing, important subjects for study.

The most important lesson we have learned about female primates since that renaissance period, and a lesson that we are still learning, is that there is no such creature as the typical female primate. Each species is different, each group within a habitat is different, and each female is different. What is even more interesting is the fact that we are gaining an appreciation for female flexibility, and that we are learning that, over a lifetime, a female might behave differently depending on her circumstances.

The chapters on female chimpanzee and bonobo behavior in this volume contain numerous examples of this behavioral flexibility. By looking at both chimpanzees and bonobos, of course, the editors recognize that there will be, by definition, differences between the species. As we know from both captive and field work, female bonobos are less dominated by males, they have swellings that continue through nonfertile periods, and they spend much more time in the company of males. Rumor has it that bonobos are also more 'sexy' than chimpanzees because they copulate more often over time and engage in other sexual behavior, such as female–female genital rubbing. The articles in this collection, however, fine-tune these broad observations.

By examining the details of cycling and sexuality, Takeshi Furuichi and Chie Hashimoto (Chapter 11) take apart the notion that bonobos are more 'sexy' than chimpanzees. Yes, bonobos do copulate more often than chimpanzees, but it seems that their rate of copulation during the peak of estrus, when it really matters in terms of passing on genes, is lower than that of chimpanzees. Bonobo females are, in fact, not even as eager to have sex with males as chimpanzee females are, and why should they be? As Furuichi & Hashimoto point out, bonobo males are always around and thus bonobo females, who are at less risk because these males are friendly, need not have sex under duress. Chimpanzee females, in contrast, have tense relationships with males. The result is that they must mate often to insure insemination and quickly to avoid harm.

Although Furuichi & Hashimoto do not broach the subject, the differences between chimpanzee and bonobo female sexual behavior echo the situation for women across cultures. In patriarchal societies where women are dominated by men, sex is a furtive affair and females live in fear. However, where women have power and freedom, they are also more free sexually. One could then hypothesize that the rate of copulation for women in patriarchal societies would be higher, but more clustered around ovulation, than in cultures where women are more free.

Richard Wrangham (Chapter 15) also looks at copulation and estrus in an attempt to explain why some female chimpanzees are simply more sexually attractive than others. Although we assume that, again, bonobos are more 'sexy,' it is the East African chimps who seem to advertise their ovulation more intensely. At Gombe, Mahale, and Kibale, female chimpanzees mate most often during their period of maximum swelling, and males fiercely guard these females during this time. Wrangham claims that females at these locations advertise their estrous state more 'loudly' than do females at other chimpanzee or bonobo locations, attributing the difference to scramble competition. Because of ecological constraints, East African females are forced to be more solitary and to avoid males; when they are fertile they need to bring males close with an intensity not necessary in other communities.

Of more interest is the observation that male chimpanzees prefer older parous females to nulliparous ones. As Blaffer Hrdy (2000) recently pointed out, this penchant for older, experienced, and obviously fertile females, should be revelatory to evolutionary psychologists who insist that preference for young, nulliparous females is adaptive for primate males. Obviously, chimpanzee males (and all other nonhuman primate males) are more aware than modern human males that they do best, in terms of passing on genes, if they favor experienced females. Viewed in this light, it is reasonable to suggest that the more recent modern human male lust for young females is socially driven, not evolutionary.

Janette Wallis (Chapter 13) also addresses female chimpanzee sexual behavior and links that behavior to particular environmental settings. Although mating is not seasonal among chimpanzees as it is in, say, some macaque species, there is still variation that makes sense relative to climate and food availability. Wallis maintains that it is not rainfall per se that has an effect, but overall changes in environmental conditions. These changes affect the fruiting and flowering of plants that may have an important impact on female hormone levels, both positively and negatively. Those levels may fluctuate, not only as abundance waxes and wanes, but also as the variety of plants utilized changes, which in turn forces females to extend their nutritional reach.

Two chapters address another major issue for chimpanzee and bonobo females – social grouping. Akiko Matsumoto-Oda (Chapter 12) asks the important question about bonds between male and female chimpanzees. In the past, the sexes have always been portrayed as clearly unfriendly, and this was in contrast to bonobos, where females and males seemed to be on better terms. Estrus is apparently the key to gender relations among chimpanzees; females and males associate more often and for longer continuous periods when females display swellings. Males who tend to hang around with estrous females also share meat with them and groom them. Females also have bonds with some males outside of estrus; in general, some males are more familiar to females than others, and those familiar males are most often of high rank.

Jennifer Williams, Hsien-Yang Liu, and Anne Pusey (Chapter 14) look at the effects of scramble and contest competition on social groups for females at Gombe Stream. Females are generally fairly solitary unless they are in estrus, but they do spend some time around others. Everybody, apparently, wants to be around high-ranking females, and low-ranking females are generally stuck with each other's company, or they are solo. Mothers with juveniles seek each other out, while mothers with infants are mostly on their own. Juveniles tagging along do not slow mothers down, but when females are accompanied by infants, they move less, which means it may be harder to find food. In many ways, having an infant costs the mother day to day – she is more alone and would probably lose out in times of food scarcity. This last chapter is especially poignant to me. Here is another study, many years after my own, that looks at the differences between females with infants and females without infants.

As all these more current studies show, chimpanzee and bonobo females are behaviorally complex creatures. Like all females, they are much more than simple reproductive machines, they are not just mothers. Also, like human females, they are flexible, able to assess options, and capable of weighing the costs and benefits of various moves over the span of different stages in their life cycles. They are, indeed, females that evolved.

REFERENCES

Blaffer Hrdy, S. (1981). *The Woman That Never Evolved.* Cambridge, MA: Harvard University Press.

Blaffer Hrdy, S. (2000). Raising Darwin's consciousness: sexual selection and the prehominid origins of patriarchy. In *Gender and Society*, ed. C. Blakemore & S. Iversen, pp. 143–99. Oxford: Oxford University Press.

Kurland, J. A. (1977). Kin selection in the Japanese monkey. *Contribution to Primatology*, 12, 1–145.

Wrangham, R. W. (1980). An ecological model of female-bonded primate groups. *Behaviour*, 75, 262–300.

NOTE

1. With apologies to Sarah Blaffer Hrdy and her book *The Woman That Never Evolved.*

11 • Why female bonobos have a lower copulation rate during estrus than chimpanzees

TAKESHI FURUICHI & CHIE HASHIMOTO

INTRODUCTION

Due to their prolonged sexual cycling period and variety of sexual behaviors, there is a prevailing impression that female bonobos are very sexually active. They copulate even during non-reproductive periods (Thompson-Handler 1990; Furuichi 1992; Kano 1992) and use sexual behaviors for various social purposes (Kuroda 1980; Goodall 1986; de Waal 1987; Furuichi 1989; Kano 1989, 1992; Idani 1991; Wrangham 1993; Parish 1994). However, Takahata and others (1996) have pointed out that the copulation rates of female chimpanzees and female bonobos over the swelling cycle, which consists of a swelling phase and a non-swelling phase, are approximately the same, although the copulation rates over the interbirth interval, which consists of a phase in which females show cyclic swelling and a phase in which females do not show cyclic swelling due to lactation or pregnancy, are higher for female bonobos. Although female bonobos copulate most frequently during the swelling phase, even during the non-swelling phase they copulate more often than female chimpanzees (Tutin 1979a; Thompson-Handler *et al.* 1984; Dahl 1986; Furuichi 1987; Kano 1992; Dixson 1998). Therefore, the findings of Takahata and others logically predict that, during the swelling phase, female bonobos copulate at a lower rate than female chimpanzees. Wrangham (Chapter 15) also points out that adult female chimpanzees copulate especially frequently during the peri-ovulatory period, whereas there is no such data for female bonobos. He suggests that the copulation rates of female chimpanzees and female bonobos differ during the peri-ovulatory period because they incur different costs when living in mixed-sex parties for mating with males.

In order to discuss chimpanzee and bonobo sexual behaviors more substantially, we need to compare the actual copulation rates of females of both species during the swelling phase. There are some studies of chimpanzee copulation rates during the swelling phase (Tutin 1979a; Hasegawa & Hiraiwa-Hasegawa 1990), but for bonobos there are only reports on copulation rates over the swelling cycle, or over the interbirth interval (Furuichi 1987; Idani 1991; Kano 1992; Takahata *et al.* 1996). In this study, we calculate the copulation rate for female bonobos both over the interbirth interval and during the swelling phase, using female sexual swelling data collected daily.

If, indeed, the copulation rate of female bonobos during the swelling phase is lower than that of female chimpanzees, as will be shown below, then what proximate factors cause such a difference? We propose two hypotheses to explain the lower copulation rate of female bonobos.

Hypothesis 1 is that female bonobos do not copulate as frequently as chimpanzees because opportunities for mating are limited; that is the number of adult males per estrous female is fewer, or the copulation rate of adult males is lower. This hypothesis assumes that either the availability of males, or the male copulation rate, or both, influence the female copulation rate. That *Pan* females sometimes show receptivity beyond their own desire (Dixson 1998) suggests that the female copulation rate may increase as the number of males per estrous female increases. Alternatively, female copulation rate could be limited by the number of adult males per estrous female, or by adult male copulation rate, or by both, because a male requires time between copulations to replenish sperm (Hashimoto, 1997).

Hypothesis 2 is that female bonobos are not as eager to copulate as are female chimpanzees. During the swelling phase, females make themselves available for copulation by ranging with males and approaching or explicitly soliciting them for copulation. Therefore, a difference in female proceptivity (Beach 1976) between chimpanzees and bonobos may explain the difference in female copulation rate between the two species. Takahata *et al.* (1996) showed that female bonobos performed pre-copulatory approach or soliciting behaviors less frequently than did female chimpanzees. In this study, we compare the female bonobos' association with males during the swelling versus non-swelling phases in order to examine the estrus-related change in attitude of female bonobos.

If female proceptivity varies between chimpanzees and bonobos, we need to identify the factors responsible. Chimpanzee and bonobo females copulate several hundred times per conception, and many hypotheses have been proposed to explain such a high number of copulations from the females' point of view, including the best-male hypothesis, the many-male hypothesis, the social-passport hypothesis, and the paternity confusion hypothesis (Boesch & Boesch-Achermann 2000; Wrangham, Chapter 15). In this chapter, we present a new hypothesis (3): that the difference in the length of the optimal period for producing next offspring explains the species difference in female proceptivity during the swelling phase.

Females generally resume ovulation cycles after weaning offspring, and resuming ovulation too early may mean that females have to nurse two offspring simultaneously, or that the older offspring receive insufficient parental care (e.g. Tutin 1979b; Trivers 1985; Furuichi *et al.* 1998). After resuming ovulation, females may suffer the costs of living and foraging in a party that includes males, as well as the costs of having fewer offspring, if it takes too long to conceive (Boesch & Boesch-Achermann 2000; Wrangham 2000; Wrangham, Chapter 15). Thus, the desirable period for females to conceive next offspring, referred to in this chapter as the 'optimal period for producing next offspring,' is likely influenced by various socio-ecological factors and may vary between chimpanzees and bonobos.

If the optimal period for producing next offspring is short, females may need to conceive quickly after postpartum amenorrhea, which may in turn encourage higher proceptivity and thus a higher frequency of copulation during the swelling phase. The shorter length of the optimal period for producing next offspring may also lead to smaller variation in interbirth intervals because more females conceive after a period similar in length to the previous birth. Therefore, in order to test *Hypothesis 3*, we compare variation in the interbirth intervals between chimpanzees and bonobos, predicting that interval variation would be greater for female bonobos, who are less proceptive than chimpanzees during the swelling phase.

In this study, we analyze reproductive parameters and sexual behaviors of parous adult females. In bonobos, nulliparous females copulate more frequently than parous females, probably because they use copulation as a means of forming new social relationships in the group into which they immigrate (Furuichi 1989; Idani 1991; Kano 1992). In contrast, nulliparous eastern chimpanzee females have a lower copulation rate than their parous female counterparts; Wrangham (Chapter 15) suggested that this difference is due to lower cost of living in mixed-sex parties for nulliparous females. Thus, the copulation rate of nulliparous females seems to be influenced by many factors other than those discussed in this study.

MATERIALS AND METHODS

This study examines the findings of previous socio-ecological studies of eastern chimpanzees (*Pan troglodytes schweinfurthii*), western chimpanzees (*Pan troglodytes verus*) and bonobos (*Pan paniscus*), and provides new data concerning the sexual behavior of eastern chimpanzees and bonobos. Eastern chimpanzees are sometimes described simply as 'chimpanzees.' In the analyses of reproductive parameters and copulation rate, we mainly examined wild eastern chimpanzees in Gombe (Tutin 1979a; Tutin & McGinnis 1981; Goodall 1986; Wallis 1997) and Mahale (Hasegawa & Hiraiwa-Hasegawa 1983, 1990; Nishida *et al.* 1990; Hasegawa 1991; Takahata *et al.* 1996), and bonobos in Wamba (Furuichi 1987; Kano 1992; Takahata *et al.* 1996; Furuichi *et al.* 1998; this study). Supplementary data of copulation rates of eastern chimpanzees were also obtained from the studies in Kibale (Wrangham, unpublished data) and Kalinzu (this study) in Uganda. In the analyses of interbirth interval, we also examined data from western chimpanzees in Bossou (Sugiyama 1994; Sugiyama, unpublished data) and Taï (Boesch & Boesch-Achermann 2000).

We observed the E1 group (equivalent to 'unit group' or 'community') of bonobos during the three observation periods, from August 1985 to January 1986, from September 1987 to February 1988, and from December 1990 to February 1991, and recorded, as far as possible, the state of sexual swelling of females on every observation day. From these data, we calculated adult sex ratio (ASR, total number adult males/total number adult females) and estrus sex ratio (ESR, total number adult males/total number adult females in swelling phase).

In 1985–86, we observed all adult females of the E1 group using focal animal sampling (Furuichi 1989). Approximately 80% of the observations occurred in natural habitat, and the remainder at provisioning sites. We observed each adult male for a total of about 500 minutes, and each adult female for about 800 minutes. From these data, we obtained the average number of males found within 3 m of each female, and the copulation rate of adult males and adult females.

In 1990–91, we observed the E1 group only at provisioning sites (Hashimoto & Furuichi 1994; Hashimoto 1997). When bonobos appeared at each site, we recorded the presence of each member at 10-minute intervals, and recorded all occurrences of sexual behaviors. The total observation time was 101.5 hours (609 observation units). From these data, we calculated the average number of adult males found with each adult female, and the copulation rates of adult males and females.

We observed the M group of chimpanzees in the Kalinzu Forest from July 1997 to March 1998 (Hashimoto *et al.* 2001). Although we only began following chimpanzees in July 1997, they were habituated by October, when we started collecting data, to the point that we could follow them at a distance of 10–20 m without disturbing them. On each observation day, we followed one party of chimpanzees for as long as possible, recording the presence of each member, the swelling state of females, and copulatory behaviors. From these data, we calculated the ESR and the copulation rate. The total observation time was 208.8 hours.

In this study, we deal with the sexual behaviors of adults of both sexes. Adult males were defined as those of 15 years old or older. Adult females were defined as parous females or females that had reached the general age for first parturition. Females were considered to be in estrus when swelling of their sexual skin was maximal (described simply as 'swelling' here) (Hasegawa & Hiraiwa-Hasegawa 1983; Dahl 1986; Furuichi 1987; Thompson-Handler 1990; Wallis 1997). Interbirth interval (IBI) was only measured in those cases in which an offspring was alive to see the birth of a sibling. In an analysis of the data in Figure 11.4 (below), we employed an operational IBI that excluded the last interbirth interval of females older than 40 years because old females can have unusually long interbirth intervals (Nishida *et al.* 1990; Sugiyama 1994). All statistical tests were two-tailed.

RESULTS

Comparison of copulation rates of females

Female bonobos seem to exhibit higher copulation rates over the IBI than female chimpanzees, though we could not test statistically due to the small number of data sets and lack of information about variances in some data sets (Figure 11.1). In Mahale, adult female chimpanzees are expected to copulate with adult males 0.04 times per hour over the IBI (Takahata *et al.* 1996). In contrast, Kano (1992) reported a copulation rate of 0.14 times per hour for all adult female bonobos at Wamba, including non-cycling nursing females, which is comparable with the copulation rate over the IBI. Takahata *et al.* (1996) reported a copulation rate of 0.28 times per hour for all adult female bonobos, except for one adult female who showed an extremely high copulation rate.

Our bonobo copulation data indicate similarly high copulation rates for all adult females; 0.18 per hour in 1985–86 and 0.08 per hour in 1990–91. This difference between the two species can be explained by female bonobos' much shorter non-cycling phase after delivery compared with that of female chimpanzees (Table 11.1).

In contrast, if calculation of copulation rate is restricted to the swelling phase of cycling females, female bonobos appear to show a lower copulation rate than that of female chimpanzees (Figure 11.1). Adult female chimpanzees in Mahale copulated 0.79 times per hour (Hasegawa 1991). For chimpanzees in Gombe, Tutin (1979a) reported that adult females copulated 0.52 times per male per hour during estrus. Tutin defined copulation rate only per male who was associating with a female, and therefore the total copulation rate of a female with all males would have been much higher. For the Kalinzu Forest chimpanzees, we calculated the copulation rate as the number of copulation events observed per total observation time of all females showing sexual swelling. This analysis yielded a conservative estimate of mean copulation rate during estrus of 0.43 times per hour; all occurence sampling probably caused some copulations to be overlooked. By all estimations, the mean copulation rate of female chimpanzees during the swelling phase is probably at least 0.5 times per hour. In contrast, female bonobos in a swelling phase copulated 0.37 times per hour in 1985–1986, and 0.11 times per hour in 1990–1991.

HYPOTHESIS 1: OPPORTUNITY FOR COPULATION IS LIMITED FOR BONOBO FEMALES

The lower copulation rate during the swelling phase for female bonobos compared to chimpanzees may be explained by fewer opportunities for copulation, that is fewer males per estrous female or a lower copulation rate of males.

In order to determine the number of adult males available for an estrous female, we compared the reproductive parameters of female chimpanzees and bonobos. Wrangham (1993) made this comparison using all available reports from different sites. Our comparison, however, was made only among the three sites at which most of the parameters were available, since some parameters may impact other measures within each site (Table 11.1).

Female chimpanzees are expected to be in the swelling phase during only 6.4% (Mahale) or 4.2% (Gombe) of the

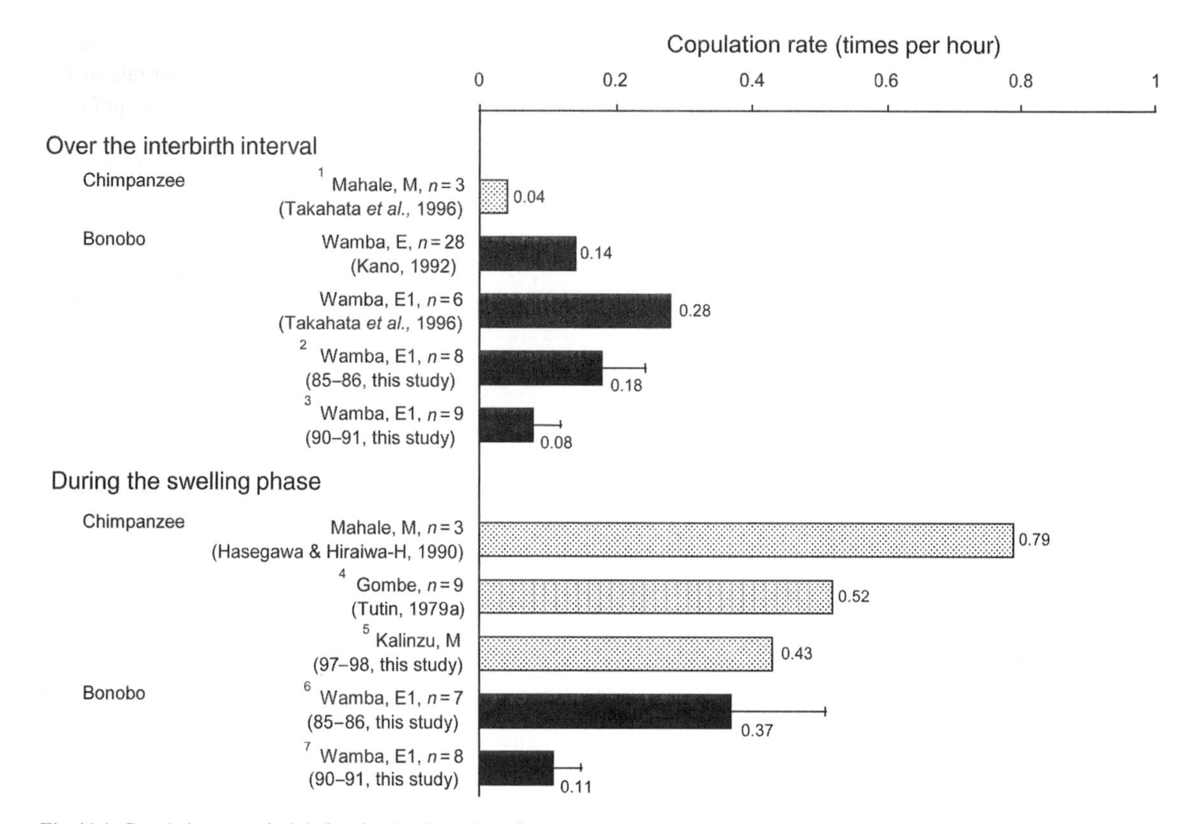

Fig. 11.1. Copulation rate of adult females (total number of copulations per hour). Name of site, name of group, and number of female subjects are shown on the left. [1]Calculated from Hasegawa 1991. [2]Based on 19 copulations for 6422 minutes of focal animal sampling of all adult females. [3]Based on 25 copulations for 10 minutes × 1982 observation units of all adult females. [4]Mean copulation rate with a male in association (see text). [5]Number of estrous females not confirmed. Based on 43 copulations for 5 minutes × 1207 scans of all observed estrous females. [6]Based on 17 copulations for 2793 minutes of focal animal sampling of females in the swelling phase. [7]Based on 17 copulations for 971 observation units of females in the swelling phase. Error bars show S.E.M.

IBI. Although the ASR of chimpanzees is largely biased toward females (0.27 in Mahale and 0.51 in Gombe), due to these small proportions of days in the swelling phase the ESR is expected to be as high as 4.2 in Mahale and 12.3 in Gombe. By contrast, female bonobos in Wamba are expected to be in the swelling phase for as much as 27% of the IBI. Although the ASR is less biased toward females (0.75), the ESR is expected to be 2.8, which is much smaller than that of chimpanzees. As for bonobos in Lomako, Gerloff *et al.* (1999) reported an ASR of 0.4, intermediate between those of chimpanzees in Mahale and Gombe. If we assume that other reproductive parameters in Lomako are similar to those in Wamba, the ESR for Lomako bonobos must be even smaller, estimated at 1.5. Thus, we predicted that the ESR would generally be much smaller for bonobos than for chimpanzees.

To test this prediction, we calculated the ESR of E1 group of bonobos over the three observation periods. As shown in Table 11.2, the observed ESR was slightly smaller than the predicted ESR.

For chimpanzees, it is difficult to monitor the swelling state of all adult females on a daily basis because many females range independently (e.g. Nishida 1979; Wrangham 1979, 2000). Therefore, we examined the ESR of chimpanzees using the composition of observed parties. In the Kalinzu Forest study, we observed the largest chimpanzee parties we could find, most of which were mixed-sex parties. Assuming adult males and estrous females were about equally likely to be in mixed-sex parties, the ESR calculated from party composition roughly represents that of the whole group. This group consisted of about 16 adult males, with

Table 11.1. *Reproductive parameters and estrus sex ratio*

	Chimpanzee (Mahale)[a]	Chimpanzee (Gombe)[b]	Bonobo (Wamba)[c]
For an interbirth interval (IBI)			
Interbirth interval (months)	72[a]	61.8[b]	57.6[c]
Anestrus after delivery (months)	55.5[a]	46.3[b]	12[d]
Cycle resumption to conception (months)	8.9[a]	4.75[b]	38
Gestation period (months)	7.6[e]	7.6[e]	7.6[e]
Cycling period after conception (months)	2.6[f]	2[g]	6.5
Anestrus before delivery (months)	5	5.1	1[d]
Proportion of cycling period before an IBI	0.16	0.11	0.77
Between ovulations			
Length of swelling cycle (days)	31.5[h]	33.6	42[i]
Duration of swelling phase (days)	12.5[h]	12.6	14.6[i]
Proportion of days in swelling phase for a swelling cycle	0.40	0.38	0.35
Calculation of estrus sex ratio (ESR)			
Proportion of days in swelling phase for an IBI	0.064	0.042	0.27
ASR (male/female)	0.27[j]	0.51[b]	0.75[c]
ESR (male/female in swelling phase)	4.2	12.3	2.8

Notes:
[a] Nishida *et al.* 1990; [b] Wallis 1997; [c] Furuichi *et al.* 1998; [d] Kano 1992; [e] Martin *et al.* 1978; [f] Takahata *et al.* 1996; [g] Tutin & McGinnis 1981; [h] Hasegawa & Haraiwa-Hasegawa 1983; [i] Furuichi 1987; [j] Goodall 1986.
Parameters without references are calculated from data presented here. All figures show mean values except figures from Nishida *et al.* 1990 that refer to mean values.

Table 11.2. *Group composition and sex ratio in E1 group of bonobos*

Observation period	No. of all members	No. of adult females	No. of cycling females	No. of estrous females[a]	No. of adult males	ASR	ESR
1985–86	29	8	7	3.1 ± 1.1 (n = 57)	7	0.88	2.3
1987–88	32	9	8	3.1 ± 1.9 (n = 48)	6	0.67	1.9
1990–91	30	9	8	4.1 ± 1.5 (n = 43)	6	0.67	1.5

Note:
[a] Mean and standard deviation of the number of females in swelling phase and the number of sample days. Sample days are days when the state of sexual skin was scored for more than 90% of all females.

an unconfirmed number of adult females and immature individuals (Hashimoto *et al.* 2001). We observed 61 parties, which included, on average, 4.43 adult males and 0.92 parous females in the swelling phase. These data yielded an ESR of 4.8 for Kalinzu, which is close to that estimated from reproductive parameters for Mahale. We also examined the ESR of the Kanyawara group in Kibale in a similar fashion (Wrangham, unpublished data). In 1998, the group had 11 adult males and 15 adult females. The average number of adult males observed in a party (15-minute observations, n = 12456) was 3.2, and that of adult females in the maximal swelling phase was 0.1. Thus the ESR calculated from the party composition was as high as 32. This was probably because the ASR of this group (0.73) was higher than those

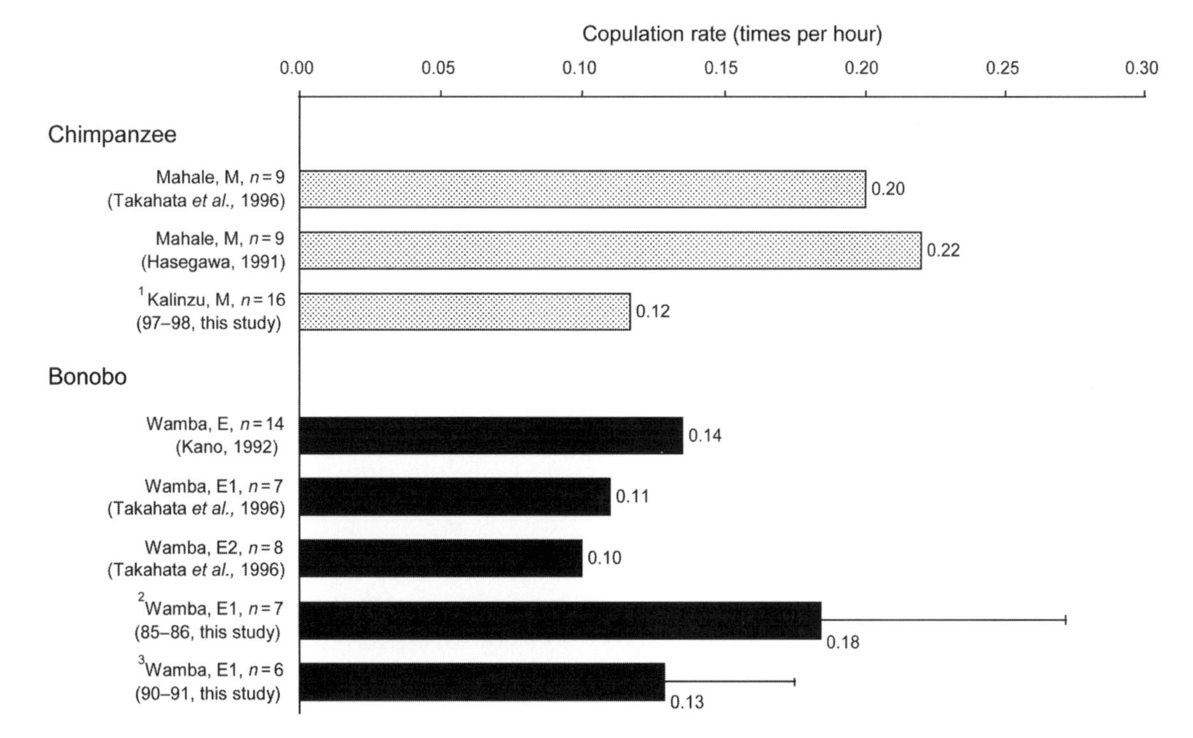

Fig. 11.2. Copulation rate of adult males. Name of site, name of group, and number of male subjects are shown on the left. [1]Based on 43 copulations for 5 minutes × 4401 scans of all adult males. [2]Based on 11 copulations for 3516 minutes of focal animal sampling of all adult males. [3]Based on 25 copulations for 10 minutes × 1175 observation units of all adult males. Error bars show S.E.M.

of Mahale or Gombe, and because females ranging in the southern part of the group's range were observed infrequently (Wrangham, personal communication).

Thus, for chimpanzees, there are considerable differences in ESR between sites, as well as differences between expected and observed ESRs. However, as we predicted, the observed ESRs of chimpanzees are much larger than those for bonobos.

As for the copulation rate of adult males, we found no apparent difference between chimpanzees and bonobos (Figure 11.2). Two data sets from male chimpanzees in Mahale showed that males copulated about 0.2 or 0.22 times per hour. A conservative estimate of the copulation rate in the Kalinzu Forest was 0.12 times per hour. In contrast, five data sets describing either different groups or periods at Wamba indicated that male bonobos copulated 0.10–0.18 times per hour.

Thus, the number of males per estrous female was smaller for bonobos than for chimpanzees, but the copulation rates of adult males were similar. These results support the hypothesis that less opportunity for copulation leads to the lower copulation rate of female bonobos compared to chimpanzees.

HYPOTHESIS 2: FEMALE BONOBOS ARE LESS EAGER TO COPULATE FREQUENTLY

Because bonobos rarely form consort pairs of a male and an estrous female, if males or females show increased attempts for association with the opposite sex when females are in estrus, then the number of males in association with estrous females will be larger in the swelling phase than in the non-swelling phase. Conversely, if there is no change in the number of males in association with females between the swelling and non-swelling phases, such a result suggests that neither males nor females show a higher inclination for association with the opposite sex due to estrus.

As shown in Figure 11.3(*a*), only two females had a significantly different number of males within 3 m of them during swelling compared with non-swelling phases, and the direction of this difference was opposite for each female. Moreover, the number of males observed in the same scanning samples with each female did not vary greatly with female estrous state, although two older females had fewer and three younger females had more males associated with

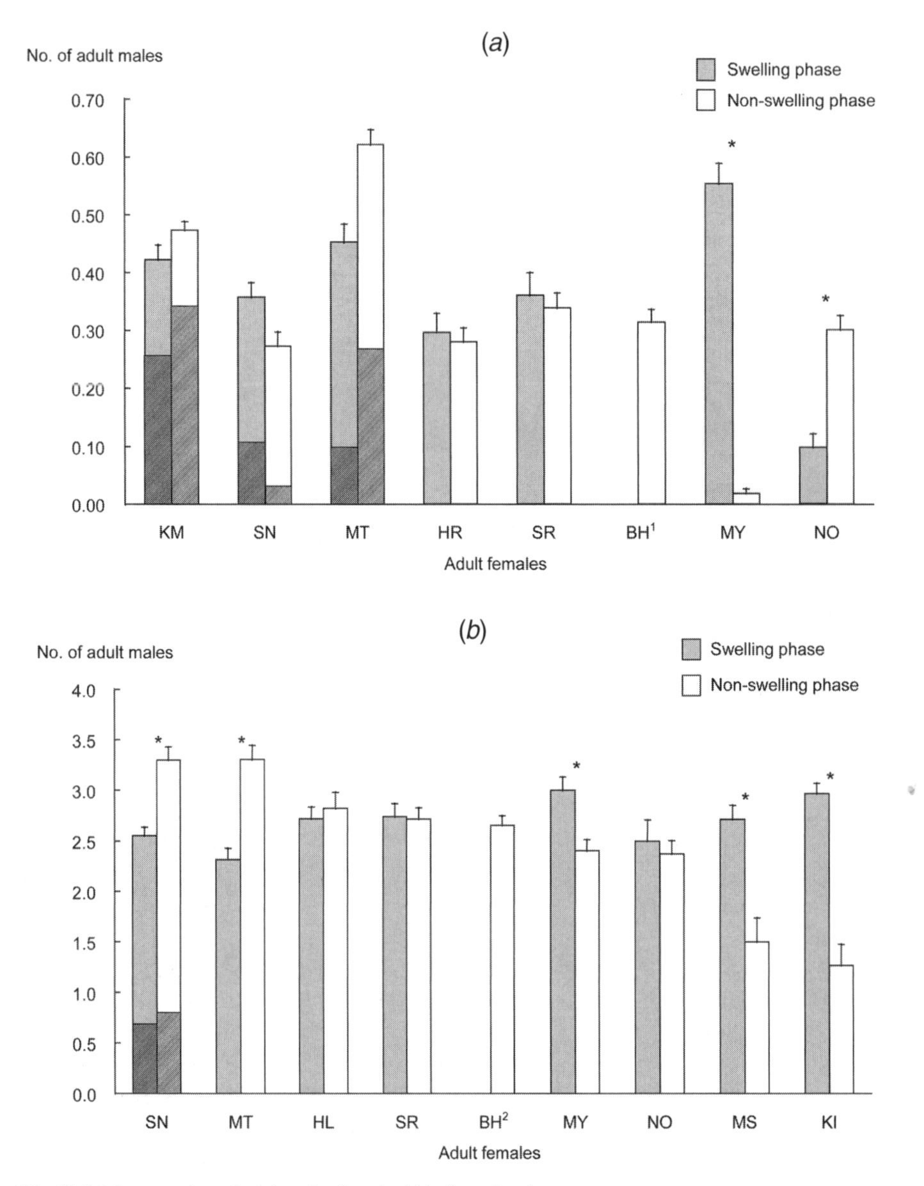

Fig. 11.3. Mean number of adult males found within 3 m of each adult female in 1985–86 (*a*), and mean number of adult males found with each adult female in provisioning sites in 1990–91 (*b*). Shaded bars refer to adult sons of each female. Error bars show S.E.M. for non-kin males. [1]BH did not exhibit swelling due to pregnancy. [2]BH did not exhibit swelling because she had an infant under a year old. *T-test, $p<0.05$.

them during the swelling phase (Figure 11.3(*b*)). Although these data sets were obtained under different conditions in different years, both suggest that female bonobos, regardless of estrous state, maintain a constant attitude about associating with males.

HYPOTHESIS 3: THE OPTIMAL PERIOD FOR PRODUCING NEXT OFFSPRING IS LONGER FOR FEMALE BONOBOS

By assuming that the shorter optimal period for producing next offspring reduces variance in the IBI, we can use bonobo and chimpanzee IBI data to examine a difference in length of the optimal period for producing next offspring. We compared the mean and variance of the operational IBI between chimpanzees (Bossou, Taï, and Mahale) and bonobos (Wamba) (Figure 11.4). Significant differences in mean IBI were found between Wamba and Taï ($t=22$, $df=36$, $p<0.05$), Wamba and Mahale ($t=4.6$, $df=38$, $p<0.001$),

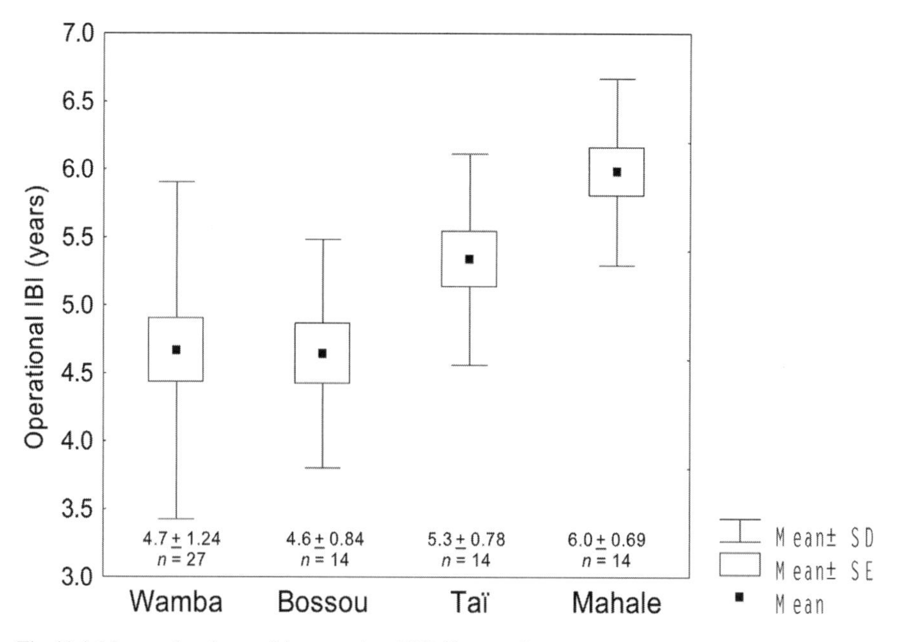

Fig. 11.4. Mean and variance of the operational IBI. Data are from Furuichi *et al.* 1998 (Wamba), Sugiyama 1994 and Sugiyama, unpublished data (Bossou), Boesch & Boesch-Achermann 2000 (Taï), and Nishida *et al.* 1990 (Mahale).

Bossou and Taï ($t = 2.3$, df $= 26$, $p < 0.05$), Bossou and Mahale ($t = 4.6$, df $= 26$, $p < 0.001$), and Taï and Mahale ($t = 2.3$, df $=$ 26, $p < 0.05$); there was no significant difference between bonobos in Wamba and chimpanzees in Bossou. Thus, there is considerable variation among chimpanzee sites, and no clear difference in mean operational IBI between species.

There was, however, a difference in variance of the operational IBI between chimpanzees and bonobos. Significant differences were found between Wamba and Taï ($F = 2.5$, $df_1 = 26$, $df_2 = 13$, $p < 0.05$) and Wamba and Mahale ($F = 3.3$, $df_1 = 26$, $df_2 = 13$, $p < 0.05$). There was also a considerable difference between Wamba and Bossou, though not statistically significant ($F = 2.2$, $df_1 = 26$, $df_2 = 13$, $p = 0.07$). In contrast, there were no significant differences among the chimpanzee sites. Thus the IBI variance is greater for bonobos, which supports the hypothesis that the longer optimal period for producing next offspring explains the lower proceptivity of female bonobos.

DISCUSSION

This study showed that the copulation rate of estrous females (i.e. females in the swelling phase) is lower for bonobos than for chimpanzees. The analyses described above support Hypotheses 1 and 2, which were generated to explain this observation.

Hypothesis 1, that opportunities for copulation are limited for female bonobos, relates to female sexual cycle and male reproductive strategy. Since the proportion of time spent in estrus during an IBI (or during adulthood) is much smaller for chimpanzees than for bonobos, the ESR for chimpanzees is much higher, although it is mitigated by a lower ASR. Given that the copulation rates of adult male chimpanzees and bonobos are similar, a high ESR results in a higher copulation rate for female chimpanzees during estrus.

Hypothesis 2 attributes the low copulation rate of estrous bonobo females to their being less eager for copulation. In both chimpanzees and bonobos, the series of interactions that leads to copulation usually begins with an initial approach or soliciting behaviors such as the raising of a hand, body swaying, or display of penile erection (de Waal 1988; Kano 1992; Nishida *et al.* 1999). By comparing the frequency of these behaviors, Takahata *et al.* (1996) suggested that female chimpanzees initiate copulation more often than do female bonobos. Even in chimpanzees, Goodall (1986) reported that copulation was almost always preceded by a male courtship display in Gombe chimpanzees. However, she did not mention whether an initial approach by females was considered to be an initiation of copulation. Study of more populations using a consistent definition of copulation initiation is needed in order to examine the interspecific difference in more detail.

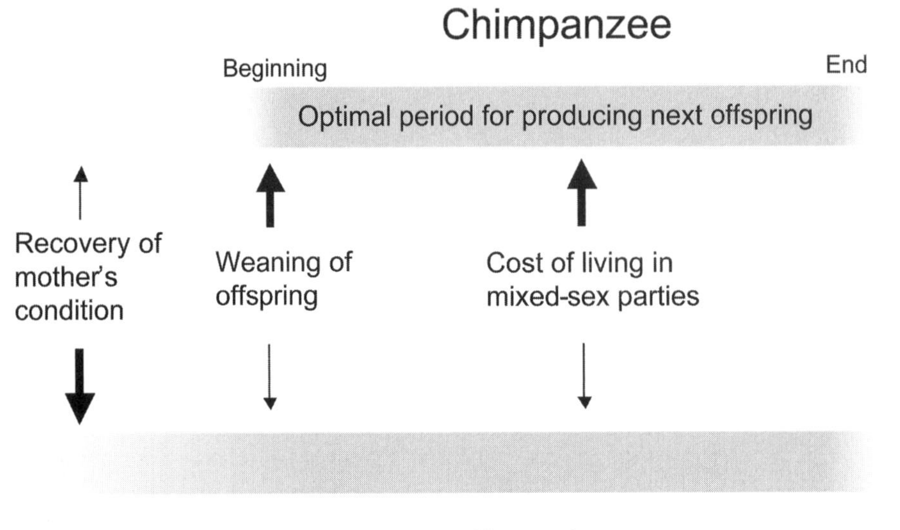

Fig. 11.5. factors influencing the length of optimal period for producing next offspring.

In chimpanzees, non-cycling nursing females tend to range alone, and even cycling females range with fewer males during non-swelling phases (e.g. Nishida 1979; Wrangham 1979; Matsumoto-Oda 1999). Thus, female chimpanzees' association with males differs due to estrus. By contrast, this study shows that the number of males in association with females does not change between the swelling and non-swelling phases. Unlike chimpanzees, female bonobos do not show a higher inclination for association with males due to estrus.

That female bonobos are less eager for copulation seems to be related to the female reproductive strategy. Since they experience a longer optimal period for producing next offspring, female bonobos do not need to copulate as frequently as chimpanzees, whose optimal period is more restricted (*Hypothesis 3*).

Several factors influence the length of the optimal period for producing next offspring (Figure 11.5). The optimal period of chimpanzees may be shorter than that of bonobos because chimpanzees incur high costs by remaining in mixed-sex parties during estrus (Wrangham, Chapter 15). When living among males, females must travel farther to forage than when they are solitary, and they are less successful because the faster-moving males get to the food first (Wrangham, 2000). Furthermore, males may dominate access to food and harass females by using agonistic displays or committing infanticide (Bygott 1979; de Waal 1982; Goodall 1986; Hamai *et al.* 1992; Arcadi & Wrangham 1999).

In comparison, living in mixed-sex parties may be less costly for female bonobos. White & Wrangham (1988) showed that the fruit patches available to bonobos in Lomako are larger than those available to chimpanzees in Gombe, although Chapman *et al.* (1994) found no such difference between bonobos in Lomako and chimpanzees in Kibale. Other studies indicate that terrestrial herbs serve as large food patches for bonobos (Wrangham 1986; Malenky & Stiles 1991). If the average food patch is larger in bonobo habitat, the ranging distance of female bonobos may not differ greatly between periods of solitary versus group living. Wrangham (2000) suggested that the existence of abundant food between fruit patches, for example terrestrial herbs, allows feed-as-you-go foraging for individuals who travel more slowly, and this could lower the cost of group living for females. In addition, due to their high social status, female bonobos are rarely attacked by males and enjoy feeding priority (Wrangham 1993; Parish 1994, 1996; Furuichi 1997; Vervaecke *et al.* 1999). These relatively lower costs of living among males are a likely explanation for why female bonobos usually live in mixed-sex parties regardless of whether they are in estrus (Kano 1982; Furuichi 1987; White 1988; Chapman *et al.* 1994; Gerloff *et al.* 1999).

Early recurrence of cyclic estrus may also be responsible for the longer optimal period for producing next offspring of bonobos. Female chimpanzees resume estrus about 4 years

Fig. 11.6. A female carrying two dependent offspring.

after giving birth (Table 11.1). In contrast, Kano (1992) suggested that female bonobos do so within a year of giving birth. In our unpublished study, we regularly checked the estrous state of six females about a year after parturition; all but one of these females were confirmed to have resumed maximal swelling in 15 or fewer months (<8, <9, <10, >12, or, in two cases, 15 months; resumption of swelling was not observed for the female recorded as >12 due to the absence of observers after twelfth month). Of course, after resuming estrus, female bonobos may not begin ovulation immediately and be able to conceive. However, female bonobos sometimes give birth after only a short interval and carry two dependent offspring simultaneously. From 1990 to 1991, two of nine females in our provisioned E1 group each had two offspring. In 1991, we also observed two females in the neighboring P group, which seldom appeared at feeding sites (Furuichi *et al.* 1998), each carrying two infants, thus suggesting that this phenomenon occurs in both wild and provisioned groups (Figure 11.6). Resumption of estrus seems to be restricted by the weaning of offspring in both chimpanzees and bonobos, since caring for two offspring simultaneously is costly. However, this cost seems to be lower for bonobos due to the availability of larger food patches or feed-as-you-go foraging.

One challenge to *Hypothesis 3* is the situation of western chimpanzees in Taï. Boesch & Boesch-Achermann (2000) reported that female western chimpanzees in the Taï Forest tend to spend more time among males than do female eastern chimpanzees, suggesting that the cost of living in a mixed-sex party is lower for those females. Furthermore, like female bonobos, more than half of the western females resumed estrus within 1 year of parturition (Boesch & Boesch-Achermann 2000). Since the dense forest habitat of Taï chimpanzees is more like the habitat of bonobos than that of studied eastern chimpanzees, the optimal period for producing next offspring may be longer for Taï chimpanzees compared with the eastern chimpanzees. If so, our assumption that the shorter optimal period for producing next offspring reduces the variance in the IBI is violated because the variance of the IBI of Taï chimpanzees is as small as those of other western and eastern chimpanzees. Although this study mainly compared bonobos and eastern chimpanzees, for which data on reproductive parameters and copulation rate of both sexes are available, further study of western chimpanzees will clarify whether differences in sexual behaviors and reproduction are due to species differences or environmental differences.

The result of this study, that is female bonobos copulate less frequently than female chimpanzees during the swelling phase, may seem to contradict the reputation that bonobos are 'hypersexual.' However, it is true that bonobos frequently do use non-copulatory sexual behaviors to regulate tension or enhance social bonds, and that female bonobos show estrus for a much longer period than eastern chimpanzees. Such prolonged, social use of sexual behaviors enhances the social status of females (Kano 1992; Wrangham 1993; Parish 1994; Furuichi 1997), which reduces their cost of living in mixed-sex parties, and makes their copulatory behaviors during the swelling phase relatively quiet.

ACKNOWLEDGEMENTS

We thank the Research Center for Ecology and Forestry (CREF) of the Democratic Republic of Congo for supporting the study of bonobos at Wamba, and the National Council for Science and Technology, the Wildlife Authority, and the Forest Department of Uganda for supporting the study of chimpanzees in the Kalinzu Forest. We also thank Dr T. Kano, Dr T. Nishida, Dr S. Uehara, and members of the Primate Research Institute, Kyoto University, for their support and for valuable discussions. Finally, we thank Dr R. W. Wrangham and Dr Y. Sugiyama for allowing us to use their unpublished data. This study was financially supported by grants under the Monbusho International Scientific Research Program to T. Nishida (58041025, 59043022, 60041020, 61043017, 62041021, 63043017), T. Kano (63041078), and T. Furuichi (01790353, 12575017), and the Monbusho COE Program to O. Takenaka (10CE2005).

REFERENCES

Arcadi, A. C. & Wrangham, R. W. (1999). Infanticide in chimpanzees: review of cases and a new within-group observation from the Kanyawara study group in Kibale National Park. *Primates*, 40, 337–51.

Beach, F. A. (1976). Sexual attractivity, proceptivity, and receptivity in female mammals. *Hormones and Behavior*, 7, 105–38.

Boesch, C. & Boesch-Achermann, H. (2000). *The Chimpanzees of the Taï Forest: Behavioural Ecology and Evolution*. Oxford: Oxford University Press.

Bygott, J. D. (1979). Agonistic behavior, dominance, and social structure in wild chimpanzees of the Gombe National Park. In *The Great Apes*, ed. D. A. Hamburg & E. R. McCown, pp. 405–27. Menlo Park: Benjamin/Cummings.

Chapman, C. A., White, F. J. & Wrangham, R. W. (1994). Party size in chimpanzees and bonobos. In *Chimpanzee Cultures*, ed. R.W. Wrangham, W. C. McGrew, F. B. M. de Waal & P. G. Heltne, pp. 41–57. Cambridge, MA: Harvard University Press.

Dahl, J. F. (1986). Cyclic perineal swelling during the intermenstrual intervals of captive female pygmy chimpanzees (*Pan paniscus*). *Journal of Human Evolution*, 15, 369–85.

Dixson, A.F. (1998). *Primate Sexuality: Comparative Studies of the Prosimians, Monkeys, Apes, and Human Beings.* Oxford: Oxford University Press.

Furuichi, T. (1987). Sexual swelling, receptivity and grouping of wild pygmy chimpanzee females at Wamba, Zaire. *Primates*, 28, 309–18.

Furuichi, T. (1989). Social interactions and the life history of female *Pan paniscus* in Wamba, Zaire. *International Journal of Primatology*, 10, 173–97.

Furuichi, T. (1992). The prolonged estrus of females and factors influencing mating in a wild group of bonobos (*Pan paniscus*) in Wamba, Zaire. In *Topics in Primatology, Vol. 2, Behavior, Ecology, and Conservation*, ed. N. Itoigawa, Y. Sugiyama, G. P. Sackett & R. K. R. Thompson, pp.179–90. Tokyo: University of Tokyo Press.

Furuichi, T. (1997). Agonistic interactions and matrifocal dominance rank of wild bonobos (*Pan paniscus*) at Wamba. *International Journal of Primatology*, 18, 855–75.

Furuichi, T., Idani, G., Ihobe, H., Kuroda, S., Kitamura, K., Mori, A., Enomoto, T., Okayasu, N., Hashimoto, C. & Kano, T. (1998). Population dynamics of wild bonobos (*Pan paniscus*) at Wamba. *International Journal of Primatology*, 19, 1029–43.

Gerloff, U., Hartung, B., Fruth, B., Hohmann, G. & Tautz, D. (1999). Intracommunity relationships, dispersal pattern and paternity success in a wild living community of bonobos (*Pan paniscus*) determined from DNA analysis of faecal samples. *Proceedings of the Royal Society of London, Series B*, 266, 1189–95.

Goodall, J. 1986. *The Chimpanzees of Gombe: Patterns of Behavior.* Cambridge, MA: Belknap Press of Harvard University Press.

Hamai, M., Nishida, T., Takasaki, H. & Turner, L. A. (1992). New records of within-group infanticide in wild chimpanzees. *Primates*, 33, 151–162.

Hasegawa, T. (1991). Rankon shakai no nazo (Sexual behavior of wild chimpanzees). In *Saru No Bunka-shi* (*The Cultural History of Primates*), ed. T. Nishida, K. Izawa & T. Kano, pp. 371–88. Tokyo: Heibon-sha (in Japanese).

Hasegawa, T. & Hiraiwa-Hasegawa, M. (1983). Opportunistic and restrictive matings among wild chimpanzees in the Mahale Mountains Tanzania. *Journal of Ethology*, 1, 75–85.

Hasegawa, T. & Hiraiwa-Hasegawa, M. (1990). Sperm competition and mating behavior. In *The Chimpanzees of the Mahale Mountains: Sexual and Life History Strategies*, ed. T. Nishida, pp. 115–32. Tokyo: University of Tokyo Press.

Hashimoto, C. (1997). Context and development of sexual behavior of wild bonobos (*Pan paniscus*) at Wamba, Zaire. *International Journal of Primatology*, 18, 1–21.

Hashimoto, C. & Furuichi, T. (1994). Social role and development of non-copulatory sexual behavior of wild bonobos. In *Chimpanzee Cultures*, ed. R. W. Wrangham, W. C. McGrew, F. B. M. de Waal & P. G. Heltne, pp. 155–68. Cambridge, MA: Harvard University Press.

Hashimoto, C., Furuichi, T. & Tashiro, Y. (2001). What factors affect the size of chimpanzee parties in the Kalinzu Forest, Uganda? Examination of fruit abundance and number of estrous females. *International Journal of Primatology*, 22, 947–59.

Idani, G. (1991). Social relationships between immigrant and resident bonobo (*Pan paniscus*) females at Wamba. *Folia Primatologica*, 57, 83–95.

Kano, T. (1982). The social group of pygmy chimpanzees (*Pan paniscus*) of Wamba. *Primates*, 23, 171–88.

Kano, T. (1989). The sexual behavior of pygmy chimpanzees. In *Understanding Chimpanzees*, ed. P. G. Heltne & L. Marquardt, pp. 176–83. Cambridge, MA: Harvard University Press.

Kano, T. (1992). *The Last Ape: Pygmy Chimpanzee Behavior and Ecology*. Stanford: Stanford University Press.

Kuroda, S. (1980). Social behavior of the pygmy chimpanzees. *Primates*, 21, 181–97.

Malenky, R. K. & Stiles, E. W. (1991). Distribution of terrestrial herbaceous vegetation and its consumption by *Pan paniscus* in the Lomako forest, Zaire. *American Journal ofPrimatology*, 23, 153–69.

Martin, D. E., Graham, C. E. & Gould, K. G. (1978). Successful artificial insemination in the chimpanzee. *Symposia of the Zoological Society of London*, 43, 249–60.

Matsumoto-Oda, A. (1999). Mahale chimpanzees: grouping patterns and cycling females. *American Journal of Primatology*, 47, 197–207.

Nishida, T. (1979). The social structure of chimpanzees of the Mahale Mountains. In *The Great Apes*, ed. D. A. Hamburg & E. R. McCown, pp. 481–9. Menlo Park: Benjamin/Cummings.

Nishida, T., Takasaki, H. & Takahata, Y. (1990). Demography and reproductive profiles. In *The Chimpanzees of the Mahale Mountains: Sexual and Life History Strategies*, T. Nishida, pp. 63–97. Tokyo: University of Tokyo Press.

Nishida, T., Kano, T., Goodall, J., McGrew, W. C. & Nakamura, M. (1999). Ethogram and ethnography of Mahale chimpanzees. *Anthropological Science*, 107, 141–88.

Parish, A. R. (1994). Sex and food control in the 'uncommon chimpanzee': how bonobo females overcome a phylogenetic legacy of male dominance. *Ethology and Sociobiology*, 15, 157–79.

Parish, A. R. (1996). Female relationships in bonobos (*Pan paniscus*). Evidence for bonding, cooperation, and female dominance in a male-philopatric species. *Human Nature*, 7, 61–96.

Sugiyama, Y. (1994). Age-specific rate and lifetime reproductive success of chimpanzees at Bossou, Guinea. *American Journal of Primatoogy*, 32, 311–18.

Takahata, Y., Ihobe, H. & Idani, G. (1996). Comparing copulations of chimpanzees and bonobos: do females exhibit proceptivity or receptivity? In *Great Ape Societies*, ed. W. C. McGrew, L. F. Marchant & T. Nishida, pp. 146–55. Cambridge: Cambridge University Press.

Thompson-Handler, N. E. (1990). Sexual cycles of pygmy chimpanzees. A Dissertation Presented to the Faculty of the Graduate School of Yale University.

Thompson-Handler. N., Malenky, R. K. & Badrian. N. (1984). Sexual behavior of *Pan paniscus* under natural conditions in the Lomako Forest Equateur Zaire. In *The Pygmy Chimpanzee*, ed. R. L. Susman, pp. 347–68. New York: Plenum Press.

Trivers, R. L. (1985). *Social Evolution*. Menlo Park: Benjamin/Cummings.

Tutin, C. E. G. (1979a). Mating patterns and reproductive strategies in a community of wild chimpanzees (*Pan troglodytes schweinfurthii*). *Behavioral Ecology and Sociobiology*, 6, 29–38.

Tutin, C. E. G. (1979b). Responses of chimpanzees to copulation, with special reference to interference by immature individuals. *Animal Behaviour*, 27, 845–54.

Tutin, C. E. G. & McGinnis, P. R. (1981). Chimpanzee reproduction in the wild. In *Reproductive Biology of the Great Apes*, ed. C. E. Graham, pp. 239–64. New York: Academic Press.

Vervaecke, H. D., Vries, H. & van Elsacker, L. (1999). An experimental evaluation of the consistency of competitive ability and agonistic dominance in different social contexts in captive bonobos. *Behaviour*, 136, 423–42.

de Waal, F. B. M. (1982). *Chimpanzee Politics: Power and Sex among Apes*. New York: Harpers & Row.

de Waal, F. B. M. (1987). Tension regulation and nonreproductive functions of sex in captive bonobos (*Pan paniscus*). *National Geographic Research*, 3, 318–35.

de Waal, F. B. M. (1988). The communicative repertoire of captive bonobos (*Pan paniscus*), compared to that of chimpanzees. *Behaviour*, 106, 183–251.

Wallis, J. (1997). A survey of reproductive parameters in the free-ranging chimpanzees of Gombe National Park. *Journal of Reproduction and Fertility*, 109, 297–307.

White, F. J. (1988). Party composition and dynamics in *Pan paniscus*. *International Journal of Primatology*, 9, 179–93.

White, F. J. & Wrangham. R. W. (1988). Feeding competition and patch size in the chimpanzee species *Pan paniscus* and *Pan troglodytes*. *Behaviour*, 105, 148–64.

Wrangham, R. W. (1979). Sex differences in chimpanzee dispersion. In *The Great Apes*, ed. D. A. Hamburg & E. R. McCown, pp. 481–89. Menlo Park: Benjamin/Cummingss.

Wrangham, R. W. (1986). Ecology and social relationships in two species of chimpanzee. In *Ecological Aspects of Social Evolution*, ed. D. L. Rubenstein & R. W. Wrangham, pp. 352–78. Princeton: Princeton University Press.

Wrangham, R. W. (1993). The evolution of sexuality in chimpanzees and bonobos. *Human Nature*, 4, 47–79.

Wrangham, R. W. (2000). Why are male chimpanzees more gregarious than mothers? A scramble competition hypothesis. In *Primate Males: Causes and Consequences of Variation in Group Composition*, ed. P. M. Kappeler, pp. 248–58. Cambridge: Cambridge University Press.

12 • Social relationships between cycling females and adult males in Mahale chimpanzees

AKIKO MATSUMOTO-ODA

INTRODUCTION

Unlike the majority of mammals, anthropoid primate societies are typically characterized by the formation of constant heterosexual groups (e.g. Hrdy & Whitten 1987; van Schaik & Kappeler 1997). However, individuals in a group do not interact with all members equally. Theoretically, it is expected that individuals cooperate according to the degree of shared genes (Hamilton 1964), and, in general, individuals form affiliative relationships with kin. After sexual maturity, individuals have no choice but to form heterosexual affiliative relationships with non-kin because the individuals of one or both sexes transfer from their natal group.

Some studies have found that affiliative relationships are formed between males and estrous females, and lasting pair bonds between the sexes also exist. To begin, it is useful to distinguish between two kinds of affiliative relationships using the terms 'short-term bonds' and 'long-term bonds.' The duration of these bonds, or the 'term,' is often determined by events that disrupt the female's reproductive state, such as births or demographic changes in the composition of the unit groups. The short-term bond refers to a sexual relationship formed by complementary interactions, typically encompassing a mating season or several estrous cycles. In contrast, long-term bonds are formed by mutual interactions, lasting through the mating season and lactating period, and often extending over several years. 'Friends' in anubis baboons (*Papio anubis*), 'particular proximate relationships' in Japanese macaques (*Macaca fuscata*), and similar relationships in rhesus macaques (*Macaca mulatta*) are well-known examples of long-term bonds (Smuts 1987; Takahata 1982a,b; Chapais 1983a,b,c).

In chimpanzees (*Pan troglodytes*), social interactions between anestrous females and males were thought to be rare (e.g. McGrew 1996) and long-term bonds between females and males may have received less attention. Only one long-term bond has been reported from Gombe. An adult female, *Fifi*, groomed two males, *Satan* and *Evered*, who had been playmates in childhood, and she played with, received meat from, and was often attacked by *Satan* (Goodall, 1986). In a recent study of bonobos (*Pan paniscus*), Hohmann *et al.* (1999) showed that long-term bonds occurred predominantly between heterosexual dyads and involved not only close kin, but also unrelated individuals. Moreover, the first-ranking male tended to be involved in bonds with non-relatives.

In contrast, in both chimpanzees and bonobos, it has been reported of short-term bonds that the associations in unrelated female–male dyads are weaker than those in related dyads or male–male dyads (Furuichi & Ihobe 1994). Although corresponding data for bonobos are not available, 'consortships' are the most typical short-term bonds of unrelated female–male dyads in chimpanzees. In a consortship, a female and male travel together, away from other members of the group, and maintain an exclusive mating relationship (Tutin 1979). After Tutin's study in Gombe, consortships were also reported in Mahale and Taï (Boesch & Boesch-Achermann 2000). The studies from Gombe, Mahale, and Taï consistently reported that not all members entered consortships: higher-ranking and younger, low-ranking males, and younger and older females were the primary participants (Tutin 1979; Hasegawa 1987; Goodall 1986; Wallis 1997; Boesch & Boesch-Acherman 2000). Tutin (1979) suggested that males who spent more time with females in estrus, shared meat, and groomed them, were the preferred partners in a consortship, but no data were reported. Although some other style of heterosexual, short-term bond might exist, there have been few studies about it.

Unlike bonobos, female chimpanzees spend a relatively short proportion of their reproductive lives in estrus (Takahata *et al.* 1996). In female bonobos at Wanba, the interbirth cycling period and percentage of time in estrus was 34.0 months and 50% of 42 days, respectively (Furuichi 1987; Kano 1992). In contrast, the interbirth cycling period of female chimpanzees at Gombe, Mahale, and Taï was 11.7, 11.2, and 37.0 months, respectively (Goodall 1986; Nishida *et al.* 1990; Takahata *et al.* 1996; Pusey *et al.* 1997; Wallis 1997; Boesch & Boesch-Achermann 2000). Estrus occupies only one-third of the 35-day estrous cycle. It has been said that cycling females only associate with other community members when in estrus, but do they switch their social relationships on and off so easily? At Gombe, Kanyawara, and Ngogo, females spend a large amount of time alone, except when in estrus, while female gregariousness is high at Taï and Mahale (Wrangham & Smuts 1980; Wrangham *et al.* 1992; Boesch 1996; Matsumoto-Oda 1999a; Pepper *et al.* 1999). At Mahale, cycling females (showing a cyclic pattern of swelling) usually range as members of heterosexual parties, regardless of their estrous state. Males and cycling females in anestrus could have opportunities for affiliative interaction because they range in the same parties. By knowing how females associate with males, we will be able to determine the social diversity of chimpanzee culture under conditions such as those at Mahale.

The purpose of this study was to investigate the short-term bonds between unrelated males and cycling females at Mahale. In order to determine whether, and how, heterosexual bonds function to benefit male and female chimpanzees, it seemed useful to study the nature of the relationships between cycling females and resident males in a well-studied population. This was done by monitoring who cycling females interacted with during their estrous and anestrous periods, and determining which sex worked to maintain the relationship. Finally, it seemed important to assess the quality of heterosexual dyads and classify them as either 'affiliative dyads' or 'non-affiliative dyads' (see below for details). The frequency of copulation, grooming, and other measures of support could then be compared between these dyads.

METHODS

Study site and subjects

The subjects were the M group of chimpanzees (*Pan t. schweinfurthii*) in the Mahale Mountains National Park, Tanzania. The chimpanzee studies at Mahale have been conducted since 1965 (Nishida 1990). Observations were made from March 1993 to February 1994. The M group, including infants, contained approximately 83 individuals during the study. There were nine adult males (≥16 years old) and 27 adult females (≥14 years old, Hiraiwa-Hasegawa *et al.* 1984).

Table 12.1. *Name and age of focal females, and observation time*

Name	Age	Observation time (h)		Observation days
		Estrus	Anestrus	
MJ	13	40.8	33.3	21
XT	19	27.7	32.4	17
NK	23	34.8	31.3	17
GW	31	44.0	43.5	21
WL	34	27.1	42.4	18
WA	39	26.7	20.2	14
Total		201.1	203.1	108

Collection and analysis of data

Ten adult females who were neither pregnant nor displaying cyclic genital swelling, were called *cycling females*. Pregnant, lactating, or post-reproductive, non-cycling females were regarded as adult females. Six out of ten cycling females were subjects and observed using focal animal sampling (Table 12.1). I followed each subject for an entire day, or for as long as possible. Of all cycling female–male dyads, one adult female, *NK*, and an adult male, *NS*, were related to each other (aunt–nephew). *NK* immigrated into M group from K group in 1981, and *NS* was born in 1973. One 13-year-old female, *MJ*, was included as a subject because she turned 14 years old during the study period and began to attract males sexually.

I recorded the name of all individuals, the turgidity of cycling females according to three classifications (flat, partial swelling, maximal swelling), and social behaviors observed while following focal individuals. Genital swelling can occur during lactation and pregnancy, but the absence of pregnancy in subjects was confirmed by monthly urine tests using 'TESTPACK PLUS (ABBOT Corp., Tokyo, Japan).' The mean duration of the estrous cycle, maximal swelling, and partial swelling was 35.3 (range = 28–39), 11.3 (range = 8–15), and 1.6 (range = 0–4.5) days, respectively. The duration of the flat period was calculated to be 20.5 days. The periods of partial swelling were excluded from the analysis because almost all copulations occurred when the females exhibited maximal swelling, and part of the partial swelling period might have been caused by injury to the sexual skins (Matsumoto-Oda 1998). The last 4 days of the period of maximal swelling were regarded as the ovulation period (Graham 1981). The number of observation hours of the ovulation periods was 48.0 h. (The daily mean focal sampling time for females was 3.7 ± 2.8 h, although all-day follows were conducted whenever possible.) An *estrous female* was defined as a cycling female with maximal swelling, while an *anestrous female* was defined as a cycling female with a flattened genital swelling. Five days on, which males behaved possessively toward females, such as a male trying to pull a female away from another male, or interrupting copulation between the subject and another male for a period exceeding one hour (Tutin 1979), were excluded from the analysis because these behaviors can clearly influence the activities and association patterns of the female.

Rank of males was determined from the occurrence of dyadic interactions, such as pant-grunt and agonistic behavior (Bygott 1979). Adult males were divided into three categories according to their rank: three high-ranking males; three middle-ranking males; three low-ranking males (Hosaka, unpublished data).

The *association index* was used as a measure of which two individuals participated in the same parties. Since chimpanzees have a fission–fusion social system, the members of a party (ranging unit) change, and multiple independent parties are often formed. In this study, parties ranging beyond auditory contact with neighboring parties during the observations were assumed not to meet during the course of an entire day, and were counted as independent parties. Each stable party that persisted for 1 day was regarded as a basic unit, and an individual was only considered involved with one party per day (cf. Matsumoto-Oda *et al.* 1998). The collected records of attendance were changed into binary data, with the basic unit being one day. I observed at least one chimpanzee per day for 237 days. The association index between individual *A* and *B* was calculated as:

number of days when both *A* and *B* were observed in the same parties/number of days when at least *A* or *B* was observed

Two adult lactating females observed on less than 10% of the total observation days were excluded from the analysis.

The *proximity index* for each dyad was calculated as:

the total time that *A* and *B* were within a 10-m radius of each other when *A* was the focal subject/the total duration of focal sampling of *A* when both *A* and *B* were in the same party

To investigate who maintained proximity, Hinde's Index of proximity maintenance for male and female dyads was calculated for each dyad (Hinde & Atkinson 1970), using the equation:

$$Af/(Af+Am) - Lf/(Lf+Lm) \qquad (12.1)$$

where Af is the number of approaches (within a 10-m radius) made to a male by a subject female, Am is the number of approaches made to a subject female by a male, Lf is the number of times a subject female leaves a male, and Lm is the number of times a male leaves a subject female. If the value is positive, the agent maintaining the dyad's proximity is the female, if the value is negative, the agent of proximity is the male, and if the value is zero, there is no direction of proximity and it is called 'neutral.' For each agent of proximity maintenance, the number of dyads was calculated per female, taking into account the rank of the male involved in that particular dyad. The distance of 10 m was chosen as the standard because the undergrowth in Mahale is so dense that in many cases chimpanzees more than 10 m from the focal individual cannot be precisely identified. For each dyad, the general coding of index values (increase, no change, and decrease) was analyzed for two conditions, estrus and anestrus.

A *grooming bout* was defined as continuous grooming without interruption for more than 2 minutes (Goodall 1986). When the partner or direction of grooming changed, the bout was considered to have ended.

Previous studies have used proximity and grooming as measures of affinity between individuals (cf. Takahata 1982a,b; Smuts 1987). A heterosexual dyad that remained in proximity, or spent more time grooming per observation hour than the standard (mean + SD), was defined as an *affiliative dyad*. An affiliative dyad overlapping both estrous and anestrous periods was called *continuously affiliative dyad (COA dyad)*.

It is difficult for an observer to confirm ejaculations in the wild. Consequently, modeling previous studies, *copulation* was defined as a heterosexual interaction that included at least one intromission; a subsequent intromission occurring within 10 minutes was considered to be the same mating (Tutin 1979; Hasegawa & Hiraiwa-Hasegawa 1983).

I recorded the number of successful hunts and the co-individuals who were fed meat during focal follows, and ad libitum. Meat sharing includes both active and passive transfer (cf. Boesch & Boesch 1989). The frequency of female meat-receiving in each dyad was calculated as the

Table 12.2. *Association indices, proximity indices and percentage of time spent grooming of female–male dyads and female–female dyads. Analyses consider two reproductive states of focal females (estrous versus anestrous). Results of a three-way repeated measure ANOVA are also shown*

	Association indices			Proximity indices			Percentage of time spent grooming		
Dyads	mean	SD	(*n*)	mean	SD	(*n*)	mean	SD	(*n*)
Estrous female–male	72.5 ±	17.8	(54)	12.0 ±	1.9	(54)	6.4 ±	11.2	(54)
Estrous female–female	37.2 ±	24.8	(143)	7.3 ±	9.4	(135)	1.1 ±	4.5	(135)
Anestrous female–male	70.9 ±	16.0	(54)	8.7 ±	9.4	(54)	5.9 ±	10.9	(54)
Anestrous female–female	44.4 ±	17.8	(143)	7.5 ±	13.2	(135)	1.2 ±	4.7	(135)
Statistics									
a. partner's sex	$F=191$, df$=1$, $p<0.01$			$F=5.9$, df$=1$, $p<0.05$			$F=40.7$, df$=1$, $p<0.01$		
b. estrous state	$F=1.6$, df$=1$, ns			$F=1.3$, df$=1$, ns			$F=0.1$, df$=1$, ns		
c. subjects	$F=5.2$, df$=5$, $p<0.01$			$F=2.7$, df$=5$, $p<0.05$			$F=5.4$, df$=5$, $p<0.01$		
a×b	$F=4.0$, df$=1$, $p<0.05$			$F=2.4$, df$=1$, ns			$F=0.1$, df$=1$, ns		
b×c	$F=0.4$, df$=5$, ns			$F=3.2$, df$=5$, $p<0.01$			$F=5.9$, df$=5$, $p<0.01$		
c×a	$F=0.5$, df$=5$, ns			$F=3.1$, df$=5$, $p<0.05$			$F=2.2$, df$=5$, ns		
a×b×c	$F=0.4$, df$=5$, ns			$F=2.8$, df$=5$, $p<0.05$			$F=1.9$, df$=5$, ns		

number of food transfers from the dyad male divided by the number of successful hunts by that male.

Direct aggression towards cycling females (e.g. chasing, hitting) and supportive behavior (e.g. interrupting aggression) from a male or a female during aggressive interactions are also recorded.

Statistics

For the association indices, proximity indices, and time spent grooming in cycling female–male dyads and in cycling female–female dyads, a three-way repeated measures analysis of variance (ANOVA) was employed for matched-pair comparisons of interactions among individual differences of subjects, estrous state (estrus or anestrus), and partner's sex (male or female). For the association indices, proximity indices, and grooming in cycling female–male dyads, three-way repeated measures ANOVA was also used for matched-pair comparisons of interactions among individual differences of subjects, estrous state, and rank of males (high, middle, or low). When there was a significant difference for rank of males by the ANOVAs, the Games-Howell post hoc test was used. The Kruskal–Wallis test was employed to test differences among the three types of proximity maintenance. To investigate proximity maintenance, changes in Hinde's index values (increase, no change, or decrease) were tested by χ^2-test and G-test. Significance of deviation from expected values was tested by Fisher's exact probability test for affiliative dyads. A t-test was used for matched-pair comparisons of interactions between affiliative and non-affiliative dyads.

RESULTS

The sex of social partners

I compared association, proximity, and grooming of cycling females with males under two different conditions, estrous and anestrous. Although the association indices showed individual differences among the cycling females, cycling females associated more often with males than with females, regardless of estrous state (Table 12.2).

Estrous state did not influence time spent alone (i.e. no other adults within a 10-m radius of the focal cycling females), but did affect time spent near any male. The mean proportion of times alone for estrous females and anestrous females was

Table 12.3. *Association indices, proximity indices, and percentage of time spent grooming of female–male dyads. Analyses consider two reproductive states of focal females (estrous versus anestrous) and three rank classes of males (high, middle, low). Results of a three-way repeated measure ANOVA are also shown*

	Association indices			Proximity indices			Percentage of time spent grooming		
Dyads	mean	SD	(n)	mean	SD	(n)	mean	SD	(n)
Estrous female–high-ranking male	81.1 ±	9.8	(18)	16.2 ±	12.3	(18)	4.1 ±	6.4	(18)
Estrous female–middle ranking male	73.0 ±	17.4	(18)	9.7 ±	8.3	(18)	2.0 ±	2.9	(18)
Estrous female–low ranking male	63.3 ±	20.5	(18)	8.0 ±	11.0	(18)	0.9 +	1.7	(18)
Anestrous female–high-ranking male	77.8 ±	11.0	(18)	15.6 ±	12.4	(18)	5.9 ±	7.5	(18)
Anestrous female–middle-ranking male	70.9 ±	14.0	(18)	5.8 ±	5.1	(18)	1.3 ±	2.8	(18)
Anestrous female–low-ranking male	64.0 ±	19.4	(18)	4.4 ±	4.2	(18)	0.7 ±	1.8	(18)
Statistics									
a. partner's rank	$F=7.4$, df$=2$, $p<0.01$			$F=13.1$, df$=2$, $p<0.01$			$F=7.6$, df$=2$, $p<0.01$		
b. estrous state	$F=0.2$, df$=1$, ns			$F=2.7$, df$=1$, ns			$F=0.04$, df$=1$, ns		
c. subjects	$F=1.5$, df$=5$, ns			$F=1.4$, df$=5$, ns			$F=2.0$, df$=5$, ns		
a×b	$F=0.1$, df$=2$, ns			$F=0.4$, df$=2$, ns			$F=0.1$, df$=2$, ns		
b×c	$F=0.1$, df$=5$, ns			$F=5.1$, df$=5$, $p<0.01$			$F=0.9$, df$=5$, ns		
c×a	$F=0.2$, df$=10$, ns			$F=1.5$, df$=10$, ns			$F=1.2$, df$=10$, ns		
a×b×c	$F=0.1$, df$=10$, ns			$F=0.5$, df$=10$, ns			$F=0.3$, df$=10$, ns		

26.4±10.1% (range=7.2–33.7, $n=6$) and 37.1±11.4% (range=20.7–49.7, $n=6$), respectively. The difference was not significant (t-test, $t=-1.6$, df=5, ns). The mean proportion of times spent near any male for estrous females and anestrous females was 60.6±11.4% (range=38.6–70.8, $n=6$) and 36.1±9.5% (range=23.1–70.8, $n=6$). Estrous females spent more time near males than anestrous females (t-test, $t=-4.1$, df=5, $p<0.01$).

Independent of estrous state, cycling females were in proximity more often with males than with females (Table 12.2). The only exception was *GW*, who had higher proximity indices with females than with males. There were 11 cycling female–female dyads with high proximity indices (>30), and *GW* participated in 7 out of 11 of them. The proximity index of *GW–WO* dyads was continuously high: 35.0 in estrus and 59.7 in anestrus.

The maximum length of a grooming bout was 45.8 minutes, while 87.7% (1025/1169) of all bouts ended in less than 5 minutes. Estrous females spent 7.2±1.5 min h^{-1} (range=4.4–8.6, $n=6$) grooming, while anestrous females spent 5.5±1.7 min h^{-1} (range=2.4–7.6, $n=6$). Cycling females spent time grooming more often with males than with females regardless of estrous state (Table 12.2). Only one female, *GW*, spent more of her time grooming with females than with males.

Rank of males

In cycling female–male dyads, the association indices did not differ with estrous state, but they differed by rank of males (Table 12.3). The association indices with high-ranking males were larger than with middle-ranking or low-ranking males ($p<0.05$). The proximity indices showed the similar results. The proximity indices with high-ranking males were larger than with middle-ranking or low-ranking males ($p<0.05$). Time spent grooming in cycling female–male dyads differed by rank of males. In cycling female–male dyads, more time was spent grooming with high-ranking males than with subdominant males ($p<0.05$, high-ranking> middle-ranking, high-ranking>low-ranking).

The changes in proximity indices between estrous and anestrous occurred in 42 dyads (increased for 29 dyads, decreased for 13 dyads). In five dyads, the proximity indices showed no change. There was a difference between observed and expected number of dyads, and it showed that female positivity to proximity maintenance increased from estrous

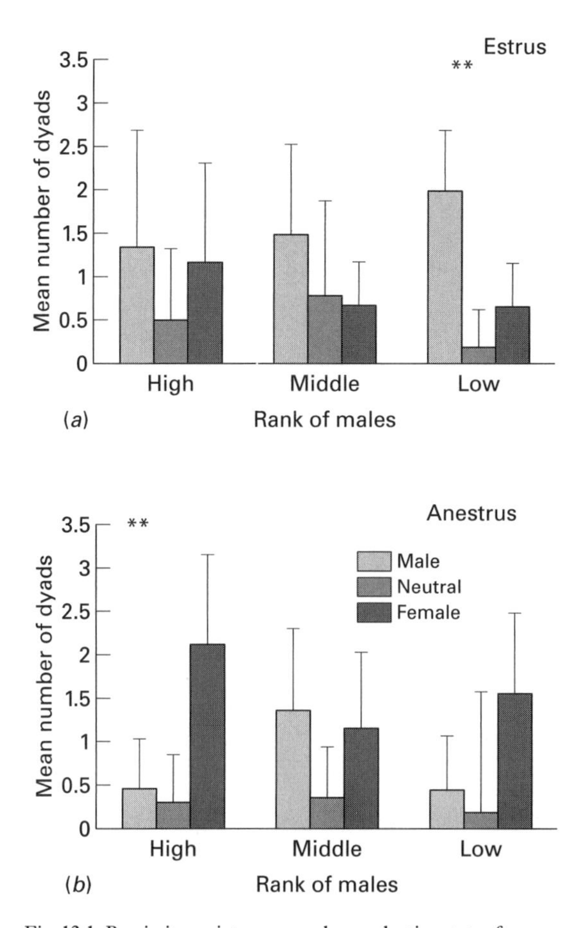

Fig. 12.1. Proximity maintenance and reproductive state of females. Mean (+SD) number of dyads by each proximity agent per male dominance rank. The number of heterosexual approaches and leaving in estrus (*a*) and in anestrus (*b*) were 538 and 266, respectively. $^{**}p<0.01$, Kruskal–Wallis test between agents of proximity per rank of males.

period to anestrous period ($\chi^2=298.7$, df=2, $p<0.01$). There was no significant difference of deviation among the dyads of high-, middle-, and low-ranking males (G=1.2, df =4, ns). Across the ranks of males, most dyads showed increasing index values. The agents of proximity maintenance varied with both the rank of the males and female estrous states (Figure 12.1). When male partners were low-ranking and focal females were in estrus, males maintained proximity more often than female partners (Kruskal–Wallis test, $H=10.06$, df=2, $p<0.01$). There was no significant difference when the dyad males were high- or middle-ranking (high-ranking: $H=1.82$, df=2, ns; middle-ranking: $H=2.37$, df=2, ns). However, when cycling females were in anestrus, there was a significant difference when dyad males were high-ranking ($H=9.74$, df=2, $p<0.01$). There was no significant difference when the dyad males were middle- or low-ranking (middle-ranking: $H=3.92$, df=2, ns; low-ranking: $H=4.03$, df=2, ns).

Social interactions with affiliative males and with non-affiliative males

Of the 54 cycling female–male dyads, the number of affiliative dyads in estrus was 12 and in anestrus was 11 (Table 12.4). Seven were COA dyads that were affiliative dyads overlapping both estrous and anestrous periods. It was greater than expected by chance alone (Fisher=0.00099). Females were more affiliative with high-ranking males. Of those seven affiliative dyads overlapping both estrous and anestrous periods, four dyads included the first-ranking male, two included other high-ranking males, and the other dyads included a single, low-ranking male.

Copulation frequencies did not differ between affiliative dyads in estrus and non-affiliative dyads in estrus (Figure 12.2(*a*), t-test, $t=1.3$, df=52, ns). However, copulation occurred more frequently in COA dyads than in non-COA dyads (Figure 12.2(*b*), t-test, $t=2.3$, df=52, $p<0.05$). Copulations during ovulation periods did not differ between affiliative dyads in estrus and non-affiliative dyads in estrus (t-test, $t=1.2$, df=52, ns), but they occurred significantly more frequently in COA dyads than in non-COA dyads (Figure 12.2(*c*), t-test, t=2.6, df=52, $p<0.05$).

I observed 39 successful hunts by adults. Males possessed the meat in 33 cases and females in six cases. When males held meat, high-ranking males were more often in possession of it than were low-ranking males: first-ranking male (17 cases); other high-ranking males (six cases); middle-ranking (six cases); and low-ranking (four cases). In the cases where cycling females were present during meat-sharing by males, the percentage of estrous females who obtained meat was 37.4±26.4% (range=0–66.7), and the percentage of anestrous females who obtained meat was 19.1 ±19.0% (range=0–41.7). Females with male partners (i.e. mixed-sex dyads) received meat more often during estrus and during anestrus (t-test, $t=2.81$, df=10, $p<0.05$). Regardless of estrous state, meat receiving from COA and non-COA males did not differ (Figure 12.3, t-test, estrus: $t=0.5$, df=18, p=0.6, anestrus: $t=1.0$, df=24, $p=0.3$).

Regardless of the estrous and anestrous state, the percentage of grooming received from COA males was longer

Table 12.4. *Affiliate male partners of each focal female*

Males		Cycling females											
		MJ		*XT*		*NK*		*GW*		*WL*		*WA*	
rank	name	estrus	anestrus	estrus	anestrus	estrus	anestrus	estrus	anestrus	estrus	anestrus	estrus	anestrus
high-ranking	*NT*	○	○	⊙	⊙	⊙	⊙	○	⊙	⊙		○	
	NS				○						⊙	○	○
	DE				○		○	○				○	○
middle-ranking	AJ												
	JI							○					
	TB												
low-ranking	BE					○							
	MU	○	○										
	MA												

Notes:
○ a dyad that remained in proximity *or* groomed more than the standard (mean + standard deviation).
⊙ a dyad that remained in proximity *and* groomed more than the standard (mean + standard deviation).
Grey box shows the continuous affiliative dyad.

than that from non-COA males (Figure 12.4). The differences were significant (t-test, estrus: $t=2.4$, df$=52$, $p<0.05$, anestrus: $t=2.4$, $p<0.05$).

There were five direct acts of aggression. All were directed towards one cycling female, *XT*. Two cases were by males (0.02 ± 0.05 times h^{-1}, $n=6$, range$=0$–0.12) and three cases by females. No support from males was observed.

DISCUSSION

The results presented here indicate that cycling females at Mahale associated with and remained near males more often than with or near other females. Of all the adult males, cycling females most often formed affiliate relationships with higher-ranking males during the estrous cycle. It is important to note that cycling females copulated much more often with males who continued affiliative relationships throughout the estrous cycle than with those who only formed affiliative relationships during estrus. Cycling females obtained meat regardless of estrous state and received more grooming from continuously affiliative males than from non-continuously affiliative males.

Males as important social partners of cycling females

In great apes, females are the emigrating sex. Except for female bonobos, great ape females associate familiarly with males more than with other unrelated females (e.g. Harcourt 1979; Nishida 1979; Wrangham & Smuts 1980; Goodall 1986; White 1989; Hasegawa 1990; Watts 1992; Pepper *et al.* 1999). In recent data from Lomako, female bonobos formed close associations not only with kin, but also with non-kin adult males during and outside their fertile period (Hohmann *et al.* 1999). The sex of association partners that I observed was similar to previous reports. In species with female dispersal, although males may be of less value than female relatives as coalition partners, males are better than nothing for females who lack such relatives (Dunbar 1984).

The gregariousness of great ape societies varies among species. In orangutans (*Pongo pygmaeus*), who are considerably less gregarious, females interact with males only when they are in estrus (Mitani *et al.* 1991; van Schaik 1999). In one-male/multi-female groups of gorillas (*Gorilla gorilla*), groups neither split nor fuse, and the time that females spend near the leading male does not change when the

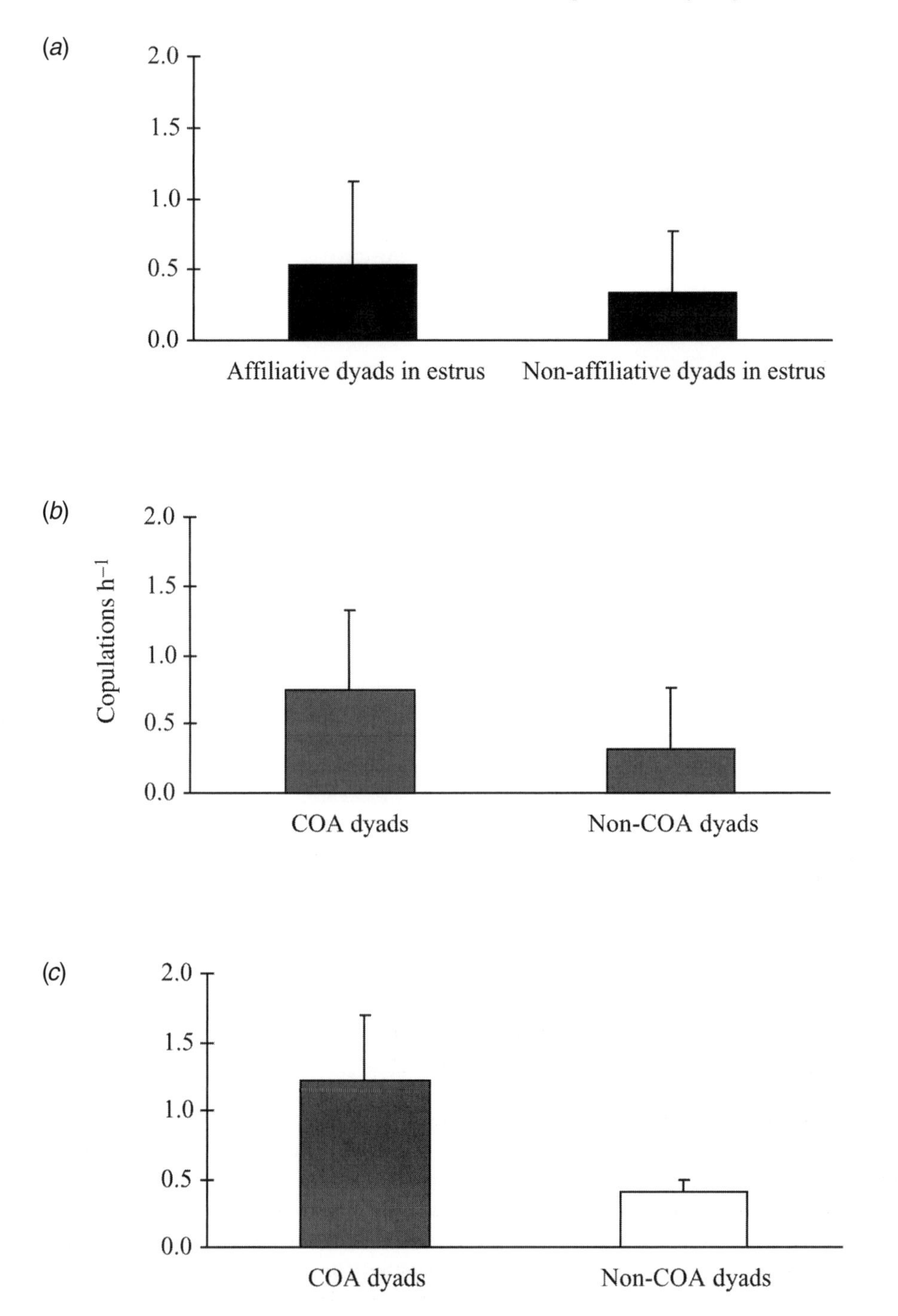

Fig. 12.2. Copulation and affiliation. Mean (+SD) copulation frequency of affiliative dyads and non-affiliative dyads. (*a*) Mean copulation frequency of affiliative dyads in estrus ($n = 12$) and that of non-affiliative dyads in estrus ($n = 42$). (*b*) Mean copulation frequency of continuous affiliative dyads (COA dyads, $n = 7$) and non-COA dyads ($n = 47$). (*c*) Mean copulation frequency in ovulation period of COA dyads ($n = 7$) and non-COA dyads ($n = 47$).

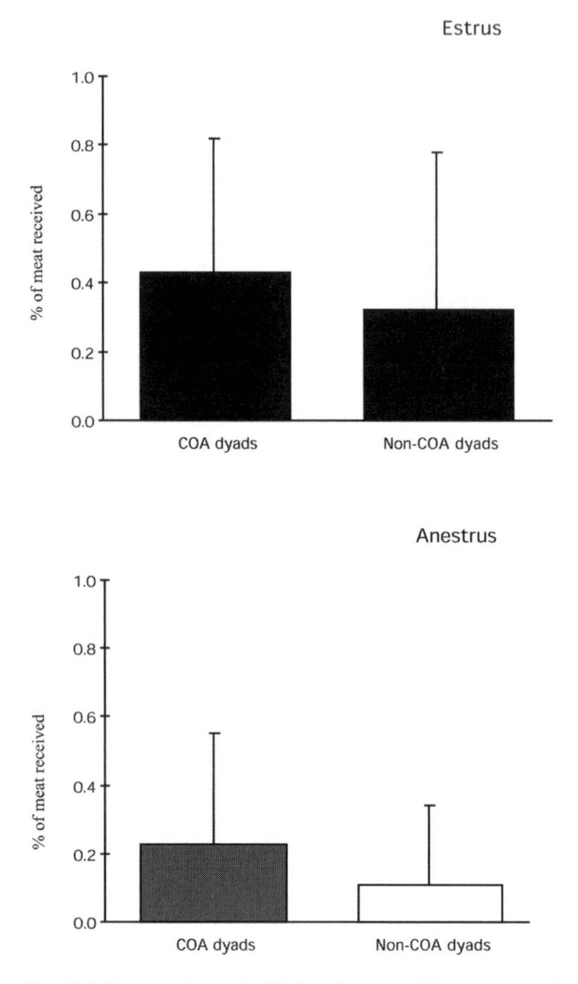

Fig. 12.3. Food sharing and affiliation. Mean (+SD) percentage of meat received from males in continuous affiliative dyads (COA dyads, $n=7$) and non-continuous affiliative dyads ($n=47$).

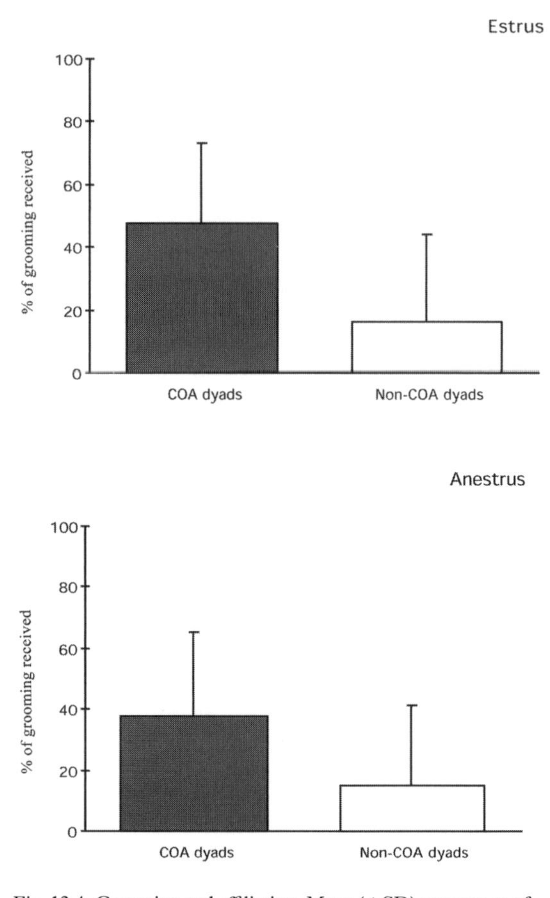

Fig. 12.4. Grooming and affiliation. Mean (+SD) percentage of grooming received from males in continuous affiliative dyads (COA dyads, $n=7$) and non-continuous affiliative dyads ($n=47$). The percentage of grooming received was calculated as follows: the time being groomed by a male was multiplied by 100 and divided by total grooming time with the partner male.

females come into estrus (Harcourt 1979). In multi-male/multi-female groups of chimpanzees, which have a fission–fusion social system, males and females reportedly interact only when food resources are abundant or when estrous females are present (Goodall 1986). The reason usually given for this is that female ranges are smaller than those of males (Wrangham & Smuts 1980; Hasegawa 1990). The image of chimpanzees reported by Goodall (1986) is more orangutan-like than gorilla-like. In contrast to chimpanzees, bonobos are gregarious (Idani 1991; Hohmann *et al.* 1999; Hohmann & Fruth, Chapter 10). In previous studies on Mahale chimpanzees, I have shown that female gregariousness in heterosexual parties changes with their reproductive state (Matsumoto-Oda 1999a). Non-cycling females (lactating, pregnant, or post-menopausal females) participated less in heterosexual parties, whereas anestrous cycling females participated as often as when they were in estrus. Using the measures of association index, proximity index, and grooming time, both anestrous and estrous cycling females associated with males. That is, the pattern of close association between cycling females and males was gorilla- or bonobo-like.

However, the gregariousness seen at Mahale might not

be characteristic of chimpanzees in general. The behavior of cycling females varies with circumstances. Among Ngogo chimpanzees in Uganda, anestrous cycling females do not participate in heterosexual parties (Pepper *et al.* 1999). It is not clear what causes the differences in female gregariousness. For example, in Mahale, disease or old age can decrease association (Huffman 1990; Matsumoto-Oda, unpublished data), and allomothering behavior can increase grooming of an infant's mother, as in the case of *GW* (Nishida 1983; this study). These differences indicate the behavioral flexibility of female chimpanzees. In a fission–fusion social system, chimpanzees can choose the size and composition of the party in which they participate based on balancing costs and benefits. One benefit for estrous females of a party with many males is that it might be a place to find a mate (Matsumoto-Oda 1999a).

Benefits for males and females from affiliative relationships

This study showed that, for the most part, males maintained relationships when cycling females were in estrus. Although the time females spent near any male and the frequency of cycling female–male approaches were greater when cycling females were in estrus, estrous state did not influence the proximity indices in dyads. This means that no dyad male increased proximity time with partner females when they were in estrus, but that several males, one after another, spent time in proximity with estrous females. Female chimpanzees copulate with multiple males several times during an estrous period, and most copulations occur opportunistically (Tutin 1979; Hasegawa & Hiraiwa-Hasegawa 1983; Matsumoto-Oda 1999b). A male may increase his chance of copulation with an estrous female by approaching when other males are not nearby (Watts 1998; Matsumoto-Oda 1999b).

Why do males interact with anestrous females? According to this study, they do so to increase the chance of mating. While cycling females tend to copulate with many males throughout the estrous period, they copulate repeatedly with higher-ranking males near the day of ovulation (Matsumoto-Oda 1999b). In this study, females in estrus copulated more often with continuously affiliative males than with males who were affiliative only during estrus. This strategy suggests that cycling females choose their mate not only during estrus but throughout the estrous cycle. Female choice might lead males to form affiliative relationships with anestrous females.

By contrast, what benefits do cycling females derive from affiliative relationships? Although the relationship between female estrous state and potential benefits from males has not been fully explored, estrous females may obtain some benefits (e.g. Yerkes 1943; Goodall 1986; de Waal 1990; Hemelrijk *et al.* 1992, 1999; Wallis 1992). Working with this premise, this study sought to determine the kinds of benefits anestrous females obtained from affiliative males in contrast with estrous females.

When cycling females were in anestrus, they were responsible for maintaining proximity more often than males. What benefits do anestrous females receive by staying close to males? To answer this question, the benefits conferred by rank of males need to be clarified because most of the affiliative males were higher ranking. In general, the benefits of higher rank are reported to be increased reproductive success and quantity of food, and dominance during aggression (e.g. Hofer & East 1993 for spotted hyena). One of the benefits conferred by a high rank is obtaining meat, which is not only highly prized, but is also only available in relatively small amounts. Stanford (1998) argues that females at Gombe may be present when meat is available, but only succeed in obtaining meat when in estrus. Conversely, estrous females in Ngogo did not consistently obtain meat through their begging efforts at hunts (Mitani & Watts 2001). In our study, estrous females obtained meat more frequently than anestrous females, and the difference occurred with meat sharing from non-continuously affiliative males. Anestrous females obtained meat less frequently from non-continuously affiliative males. However, females constantly obtained meat from continuously affiliative males regardless of estrous state, and a greater portion of meat sharing was performed by the first-ranking male. As far as meat sharing in Mahale is concerned, it seems to be a 'service' provided by non-continuously affiliative males to estrous females. Grooming is another potential benefit both for the one who is groomed and the one who provides the grooming, for example by reducing stress or in providing skin care (Goodall 1986). In this study, cycling females in affiliative relationships received the same level of grooming from males during both the estrous and anestrous periods. The remaining benefit is protection from threat or attack. In this study, all of the aggressive behavior was directed at one estrous female, *XT*, and there is no clear explanation as to why she was the subject of this aggression. In the few instances of aggression observed, she never received support. However, protection from aggression is usually considered one reason for affiliative relationships, because support from the higher-ranking males is observed, albeit rarely (e.g. Goodall 1986). Female gregariousness is considered to balance cost-of-grouping for the female (Wrangham, Chapter 15). Perhaps aggression is so rare at Mahale because female gregariousness is high.

Short-term bonds in chimpanzees

Speculations have been made that copulation and affiliative interactions should lead to a simple form of heterosexual bond (Wrangham 1993). We are apt to link short-term bonds with a sexual relationship formed by complementary interactions that endure over a few estrous cycles. Regardless of estrous state, cycling females formed short-term bonds with some males continuously over the course of a year in a gregarious unit-group at Mahale. There were nine adult males in the group during the study, but cycling females did not interact equally with each male. The weight given to the first-ranking male, *NT* (*Ntologi*), was disproportionately large. Preference for the higher-ranking males was seen in a previous study at Mahale: In 1981, five out of ten dyads that were frequently in close proximity were formed with three higher-ranking males, including *NT* (Takahata 1990). This suggests that cycling females do not prefer specific males, but prefer higher-ranking males as bond partners. The extent to which this is or is not a special feature of dyads awaits further investigation.

ACKNOWLEDGEMENTS

I am grateful to the Tanzania Commission for Science and Technology, Tanzania National Parks, and the Tanzanian Wildlife Research Institute for permission to conduct the present research. I am indebted to C. Boesch, L. F. Marchant and G. Hohmann for discussion and valuable comments on a draft of this manuscript. I thank T. Nishida, T. Kano, T. Enomoto, J. Yamagiwa, Y. Takahata, H. Ihobe, and R. Oda for their comments on earlier drafts; and K. Kawanaka, M. A. Huffman, K. Hosaka, and R. Hawazi for their cooperation in the field. This study forms part of my PhD thesis at Kyoto University. This research was financed by a grant under the Monbusho International Scientific Research Program (no. 03041046 to Prof. T. Nishida), and by a Monbusho Scientific Research Fellowship from the Japan Society for the Promotion of Science for Japanese Junior Scientists (no. 2443 to A.M.O.).

REFERENCES

Boesch, C. (1996). Social grouping in Taï chimpanzees. In *Great Ape Societies*, ed. W. C. McGrew, L. F. Marchant & T. Nishida, pp. 101–13. Cambridge: Cambridge University Press.

Boesch, C. & Boesch, H. (1989). Hunting behavior of wild chimpanzees in the Taï National Park. *American Journal of Physical Anthropology*, 78, 547–73.

Boesch, C. & Boesch-Achermann, H. (2000). *The Chimpanzees of Taï Forest: Behavioural Ecology and Evolution*. Oxford: Oxford University Press.

Bygott, J. D. (1979). Agonistic behavior, dominance, and social structure in wild chimpanzees of the Gombe National Park. In *The Great Apes*, ed. D. A. Hamburg & E. R. McCown, pp. 405–27. Menlo Park: Benjamin/Cummings.

Chapais, B. (1983a). Matriline membership and male rhesus reaching high rank in their natal troops. In *Primate Social Relationships*, ed. R. A. Hinde, pp. 171–5. Oxford: Blackwell.

Chapais, B. (1983b). Structure of the birth season relationship among adult male and female rhesus monkeys. In *Primate Social Relationships*, ed. R. A. Hinde, pp. 200–8. Oxford: Blackwell.

Chapais, B. (1983c). Adaptive aspects of social relationships among adult rhesus monkeys. In *Primate Social Relationships*, ed. R. A. Hinde, pp. 286–9. Oxford: Blackwell.

de Waal, F. B. M. (1990). Sociosexual behavior used for tension regulation in all age and sex combinations among bonobos. In *Pedophilia: Biosocial Dimensions*, ed. J. R. Feierman, pp. 378–93. New York: Springer.

Dunbar, R. I. M. (1984). *Reproductive Decisions: An Economic Analysis of Gelada Baboon Social Strategies*. Princeton, NJ: Princeton University Press.

Furuichi, T. (1987). Sexual swelling receptivity, and grouping of wild pygmy chimpanzee females at Wamba, Zaire. *Primates*, 28, 309–18.

Furuichi, T. & Ihobe, H. (1994). Variation in male relationships in bonobos and chimpanzees. *Behaviour*, 130, 211–28.

Goodall, J. (1986). *The Chimpanzees of Gombe: Patterns of Behavior*. Cambridge, MA: Harvard University Press.

Graham, C. E. (1981). Menstrual cycle physiology of the great apes. In *Reproductive Biology of the Great Apes*, ed. E. Graham, pp. 286–303. New York: Academic Press.

Hamilton, W. D. (1964). The genetical evolution of social behavior, *Journal of Theoretical Biology*, 7, 1–52.

Harcourt, A. H. (1979). Social relationships between adult male and female mountain gorillas in the wild. *Animal Behavior*, 27, 325–42.

Hasegawa, T. (1987). Sexual behaviors in wild chimpanzees. PhD thesis, University of Tokyo (in Japanese).

Hasegawa, T. (1990). Sex differences in ranging patterns. In *The Chimpanzees of the Mahale Mountains*, ed. T. Nishida, pp. 99–114. Tokyo: University of Tokyo Press.

Hasegawa, T. & Hiraiwa-Hasegawa, M. (1983). Opportunistic and restrictive mating among wild chimpanzees in the Mahale Mountains, Tanzania. *Journal of Ethology*, 1, 75–85.

Hemelrijk, C. K., van Laere, G. J. & van Hooff, J. A. R. A. M. (1992). Sexual exchange relationships in captive chimpanzees? *Behavioral Ecology and Sociobiology*, 30, 269–75.

Hemelrijk, C. K., Meier, C. & Martin, R. (1999). 'Friendship' for fitness in chimpanzees? *Animal Behaviour*, 58, 1223–9.

Hinde, R. A. & Atkinson, S. (1970). Assessing the roles of social partners in maintaining mutual proximity, as exemplified by mother–infant relations in rhesus monkeys. *Animal Behaviour*, 18, 169–76.

Hiraiwa-Hasegawa, M., Hasegawa, T. & Nishida, T. (1984). Demographic study of a large-sized unit-group of chimpanzees in the Mahale Mountains, Tanzania. *Primates*, 25, 401–13.

Hofer, H. & East, M. L. (1993). The commuting system of Serengeti spotted hyaenas: how a predator copes with migratory prey. I. Social organization. *Animal Behaviour*, 46, 547–58.

Hohmann, G., Gerloff, U., Tautz, D. & Fruth, B. (1999). Social bonds and genetic ties: kinship, association and affiliation in a community of bonobos (*Pan paniscus*). *Behaviour*, 136, 1219–35.

Hrdy, S. B. & Whitten, P. L. (1987). Patterning of sexual activity. In *Primate Societies*, ed. B. B. Smuts, D. L. Cheney, R. M. Seyfarth, R. W. Wrangham & T. T. Struhsaker, pp. 370–84. Chicago: University of Chicago Press.

Huffman, M. A. (1990). Some socio-behavioral manifestations of old age. In *The Chimpanzees of the Mahale Mountains*, ed. T. Nishida, pp. 237–56. Tokyo: University of Tokyo Press.

Idani, G. (1991). Social relationships between immigrant and resident bonobo (*Pan paniscus*) females at Wamba. *Folia Primatologica*, 57, 83–95.

Kano, T. (1992). *The Last Ape: Pygmy Chimpanzee Behavior and Ecology*. Stanford: Stanford University Press.

Matsumoto-Oda, A. (1998). Injuries to the sexual skin of cycling female chimpanzees at Mahale and their effect on behaviour. *Folia Primatologica*, 69, 400–4.

Matsumoto-Oda, A. (1999a). Mahale chimpanzees: grouping patterns and cycling females. *American Journal of Primatology*, 47, 197–207.

Matsumoto-Oda, A. (1999b). Female choice in the opportunistic mating of wild chimpanzees at Mahale. *Behavioral Ecology and Sociobiology*, 46, 258–66.

Matsumoto-Oda, A., Hosaka, K., Huffman, M. A. & Kawanaka, K. (1998). Factors affecting party size in chimpanzees of the Mahale Mountains. *International Journal of Primatology*, 19, 999–1011.

McGrew, W. C. (1996). Dominance status, food sharing, and reproductive success in chimpanzees. In *Food and the Status Quest*, ed. P. Wiessner & W. Schiefenhövel, pp. 39–46. Oxford: Berghahn Books.

Mitani, J. C. & Watts, D. P. (2001). Why do chimpanzees hunt and share meat? *Animal Behaviour*, 61, 915–24.

Mitani, J. C., Grether, G. F., Rodman, P. S. & Priatna, D. (1991). Associations among wild orang-utans: sociality, passive aggregations or chance? *Animal Behaviour*, 42, 33–46.

Nishida, T. (1979). The social structure of chimpanzees of the Mahale Mountains. In *The Great Apes*, ed. D. A. Hamburg & E. R. McCown, pp. 73–121. Menlo Park: Benjamin/Cummings.

Nishida, T. (1983). Alloparental behavior in wild chimpanzees of the Mahale Mountains, Tanzania. *Folia Primatologica*, 41, 1–33.

Nishida, T. (1990) A quarter century of research. In *The Chimpanzees of the Mahale Mountains*, ed. T. Nishida, pp. 3–36. Tokyo: University of Tokyo Press.

Nishida, T., Takasaki, H., and Takahata, Y. (1990). Demography and reproductive profiles. In *The Chimpanzees of the Mahale Mountains*, ed. T. Nishida, pp. 63–98. Tokyo: University of Tokyo Press.

Pepper, J. W., Mitani, J. C. & Watts, D. P. (1999). General gregariousness and specific social preference among wild chimpanzees. *International Journal of Primatology*, 20, 613–32.

Pusey, A., Williams, J. & Goodall, J. (1997). The influence of dominance rank on reproductive success of female chimpanzees. *Science*, 277, 828–31.

Smuts, B. B. (1987). *Sex and Friendship in Baboons*. New York: Aldine.

Stanford, C. (1998). *Chimpanzee and Red Colobus: The Ecology of Predator and Prey*. Cambridge, MA: Harvard University Press.

Takahata,Y. (1982a). Social relations between adult males and females of Japanese monkeys in the Arashiyama B troop. *Primates*, 23, 1–23.

Takahata,Y. (1982b). The socio-sexual behavior of Japanese monkeys. *Zeitschrift für Tierpsychologie*, 59, 89–108.

Takahata,Y. (1990). Adult males' social relations with adult females. In *The Chimpanzees of the Mahale Mountains*, ed. T. Nishida, pp. 133–48. Tokyo: University of Tokyo Press.

Takahata, Y., Ihobe, H. & Idani, G. (1996). Comparing copulations of chimpanzees and bonobos: do females exhibit proceptivity or receptivity? In *Great Ape Societies*, ed. W. C. McGrew, L. F. Marchant & T. Nishida, pp. 146–58. Cambridge: Cambridge University Press.

Tutin, C. E. G. (1979). Mating patterns and reproductive

strategies in a community of wild chimpanzees (*Pan troglodytes schweinfurthii*). *Behavioral Ecology and Sociobiology*, 6, 29–38.

van Schaik, C. P. (1999). The socioecology of fission–fusion sociality in orangutans. *Primates*, 40, 69–86.

van Schaik, C. P. & Kappeler, P. M. (1997). Infanticide risk and the evolution of male–female association in primates. *Proceedings of the Royal Society of London, Series B*, 264, 1687–94.

Wallis, J. (1992). Chimpanzee genital swelling and its role in the pattern of sociosexual behavior. *American Journal of Primatology*, 28, 101–13.

Wallis, J. (1997). Chimpanzee consortships: new information on conception rate, seasonality, and individual preference. *American Journal of Primatology*, 42, 152 (abstract).

Watts, D. P. (1992). Social relationships of immigrant and resident female mountain gorillas. I. Male–female relationships. *American Journal of Primatology*, 28, 159–81.

Watts, D. P. (1998). Coalitionary mate guarding by male chimpanzees at Ngogo, Kibale National Park, Uganda. *Behavioral Ecology and Socobiology*, 44, 43–55.

White, F. J. (1989). Social organization of pygmy chimpanzees. In *Understanding Chimpanzees*, ed. P. G. Heltne & L. A. Marquardt, pp. 194–207. Cambridge, MA: Harvard University Press.

Wrangham, R. W. (1993). The evolution of sexuality in chimpanzees and bonobos. *Human Nature*, 4, 47–79.

Wrangham, R. W. & Smuts, B. B. (1980). Sex differences in the behavioral ecology of chimpanzees in the Gombe National Park, Tanzania. *Journal of Reproduction and Fertility*, 28, 13–31.

Wrangham, R. W., Clark, A. & Isabiryre-Basuta, G. (1992). Female social relationships and social organization of Kibale Forest chimpanzees. In *Topics in Primatology. Volume I. Human Origins*, ed. T. Nishida, W. McGrew, P. Marler, M. Pickford & F. de Waal, pp. 81–98. Tokyo: Tokyo University Press.

Yerkes, R. M. (1943). *Chimpanzees: A Laboratory Colony*. New Haven: Yale University Press.

13 • Seasonal aspects of reproduction and sexual behavior in two chimpanzee populations: a comparison of Gombe (Tanzania) and Budongo (Uganda)

JANETTE WALLIS

INTRODUCTION

> Natural selection ensures that reproduction will occur in harmony with existing environmental conditions.
>
> Bronson 1988, p. 1831

The study of seasonal variation in the environment and the role this plays in influencing behavior and physiology of chimpanzees has been the focus of numerous studies in recent years. Chimpanzee habitat shows a highly seasonal pattern in food abundance and distribution, which has long been known to influence many aspects of social behavior and ranging patterns.

Several social behaviors of chimpanzees are seasonal. For example, Stanford *et al.* (1994) found that hunting activity at Gombe peaked during the dry season. Chimpanzee party size varies seasonally (Wrangham 1977; Wallis & Bettinger, unpublished data; Chapman *et al.* 1995; Matsumota-Oda *et al.* 1998). At Gombe, there are significantly more adult males and estrous females present in a party during the dry season than during the wet season (Wallis & Bettinger, unpublished data).

It was previously assumed that within-group feeding competition drives the correlation between party size and the availability of food in chimpanzees (Wrangham 1977; Chapman *et al.* 1995) and may thus explain the seasonal trend in party size. However, the presence of sexually receptive females influences party size much more than do the local conditions of food abundance or distribution (Anderson 2001; Anderson *et al.*, Chapter 6). Several investigators have found a correlation between party size and the presence of estrous females (Matsumoda-Oda 1999; Wallis & Reynolds 1999; Anderson 2001; Anderson *et al.*, Chapter 6). Thus, the key to understanding some seasonal social patterns in chimpanzees may lie in further investigation of female reproductive physiology.

CHIMPANZEE REPRODUCTION: THE IMPORTANCE OF GENITAL SWELLING CYCLES

Female chimpanzees are most sexually receptive when they exhibit a maximal anogenital swelling, which lasts for approximately 10-12 days (Goodall 1986; Wallis 1997). These genital swellings enhance sexual activity and thus promote intermale competition (Clutton-Brock & Harvey 1976), confuse paternity (Hrdy 1977), and may increase male parental care by decreasing the chance of possible infanticide (Hamilton 1984).

For regularly cycling females, as well as pregnant females, a genital swelling serves as a passport to intercommunity transfer (Nishida 1979; Pusey 1979; Wallis 1992a); female chimpanzees attempting to transfer without a swelling may be fatally attacked (Nishida & Hiraiwa-Hasegawa 1985). Review of the Gombe chimpanzee records revealed that at least two conceptions occurred when the mothers were away from their home community and no community males were coincidentally absent (suggesting consortship). This finding implies that these females moved between communities while pregnant (A. Pusey, personal communication; Wallis, unpublished data). Boesch & Boesch-Achermann (2000) suggest that a genital swelling may also serve as a social passport by way of which females in estrus may gain the support of males during competition with other females.

Large party sizes may be formed by males seeking sexually receptive females, or by sexually receptive females seeking males for copulation. Indeed, parties with more estrous females contain more males (Matsumoto-Oda 1999; Wallis & Reynolds 1999; Anderson *et al.*, Chapter 6). Additionally, non-sexually-receptive females may seek the company of estrous females, an action shown to stimulate the resumption of postpartum cycles (Wallis 1985, 1992b) and initiate the first full anogenital swelling in adolescent females (Wallis 1994).

Many aspects of chimpanzee reproduction show a strong

seasonal pattern. The onset of postpartum cycles is highly seasonal at Gombe National Park (Wallis 1992b, 1995), Mahale National Park (Nishida 1990), and Budongo Forest Reserve (Wallis & Reynolds 1999). At these locations, the key time of year for resuming cycles is the late dry season. In addition, young female chimpanzees at Gombe show a tendency to exhibit their first adult anogenital swelling during the same period of the year – the late dry season (Wallis 1994). In a review of 19 years' data from Gombe, Wallis (1995) found seasonal patterns in the presence of females with estrous swellings and, consequently, conceptions.

Predictably, the pattern of observed copulation is also highly seasonal at Gombe (Wallis and Bettinger, unpublished data) and Budongo (Wallis & Reynolds 1999). In fact, as long-term records are analyzed at field sites across Africa, additional researchers are finding evidence of a strong seasonal influence on chimpanzee sexual activity and conception/birth cycles (Gombe: Wallis 1995; Kibale: Wrangham, personal communication; Mahale: Nishida, personal communication; Budongo, Wallis & Reynolds 1999; Taï: Boesch & Boesch-Achermann 2000; Anderson 2001).

THE POTENTIAL ROLE OF DIET

Whereas the availability of food may *directly* affect seasonal party size and hunting in chimpanzees, as noted above, current evidence suggests that party size is much more influenced by the presence of sexually receptive females. However, food availability, nutrient content, and/or chemical content may have an important *indirect* effect on seasonal behavior patterns by first influencing the menstrual cycle, and therefore the timing of sexual receptivity.

In previously published work, I first proposed the theory that the seasonal variations in diet – specifically the possible presence of phytoestrogens – might be the driving force producing seasonal patterns in reproductive parameters (Wallis 1995). This idea was further discussed in relation to evidence that particular flowers are eaten by Gombe chimpanzees most often during the months just prior to the peak in anogenital swellings, sexual activity, and conceptions (Wallis 1997).

As in humans, chimpanzee reproduction is highly rhythmical, highly cyclical, and anything that disrupts (or stimulates) this rhythm can have a great impact on reproductive function. This is particularly true for events such as the resumption of postpartum estrus. The interbirth interval for chimpanzees is approximately 5.5 years (Wallis 1997); when a female begins to exhibit her first genital swellings after delivery, the sex skin has been 'dormant' for more than 4 years. Any external influence on the timing of this and other events – whether inhibitory or stimulatory – is of key importance to an individual chimpanzee's reproductive success. This chapter examines the seasonal patterns of some reproductive parameters in two populations of chimpanzees in an effort to learn more about potential influences on reproductive timing, and discusses the possible link between the annual variation in diet, the chemical content of this diet, and the seasonal patterns of reproductive behavior and biology.

METHODS

This study examined the seasonal influence on reproductive behavior of chimpanzees (*Pan troglodytes schweinfurthii*). It compared the long-term data collected from observations of chimpanzee populations living in Gombe National Park, Tanzania, and the Budongo Forest Reserve, Uganda. Portions of the Gombe data have been published previously (Wallis 1995, 1997).

The study populations

SONSO COMMUNITY, BUDONGO FOREST RESERVE, UGANDA

The Budongo Forest Reserve is located in western Uganda (1°44′N, 31°33′E). It has a mean altitude of 1100 m and covers 793 km^2 of moist, semi-deciduous forest and grassland, of which 428 km^2 is forested (Plumptre 1996). The Sonso region lies in the south central portion of the forested area of Budongo, which is a mosaic of forest types as a result of selective logging prior to 1952 (Reynolds 1992; Plumptre 1996). The chimpanzees are fully habituated and have not been provisioned (Reynolds 1998). Throughout the period under review, the Sonso Community consisted of approximately 40–45 chimpanzees, with 16 sub-adult or adult females and 15 sub-adult or adult males.

KASAKELA COMMUNITY, GOMBE NATIONAL PARK, TANZANIA

The Gombe National Park is located in western Tanzania (4°40′S, 29°38′E). The Park is bordered by Lake Tanganyika (772 m above sea level) on the west, the rift escarpment on the east, and villages on both the north and south. It is the smallest national park in Tanzania, covering approximately 32 km^2 of mountainous terrain, although actual surface area is estimated to be closer to 60 km^2. Gombe is composed of alternating thick riverine forests, deciduous woodland, and hilltop grassland. The Kasakela Community has been studied since the early 1960s (see Goodall 1986, for complete details) and ranges in the central portion of the Park. The chimpanzees were provisioned for habituation in the early days of the research program. Limited provisioning continued during portions of the present study. Throughout the period under review, the Kasakela Community consisted of approximately 40–45 chimpanzees, with 16 sub-adult or adult females and 12 sub-adult or adult males.

Rainfall data

Because they are located on opposite sides of the equator, the rainfall patterns of Budongo and Gombe are not synchronous. Budongo receives approximately 1500 mm of rain annually (Plumptre 1996; Newton-Fisher 1999), with a single dry season from December to February. Gombe, too, has a distinct dry season, typically lasting from May to October.

Rainfall data were collected daily at each field site, using standard rain gauges. The Budongo data were collected by Field Assistants employed by the Budongo Forest Project. The Gombe data were collected by Gombe National Park employees.

Figure 13.1 provides the annual rainfall patterns for the two field sites. The graphs on the left show the pattern as traditionally illustrated in a calendar year – from January to December. To compare the two sites better, the graphs on the right realign the data to begin with the first month of the rainy season (i.e. when average monthly rainfall is approximately 100 mm or more). Thus, Gombe's data are depicted from November through October, Budongo from March through February.

Several important features appear when the rainfall data are aligned in this fashion (Figure 13.1). Both sites show a rainy season that is characterized by two peaks. Although in some years the period between these two peaks may be referred to as a 'short dry season,' in neither location is this true; the average monthly rainfall does not go below 100 mm as in a normal dry season. Overall, rainfall is greater at Gombe, but Budongo remains wet throughout most of the year; Gombe's dry season lasts for approximately half of each year, while Budongo's dry season is confined to only 3 months.

The data sets

REPRODUCTIVE EVENTS

The comparison of reproductive events such as the onset of postpartum estrus and estimated time of conception involved the review of long-term records of both field sites. For Gombe, this included the review of data from 1964–94 (see Wallis 1997). For Budongo, the data set ranged from 1990 to 2000 (Wallis & Reynolds, unpublished data).

Reproductive events were defined as in Wallis (1997), that is the last day of maximal anogenital swelling during the last normal swelling cycle (the fertile cycle) was considered the day of conception (Graham 1982). Postpartum cycle assessment included only those female subjects for whom sufficient observations were made during the late stages of postpartum amenorrhea (i.e. no swelling cycles were missed).

Copulation, party size, and estrous females

A detailed review of the Field Assistants' records for Budongo and Gombe produced information regarding party size, the presence of estrous females, and occurrence of sexual behavior (Wallis & Bettinger, unpublished data; Wallis & Reynolds, unpublished data). In addition, the time of day for all behavioral interactions was recorded to allow assessment of diurnal activity.

Party size is defined as the number of chimpanzees traveling together in a sub-group. 'Estrus' is a term typically used to describe a period of heightened sexual arousal, resulting in increased sexual attractivity and proceptivity. For chimpanzees, the cyclical anogenital swelling of the female is highly associated with sexual activity, even if the female happens to be pregnant at the time (Wallis & Lemmon 1986; Wallis 1992a; Wallis & Goodall 1993). Thus, the term 'estrous female' is used here to refer to female chimpanzees with anogenital swellings.

Under field conditions, some details of copulatory activity are difficult to determine. For example, it is often impossible to detect evidence of ejaculation, particularly if the mating pair is high in a tree. Thus, for the purpose of this analysis, 'copulation' is minimally defined as the male mounting the female, achieving intromission, and thrusting.

The data from Gombe were collected by Field Assistants from January 1990 to December 1992. A total of 6419 hours of data were analyzed from that 3-year period and used in this analysis. The data yielded 1146 recorded bouts of copulation in the Kasakela Chimpanzee Community. The

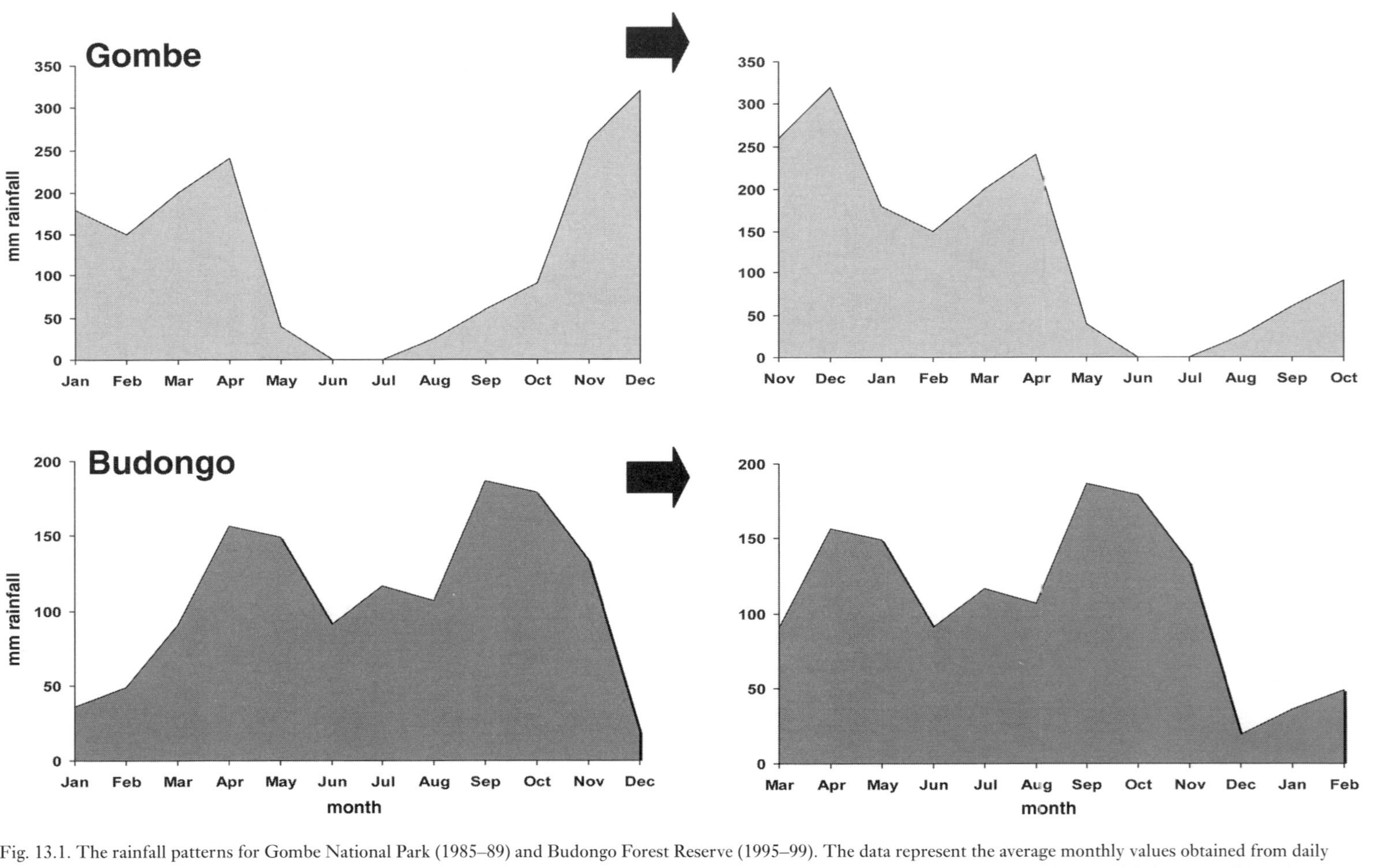

Fig. 13.1. The rainfall patterns for Gombe National Park (1985–89) and Budongo Forest Reserve (1995–99). The data represent the average monthly values obtained from daily record keeping. Graphs on the left present the data in the normal January to December fashion; graphs on the right shift the data to begin the year with the onset of the rainy season.

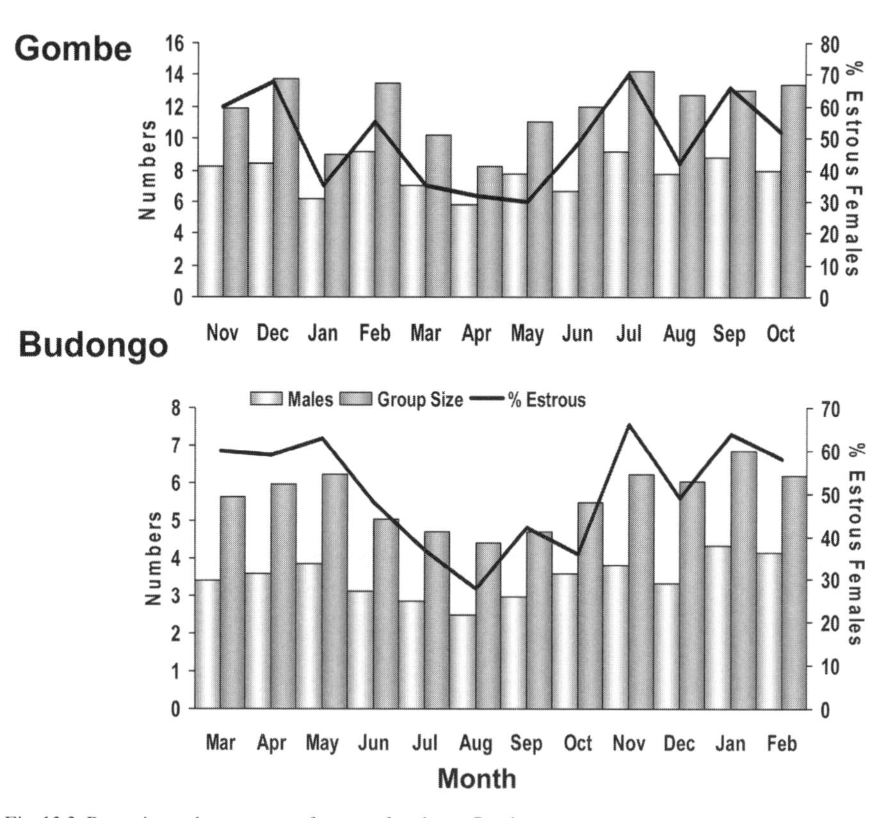

Fig. 13.2. Party size and percentage of estrous females at Gombe and Budongo. Data are aligned as in Figure 13.1, with the graphs beginning at the wet season's onset. The bars represent mean numbers per month. The solid lines represent the percentage of adult females present that are currently in estrus.

Budongo data set included 3515 hours of observation by Field Assistants. From November 1995 to February 2000, the Budongo Field Assistants recorded 966 copulatory bouts in the Sonso Chimpanzee Community. Owing to different data collection techniques at the two field sites, the copulatory rates are not directly comparable for this study. Gombe data collection uses focal animal observations with ad libitum notation of non-focals' behavior. Budongo data collection uses all occurrence sampling of key social interactions of any individual in the party. Although the data cannot be directly compared, the overall patterns of rates and monthly frequencies can be compared.

RESULTS

At both Gombe and Budongo, party size was associated with the presence of estrous females. Moreover, at both sites, these parameters show a strong seasonal variation. Figure 13.2 presents the mean monthly figures for party size, number of sub-adult and adult males, and percentage of adult females that are in estrus. When the data are modified to align the rainfall patterns, there is a similar pattern for the two sites. Budongo has a more striking pattern; however, at both sites there are smaller party sizes and fewer estrous females during the period coinciding with the waning months of the wet season.

At both field sites, other important reproductive events show a similar seasonally influenced pattern. Figure 13.3 provides the realigned data for the onset of postpartum estrous cycles of females at Gombe and Budongo. At both sites, there is a tendency for cycle onset to occur in the dry part of the year. Very few females at either site exhibit their first postpartum cycle during the wet season. Similar seasonal patterns occur for conception, showing a concentration during the dry season at both sites.

Not surprisingly, at both Gombe and Budongo, there is a seasonal pattern of sexual activity (Figure 13.4). During the period when the wet season is fully under way, there were fewer copulations recorded and a lower copulatory rate for

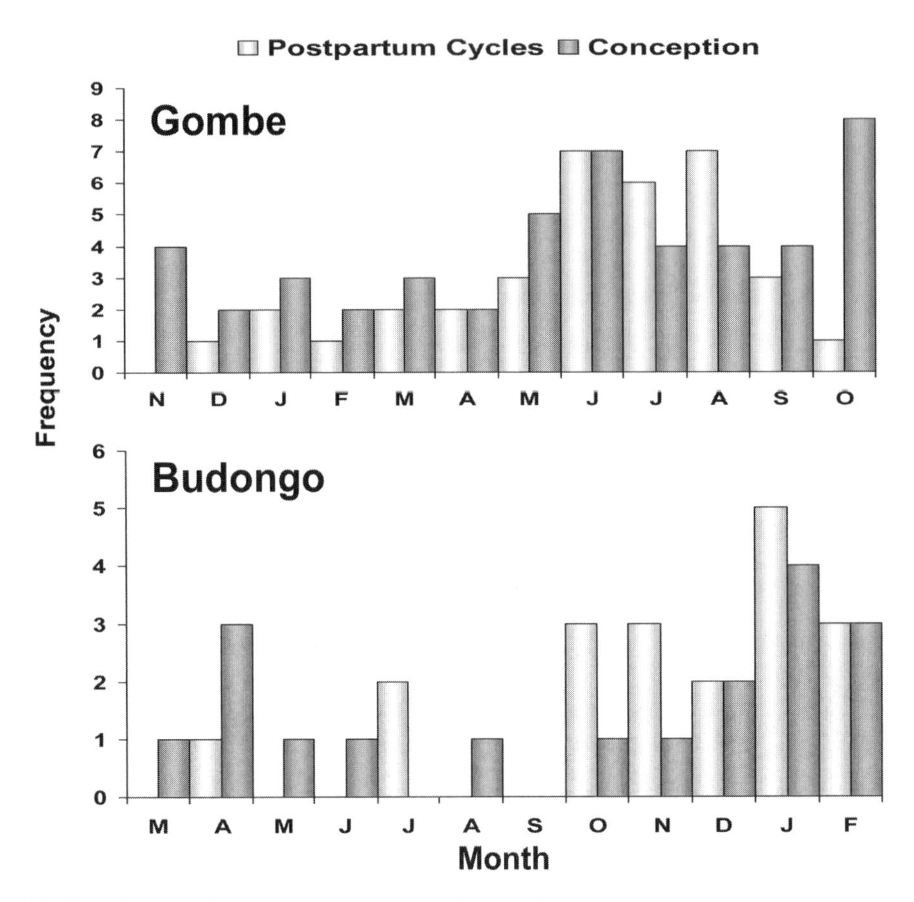

Fig. 13.3. The onset of postpartum estrous cycles and timing of conception in Gombe and Budongo chimpanzees. Data are aligned as in Figure 13.1, with the graphs beginning at the wet season's onset.

chimpanzees at both sites. Whereas the distinction can be made that Gombe's pattern precisely reflects the rainfall patterns (e.g. compare the six wet months versus the six dry months), the pattern at Budongo appears to be related to a peak specifically marking the end of the wet season.

Figure 13.5 provides the diurnal pattern of the copulatory bouts examined in this study. As in captive chimpanzee populations, the chimpanzees at both Gombe and Budongo tend to concentrate their sexual activity in the morning hours.

DISCUSSION AND CONCLUSIONS

The data comparison presented here confirms a seasonal influence on reproduction in two chimpanzee communities. As stated above, previous reports of the seasonal influence on chimpanzee reproductive parameters suggested that dry season, per se, plays a role in the timing of reproductive events. However, by aligning the data according to rainfall patterns at the two sites, each parameter presented shows a seasonal pattern, but they are less clearly linked to rainfall than previously thought. Budongo is a wetter habitat, yet the patterns of sexual behavior and onset of postpartum estrus (among other things) are remarkably similar to those at Gombe. Rather than being specifically aligned to wet versus dry season, perhaps these patterns are more closely related to *changes* in the environment. For example, at Budongo the copulatory activity is lowest during the period during the wet season characterized by a brief respite from heavy rainfall (see Figure 13.1). Anderson (2001) also found that change was important: the number of estrous females at Taï was negatively related to the mean rainfall specifically in the two preceding months. At Taï, the season with the fewest females with sexual swellings falls during the dry months of June–August (Boesch & Boesch-Achermann 2000;

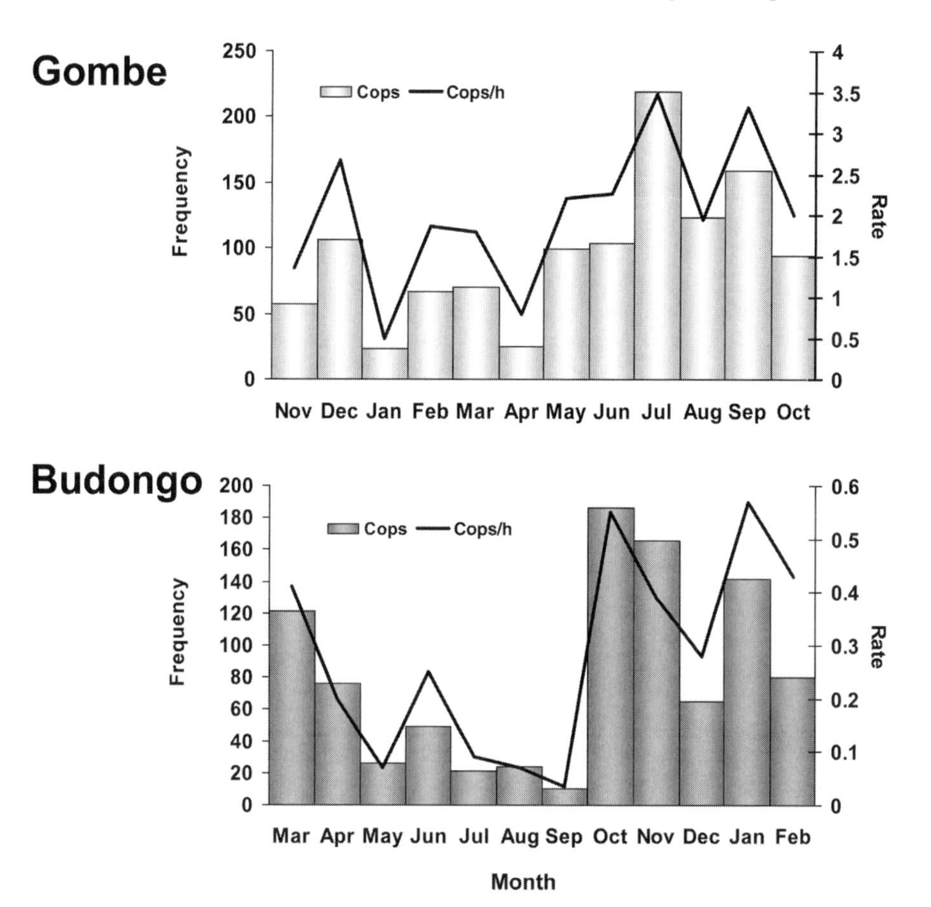

Fig. 13.4. Chimpanzee copulatory behavior at Gombe and Budongo. Data are aligned as in Figure 13.1, with the graphs beginning at the wet season's onset. The bars represent the total number of copulations observed; the solid line represents the rate of copulations per hour.

Anderson *et al.*, Chapter 6). The number of sexual swellings increases in September through May, when the rains begin and fruit productivity increases. Conversely, at Gombe, the dry season corresponds with the highest number of sexually receptive females (Goodall 1986; Wallis 1995).

The role of endogenous hormones

Key to any seasonal fluctuations in reproductive activity is, of course, the endogenous steroids driving the females' reproductive cycles. Without the rise and fall of circulating estrogens and progesterones, the anogenital skin of the female chimpanzee will not exhibit its 36-day cyclical swelling patterns. The 'estrous swelling' serves to attract males and likely provides stimulation to the female, as it almost completely engulfs the clitoris. The role of circulating steroids on genital swelling and sexual activity has been well documented in the literature (see Graham 1982, for a review).

We often take for granted the lead role that *female* endogenous hormones play in chimpanzee reproduction. Rarely, however, do we consider the role of the *male's* hormonal milieu on chimpanzee sexual behavior. Figure 13.5 illustrates the importance of the male endocrine state. At both sites, copulatory activity is greater in the morning, leveling off by about mid-day. These findings are not surprising. Captive chimpanzees have long been known to exhibit heightened sexual activity during the morning (personal observation). Recent study of urinary samples from adult male wild chimpanzees indicated a peak in testosterone during the early morning hours (M. Muller, personal communication). This finding may explain the heightened sexual arousal during morning hours seen in male chimpanzees, particularly high ranking ones. For example, at

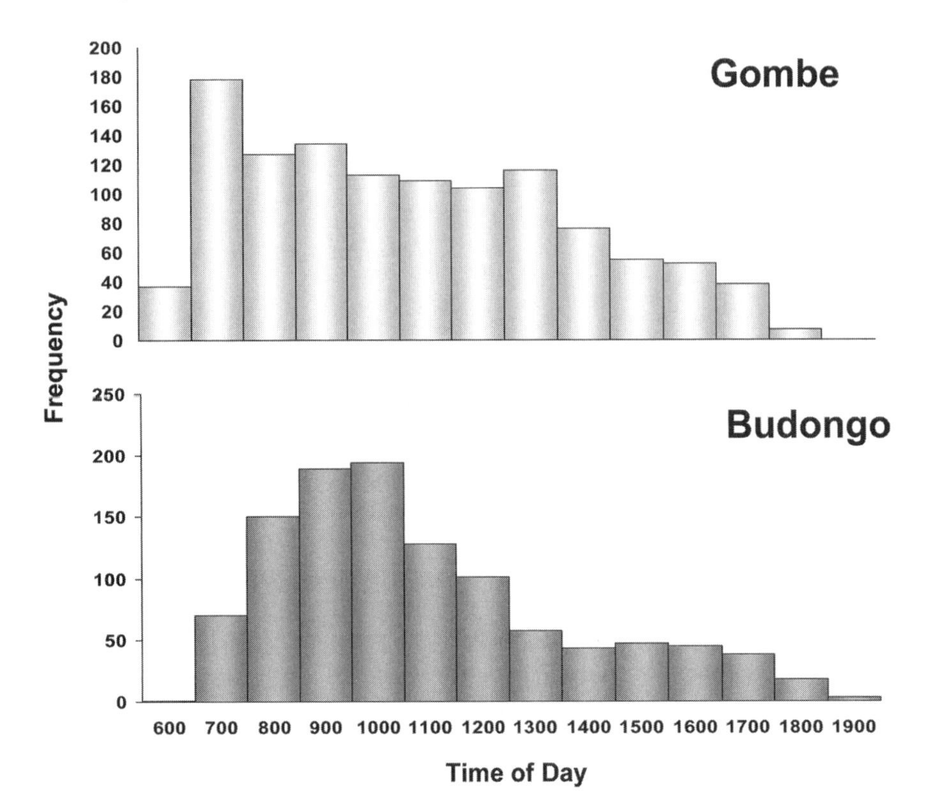

Fig. 13.5. Chimpanzee copulation by time of day at Gombe and Budongo. [Note: The slight difference between the two sites' diurnal mating patterns is probably related to the difference in data collection techniques; Gombe Field Assistants collect data starting at the time of unnesting, whereas Budongo Field Assistants typically begin work at a later time of day.]

Gombe, the alpha male *Wilkie* typically mated with an estrous female as soon as he descended from the nest each morning (personal observation).

The role of exogenous hormones

Logically, we may look to other seasonally fluctuating factors in the chimpanzees' environment to help explain the seasonal pattern in sex and reproduction. Specifically, the annual variability of diet for wild chimpanzees may well be involved. Species composition and availability of the diet vary significantly from month to month at every chimpanzee field site studied (see Wrangham 1977; Newton-Fisher 1999).

Reproductive cycles may be influenced directly by the nutritional content of the foods consumed. Chimpanzees are ripe fruit specialists (Wrangham *et al.* 1998; Newton-Fisher 1999), and they prefer food items that have high carbohydrate and low fiber levels (Conklin-Brittain *et al.* 1998; Reynolds *et al.* 1998). As a consequence, females may not be able to meet the physiological and behavioral demands of a sexual swelling during seasons in which few or no foods have relatively high carbohydrate and low fiber levels (Anderson 2001). For example, the dry season at Gombe has less food available than the wet season (Wrangham 1977). As a result, the chimpanzees respond to the reduction in preferred foods by eating a wider variety of food items at both Gombe (Wallis *et al.* 1995) and Taï (Doran 1997). It is toward the end of this dry period, with a wide range of foods and imminent rainfalls, that Gombe chimpanzees exhibit more female genital swellings and related events. The pattern at Gombe, therefore, suggests that it is not the abundance and distribution of food, but rather the content of the foods that may influence the timing of female sexual receptivity (Wallis 1995, 1997).

In addition to the nutritional content of plant foods, however, there is a more subtle way the diet of chimpanzees may influence reproductive and behavioral biology. Although nutrients are the most important factor in plant choice by primates, the presence of secondary compounds in the diet clearly influences primate feeding behavior (Glander 1982). Animals do not feed randomly on the plants available to them; they feed as though the secondary compounds occur in predictable patterns (Glander 1978). These patterns may fluctuate annually, though not always consistently from year to year (Glander 1981, 1982). Clearly, this pattern is related to rain; rainfall provides trees with a cue to begin leaf-drop, flower growth, and new leaf production. Typically, it is only a portion of the plant that contains a secondary compound (Sim 1965), and several primate studies have found a strong preference for or avoidance of select plants – presumably based on the presence of secondary compounds.

Most flower and new leaf production occurs during the dry season (Glander 1978). Plants concentrate nutrients in their stems and young leaves (Fraenkel 1953). These young leaves contain more protein and more total phenolics (Milton 1979), with significantly higher concentrations of zinc, phosphorus, and potassium, than mature leaves (Yeager *et al.* 1997). Not surprisingly, chimpanzees prefer young leaves over mature leaves in the species they eat (Reynolds *et al.* 1998). At the Taï Forest, more new leaves are available at the end of the dry season and early in the wet season, consistent with the suggestion that estrus in chimpanzees at that site is related to the seasonal availability of new and less defended leaves (Anderson 2001).

Several plant compounds have been found to stimulate or inhibit reproduction when ingested (see Whitten & Naftolin 1998). Phytoestrogens may alter ovarian cycles through ovarian, pituitary, or hypothalamic actions. Timing is the key here; phytoestrogens can inhibit estradiol action when ingested in the early stages of follicular development, but augment it during the luteal phase. In fact, flavones can compete with existing steroids and alter their normal metabolism (Kellis & Vickery 1984). Other plant compounds, such as phenolic acids, can produce an inhibitory effect on gonadotropin release, thus blocking ovulation.

There is a noticeable delay in the outward effects of hormones once they are ingested. For example, human patients prescribed oral contraceptives are advised to use 'backup' protective measures for the first 2 or 3 months until the hormone supplement can be considered sufficient for birth control. Thus, any action of phytochemicals in chimpanzee diet may result in behavioral and physiological changes that appear 3 or 4 months after ingestion. The rhythms of the primate body – whether human or nonhuman – take time to adjust to environmental influence.

FUTURE DIRECTIONS AND PREDICTIONS

The relationship between food, social contact, and reproduction are clearly very complex. A thorough investigation of chimpanzee behavior, food availability, dietary content, and reproductive hormones may lead to a better understanding of this relationship. Current efforts are under way at Budongo to investigate the phytochemical content of the chimpanzees' diet.

As we examine this topic further, we should expect *between site* variation. Chimpanzees occupy a wide range of habitats, located at varying distances from the equator, with different dietary content and availability at each site. Thus, no two sites will be the same. Moreover, geographical variation and climatic factors affect secondary compounds (Swain 1972). We should also expect *within site* variation; analysis of the overall seasonal trend in conception at Gombe revealed that this trend was not present in the late 1960s/early 1970s (Wallis 1997). Fluctuation in a variety of environmental factors may play a role in shifting behavior or physiology patterns over time at any one field site.

If we confirm that phytochemicals indeed play a major role in the diet of chimpanzees, we will likely find combined effects of several plant chemicals. In other words, there will be no easy answer. In 1977, Swain reported that there have been between 10000 and 400000 total secondary compounds identified. We should not expect to find one compound that works as a chimpanzee contraceptive, another that works as a chimpanzee aphrodisiac. There is likely a highly complex system that influences chimpanzee reproductive behavior and biology.

Even if we confirm the role of phytochemicals for reproductively active female chimpanzees, this will no doubt lead to additional research questions. For example, although males and females of varying ages and reproductive states select a slightly different menu than cycling females, all chimpanzees in a community are exposed to the same foods. What is the effect of phytochemical ingestion on males? What is the potential effect on the timing of puberty for members of either sex? What is the effect on pregnant females? These and related questions will be addressed as we learn more about the presence and potential action of plant steroids in the chimpanzees' diet.

As chimpanzee researchers we are, after all, conservationists. Conservation activities typically focus on the bushmeat trade, habitat destruction, and disease transmission. Each of these is a human-controlled activity. Yet, a key aspect of the chimpanzee's future may not be under human control at all – reproduction. When and how frequently a species reproduces clearly plays a role in its future. With long interbirth intervals, rare twinning, and relatively high infant mortality, chimpanzee population growth may be especially vulnerable to environmental change.

This comparative assessment of data from two field sites serves to highlight the importance of using long-term data for assessing trends in chimpanzee behavioral ecology. As we seek to expand our knowledge of chimpanzees – and work to assure their survival in nature – it is imperative that we strengthen our collaborations and develop additional multi-site comparisons of data. Only then can we come to understand fully how environmental changes influence chimpanzee reproduction and the long-term viability of individual populations.

ACKNOWLEDGEMENTS

I wish to thank the Field Assistants working for the Budongo Forest Project and the Gombe Stream Research Centre, whose hard work and dedication provided the data presented here. In addition, I am very grateful to Vernon Reynolds and Jane Goodall for so generously allowing access to these data.

REFERENCES

Anderson, D. P. (2001). Tree phenology and distribution, and their relation to chimpanzee social ecology in the Taï National Park, Côte d'Ivoire. Dissertation, University of Wisconsin.

Boesch, C. & Boesch-Achermann, H. (2000). *The Chimpanzees of the Forest: Behavioural Ecology and Evolution*. Oxford: Oxford University Press.

Bronson, F. H. (1988). Seasonal regulation of reproduction in mammals. In *The Physiology of Reproduction*, ed. E. Knobil & J. Neill, pp. 1831–71. New York: Raven Press.

Chapman, C. A., Wrangham, R. W. & Chapman, L. J. (1995). Ecological constraints on group size: an analysis of spider monkey and chimpanzee subgroups. *Behavioral Ecology and Sociobiology*, 36, 59–70.

Clutton-Brock, T. & Harvey, P. H. (1976). Evolutionary rules and primate societies. In *Growing Points in Ethology*, ed. P. P. G. Bateson & R. Hinde, pp. 201–14. Cambridge: Cambridge University Press.

Conklin-Brittain, N., Wrangham, R. & Hunt, K. (1998). Dietary response of chimpanzees and cercopithecines to seasonal variation in fruit abundance. II. Macronutrients. *International Journal of Primatology*, 19, 971–98.

Doran, D. (1997). Influence of seasonality on activity patterns, feeding behavior, ranging, and grouping patterns in Taï chimpanzees. *International Journal of Primatology*, 18, 183–206.

Fraenkel, G. (1953). The nutritional value of green plants for insects. *Symposium of the 9th International Congress of Entomology*, 1951, 90–100.

Glander, K. E. (1978). Howling monkey feeding behavior and plant secondary compounds: a study of strategies. In *The Ecology of Arboreal Folivores*, ed. G. G. Montgomery, pp. 561–74. Washington, DC: Smithsonian Institution Press.

Glander, K. E. (1981). Feeding patterns in mantled howling monkeys. In *Foraging Behavior: Ecological, Ethological, and Psychological Approaches*, ed. A. Kamil & T. D. Sargent, pp. 231–59. New York: Garland Press.

Glander, K. E. (1982). The impact of plant secondary compounds of primate feeding behavior. *Yearbook of Physical Anthropology*, 25, 1–18.

Goodall, J. (1986). *The Chimpanzees of Gombe: Patterns of Behavior*. Cambridge, MA: Belknap Press of Harvard University Press.

Graham, C. E. (1982). Ovulation time: a factor in ape fertility assessment. *American Journal of Primatology*, Supplement, 1, 51–5.

Hamilton, W. (1984). Significance of paternal investment by primates to the evolution of male–female associations. In *Primate Paternalism*, ed. D. Taub, pp. 309–35. New York: Van Nostrand Reinhold.

Hrdy, S. (1977). *The Langurs of Abu: Female and Male Strategies of Reproduction*. Cambridge, MA: Harvard University Press.

Kellis, J. T. & Vickery, L. E. (1984). Inhibition of human estrogen synthetase (Aromatase) by flavones. *Science*, 225, 1032–34.

Matsumoto-Oda, A. (1999). Mahale chimpanzees: grouping patterns and cycling females. *American Journal of Primatology*, 47, 197–207.

Matsumoto-Oda, A., Hosaka, K., Huffman, M. A. & Kawanaka, K. (1998). Factors affecting party size in chimpanzees of the Mahale Mountains. *International Journal of Primatology*, 19(6), 999–1011.

Milton, K. (1979). Factors influencing leaf choice by howler monkeys: a test of some hypotheses of food selection by generalist herbivores. *American Naturalist*, 114, 362–78.

Newton-Fisher, N. (1999). The diet of chimpanzees in the Budongo Forest Reserve, Uganda. *African Journal of Ecology*, 37, 344–54.

Nishida, T. (1979). The social structure of chimpanzees of the

Mahale Mountains. In *The Great Apes*, ed. D. A. Hamburg & E. R. McCown, pp. 73–121. Menlo Park, CA: Benjamin/Cummings.

Nishida, T. (1990). *The Chimpanzees of the Mahale Mountains: Sexual and Life-History Strategies.* Tokyo: University of Tokyo Press.

Nishida, T. & Hiraiwa-Hasegawa, M. (1985). Responses to a stranger mother–son pair in the wild chimpanzee: a case report. *Primates*, 26, 1–13.

Plumptre, A. (1996). Changes following 60 years of selective timber harvesting in the Budongo Forest Reserve, Uganda. *Forest Ecology and Management*, 89, 101–13.

Pusey, A. (1979). Inter-community transfer of chimpanzees in Gombe National Park. In *The Great Apes*, ed. D. A. Hamburg & E. R. McCown, pp. 465–79. Menlo Park, CA: Benjamin/Cummings.

Reynolds, V. (1992). Chimpanzees in the Budongo Forest, 1962–1992. *Journal of Zoology*, 228, 695–9.

Reynolds, V. (1998). Demography of chimpanzees *Pan troglodytes schweinfurthii* in Budongo Forest, Uganda. *African Primates*, 3, 25–8.

Reynolds, V., Plumptre, A., Greenham, J. & Harborne, J. (1998). Condensed tannins and sugars in the diet of chimpanzees (*Pan troglodytes schweinfurthii*) in the Budongo Forest, Uganda. *Oecologia*, 115, 331–6.

Sim, S. K. (1965). *Medicinal Plant Alkaloids*. Toronto: University of Toronto Press.

Stanford, C. B., Wallis, J., Mpongo, E. & Goodall, J. (1994). Hunting decisions in wild chimpanzees. *Behaviour*, 131, 1–18.

Swain, T. (1972). The significance of comparative phytochemistry in medical botany. In *Plants in the Development of Modern Medicine*, ed. T. Swain, pp. 125–59. Cambridge, MA: Harvard University Press.

Wallis, J. (1985). Synchrony of estrous swelling in captive group-living chimpanzees (*Pan troglodytes*). *International Journal of Primatology*, 6(4), 335–50.

Wallis, J. (1992a). Chimpanzee genital swelling and its role in the pattern of sociosexual behavior. *American Journal of Primatology*, 28, 101–13.

Wallis, J. (1992b). Socioenvironmental effects on timing of first postpartum cycles in chimpanzees. In *Topics in Primatology, Vol. 1: Human Origins*, ed. T. Nishida, W. C. McGrew, P. Marler, M. Pickford & F. de Waal, pp. 119–30. Tokyo: University of Tokyo Press.

Wallis, J. (1994). Socioenvironmental effects on first full anogenital swellings in adolescent female chimpanzees. In *Current Primatology Vol. II: Social Development, Learning, and Behavior*, ed. J. J. Roeder, B. Thierry, J. R. Anderson & N. Herrenschmidt, pp. 25–32. Strasbourg: University of Louis Pasteur.

Wallis, J. (1995). Seasonal influence on reproduction in chimpanzees of Gombe National Park. *International Journal of Primatology*, 16(3), 435–51.

Wallis, J. (1997). A survey of reproductive parameters in the free-ranging chimpanzees of Gombe National Park. *Journal of Reproduction & Fertility*, 109, 297–307.

Wallis, J. & Goodall, J. (1993). Genital swelling patterns of pregnant chimpanzees in Gombe National Park. *American Journal of Primatology*, 31, 89–98.

Wallis, J. & Lemmon, W. B. (1986). Social behavior and genital swelling in pregnant chimpanzees (*Pan troglodytes*). *American Journal of Primatology*, 10(2), 171–83.

Wallis, J. & Reynolds, V. (1999). Seasonal aspects of sociosexual behavior in two chimpanzee populations: a comparison of Gombe (Tanzania) and Budongo (Uganda). *American Journal of Primatology*, 49(1), 111 (abstract).

Wallis, J., Mbago, F., Mpongo, E. & Chepstow-Lusty, A. (1995). Sex differences and seasonal effects in the diet of chimpanzees. *American Journal of Primatology*, 36, 162 (abstract).

Whitten, P. & Naftolin, F. (1998). Reproductive actions of phytoestrogens. *Clinical Endocrinology & Metabolism*, 12(4), 667–90.

Wrangham, R. W. (1977). Feeding behaviour of chimpanzees in Gombe National Park, Tanzania. In *Primate Ecology*, ed. T. Clutton-Brock, pp. 503–38. London: Academic Press.

Wrangham, R., Conklin-Brittain, N. & Hunt, K. (1998). Dietary response of chimpanzees and cercopithecines to seasonal variation in fruit abundance. I. Antifeedants. *International Journal of Primatology*, 19, 949–70.

Yeager, C. P., Silver, S. C. & Dierenfeld, E. S. (1997). Mineral and phytochemical influences on foliage selection by the proboscis monkey (*Nasalis larvatus*). *American Journal of Primatology*, 41, 117–28.

14 • Costs and benefits of grouping for female chimpanzees at Gombe

JENNIFER M. WILLIAMS, HSIEN-YANG LIU & ANNE E. PUSEY

INTRODUCTION

Most diurnal primates live in permanent social groups. Chimpanzees, however, exhibit a fission–fusion social structure in which individuals of the social group or community spend some time alone and the rest of the time in parties of varying composition. These, in turn, comprise subsets of the total community. In this regard, chimpanzees resemble several other species of large-bodied, frugivorous primates. A common explanation is that in such species, predation risk, a major factor in producing group living, is relaxed because of large body size. Group size then adjusts to the distribution and abundance of food sources, which vary over time (van Schaik & van Hooff 1983; Terborgh & Janson 1986; Dunbar 1988). These explanations regard plentiful food as a permissive factor for group formation. The possibility that individuals actually gain more food by joining groups through information transfer or better exploitation of resources has received less attention in primate than in other animal studies (Wilson 1975; Rodman 1988) and deserves further study.

A number of researchers have investigated variables important in controlling party formation in chimpanzees. Besides grouping at plentiful food sources, the benefits of grouping they have considered include mating prospects, measured by the presence of sexually receptive females (Goodall 1986; Sakura 1994; Boesch 1996; Wrangham 2000; Anderson *et al.*, Chapter 6; Mitani *et al.*, Chapter 7; Wrangham, Chapter 15), cooperative hunting (Boesch 1996) and safety in numbers, either from predation (Boesch 1991), conspecific threat (Bauer 1980), or human disturbance (Sakura 1994). Research investigating party size in chimpanzees has focused on limits to party size, assuming that benefits of grouping are always high enough that observed patterns are a product of the costs (Wrangham 1979, 2000). Feeding competition is perhaps the most important cost of grouping (Macdonald 1983), and can be broken down into scramble and contest competition (van Schaik 1989). Scramble competition occurs when animals, without interacting with each other directly, attempt to obtain resources before their competitors can. Thus, when there are more individuals in a food patch, each individual obtains less food. In contest competition, competitors interact directly to obtain disputed resources, and the winner generally has some kind of advantage over the other, such as dominance rank, territory ownership, size, strength, or motivation (Milinski & Parker 1991).

In most chimpanzee populations, females spend less time in parties than males do, but female sociability varies considerably, both within and between populations. At Gombe, females spend the majority of their time alone or with only their family (Wrangham & Smuts 1980; Goodall 1986), but female chimpanzees at Taï and Bossou, and female bonobos are far more sociable (Wrangham 1986; Sakura 1994; Boesch 1996; Hohmann & Fruth, Chapter 10). In this chapter, by examining some of the factors that affect female sociability at Gombe, we search for ways to explain this variability. We consider the effects of costs to grouping from feeding competition, and benefits to grouping from social factors, but note that these effects may vary depending on the female's reproductive state and the age of her infants.

Costs of grouping

SCRAMBLE COMPETITION

In scramble competition, the assumption is that each area or patch contains only a limited amount of food. The rate at which food is depleted depends on the number of individuals feeding there. Individuals in larger groups will deplete the food more quickly than individuals in small groups. Therefore, individuals in large groups will have to travel further in a day, and an increase in time traveling is assumed to be costly (Janson & Goldsmith 1995; Wrangham 2000). Scramble competition for food occurs in all groups and, in a comparative study, Janson and Goldsmith (1995) provide evidence that it is particularly important in limiting group size in frugivorous primates. Scramble competition and the subsequent increase in travel with party size has been the focus of a great deal of work in chimpanzees (Wrangham & Smuts 1980; Wrangham *et al.* 1993; Chapman *et al.* 1995; Wrangham 2000). Janson & Goldsmith (1995) and Wrangham (2000) suggest that the main cost of scramble competition while foraging in groups is the increase in distance traveled with group size. Females with dependent offspring are thought to have higher travel costs, due to the cost of carrying infants or traveling with juvenile offspring (Wrangham 2000). This could explain the difference between male and female chimpanzee grouping patterns; males, who pay lower costs of grouping, can afford to spend more time in parties than can females (Wrangham 2000).

We address four related issues to test the effect of scramble competition on sociability of female chimpanzees. First, Janson & Goldsmith (1995) and Wrangham (2000) found that larger parties traveled greater distances, so we test whether this is true at Gombe as well. Second, to determine whether females travel more slowly than males, we examine the relationship between party size and daily travel distance separately for males, sexually receptive females, and non-receptive females. We expect females, in general, to travel shorter distances than males, and sexually receptive females to travel further than non-receptive females since sexually receptive females generally are not carrying infants and may need to travel to find or stay with mates. Third, we test whether offspring age is related to mother's travel rate. If infants are costly to carry or juveniles slow a mother down (Altmann & Samuels 1992), we would expect the mother to travel more slowly with offspring. Finally, by looking at the relationship between offspring age and the amount of time females spend in parties of different compositions, we test whether any differences in female travel rates result in different association patterns.

CONTEST COMPETITION

It is likely that contest competition is a major factor in feeding competition among female chimpanzees (Pusey *et al.* 1997; Wrangham 2000). Although the frequency of contest competition between females has not yet been measured systematically, Goodall (1986) found that most aggressive interactions occurred over food and that dominant individuals often occupied prime feeding sites. At Gombe, the infants of high-ranking females survive better, their daughters reach sexual maturity more quickly (Pusey *et al.* 1997), and high-ranking females have higher body weight (Pusey, Williams & Oerlert, unpublished data). This suggests that high-ranking females obtain better access to high-quality food, and it is likely that this is at least partly because of their success in contest competition. If dominant females win contests over food, low-ranking females should avoid traveling with higher-ranking females, while high-ranking females should not be as constrained by contest competition. We explore this issue by examining the relationship between rank and dyadic association patterns.

Benefits of grouping

SOCIALIZING OFFSPRING

Chimpanzee societies are particularly complex because the constant fissioning and fusing of parties means that the outcome of dominance interactions and the likelihood of aggressive or affiliative behaviors depends on which individuals are in the party. As in other social species, young chimpanzees must learn to interact effectively with others (Goodall 1986). Thus, although female chimpanzees may benefit by feeding alone, mothers may benefit by spending time with other chimpanzees in order to socialize their offspring.

We explore two aspects of this hypothesis. First, we consider the extent to which mothers provide opportunities for social play. Play is a frequent activity of infant and juvenile primates, which probably contributes to the development of skilled movement, communication, and social relationships (Fagen 1993), and allows juveniles to learn adult social behavior in a safe setting (Watts & Pusey 1993). Watts & Pusey (1993) found that great ape juveniles played with like-aged companions more than they played with older or younger juveniles, and Pusey (1990) found that juvenile chimpanzees spent at least 20% of their time playing with other juveniles when they were all in the same party. Thus, by analyzing the effect of offspring in two age classes on the mother's rate of association with other females, we investigate whether mothers spend more time with each other when their offspring are the same age.

Second, if sons need to learn how to compete effectively for rank by watching adult males doing so, the sons of females who spend more time with adult males should attain higher rank. To test this, we examine the relationship between a mother's association with males and the lifetime maximum rank of her sons.

SEXUAL RECEPTIVITY

In all study sites, females are more social when sexually receptive (Goodall 1986; Boesch 1996; Wrangham 2000; Mitani *et al.*, Chapter 7; Wrangham, Chapter 15), despite the scramble competition costs of being in a party. Sexually receptive females might join and stay with other chimpanzees to ensure conception, facilitate sperm competition, increase paternity uncertainty, or establish their identities as community members. While it is beyond the scope of this chapter to investigate the relative importance of these benefits, by including sexual state in all analyses we can document the different sociability and travel patterns when a female is sexually receptive.

METHODS

Study population, site, and field methods

We use data on the Kasakela chimpanzee community collected at Gombe National Park, Tanzania between 1975 and 1992 (see Goodall 1986 for details). Jane Goodall started collecting data on this community in 1960, and data have been collected in the same format since 1972. Continuous party membership and status of sexual swellings (coded as flat, quarter swollen, half swollen, three-quarters swollen, or full) of females were recorded on checksheets on all-day focal follows, while location was recorded every 15 minutes and behaviors were recorded in longhand notes. Travel distance was computed by adding together the linear distance between each 15-minute sample. Mean travel distance on nest-to-nest follows was 2.6 km, standard deviation 1.4. In follows that were not nest-to-nest but were at least 5 hours long, we divided the travel distance by the length of time the follow lasted to get overall travel rate per hour. We excluded follows less than 5 hours long since distance traveled during that time was generally not representative of the entire day.

Pant-grunts were extracted from the notes by students to determine social rank of both females and males. We assigned female dominance rank using directions of pant-grunts for 2-year periods since pant-grunts were too rare to assign annual rank values (Pusey *et al.* 1997). All females for whom we could assign rank were assessed as high-, medium-, or low-ranking. Each dyad was characterized by the rank of both members; that is, high–high, high–medium, high–low, and so on. This analysis included a subset of all possible dyads since we were not able to assign dominance rank to all females ($n = 26$ individuals, 5 of whom were high-ranking, 16 medium-ranking, and 15 low-ranking for at least 1 year, some counted in more than one rank due to ranks change; this resulted in 1394 dyad–year pairs).

We categorized sons' rank as alpha, medium-, or low-ranking each year using directions of pant-grunts when possible (1975–89), and estimations by field researchers during years when the field notes were not yet translated into English (1990–98). Even in the absence of pant-grunt data, observers agreed on these categories based on patterns of challenging and subservient behavior. We only included sons who either reached alpha status or started to decrease in rank before their death or the end of the study. Since males generally rise and then fall in rank over time (Goodall 1986, figure 15.2), we excluded males who were still rising (or could have risen) in rank at the time of their death or at the end of the study. We also excluded one male (Pax), who as an infant was castrated during an attack, shows no interest in attaining rank, and exhibits a variety of juvenile behaviors.

Demographic records were kept throughout the period. Kin relations between mothers and offspring were known since all offspring were observed with the mother within days or weeks of birth. All check sheet, location, and demographic data were transferred to computer for analysis (see Williams *et al.* 2002).

Gombe National Park is made up of a mosaic of semi-deciduous forest types and grassland (Collins & McGrew 1988). It also experiences extreme seasonality, with 7 months of rain daily, and 5 months with few or no days of rain (unpublished rain data, 1964, 1969–82, 1993–97). Thus, food availability is extremely patchy both in time and space.

Female space use at Gombe appears to be a result of both feeding competition among females and male territorial behavior. Females of the Kasakela community at Gombe spend most of their time alone in overlapping core areas that are clustered into two neighborhoods in the center of the community range (Williams *et al.* 2002). About half the females transfer to other communities at adolescence (Pusey *et al.* 1997). During the 18-year study period, the location (and identity) of the alpha female changed, and the number of females in each neighborhood changed as well. Immigrant females and natal females that remain in their community tend to settle in the least crowded neighborhood away from the highest ranking female (Williams 2000). Once they have settled in a core area, most females show strong site fidelity, although they may adjust their ranges according to changes in the community range (Williams *et al.* 2002).

Measures of association

We use three measures of association patterns:

1. The proportion of time a focal chimpanzee spent in parties of different compositions (the *focal association measure*): This samples only a subset of all chimpanzees

since many individuals were never the subjects of focal follows (20 of 24 males and 32 of 44 females over age 7 years were followed during the study period). We classified parties in three categories: female-only parties, which included at least two maternally unrelated adult females and any of their offspring, but no adult males (8.5% of all 15-minute samples); male-only parties, which included at least two adult males and no females (8.4%); and mixed-sex parties, which included at least one adult male and one unrelated adult female (49% of 15-minute samples). Parties that were composed of only maternally related individuals, we categorized as alone, following Goodall (1986) and Wrangham & Smuts (1980). Females were alone 24% of all 15-minute samples, while males were alone 9.9%. We sampled party size and composition every 15 minutes, which allowed us to calculate the proportion of time spent in parties of different compositions, or mean party size over time for all parties and for each party type. Averages are calculated using 15-minute sample points as the unit. Party size excludes dependent juveniles and infants.

2. Each female's level of association with all adult males (the *male association measure*): We recorded the proportion of male follows in which each female was encountered each year.
3. A modified version of the *dyadic association measure* described by Boesch & Boesch-Achermann (2000), which evaluates the level of association between individual females: This measure is equivalent to the simple ratio association index described by Cairns & Schwager (1987). Rather than calculating the time dyads spent together, we followed the suggestion of Cairns & Schwager (1987) in considering just the first sighting of a given pair each day to reduce autocorrelation. We extracted the first time in a day when an individual was encountered by the focal animal (first arrival). Then, for each pair of females, we determined the proportion of all of their first arrivals that were within 5 minutes of each other, taking care to avoid double counting (Cairns & Schwager 1987; see Williams 2000 for details). We assumed that arrivals occurring within this defined time period represented individuals who were already together when encountered by the focal animal. This measure uses data from all follows and thus reduces bias towards females who were followed frequently. Note that while the measure is dyadic, the associations are not – that is, females who are recorded as arriving together may have arrived with other individuals, both male and female. Overall, the mean dyadic association rate among females was 0.10 (SD = 0.11, sample size = 11868 dyad–year pairs).

We calculated the dyadic association measure annually from 1975 to 1992 for each pair of females who had experienced their first adult sexual swellings, producing a dyadic association measure for each pair of adult females for each year. Unlike the measure of associations from focal data, this measure includes mother and adult daughter dyads since without those dyads the statistical design would be unbalanced. We classified each dyad by whether both females, one female, or neither of the females in the pair were observed as being sexually receptive more than 1 day that year. We did the same for the number of females in each dyad who had offspring ages 0–3.5 years (infants) or 3.5–7 years (juveniles) or either (classified as mothers) each year. We separated infants and juveniles since offspring under 3.5 years are generally carried during travel, while offspring over that age are not (Gardner-Roberts 1998). Thus, offspring in the two age classes should impose different costs on their mothers during travel. We also classified dyads by the number of mothers in order to determine the effect of offspring, irrespective of age.

Analysis

To assess grouping patterns of *focal* females, we conducted analyses of covariance, with a categorical variable to take variation among individuals into account. We also used analyses of covariance to examine associations among individual females, using the *dyadic* measure of first arrivals together and a categorical variable to identify each unique dyad in all tests. In the dyadic analyses, we treated all predictors as categorical rather than continuous variables to test for non-linear relationships between them and associations among females. As mentioned earlier, we included sexual receptivity as a predictor in all analyses.

RESULTS

Scramble competition

DO LARGER PARTIES TRAVEL GREATER DISTANCES?

Overall, daily travel distance is positively correlated with average daily party size (linear regression, $p < 0.0000$, $r^2 = 0.20$, $n = 2706$ days), with the focal individual traveling

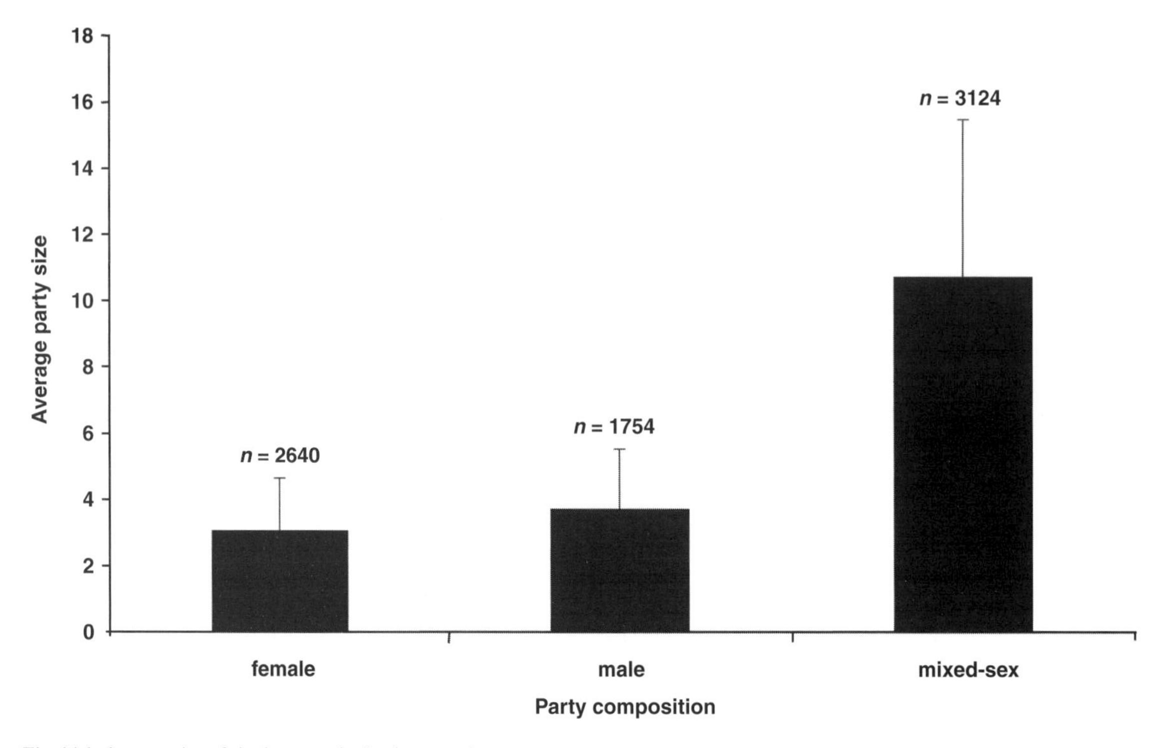

Fig. 14.1. Average size of single-sex and mixed-sex parties, averaged over daily averages of 15-minute samples. Sample size refers to the number of follows.

108 m farther for every additional member in the average party.

Party size is related to the party composition. Single-sex parties are significantly smaller than mixed-sex parties (Figure 14.1: female-only mean party size = 3.51, SD = 1.59, n = 2640 focal follows; male-only mean party size = 4.02, SD = 1.83, n = 1754; mixed-sex mean party size = 10.7, SD = 4.76, n = 3124). Thus, we looked at membership in female-only and mixed-sex parties separately since scramble competition should be more costly in mixed-sex parties.

DO MALE AND FEMALE CHIMPANZEES TRAVEL DIFFERENT DISTANCES DAILY?

If the relationship between average party size and daily travel distance discussed above is examined separately for male and both sexually receptive and non-receptive female focal animals, travel distance increases for all categories (Figure 14.2), but does so at a much higher rate for sexually non-receptive females (143.3 m farther for every additional party member) than for sexually receptive females or males (65.4 m and 71.4 m farther for each additional party member, respectively). In addition, non-receptive females travel a much shorter distance each day when alone than either receptive females or males (Figure 14.2).

DO FEMALES WITH OFFSPRING OF DIFFERENT AGES TRAVEL AT DIFFERENT RATES?

To test this question, we used travel rate rather than travel distance in order to maximize our sample size, in order to utilize data on females who were never followed for an entire day. We then determined whether or not the focal female was sexually receptive, had an infant or a juvenile offspring, or had an adult daughter with her. Infant but not juvenile offspring had significant effects on the travel rate of their mothers (Table 14.1). Females covered smaller distances over time with a small infant, greater distances when sexually receptive, and may have traveled farther when accompanied by an adult daughter.

DOES OFFSPRING AGE CORRELATE WITH THE AMOUNT OF TIME A FEMALE SPENDS IN PARTIES OF DIFFERENT COMPOSITIONS?

In female focal follows, focal females spent less time in mixed-sex parties and more time in female-only parties

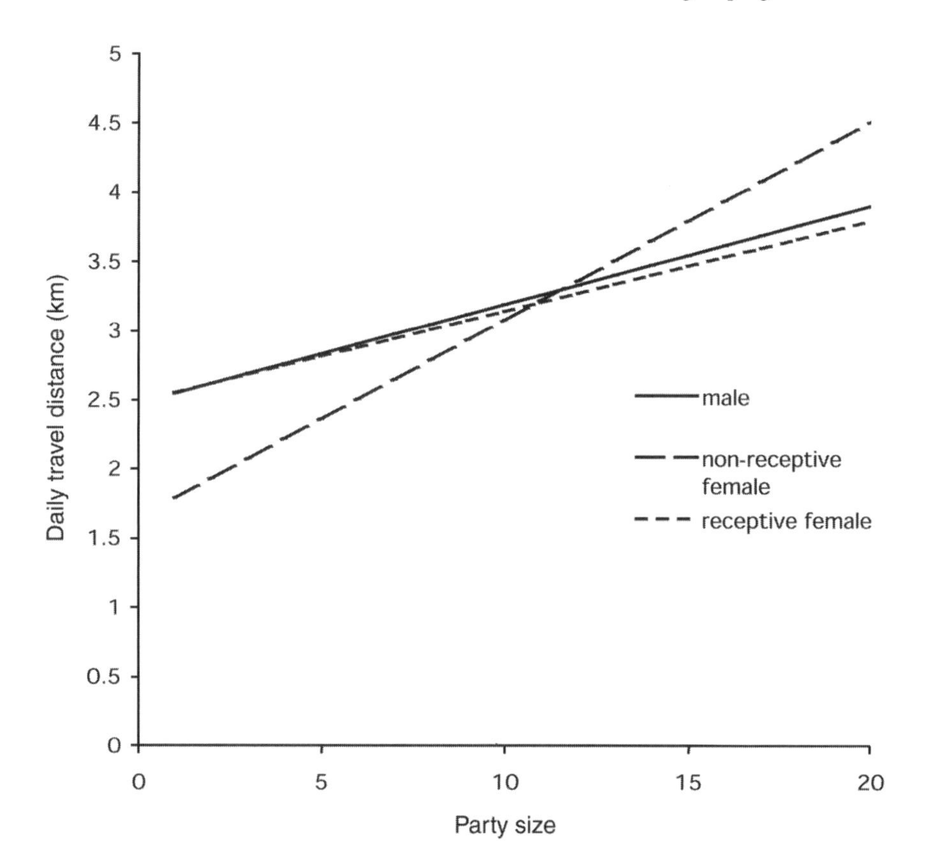

Fig. 14.2. Relationship between daily travel distance and daily mean party size for males, sexually non-receptive females, and sexually receptive females. This graph shows the regression lines for 1185 follows of 20 males, 1267 follows of 32 non-receptive females, and 123 follows of 29 receptive females.

when they had an infant, but adult daughters decreased the amount of time in female-only parties (i.e. parties made up of unrelated adult females) (Table 14.1). Additionally, females spent more time in all parties when sexually receptive. Again, juveniles had no effect on the amount of time their mothers spent in different party types.

Contest competition

DOES RANK INFLUENCE FEMALE–FEMALE ASSOCIATIONS?

Female dominance rank had a strong effect on dyadic associations among females (Table 14.2). According to the coefficients, both high-ranking and medium-ranking females associated most with high-ranking, then medium-ranking, then low-ranking females. Low-ranking females had exactly the reciprocal pattern, associating most with each other, and least with high-ranking females. High-ranking females were the most social with other females, while low-ranking were the least. Even high-ranking females associated at relatively low rates, however, as the mean dyadic association index among high-ranking females is 0.12, including one mother–adult daughter pair (SD = 0.15, $n = 5$ high-ranking females). In contrast, mothers and adult daughters associated at much higher rates regardless of rank differences. The mean mother–daughter dyadic association index was 0.57 (SD = 0.13, 12 mother–daughter pairs), while in only two of seven mother–daughter pairs in which both were ranked were the mother and daughter the same rank for any 2-year period.

Socializing offspring

DOES OFFSPRING AGE INFLUENCE FEMALE ASSOCIATION PATTERNS?

In order to test whether mothers with like-aged offspring spend more time together, we utilized the yearly dyadic

Table 14.1. *Effect of offspring on party membership and travel speed of females*

Variable	Infant	Adult daughter	Sexual receptivity
Model: Travel speed = Individual + Infant + Adult daughter + Sexual receptivity			
P-values	<0.01	0.12	<0.0000
Coefficients	−1.59	3.92	13.97
Model: Time in mixed-sex parties = Individual + Sexual receptivity + Infant + Adult daughter			
P-values	<0.0000	0.16	<0.0000
Coefficients	−0.082	0.033	0.277
Model: Time in female parties = Individual + Infant + Adult daughter + Sexual receptivity			
P-values	<0.005	<0.02	<0.0000
Coefficients	0.0047	−0.0060	0.0063

Note:
Samples are follows of female focal individuals: 2087 follows of 27 females, who had 55 infants or juveniles and 8 adult daughters when followed. The factor 'Individual' is the ID code of the focal individual, which is included in the model to take individual differences into account. Predictor variables refer to the state of the focal individual; whether she had an infant or an adult daughter, or was sexually receptive. Coefficients indicate the relationship between predictor and dependent variables; negative values correspond to a negative relationship, positive values indicate a positive relationship. Juveniles had no significant effect in any analysis, and were dropped from the models.

association measure. Females with juvenile offspring were most likely to spend time with other mothers of juvenile offspring (Table 14.3). In addition, females who did not have a juvenile offspring spent more time with mothers of juvenile offspring than they did with females without juvenile offspring. This relationship suggests that females with juveniles were attracted to any other females, not just other mothers with juvenile offspring. The opposite was true for mothers of infants, who were less likely to be with any other females. Thus, mothers may join other females to socialize juveniles but not infants.

DOES A FEMALE'S SOCIABILITY WITH MALES RELATE TO HER SON'S RANK?

We examined the relationship between the amount of time a female spent with adult males, and the eventual rank of her sons. We averaged the yearly male association measures for each female. Sons of females who were encountered on male follows more frequently appeared to attain higher rank in their lifetime (Figure 14.3; in a linear regression $p=0.07$, $r^2=0.52$, $n=7$ sons, 6 mothers). Pusey (1983) found that once juvenile males reached the age of about 7 years, they often led their mothers into groups, and could significantly influence the association patterns of their mothers. Even if we restricted the analysis to the mother's association with males before the son reached age 7 years, the same trend held (son's maximum rank vs. average yearly proportion of male follows in which mother was encountered from 1974 or birth of son, to the year he turned seven; $p=0.11$, $r^2=0.62$, $n=5$ sons, 4 mothers). At the same time, there was no relationship between mother's average rank and son's maximum rank (in a regression $p=0.37$, $r^2=0.16$, $n=7$ sons of 6 mothers), or between a female's rank and the amount of time she spent with adult males ($p=0.97$, $r^2=0.0004$, $n=6$ mothers).

DISCUSSION

Our results suggest that both feeding competition and social factors influence the associations of females at Gombe.

Scramble competition

Several of the patterns we observed suggest that scramble competition for food may be an important influence on female grouping patterns. Overall, daily travel distance increases with party size, in particular for females who are not sexually receptive. This result is consistent with Janson & Goldsmith's (1995) model of the effects of scramble com-

Table 14.2. *Effect of rank on dyadic associations among females*
Model **Arrivals together** = Individual + Dominance rank + Number of cycling females

Rank P-value < 0.01	Coefficient
High–high	0.033
High–medium	0.0082
Medium–medium	−0.0048
Low–low	−0.0053
Medium–low	−0.0053
High–low	−0.023

Note:
Again, we measured arrivals together of each female with every other female in the community each year ($n=26$ females, of whom 5 were high-ranking, 16 were medium-ranking, and 15 were low-ranking; some females changed rank over time). 'Individual' is an identifying variable for each specific combination of individuals. The variable 'rank' indicates the rank of both individuals in the dyad. The coefficients show that dyads in which both females are high-ranking arrive mostly together, followed by a high- and a medium-rank pair, two medium-rank females, two low-rank females, a medium- and a low-rank pair, and finally a high- and a low-rank pair. As in other analyses, number of sexually receptive females was a significant variable ($p<0.0001$), with a dyad of two sexually receptive females the most social and two non-receptive females the least social.

petition on group size. In addition, in their own focal follows, females who carry infants travel shorter distances over time, spend less time in mixed-sex parties, and more time in female-only parties, which are smaller than mixed-sex parties and are thus likely to impose lower scramble competition costs. These patterns of behavior suggest that scramble competition may restrict travel and sociability of females with infants. There is no such effect of juvenile offspring, who generally travel under their own power. The patterns of association among individual females also suggest that infants increase scramble competition costs since females are less likely to arrive together if they have infants, but more likely to arrive together if they have juveniles. Note that dyadic associations among individual females include time in both female-only and mixed-sex parties, so this result suggests that although females with infants spend more time during their focal follows in female-only parties, this does not offset the decrease in time they spend in mixed-sex parties. Thus, scramble competition appears to cost more for mothers with infants than mothers with juveniles.

It should be noted that in these models of scramble competition, increase in travel distance is taken as a proxy for feeding competition, and food intake is not measured directly. The assumption is that individuals do not completely make up the costs of increased travel between food sources by increased food intake in each patch as, for example, might occur if large groups only form when there is much more food available in each patch. Rates of food intake in different-sized chimpanzee parties have yet to be measured, but Wrangham & Smuts (1980) found that both male and female chimpanzees spent less time feeding in parties than they did when alone.

Our results from 18 years at Gombe indicating a strong effect of scramble competition are consistent with those from shorter periods at Gombe and at Kibale (Wrangham 2000). The fact that female chimpanzees at Taï and Bossou and bonobo females spend more time in parties has been attributed both to higher levels of food availability (Boesch & Boesch Achermann 2000) and to differences in the diet and distribution of food (Wrangham 2000). These ideas can only be tested by detailed comparisons of food availability in patches, and travel rate and distance between patches in each population.

Contest competition

The fact that females with high dominance rank arrive more frequently with other females, and low-ranking females spend little time with females who outrank them suggests that contest competition is an important aspect of female association patterns. This is consistent with the idea that, for high-ranking females, the cost of contest competition is low since they are likely to win encounters with other females. Another possibility is that high-ranking females are in such good condition that they can afford to associate with other females, despite feeding costs.

Association rates at Gombe are still fairly low, even between high-ranking females. Only mother and adult daughter pairs associate at high rates at Gombe. These include pairs in which the daughter never emigrated, and those in which the daughter emigrated a few years after her first estrus. In these pairs, daughters generally had different

Table 14.3. *Effect of offspring on dyadic associations among females*
Model **Arrivals together** = Individual + number of mothers + number of juveniles + number of infants + number of cycling females

Variable	Mothers	Juveniles	Infants	Cycling females
P-values	<0.01	<0.0000	<0.0000	<0.0000
Coefficient = 0	−0.0012	−0.0062	0.0012	−0.016
Coefficient = 1	−0.00049	−0.00088	0.0000093	−0.0055
Coefficient = 2	0.0017	0.0071	−0.0012	0.021

Note:
We measured arrivals together of each female with every other female in the community each year. 'Individual' is an identifying variable for each specific combination of individuals, and all other variables refer to the number of individuals in the dyad who are mothers (e.g. have juveniles or infants), or cycle that year. Coefficients are listed for each value the predictor variables can have: for example, to understand the effect of mothers, coefficient = 0 means that the dyad includes no mothers, coefficient = 1 indicates that one of the females in the pair is a mother, coefficient = 2 means that both females are mothers. The same is true for other predictors.

ranks from their mothers, at least while their mothers were still alive. Only one pair, Passion and Pom, were both high-ranking. In populations in which all females associate at higher rates, such as Taï chimpanzees and bonobos, contest competition may have different effects on female association patterns. Boesch & Boesch-Achermann (2000) observed high-ranking female pairs at Taï who spent over 50% of their time together for years, just as Gombe mothers and adult daughters do, and they suggest that these dyads support each other in competing for food with other chimpanzees. Among bonobos, unlike Gombe chimpanzees (Williams *et al.* 2002), young immigrant females quickly form friendly relationships with older resident females that are thought to help them integrate into the female group (Idani 1991). In captive bonobos, alliances between high-ranking females allow them to exclude males and lower-ranking females from food (Parish 1996). Perhaps, when females spend more time in groups, either because scramble competition is relaxed or predation is a greater danger, the higher frequency of contest competition encourages alliances among unrelated females.

In a general discussion of the evolution of female social relationships in primates, Sterck *et al.* (1997) classified chimpanzees as a 'dispersal egalitarian' species. In such species, they suggest, contest competition is low between females in the same group so dominance relationships are unimportant. Therefore, females can afford to disperse away from female kin because they do not need allies in competition (Sterck *et al.* 1997). Data from Gombe showing strong effects of female dominance rank on reproductive success (Pusey *et al.* 1997), intragroup infanticide or attempted infanticide by coalitions of females (Goodall 1977; Pusey *et al.* 1997), and the data we present here do not support the idea that female chimpanzees are egalitarian. Rather, it appears that female competition over space (and probably food sources) is very strong (Williams *et al.* 2002). We are currently exploring the possibility that daughters are more likely to stay in their natal groups at Gombe because female competition and opposition to immigrant females is particularly strong at this site. If we accept that female competition is an important force in chimpanzees, we need to seek other explanations to account for female dispersal in chimpanzees, such as the avoidance of inbreeding with philopatric males (Pusey 1979).

Socializing offspring

Our results suggest that females may spend time together to socialize their juvenile offspring. Two females are most likely to associate with each other if they both have juvenile off-

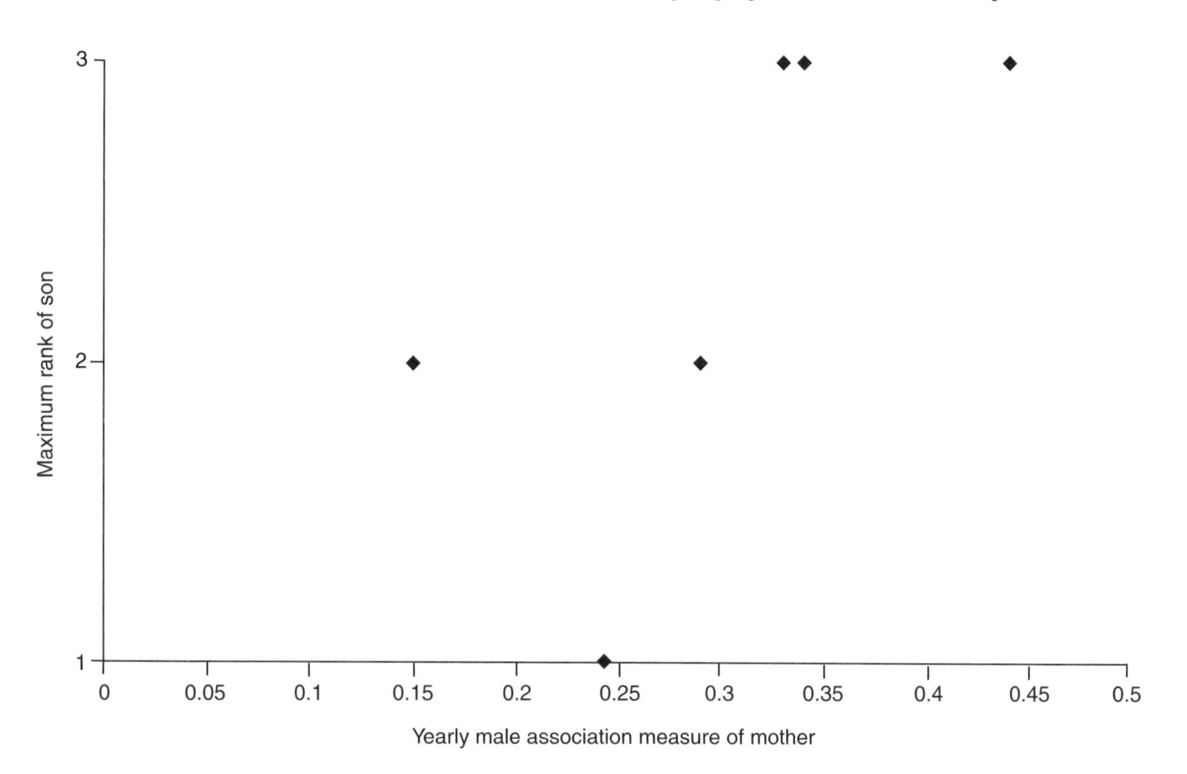

Fig. 14.3. Relationship between mother's average yearly proportion of encounters during follows of males and maximum rank of her sons. Highest (alpha) rank = 3, middle rank = 2, low rank = 1. Seven sons of six mothers are sampled (the two sons from one mother overlap on the graph).

spring or if neither have infants. This pattern best supports the importance of socializing juvenile offspring, but not infants. In addition to the benefits of play, older juveniles also interact quite extensively with adult females and may benefit by interactions with the juveniles' mothers and other adult females as well (Pusey 1990).

Additionally, mothers and adult daughters may be less likely to be in a female-only party with non-related females. This is not surprising since mothers and adult daughters have the benefits of a female-only party whenever they are together.

Females appear to benefit from socializing with adult males via the rank of their sons since sons of mothers who frequently encounter resident males may attain higher rank as adults. While dominant females socialize more with other females, they are not more sociable with males, and unlike at Taï (Boesch & Boesch-Achermann 2000) or in bonobos (Furuichi 1989), at Gombe there is no evidence that a female's rank is related to her son's rank. Thus, sons do not benefit from their mother's rank, but rather from her sociability with males. Mothers likely benefit reproductively from producing high-ranking sons since male rank increases mating success (Nishida & Hiraiwa-Hasegawa 1987) and reproductive success (Constable *et al.* 2001). So, why do not all mothers of sons associate frequently with adult males? Perhaps some mothers cannot afford the increase in scramble competition associated with party membership. Alternatively, some mothers might not have spent much time with adult males when they were young, and therefore do not have the social skills necessary to interact with them effectively. Goodall (1986) noted that some females are very uncomfortable around adult males and behave in ways that incite male aggression towards them. If this is the result of having an unsociable mother, socializing daughters may be as important as socializing sons since socially adept daughters may also be able to produce higher-ranking sons. Our sample is still too small to examine the effect of sociability of mothers on their daughters' reproductive success because at least half of adolescent females transfer to other communities in which their success cannot be measured.

Sexual receptivity

As in other studies, we found that sexually receptive females are more sociable and travel more than non-receptive females. Females who are sexually receptive shoulder the costs of faster travel even when alone (see also Wrangham, Chapter 15), more time in parties of all types, and male coercion (Smuts & Smuts 1993). Female chimpanzees cycle far more than appears necessary for conception (Goodall 1986), so what benefits counteract the above costs? Boesch & Boesch-Achermann (2000) suggest that frequent sexual cycles among nulliparous and recent immigrant females at Taï act as a 'social passport,' allowing females to stay with males, who protect them from aggressive resident females. Wrangham (Chapter 15) proposes that frequent and long, but unattractive cycles in nulliparous female chimpanzees and female bonobos reduce the costs of sexual coercion. We suggest that both nulliparous and parous female chimpanzees would benefit from increased sociability during sexual cycles via paternity confusion and maintenance of community membership in the eyes of the males in their community. Both reduce the chance of even greater costs, namely, loss of a female's own life via unclear community identity, and the lives of her offspring via paternity uncertainty (Goodall *et al.* 1979; Nishida *et al.* 1985; Williams 2000).

In sum, these results suggest that both scramble and contest competition are important variables for determining female chimpanzee association patterns. In particular, females with infants may be strongly influenced by scramble competition, and low-ranking females are most influenced by contest competition. At the same time, sociality of a mother could have far-reaching effects on the success of her offspring. Females with juvenile offspring are most likely to spend time with each other, suggesting that juveniles may benefit from spending time with others their own age, and sons of mothers who frequently encounter resident males may attain higher rank as adults. Finally, sexually receptive females are more social across all measures and travel farther, suggesting that sociability while sexually receptive is a substantial benefit.

REFERENCES

Altmann, J. & Samuels, A. (1992). Costs of maternal care: infant carrying in baboons. *Behavioral Ecology and Sociobiology*, 29, 391–8.

Bauer, H. R. (1980). Chimpanzee society and social dominance in evolutionary perspective. In *Dominance Relations*, ed. D. R. Omark, F. F. Strayer & D. G. Freedman, pp. 97–119. New York: Garland STPM Press.

Boesch, C. (1991). The effects of leopard predation on grouping patterns in forest chimpanzees. *Behaviour*, 117, (3–4), 220–42.

Boesch, C. (1996). Social grouping in Taï chimpanzees. In *Great Ape Societies*, ed. W. C. McGrew, L. F. Marchant & T. Nishida, pp. 101–13. Cambridge: Cambridge University Press.

Boesch, C. & Boesch-Achermann, H. (2000). *The Chimpanzees of the Taï Forest*. New York: Oxford University Press.

Cairns, S. J. & Schwager, S. J. (1987). A comparison of association indices. *Animal Behaviour*, 35, 1454–69.

Chapman, C. A., Wrangham, R. W. & Chapman, L. J. (1995). Ecological constraints on group size: an analysis of spider monkey and chimpanzee subgroups. *Behavioral Ecology and Sociobiology*, 36, 59–70.

Collins, D. A. & McGrew, W. C. (1988). Habitats of three groups of chimpanzees (*Pan troglodytes*) in western Tanzania compared. *Journal of Human Evolution*, 17, 553–74.

Constable, J. L., Ashley, M. V., Goodall, J. & Pusey, A. E. (2001). Noninvasive paternity assignment in Gombe chimpanzees. *Molecular Ecology*, 10, 1279–300.

Dunbar, R. I. M. (1988). *Primate Social Systems*, Ithaca, NY: Comstock, Cornell University Press.

Fagen, R. (1993). Primate juveniles and primate play. In *Juvenile Primates*, ed. M. E. Pereira & L. A. Fairbanks, pp. 182–96. Oxford: Oxford University Press.

Furuichi, T. (1989). Social interactions and the life history of female *Pan paniscus* in Wamba, Zaire. *International Journal of Primatology*, 10, 173–97.

Gardner-Roberts, D. (1998). From birth to five: developmental changes in chimpanzee behavior. PhD dissertation, Cambridge University, Cambridge.

Goodall, J. (1977). Infant killing and cannibalism in free-living chimpanzees. *Folia Primatologica*, 28, 259–82.

Goodall, J. (1986). *The Chimpanzees of Gombe*. Cambridge, MA: Harvard University Press.

Goodall, J., Bandora, A., Bermann, E., Busse, C., Matama, H., Mpongo, E., Pierce, A. & Riss, D. (1979). Intercommunity interactions in the chimpanzee population of the Gombe National Park. In *The Great Apes*, ed. D. A. Hamburg & E. R. McCown, pp. 13–53. Menlo Park, CA: Benjamin/Cummings.

Idani, G. (1991). Social relationships between immigrant and resident bonobo (*Pan paniscus*) females at Wamba. *Folia Primatologica*, 57, 83–95.

Janson, C. H. & Goldsmith, M. L. (1995). Predicting group size in primates: foraging costs and predation risks. *Behavioral Ecology*, 6, 326–36.

Macdonald, D. W. (1983). The ecology of carnivore social behaviour. *Nature*, 301, 379–84.

Milinski, M. & Parker, G. A. (1991). Competition for resources. In *Behavioral Ecology: An Evolutionary Approach*, ed. J. R. Krebs & N. B. Davies, pp. 137–68. Cambridge: Blackwell Scientific.

Nishida, T. & Hiraiwa-Hasegawa, M. (1987). Chimpanzees and bonobos: cooperative relationships among males. In *Primate Societies*, ed. B. B. Smuts, D. L. Cheney, R. M. Seyfarth, T. T. Struhsaker & R. W. Wrangham, pp. 165–80. Chicago: University of Chicago Press.

Nishida, T., Hiraiwa-Hasegawa, M., Hasegawa, T. & Takahata, Y. (1985). Group extinction and female transfer in wild chimpanzees in the Mahale National Park, Tanzania. *Zeitschrift für Tierpsychologie*, 67, 284–301.

Parish, A. R. (1996). Female relationships in bonobos. *Human Nature*, 7, 61–96.

Pusey, A. E. (1979). Intercommunity transfer of chimpanzees in Gombe National Park. In *The Great Apes*, ed. D. A. Hamburg & E. R. McCown, pp. 465–79. Menlo Park, CA: Benjamin/Cummings.

Pusey, A. E. (1983). Mother-offspring relationships in chimpanzees after weaning. *Animal Behaviour*, 31, 363–77.

Pusey, A. E. (1990). Behavioral changes at adolescence in chimpanzees. *Behaviour*, 115, 204–46.

Pusey, A. E., Williams, J. M. & Goodall, J. (1997). The influence of dominance rank on the reproductive success of female chimpanzees. *Science*, 277, 828–31.

Rodman, P. S. (1988). Resources and group sizes in primates. In *The Ecology of Social Behavior*, ed. Slobodchikoff, pp. 83–108. New York: Academic Press.

Sakura, O. (1994). Factors affecting party size and composition of chimpanzees (*Pan troglodytes verus*) at Bossou, Guinea. *International Journal of Primatology*, 15, 167–83.

Smuts, B. B. & Smuts, R. W. (1993). Male aggression and sexual coercion of females in nonhuman primates and other mammals: evidence and theoretical implications. *Advances in the Study of Behavior*, 22, 1–63.

Sterck, E. H. M., Watts, D. P. & van Schaik, C. P. (1997). The evolution of female social relationships in nonhuman primates. *Behavioral Ecology and Sociobiology*, 41, 291–309.

Terborgh, J. & Janson, C. H. (1986). The socioecology of primate groups. *Annual Review of Ecology and Systematics*, 17, 111–36.

van Schaik, C. P. (1989). The ecology of social relationships amongst female primates. In *Comparative Socioecology*, ed. V. Standen & R. A. Foley, pp. 195–218. Oxford: Blackwell Press.

van Schaik, C. P. & van Hooff, J. A. R. A. M. (1983). On the ultimate causes of primate social systems. *Behaviour*, 85, 91–117.

Watts, D. P. & Pusey, A. E. (1993). Behavior of juvenile and adolescent Great Apes. In *Juvenile Primates: Life History, Development, and Behavior*, ed. M. E. Pereira & L. A. Fairbanks, pp. 148–67. New York: Oxford University Press.

Williams, J. M. (2000). Female strategies and the reasons for territoriality in chimpanzees: lessons from three decades of research at Gombe. PhD thesis, University of Minnesota.

Williams, J. M., Pusey, A. E., Carlis, J. V., Farm, B. E. & Goodall, J. (2002). Female competition and male territorial behavior influence female chimpanzees' ranging patterns. *Animal Behaviour*, 63, 347–60.

Wilson, E. O. (1975). *Sociobiology*. Cambridge, MA: Belknap Press of Harvard University Press.

Wrangham, R.W. (1979). On the evolution of ape social systems. *Social Science Information*, 18(3), 335–68.

Wrangham, R. W. (1986). Ecology and social relationships in two species of chimpanzee. In *Ecology and Social Evolution*, ed. D. I. Rubenstein & R. W. Wrangham, pp. 352–78. Princeton: Princeton University Press.

Wrangham, R. W. (2000). Why are male chimpanzees more gregarious than mothers? A scramble competition hypothesis. In *Male Primates*, ed. P. M. Kappeler, pp. 248–58. Cambridge: Cambridge University Press.

Wrangham, R. W., Gittleman, J. L. & Chapman, C. A. (1993). Constraints on group size in primates and carnivores: population density and day-range as assays of exploitation competition. *Behavioral Ecology and Sociobiology*, 32, 199–209.

Wrangham, R. W. & Smuts, B. B. (1980). Sex differences in the behavioral ecology of chimpanzees in the Gombe National Park, Tanzania. *Journal of Reproduction and Fertility*, Supplement, 28, 13–31.

15 • The cost of sexual attraction: is there a trade-off in female *Pan* between sex appeal and received coercion?

RICHARD WRANGHAM

INTRODUCTION

Chimpanzees and bonobos are generally similar in their adult heterosexual behavior. In both species, copulations are largely restricted to females with maximally tumescent swellings (Furuichi 1987; Wallis 1992; Takahata *et al.* 1996). 'Group mating' is the predominant pattern. Rates of copulation vary, but there tend to be several hundred copulations per conception (Goodall 1986; Kano 1992). Furthermore, in both species the relationship between male rank and copulation rate varies in similar ways depending on group composition. Sometimes there is no relationship between a male's dominance rank and his copulation rate (e.g. Gombe chimpanzees: Tutin 1979; Mahale chimpanzees (1980–82,1991): Hasegawa & Hiraiwa-Hasegawa 1990; Nishida 1997; Taï chimpanzees: Boesch & Boesch-Achermann 2000; Wamba bonobos, E1 and E2 1986, P 1988: Takahata *et al.* 1996). However when a community or party contains few adult males in a stable hierarchy, higher-ranking adult males can achieve more copulations than lower-ranking males (Gombe chimpanzees in 1973–74: Goodall 1986; Mahale chimpanzees in 1992, Nishida 1997; Wamba E-group bonobos 1979: Kano 1996; Lomako bonobos: Gerloff *et al.* 1999). Dominant males then gain their advantage partly by interfering, at least occasionally, with subordinates' copulations (e.g. in 7% of bonobo copulations, Kano 1992).

Despite such similarities, heterosexual activity also varies in significant ways. A prominent example is the number of sexual cycles per conception, which ranges from an average of 3.6 for Gombe mothers to an average of 29.2 for Taï mothers. Such differences in sexual cycling have no relation to reproductive rates since the population interbirth intervals are similar in different sites (Takahata *et al.* 1996; Boesch and Boesch-Achermann 2000; Knott 2001). To account for this variation, it has been argued that a high number of sexual cycles increases the total benefits that a female derives from being sexually attractive. For chimpanzees, proposed benefits include social support from sexually attracted males (Boesch & Boesch-Achermann 2000), development of affiliative relationships (Wallis 1992), or the acquisition of mating skills (Wrangham 1993). For bonobos, they include alliances with sexually attracted females (Takahata *et al.* 1996), and facilitation of reconciliation and tension regulation (Hohmann & Fruth 2000).

In this chapter, I ask why such benefits are not available to parous East African chimpanzees. Compared to other females, East African mothers have a very low number of sexual cycles per conception, affording them few benefits from nonconceptive sex. They are also unusually attractive to males, so much so that they often experience intense sexual coercion, increased travel, and reduced feeding time (Wrangham & Smuts 1980; Goodall 1986; Matsumoto-Oda & Oda 1998). Their heightened appeal to males could be explained by the fact that a shorter time spent maximally tumescent gives females a greater probability of conception per mating day, so that their copulations are therefore more valuable to males (Goodall 1986; Nunn 1999; Boesch & Boesch-Achermann 2000). However, even though it is an understandable phenomenon, the strong attraction shown by males only exacerbates the theoretical problem of why East African mothers have such few cycles. If they increased the number of their sexual cycles per conception from 4 to 40, wouldn't they gain social benefits and also experience a lower level of sexual coercion?

To solve this problem, I propose a cost-of-sexual-attraction hypothesis. The hypothesis suggests that because of an ecological constraint (their relatively high cost-of-grouping), parous East African chimpanzees cannot afford to spend much time with males. Yet having a sexual swelling tends to increase a female's party size (Wrangham 2000a). Therefore, East African mothers cannot afford to have a high number of sexual cycles. This means that their total time available for mating is limited. During their few cycles, females must therefore mate at a relatively high rate. This requires them to intensify their sexual attractiveness; this, in turn, leads to elevated levels of coercion by males. For nulliparous East African chimpanzees, Taï chimpanzees, and bonobos, by contrast, I suggest that a lower cost-of-

Table 15.1. *Age–sex composition of the Kanyawara chimpanzee community from 1993 to 1999*

	Females				Males			
	Adult	Adol	Juv	Inf	Adult	Adol	Juv	Inf
1993	13	5	1	7	12	1	2	3
1994	14	4	0	8	11	1	4	3
1995	14	5	3	6	11	1	4	6
1996	15	5	4	6	11	2	3	7
1997	16	3	5	4	10	2	2	7
1998	15	2	6	4	11	1	3	8
1999	15	0	5	5	10	3	1	7

Notes:
Ages are: adult, >15 years; adolescent 11–15 years; juvenile 6–10 years; infant 0–5 years. The oldest individual whose birth-year was recorded was born in 1985 (G. Isabirye-Basuta, personal communication).

grouping affords females the luxury of many sexual cycles, a low intensity of sexual attractiveness to males, and a release from sexual coercion.

The proposal that the cost-of-grouping is a driving variable comes from the observation, which is central to this study, that East African mothers are the least gregarious of female *Pan* (Wrangham 2000a). The putative effect of scramble competition is that as traveling parties increase in size, all individuals have to spend more time traveling. If the rate of increase of travel cost is sufficiently steep, individuals leave parties and travel alone. Cost-of-grouping theory points to two particularly important influences on how rapidly travel costs rise with party size.

First, the travel cost rises more quickly for low-velocity individuals. East African chimpanzee mothers have a lower velocity than males or nulliparous females. This can therefore explain why they spend much of their time alone or in relatively small parties (Wrangham 2000a).

Second, the travel cost rises more quickly in patchier environments, for example for parties that travel further, or that have fewer foraging opportunities between major meals. This may explain why East African populations form smaller parties than other *Pan* (Fruth *et al.* 1999; Wrangham 2000a; Boesch & Boesch-Achermann 2000).

Thus, in this chapter, I draw attention to the relatively solitary nature of East African mothers, and I assume that this uniquely low level of gregariousness among female *Pan* is the result of ecological factors. In order to test the cost-of-sexual-attraction hypothesis, I review relevant data from the wild, and present new information comparing sexual behavior of nulliparous and parous chimpanzees in an East African population. My focus is on the wild because the adaptive significance of sexual patterns observed in captivity is often unclear.

METHODS

Chimpanzees were searched for in the Kanyawara community of Kibale National Park (Uganda) daily from January 1993 to June 1999. The study site was described by Wrangham *et al.* (1996). The age–sex class composition of the Kanyawara community is shown in Table 15.1. Monthly observation hours varied from 8 to 398, totaling 10913 hours over the 78 months. Statistical tests are two-tailed.

Copulations were recorded by 'all-event sampling', that is recording all copulations observed by myself and other observers. This undoubtedly underestimates the rate of copulation because individuals were sometimes hidden behind vegetation despite being recorded as present together. Maximal tumescence, following standard definitions (e.g. Wallis 1992), was defined visually as a fully expanded sexual swelling, that is tense and shiny. Male rank was assessed from dyadic aggressive and greeting interactions (see Muller, Chapter 8). Copulations were defined as mounting with intromission and pelvic thrusting, followed by separation. Though ejaculation was often confirmed, it could not always be seen. The duration of copulations averaged 6.6 s (SD = 0.5 s) across 13 males for whom at least 45 copulations were observed. Copulations normally averaged about one thrust per second. After the first two to three pelvic thrusts, a sperm plug left by a prior male was sometimes dislodged and fell to the ground. A total of 1898 copulations was recorded (24.3 per month; SD = 24.9).

HYPOTHESES AND RESULTS

Variations in the sexual attractiveness of female *Pan*

Unlike bonobos or nulliparous chimpanzees (Wallis 1992), the sexual attractiveness of parous East African chimpanzees (best known from Gombe and Mahale) varies markedly within the period of maximal tumescence. The peak of parous chimpanzee sexual attractiveness is reached during the 'peri-ovulatory period' or 'POP,' defined by Goodall (1986) as the last 4 days of maximal tumescence, that is terminating on the presumed day of ovulation (Graham 1981). Evidence for this temporal variation in attractiveness of East African females is as follows.

First, during the POP, parous chimpanzees in Gombe and Mahale experience a change in copulation rate. If no males are sufficiently dominant to mate-guard effectively, copulation rate increases significantly compared to the preceding days (Hasegawa & Hiraiwa-Hasegawa 1990). Alternatively, if some males are sufficiently dominant, a decreased rate of copulation by the female results from her limited access to lower-ranking males (Goodall 1986). Either way, male interest intensifies during the POP (Wallis 1992).

Second, mate-guarding by male chimpanzees reaches its highest frequency in the POP, for example in the last 3 days of the swelling period (Tutin & McGinnis 1981). In Kibale (Ngogo), POP females were guarded by top-ranking males, either alone or in coalitions of two to three males, unlike females that were maximally tumescent but outside the POP (Watts 1998). Intensified aggression during the POP means that during this period, a male's mating success is most likely to be related to his dominance rank (Tutin 1979; Goodall 1986; Hasegawa & Hiraiwa-Hasegawa 1990; Nishida 1997).

By contrast, among bonobos and nulliparous chimpanzees, and perhaps among Taï chimpanzees, there does not appear to be an equivalent change in female attractiveness during the POP. Thus, although mating rights can be contested, there are no reports of increased attractiveness during any particular period (Kano 1996; Fruth *et al.* 1999; Takahata *et al.* 1999; Boesch & Boesch-Achermann 2000). Taï females are sometimes the objects of male–male aggression, but they experience low rates of copulation (40 copulations per 10-day period), and are never monopolized by high-ranking males (Boesch & Boesch-Achermann 2000). Boesch & Boesch-Achermann (2000, p. 87) state that 'Ovulation in (Taï) females is hidden by prolonged sexual swellings . . .' In Gombe, by contrast, 'male chimpanzees . . . seem able to determine the peri-ovulatory period with some precision' (Goodall 1986, p. 445). This difference occurs even though the pattern of sexual swelling appears to be similar in the two sites (10 days of maximal tumescence in a 36-day cycle, Boesch & Boesch-Achermann 2000).

These patterns can be summarized as parous East African chimpanzees giving louder signals of their peri-ovulatory period than other females. In other words, parous East African chimpanzees have a loud POP, whereas bonobos, Taï chimpanzees, and nulliparous chimpanzees have a quiet POP. The chimpanzees currently known to have a loud POP, therefore, are all from East Africa (Gombe, Mahale, and Kibale).

The cost-of-sexual-attraction hypothesis

The cost-of-sexual-attraction hypothesis is presented in Figures 15.1 and 15.2. It proposes that the factor ultimately responsible for the intensity of female sexual attractiveness during the POP is the level of within-group scramble competition. Parous East African chimpanzees appear to experience relatively intense within-group scramble competition, which is responsible for their being less gregarious than nulliparous chimpanzees, bonobos, or Taï chimpanzees (Wrangham 2000a).

The consequences of the varying levels of scramble competition are portrayed in Figure 15.1 as a logical flow. All females are assumed to require several hundred copulations per conception. This assumption is examined below. From the female's perspective, the fastest way to achieve this is to be as attractive as possible. However, a high level of attractiveness has the disadvantage of inducing an elevated intensity of male coercion. Females accept this high level of male coercion only if it enables them to achieve their required number of copulations in a short time – that is, if their cost of scramble competition is high. Figure 15.2 presents this argument graphically.

The following sections examine and test the assumptions in this model.

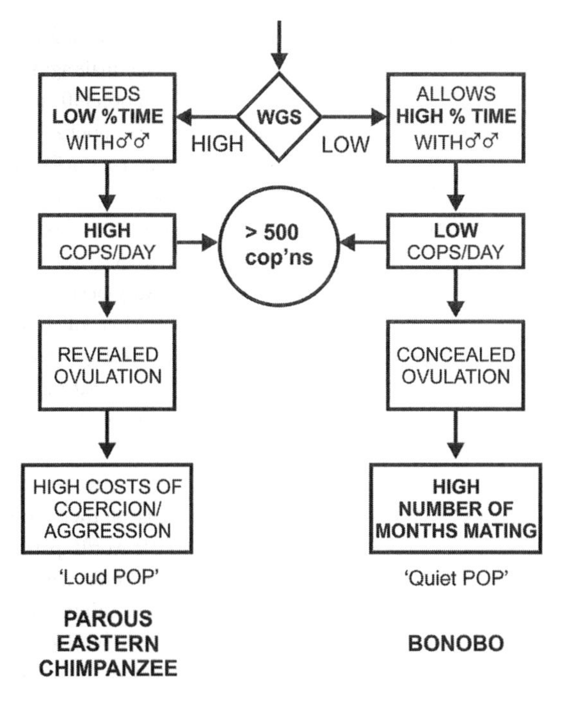

Fig. 15.1. The cost-of-sexual-attraction hypothesis is illustrated by a comparison of bonobos with Eastern parous chimpanzees. For females in both species, it is assumed that at least 500 copulations are required per conception. This can be achieved either by mating at a high rate (copulations per day) for a few cycles, or at a low rate for many cycles. The starting-point for deciding between these strategies is to ask whether the intensity of within-group scramble competition (WGS) is High or Low. For parous (Eastern) chimpanzees, it is considered to be high (a result of their traveling with an infant, and having no important foods to eat between fruit trees). Consequently, it pays these mothers to minimize the time they spend with males. This means that in order to achieve the requisite number of copulations, parous Eastern chimpanzees must attract copulating males at a high rate per day. The mechanism for attracting males (a loud POP, i.e. a peri-ovulatory period with chemical, behavioral, and/or other signals that attract intense male sexual interest) has unfortunate consequences, including both male aggression toward the female, and disturbance of the female's activity budget. By contrast, bonobos experience low WGS. A low WGS affords these females a low rate of copulations, and therefore a quiet POP (a peri-ovulatory period without chemical, behavioral, and/or other signals that attract intense male sexual interest). Taï chimpanzees and nulliparous chimpanzees also fall into this category (see text). The cost is that in order to achieve the high copulation number, they must be sexually active for relatively more days than loud-POP chimpanzees. The increased number of sexually active days is achieved mainly by an increase in the number of mating cycles prior to conception (see Figure 15.2.).

Total number of copulations required per birth

Both chimpanzees and bonobos normally copulate several hundred times per conception. Table 15.2 gives estimates for total copulation number per birth. They vary from a minimum of at least 400 (parous chimpanzees, Kanyawara) to more than 13000 (nulliparous bonobos, Wamba). In the cost-of-sexual-attraction model developed here, I assume that a high number of copulations per conception benefits females, even though it is doubtless not needed purely for the physiological act of conception. I therefore briefly consider possible reasons for the high copulation number.

In some mammals, the probability of conception increases with the number of copulations (Hoogland 1998). However, not only is the total copulation number in chimpanzees and bonobos clearly excessive for ensuring female choice of good sperm or good genes, but many copulations are also solicited by females in periods when they are unlikely to be fertile (Goodall 1986; Wrangham 1993).

Paternity confusion is therefore the predominant hypothesis accounting for multiple copulation by female primates. Paternity confusion is normally considered to be an anti-infanticide strategy (Hrdy 1979; Nunn 1999; van Schaik *et al.* 1999), an idea that has been well-supported in some species (savanna baboons *Papio cynocephalus*, Palombit *et al.* 1997; langurs *Presbytis entellus*, Borries *et al.* 1999; Japanese macaques *Macaca fuscata*, Soltis *et al.* 2000). Infanticide-inhibition is plausible for chimpanzees because infanticide by adult males is an important risk factor (recorded in Gombe, Mahale, Kibale (Kanyawara and Ngogo), Budongo and Taï: Arcadi & Wrangham 1999; Boesch & Boesch-Achermann 2000). Furthermore, the majority of male infanticides are by individuals unlikely to have copulated with the mother. However, there is no direct evidence that within communities, males are more likely to kill a female's infant if they did not mate with her.

In contrast, infanticide has never been recorded among bonobos. This difference could, in theory, result from small sample size, but observation time for the two species suggests that infanticides should have been seen among bonobos by now if they occur at rates equivalent to those in chimpanzees (Wrangham 2000b). Low infanticide rates in bonobos could come from reduced male interest in infanticide, or from females having particularly effective anti-infanticide copulation strategies such as social dominance over males. Either way, it seems unlikely that infanticide can explain multiple-male mating in bonobos. Instead, multiple-male mating might induce males to invest in a female's

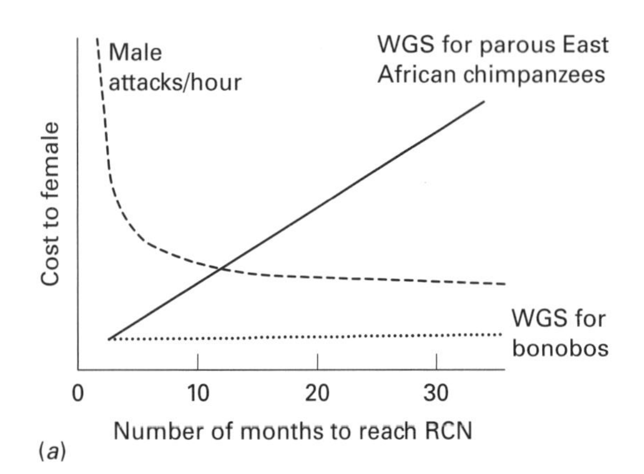

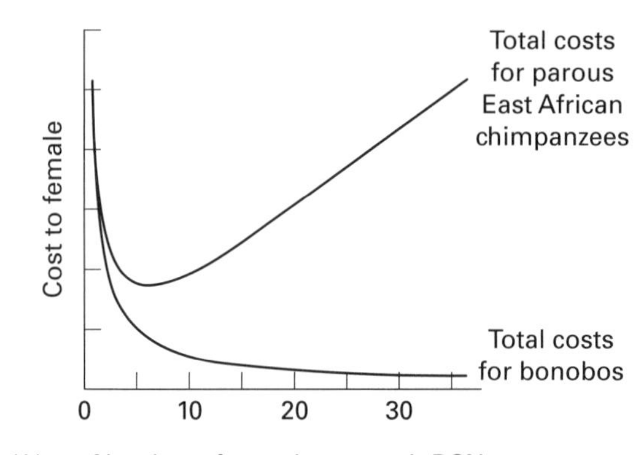

Fig. 15.2. Optimal number of mating cycles: minimizing total mating costs for female chimpanzees. 'Cost to female' means cost to the female's reproductive potential. 'Number of months to reach RCN' refers to the time to reach the female's required copulation number, assumed here to be several hundred copulations. This figure is a graphical version of Figure 15.1. As in Figure 15.1, the cost-of-sexual attraction hypothesis is illustrated by a comparison of bonobos with Eastern parous chimpanzees.

In (*a*), females experience two kinds of cost. First, all females experience costs of male coercion, here labeled 'Male attacks/hour' (dashed curve). These costs include physical wounds, physiological stress, loss of feeding time as a result of social interactions, and reduced social freedom. 'Male attacks/hour' are high for a female who takes only a few months to reach her RCN because the only way she can do this is to be highly attractive. By contrast, 'Male attacks/hour' becomes increasingly low for females who take a long time to reach their RCN. However, they do not decline to zero.

The second kind of cost results from traveling with a party of males, that is the cost of within-group scramble competition (WGS). Costs of WGS increase in direct proportion to the number of mating months. However, costs increase more steeply for parous Eastern chimpanzees than for bonobos because of ecological differences (see text).

In (*b*), the costs of coercion and WGS are combined. The total cost for parous Eastern chimpanzees is minimized at a low number of copulation months (between five and ten). Compared to parous chimpanzees, bonobos have lower costs, and a longer optimal number of mating months (i.e. as many as are compatible with lactation and conception).

young, for example by defending the community range against neighbors. If so, this kind of explanation could also apply to chimpanzees (Wrangham 1979).

Thus no hypothesis for multiple-male mating has yet been strongly supported. This means that it is unclear what a female's total copulation number should be. Furthermore, even if we were confident that a high copulation number is adapted to inhibiting male infanticide, we would still find it difficult to predict how often a female should mate. Her total copulation number would depend, for example, on how the inhibitory effect of copulation works. Is the male's infanticidal tendency inhibited as effectively by a single copulation as by several? Is the inhibition stronger as a result of copulations that are more likely to be successful, for example those nearer to the time of ovulation? Without such information, the female's optimal copulation number is unpredictable.

Could infanticide be adequately inhibited by a single, pre-conception copulation? To test this, the distribution of matings is illustrated in Figure 15.3. Figure 15.3 shows that the proportion of a community's adult males that mated a given female at least once rose very sharply as her total copulations increased. Each male's probability of mating rises so quickly with the female's total that the typical number of female copulations per birth (i.e. >400, Table 15.2) substantially exceeds the number needed to ensure a single copulation per male. Possibly she is adapted to achieving many copulations in order to ensure mating the most elusive males, for example those temporarily absent due to consortships or distant foraging. More likely, however, in view of the female's persistent proceptivity towards males whom she has already mated, the requirement of a single, preconception copulation per male cannot account for her high copulation number.

In sum, current hypotheses do not predict a female's optimal copulation total. Therefore, I make the simplifying assumption that observed behavior reflects a female's strategic goals. Accordingly, her optimal strategy is to distribute several hundred copulations rather evenly among the males in her community (see Table 15.2).

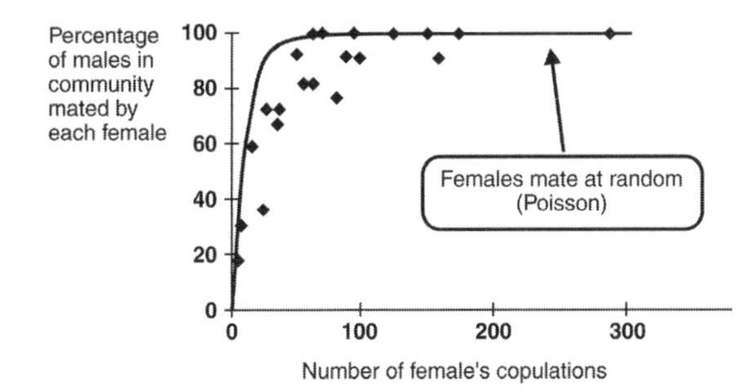

Fig. 15.3. Probability of a female Kanyawara chimpanzee copulating a male at least once, in relation to her total number of copulations. Each data point represents one conception. 'Number of female's copulations' shows her observed number of copulations during a conception period, that is starting with the resumption or initiation of copulations, and ending with the birth of an offspring. No allowance is made for periods when the female was not observed. 'Percentage of males in community mated by each female' refers to adult males present in the community at the time of the copulation. For most of the data collection period, there were 12 adult males present. The solid line shows the percentage of males mated if a female mated all males at random, that is if her mate choice followed a Poisson distribution. The actual distribution is close to random. Thus, after about 60 copulations, the probability of a female having copulated every male in the community at least once rises to >99%. Data are from 24 conception periods, 1652 copulations in Kanyawara, 1993–99.

Table 15.2. *Estimated number of copulations per birth for female chimpanzees and bonobos*

	Chimpanzee		Bonobo	
	Site	*n*	Site	*n*
Nulliparous	Kanyawara	1000	Wamba	>13100
Parous	Kanyawara	>400	Wamba	>1800
Parous	Taï	760		
Parous	Mahale	700		
Parous	Gombe	3000		

Notes:
Estimated number of copulations per birth was calculated from data on female copulation rates per hour or day, and on the number of mating days per conception. Data sources and observation methods are: Kanyawara 1991–99, all-event sampling; Mahale 1988 M-group, focal sampling (Hasegawa 1991, 1992); Gombe 1972–1975, focal sampling (Tutin 1975); Taï all-event sampling (Boesch & Boesch-Achermann 2000); Wamba 1978–79, all-event sampling (Kano 1992). 'All-event sampling' is assumed to underestimate total copulation number because compared to focal sampling, it increases the chance that some copulations occur without being recorded.

Variation in copulation rate

The cost-of-sexual-attraction model predicts that as a result of their higher level of sexual attractiveness, parous East African chimpanzees will have a higher daily rate of copulation than other females. Different studies have had different results when comparing copulation rates for parous chimpanzees and bonobos. For example, Takahata *et al.* (1999) found a trend for parous chimpanzees to copulate at higher rates than nulliparous chimpanzees, in support of the model.

Since copulation rates are likely to be affected by the community sex ratio, the most effective test is made within communities. To test the effects of parity among Kanyawara chimpanzees, I calculated copulation rates for maximally tumescent females per 100 hours of time together with specific males. Overall there was no discernible effect of parity. For example when copulating with adult males, the median copulation rate (across 12 adult males) was 4.7 copulations/100 hours for six parous females, vs. 4.3 copulations/100 hours for seven nulliparous females. This contradicts the prediction of the cost-of-sexual-attraction model.

However, the model could be correct if a goal of female sexual activity is to copulate with higher-ranking males, and copulation rates with higher-ranking males are lower for nulliparous than for parous females. To test this possibility, I hypothesized that, compared to nulliparous females, parous females would have a relatively high rate of copulation with high-ranking males. This prediction was supported. Thus, males that were adult throughout the central part of the study period ($n=10$, 1994–98) were divided into five high-rank males and five low-rank males based on their mean dominance ranks. Parous females had significantly higher copulation rates with high-rank males (7.7 copulations/100 hours) than did nulliparous females (2.5 copulations/100 hours) (Sign test, $t=0$, $p<0.05$, 1-tailed). By contrast, there was no effect of parity on copulation rates with low-rank males (parous 3.5, nulliparous 4.4 copulations/100 hours). This result suggests that parous females were more attractive than nulliparous females. If so, the

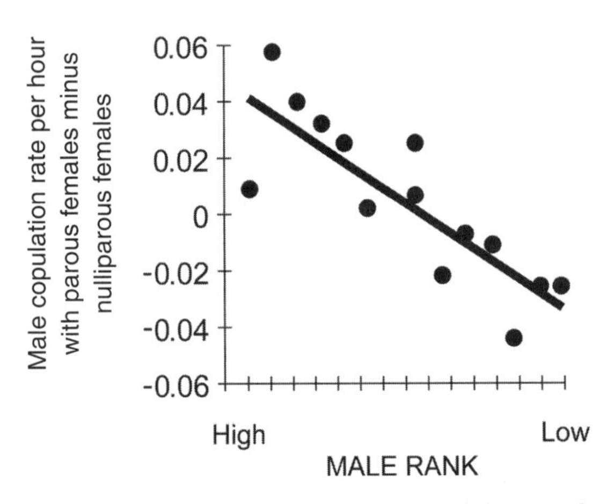

Fig. 15.4. Effects of male rank on the relative copulation rates of parous and nulliparous females. Copulation rate was calculated as the number of copulations observed per hour that each maximally tumescent female spent together with each male. Median copulation rates were then calculated for each of 14 adult and adolescent males across 12 parous and nine nulliparous females. The figure shows how the difference in copulation rates with parous vs. nulliparous females varied with male dominance rank. High-ranking males copulated with parous females at higher rates than with nulliparous females, whereas low-ranking males did the opposite. Middle-ranking males copulated with parous and nulliparous females at similar rates; $r = -0.87$, $p < 0.01$, $n = 24$ conception periods (1843 copulations), Kanyawara 1993–99.

reason that parous females experienced overall rates of copulation similar to those of nulliparous females may have been that low-rank males were inhibited from mating parous females at high rates.

To assess the effect of male rank on mating success with parous vs. nulliparous females, I compared the median number of copulations/100 hours together, for 12 parous and 9 nulliparous females respectively, with the dominance ranks of 14 adult and adolescent males. For parous females, higher-ranking males had somewhat higher copulation rates (Spearman $r = 0.58$, $n = 14$, $p = 0.07$), whereas for nulliparous females, higher-ranking males had significantly lower copulation rates (Spearman $r = -0.68$, $n = 14$, $p < 0.05$). These results are combined in Figure 15.4, which shows the difference in copulation rates between parous and nulliparous females in relation to male rank. Higher-ranking males had significantly higher copulation rates with parous females, compared to nulliparous females, than did lower-ranking males (Spearman $r = -0.87$, $n = 14$ males, $p < 0.01$). In view of evidence that parous females induced more contest competition than did nulliparous females (below), this result supports the idea that relatively high copulation rates by parous females with high-ranking males are the consequence of their sexual attractiveness. In this respect, the cost-of-sexual-attraction hypothesis is supported.

Among bonobos, adolescent males have lower copulation rates than adult males because adult males inhibit the copulation attempts of these younger males (Takahata *et al.* 1999). In contrast to the expectations of the cost-of-sexual-attraction hypothesis, this means that parous bonobos are sufficiently attractive to induce at least one form of male coercion.

Number of mating days per conception

The cost-of-sexual-attraction model proposes a trade-off between sexual attractiveness and the number of mating days per conception, with more mating days for parous East African chimpanzees than for other females. In contrast to early reports from captivity (Dahl 1986; Dahl *et al.* 1991), recent evidence suggests no consistent differences between bonobos and chimpanzees in the lengths of intermenstrual intervals or the duration of maximal tumescence (Stanford 1998; Vervaecke *et al.* 1999; Hohmann & Fruth 2000). Similarly, there appears to be no difference between Taï and East African chimpanzees in these variables (Boesch & Boesch-Achermann 2000).

However the number of mating cycles varies widely (Table 15.3). As discussed earlier, parous chimpanzees in East Africa have fewer cycles per birth (median 5) than nulliparous East African chimpanzees (median 19), parous Taï chimpanzees (29.2) or bonobos (34.8). Nevertheless interbirth intervals are similar across populations for parous females with surviving young (Table 15.3). Thus, there is no indication that an increased number of cycles delays the expected time of conception. Accordingly, the fact that some females have many cycles is due to social, not reproductive, considerations. Females should not be thought of as taking a long or short time to conceive in relation to the resumption of cycling. Instead, they should be viewed as resuming cycling early or late in relation to the anticipated conception time.

Sometimes the ‘decision’ about when to resume cycling is taken out of the females’ control. For example, Table 15.3 shows that mothers whose infants have died have few cycles per birth (4.5–5, Mahale, Taï). In these cases, differences in the intensity of scramble competition are irrelevant because, unlike mothers of surviving infants, an increased number of cycles would extend their interbirth interval. The fact that she cycles as quickly as an East African mother is under-

Table 15.3. *Number of sexual swelling cycles per conception and interbirth intervals for chimpanzees and bonobos*

Site/species	Nulliparous	Parous	Interbirth interval (y)	Reference
Gombe chimpanzee	19 (10–27)	3.6 (1–11)	5.5	Tutin and McGinnis (1981)
Mahale chimpanzee	24 (0.5–82)	9.0 (1.5–32.5) (L), 4.5 (0–34) (D)	6.0	Nishida *et al.* (1990)
Kibale (Kanyawara) chimpanzee	16 (2–25)	5 (2–10)	6.0	This study
Taï chimpanzee	26	29.2 (L), 5 (D)	5.4	Boesch and Boesch-Achermann (2000)
Wamba bonobo		34.8	4.5	Calculated by Wrangham (1993) from Kano (1992)

Notes:
Number of sexual swelling cycles per conception is shown as a mean or median prior to estimated date of conception, with the range in parentheses. Sample size: Kanyawara (medians) five nulliparous, ten parous conceptions; Mahale (medians) ten nulliparous (median calculated from data in Nishida *et al.* (1990, p. 76, table III.6), L = parous conceptions following a surviving infant, D = parous conceptions following a dead infant; Taï (medians) seven nulliparous conceptions, L = 26 parous conceptions following a surviving infant (calculated by subtracting median 22.5-month period of lactational amenorrhea and 7.5-month gestation from 65-month interbirth intervals), D = 8 parous conceptions following a dead infant (calculated from table III.3 in Boesch & Boesch-Achermann 2000, based on a 36-day sexual cycle; calculations assume continuous cycling following resumption). Interbirth intervals are from Knott (2001).

standable, therefore, as a way to avoid a delay in conception. The similarity in the number of cycles between Taï females whose infants have died, and East African mothers whose infants survive, suggests that four to five cycles may be the minimum number that chimpanzees need to conceive. This is a relatively long time compared to bonobos and gorillas (Knott 2001). An interesting possibility is that this results from the stress of being coerced by males.

The number of cycles possible for nulliparous females is probably also constrained, but in a different way. In order for nulliparous females to increase their number of cycles without postponing their conception, they would have to immigrate into their new community earlier, that is as late juveniles. This would likely be costly.

Consideration of such constraints raises the possibility that where the costs-of-grouping are low, the number of sexual cycles is generally maximized. Figure 15.2 shows one way that maximization of the number of sexual cycles may be favored.

Variation in male coercion

Coercive behavior by males includes interfering with copulations, restricting a female's access to other males, herding a female, forced copulation, and forced consortships. In addition, estrous females experience more male aggression than anestrous females (in Mahale, Matsumoto-Oda & Oda 1998), suggesting that male aggression toward females is a sexually coercive behavior. Sometimes male–male interference also occurs, both among bonobos and chimpanzees (Goodall 1986; Kano 1996). So although male coercion is thus widespread, it also appears to be more intense for parous East African chimpanzees than for other females.

First, persistent mate-guarding by one or more males occurs regularly in East African chimpanzees (i.e. a male repeatedly interposing himself between a female and a male trying to mate with her, see Watts 1998). This behavior has not been reported for bonobos or Taï chimpanzees.

Second, I have found no reports of persistent mate-guarding of nulliparous chimpanzees. For example, coalitionary mate-guarding always appears to be associated with parous rather than nulliparous females (Mahale: Nishida 1997, p. 387, NT/MU herding ZP; Gombe: Tutin & McGinnis 1981, p. 254, FG/FB twice possessing females; Kibale (Ngogo): Watts 1998). Boesch & Boesch-Achermann (2000) never saw competition between males over nulliparous females, but noted that later (i.e. after they were parous), males started to compete and guard these same females.

The difference in the level of mate-guarding directed towards parous vs. nulliparous females was clear in Kanyawara. On mating days, nulliparous females were observed much more frequently than parous females (e.g. Table 15.3), yet aggressive mate-guarding by males occurred only with parous females (12 out of 13 conceptions of parous females (92%), none out of five conceptions of nulliparous females (0%), Fisher's Exact test, $p<0.01$). Strikingly, three females were observed both when nulliparous and when parous (OU, TG, and AL). In each case, mate-guarding was absent during nulliparous mating cycles, then emerged with parous mating cycles.

Third, consortships have been described in chimpanzees but not in bonobos (Takahata *et al.* 1999). Little is known about the factors responsible for their variation within and between populations (Boesch & Boesch-Achermann 2000, Muller & Wrangham 2001). Since consortships presumably reflect relatively intense male attraction, parous females are expected to form consortships more than nulliparous females. This appears to have been true in Gombe. Thus, consortships occurred in 14 out of 33 conception cycles of parous females (42%), compared to none in 13 conception cycles of nulliparous females (0%) (Goodall 1986; table E2; Fisher's Exact test, $p<0.01$). In Kanyawara, only a single clear consortship has been recorded, involving a parous female for three months. In Taï, by contrast, consortships occurred independent of parity (Boesch & Boesch-Achermann 2000, p. 79).

DISCUSSION AND CONCLUSIONS

Chimpanzees and bonobos, whether parous or nulliparous, largely confine their copulations to periods with a maximally tumescent sexual swelling. It is sometimes assumed that sexual swellings reveal the timing of ovulation. However, for female *Pan* this assumption is clearly wrong (Boesch & Boesch-Achermann 2000). First, a maximally tumescent sexual swelling often occurs without ovulation, for example in parous bonobos and nulliparous chimpanzees during their first year of cycling. Second, in conception cycles of nulliparous chimpanzees, ovulation clearly occurs, but there is no increase in male sexual interest around the time of ovulation. Third, even in cycles when males do compete intensely around the time of ovulation, as in parous East African females, males show relatively low interest during the early (mid-follicular) phase of maximal tumescence (Tutin & McGinnis 1981). These three points all imply that the intense attraction experienced by males during the loud POP of parous East African females results from something more than simply the correlation of a sexual swelling with ovulation. We can only speculate about what this second cue, or set of cues, might be. It could, in theory, be behavioral, or chemical, or due to males estimating the probability of conception. Yet no relevant behaviors have been reported; the idea of a chemical cue is challenged by male olfactory inspections of females peaking at the wrong time (during the early follicular phase, before swelling has even started, Nishida 1997); and a cognitive calculation by males is a demanding concept.

For these reasons, the question of how ovulation is revealed or concealed is a problem both for copulating males and for evolutionary biologists. Nevertheless, evidence reviewed in this chapter suggests that in bonobos, Taï chimpanzees and nulliparous chimpanzees ovulation is effectively concealed, even though their sexual swellings offer prominent signals of the period during which mating is generally welcomed. The advantage of this quiet POP system for females is that because males cannot precisely assess the time of ovulation, males treat each copulation as having relatively low value. Therefore, males reduce (though they do not eliminate) their coercive behavior. As a consequence, a wide variety of novel and exaggerated forms of nonconceptive sexual interaction become possible (Wrangham 1993; Hohmann & Fruth 2000).

In East African mothers, by contrast, ovulation is effectively revealed. These females therefore pay the price of 'the occasional undesired consortship, forcible recruitment by neighboring males, the risk of injury during forcible group mating which, in turn, may delay fertilization – and a good deal of minor discomfort' (Goodall 1986, pp. 483–4).

The cost-of-sexual-attraction hypothesis thus appears capable of explaining the two extreme types of POP. To judge from current evidence, East African mothers experience relatively intense within-group scramble competition, and as a consequence, are driven towards having few cycles per conception, hence high attractivity, and more intense sexual coercion, that is a loud POP. By contrast, species, populations, or reproductive stages that experience low within-group scramble competition are those with many mating cycles and low levels of male coercion (bonobos, Taï chimpanzees, and nulliparous chimpanzees), that is a quiet POP.

This argument therefore suggests that nonconceptive swellings evolved because they enabled females to reduce the costs of coercion; and that the various nonconceptive benefits of extended cycling evolved subsequently. Even if the nonconceptive benefits are thus seen as evolutionarily secondary, however, they are expected to influence the trade-off between number of sexual cycles and intensity of attraction, since some types of benefits are doubtless greater than others.

For example, Boesch & Boesch-Achermann (2000) found that high-ranking Taï females cycle less often than low-ranking females, and suggested that this difference occurs because high-ranking females have less need for male social support. Thus, low-ranking females are better able to use extended cycling for social benefits. Similar considerations might explain the otherwise puzzling observation that, among bonobos, nulliparous females are more attractive than parous females (Takahata *et al.* 1996, 1999). Again, among bonobos and parous Taï females, adolescent males have lower copulation rates than adult males because adult males inhibit the copulation attempts of low-ranking males (Takahata *et al.* 1999; Boesch & Boesch-Achermann 2000). This means that in contrast to the expectations of the cost-of-sexual-attraction hypothesis, these quiet-POP females are sufficiently attractive to induce at least one form of male coercion. These observations suggest that it is necessary to consider not only the costs but also the benefits that females derive from extending or intensifying their sexual attractiveness. However, the essential principle argued here is that such benefits are possible only because of the relatively low cost-of-grouping for certain classes of female.

The cost-of-sexual-attraction hypothesis can be tested further in other primates with prominent sexual swellings where ovulation can either be revealed (e.g. *Papio cynocephalus*, Hausfater 1975) or concealed (e.g. *Macaca sylvanus*, Small 1993). In these group-living species, the essential expectation is that in loud-POP species, females benefit from intense male attraction, whereas in quiet-POP species they do not. It is only in fission–fusion species that the economics of sexual attractivity include the time spent with males, and therefore that concealed ovulation is expected to be limited to females whose cost-of-grouping is relatively low.

If the cost-of-sexual-attraction hypothesis is upheld, an interesting question is whether we can use it to reconstruct the phylogeny of loud or quiet POPs. East African chimpanzees occupy relatively peripheral and seasonally dry habitats compared to those in Taï and most of the chimpanzee range, and they have occupied those habitats relatively recently (Goldberg & Ruvolo 1997). If the East African populations are recently derived from the central forest populations, the pattern of loud POPs ('revealed ovulation') may also be recently derived. However, if the Central African chimpanzees also prove to have loud POPs, the West African populations will likely be found to be the more derived form. This question therefore awaits study of Central African chimpanzees.

In conclusion, I suggest that in *Pan* a low cost-of-grouping is a necessary condition for the evolution of concealed ovulation, and therefore for the evolution of various kinds of nonconceptive sexuality. As in other primates, the evolution of concealed ovulation in humans is often considered solely from the perspective of the benefits that it brings. The logic developed in this study, however, suggests that it is just as important to ask how females could afford a quiet POP (equivalent to concealed ovulation) as to ask how they would benefit from it. If hominid females benefited from multiple matings, as chimpanzees and bonobos apparently do, they would face the same trade-off as chimpanzees and bonobos. Accordingly, concealed ovulation in the human lineage is expected to have evolved only after our ancestors developed a way of life with a low cost-of-grouping for females (Wrangham *et al.* 1999).

ACKNOWLEDGEMENTS

C. Boesch, M. Gerald, G. Hohmann, S. Kahlenberg, L. Marchant, and D. Zinner made valuable comments. Fieldwork was supported by the Leakey Foundation, the National Geographic Society (#5626–96), and the National Science Foundation (NSF SBR-9120960, BCS-9807448). I thank Makerere University and the Government of Uganda for permission to conduct research at the Makerere University Biological Field Station at Kanyawara, and the Uganda Wildlife Authority for permission to work in Kibale National Park. The support of G. Isabirye-Basuta and J. Kasenene is much appreciated. Long-term data were collected by J. Barwogeza, the late J. Basigara, K. Clement, K. Deo, C. Katongole, F. Mugurusi, D. Muhangyi, M. Muller, C. Muruuli, P. Tuhairwe, and M. Wilson. Other field assistance was provided by M. Becker, C. Hooven, the late J. Obua, S. Mugume and K. Pieta.

REFERENCES

Arcadi, A. C. & Wrangham, R. W. (1999). Infanticide in chimpanzees: review of cases and a new within-group observation from the Kanyawara study group in Kibale National Park. *Primates*, 40, 337–51.

Boesch, C. & Boesch-Achermann, H. (2000). *The Chimpanzees of the Taï Forest: Behavioral Ecology and Evolution*. New York: Oxford University Press.

Borries, C., Launhardt, K., Epplen, C., Epplen, J. T. & Winkler, P. (1999). Males as infant protectors in Hanuman langurs (*Presbytis entellus*) living in multimale groups – defence pattern, paternity and sexual behavior. *Behavioral Ecology and Sociobiology*, 46, 350–6.

Dahl, J. F. (1986). Cyclic perineal swelling during the intermenstrual intervals of captive female pygmy chimpanzees. *Journal of Human Evolution*, 15, 369–85.

Dahl, J. F., Nadler, R. D. & Collins, D. C. (1991). Monitoring the ovarian cycles of *Pan troglodytes* and *P. paniscus*: a comparative approach. *American Journal of Primatology*, 24, 195–209.

Fruth, B., Hohmann, G. & McGrew, W. C. (1999) The *Pan* species. In *The Nonhuman Primates*, ed. P. Dolhinow & A. Fuentes, pp. 64–72. Mountain View, CA: Mayfield.

Furuichi, T. (1987). Sexual swelling, receptivity, and grouping of wild pygmy chimpanzee females at Wamba, Zaïre. *Primates*, 28, 309–18.

Gerloff, U., Hartung, B., Fruth, B., Hohmann, G. & Tautz, D. (1999). Intracommunity relationships, dispersal pattern and paternity success in a wild living community of Bonobos (*Pan paniscus*) determined from DNA analysis of faecal samples. *Proceedings of the Royal Society of London, Series B*, 266, 1189–95.

Goldberg, T. & Ruvolo, M. (1997). The geographic apportionment of mitochondrial genetic diversity in East African chimpanzees, *Pan troglodytes schweinfurthii*. *Molecular Biology and Evolution*, 14, 976–84.

Goodall, J. (1986). *The Chimpanzees of Gombe: Patterns of Behavior*. Cambridge, MA: Harvard University Press.

Graham, C. E. (1981). Menstrual cycle of the great apes. In *Reproductive Biology of the Great Apes: Comparative and Biomedical Perspectives*, ed. C. E. Graham, pp. 1–43. New York: Academic Press.

Hasegawa, T. (1991). Sexual behavior of wild chimpanzees. In *The Cultural History of Primates*, ed. T. Nishida, K. Igawa & T. Kano, pp. 37, 1–88. Yokyo: Heibonsh. (In Japanese.)

Hasegawa, T. (1992). The evolution of female promiscuity: chimpanzees and Japanese macaques. In *Aggression and Cooperation among Animal Societies*, ed. Y. Ito, pp. 223–50. Tokyo: Tokaidaigaku-Shuppankai. (In Japanese.)

Hasegawa, T. & Hiraiwa-Hasegawa, M. (1990). Sperm competition and mating behavior. In *The Chimpanzees of the Mahale Mountains: Sexual and Life History Strategies*, ed. T. Nishida, pp. 115–32. Tokyo: University of Tokyo Press.

Hausfater, G. (1975). *Dominance and reproduction in baboons (*Papio cynocephalus*). Contributions to Primatology*, Vol. 7. Basel: S. Karger.

Hohmann, G. & Fruth, B. (2000). Use and function of genital contacts among female bonobos. *Animal Behaviour*, 60, 107–20.

Hoogland, J. (1998). Why do female Gunnison prairie dogs copulate with more than one male? *Animal Behaviour*, 55, 351–9.

Hrdy, S. B. (1979). Infanticide among animals: a review, classification, and examination of the implications for the reproductive strategies of females. *Ethology and Sociobiology*, 1, 13–40.

Kano, T. (1992). *The Last Ape: Pygmy Chimpanzee Behavior and Ecology*. Stanford, CA: Stanford University Press.

Kano, T. (1996). Male rank order and copulation rate in a unit-group of bonobos at Wamba, Zaïre. In *Great Ape Societies*, ed. W. C. McGrew, L. F. Marchant and T. Nishida, pp. 135–145. Cambridge: Cambridge University Press.

Knott, C. (2001). Female reproductive ecology of the apes: implications for human evolution. In *Reproductive Ecology and Human Evolution*, ed. P. Ellison, pp. 429–63. New York: Aldine.

Matsumoto-Oda, A. & Oda, R. (1998). Changes in the activity budgets of cycling female chimpanzees. *American Journal of Primatology*, 46, 157–166.

Muller, M. & Wrangham, R. W. (2001) The reproductive ecology of male hominoids. In *Reproductive Ecology and Human Evolution*, ed. P. Ellison, pp. 397–427. New York: Aldine.

Nishida, T. (1997). Sexual behavior of adult male chimpanzees of the Mahale Mountains National Park, Tanzania. *Primates*, 38, 379–98.

Nishida, T., Takasaki, H. & Takahata, Y. (1990). Demography and reproductive profiles. In *The Chimpanzees of the Mahale Mountains: Sexual and Life History Strategies*, ed. T. Nishida, pp. 63–97. Tokyo: Tokyo University Press.

Nunn, C. (1999). The evolution of exaggerated sexual swellings in primates and the graded-signal hypothesis. *Animal Behavior*, 58, 229–46.

Palombit, R. A., Seyfarth, R. M. & Cheney, D. L. (1997). The adaptive value of 'friendship' to female baboons: experimental and observational evidence. *Animal Behaviour*, 54, 599–614.

Small, M. (1993). *Female Choices*. Ithaca: Cornell University Press.

Soltis, J., Thomsen, R., Matsubayashi, K. & Takenaka, O. (2000). Infanticide by resident males and female counter-strategies in wild Japanese macaques (*Macaca fuscata*). *Behavioral Ecology and Sociobiology*, 48, 195–202.

Stanford, C. B. (1998). The social behavior of chimpanzees and bonobos: empirical evidence and shifting assumptions. *Current Anthropology*, 39, 399–420.

Takahata, Y., Ihobe, H. & Idani, G. (1996). Comparing copulations of chimpanzees and bonobos: do females exhibit proceptivity or receptivity? In *Great Ape Societies*, ed. W. C.

McGrew, L. F. Marchant and T. Nishida, pp. 146–58. Cambridge: Cambridge University Press.

Takahata, Y., Ihobe, H. & Idani, G. (1999). Do bonobos copulate more frequently and promiscuously than chimpanzees? *Human Evolution*, 14, 159–67.

Tutin, C. E. G. (1975). Sexual behaviour and mating patterns in a community of wild chimpanzees (*Pan troglodytes schweinfurthii*). PhD dissertation, University of Edinburgh.

Tutin, C. E. G. (1979). Mating patterns and reproductive strategies in a community of wild chimpanzees (*Pan troglodytes schweinfurthii*). *Behavioral Ecology and Sociobiology*, 6, 29–38.

Tutin, C. E. G. & McGinnis, P. R. (1981). Sexuality of the chimpanzee in the wild. In *Reproductive Biology of the Great Apes: Comparative and Biomedical Perspectives*, ed. C. E. Graham, pp. 239–264. New York: Academic Press.

van Schaik, C. P., van Noordwijk, M. A. & Nunn, C. L. (1999). Sex and social evolution in primates. In *Comparative Primate Socioecology*, ed. P. C. Lee (ed), pp. 204–40. Cambridge: Cambridge University Press.

Vervaecke, H., van Elsacker, L., Möhle, U., Heistermann, M. & Verheyen, R. F. (1999). Inter-menstrual intervals in captive bonobos (*Pan paniscus*). *Primates* 40, 283–9.

Wallis, J. (1992). Chimpanzee genital swelling and its role in the pattern of sociosexual behavior. *American Journal of Primatology*, 28, 101–13.

Watts, D. P. (1998). Coalitionary mate-guarding by male chimpanzees at Ngogo, Kibale National Park, Uganda. *Behavioral Ecology and Sociobiology*, 44, 43–55.

Wrangham, R. W. (1979). On the evolution of ape social systems. *Social Science Information*, 18, 335–68.

Wrangham, R. W. (1993). The evolution of sexuality in chimpanzees and bonobos. *Human Nature*, 4, 47–79.

Wrangham, R. W. (2000a). Why are male chimpanzees more gregarious than mothers? A scramble competition hypothesis. In *Male Primates*, ed. P. Kappeler, pp. 248–58. Cambridge: Cambridge University Press.

Wrangham, R. W. (2000b). Evolution of coalitionary killing. *Yearbook of Physical Anthropology*, 42, 1–30.

Wrangham, R. W. & Smuts, B. B. (1980). Sex differences in the behavioural ecology of chimpanzees in Gombe National Park, Tanzania. *Journal of Reproduction and Fertility* 28 (Suppl.), 13–31.

Wrangham, R. W., Chapman, C. A., Clark, A. P. & Isabirye-Basuta, G. (1996). Social ecology of Kanyawara chimpanzees: implications for understanding the costs of great ape groups. In *Great Ape Societies*, ed. W. C. McGrew, L. F. Marchant and T. Nishida, pp. 45–57. Cambridge: Cambridge University Press.

Wrangham, R. W., Jones, J., Laden, G., Pilbeam, D. & Conklin-Brittain, N. L. (1999). The raw and the stolen: cooking and the ecology of human origins. *Current Anthropology*, 40, 567–94.

Part IV
Hunting and food sharing

Introduction

LINDA F. MARCHANT

The knowledge that chimpanzees hunt vertebrates and share meat continues to fuel the imagination and grant proposals of many field researchers who study *Pan*. Hunting and food sharing have played an important role in constructing models of human origins (Dart 1953; Isaac 1978; Lovejoy 1981; Stanford & Bunn, 2001). The presence of sharing in *Pan* helps us to speculate about how the last common ancestor of living apes and humans may have dealt with scarce and contested food resources (Marchant, in press).

It was in 1963 that Jane Goodall published her findings revealing how powerfully meat figures in the lives of chimpanzees. More recently, we have also learned that bonobos are intensely interested in eating vertebrate flesh and share the prey (Ihobe 1992; Hohmann & Fruth 1993). The three chapters in this section provide new insights into both *Pan* species and enlarge our understanding of species differences and similarities.

Boesch *et al.* discuss chimpanzee predation on red colobus monkeys (Chapter 16), sharpening our understanding of this predator–prey system, and adding more complexity to how we might model ancestral hominid hunting patterns. The chapter is unusual in providing a three-site comparison (Gombe and Mahale in East Africa, and Taï in West Africa) by the same observer (C. Boesch). Boesch (1994) studied chimpanzee hunting behavior at Gombe and Taï and here extends the analysis, while Uehara and Ihobe provide an update and further analysis of prey availability at Mahale (Uehara *et al.* 1992). We learn that demographic changes in either predator or prey may have a significant impact on hunting frequency, species targeted, and number of monkeys taken annually. The cross-site comparison also reveals differences between red colobus populations in their rate of calling in the presence or absence of chimpanzees. It also appears that red colobus in eastern Africa are far more aggressive toward their ape predators than are those in western Africa. This may explain the Gombe and Mahale hunters' preference for immature red colobus, while those at Taï target the adults.

Fruth & Hohmann's work (Chapter 17) is a systematic and detailed analysis of how the rainforest-dwelling bonobos of Lomako (Democratic Republic of Congo) obtain and share two different but much sought-after food items. Fruth & Hohmann test four hypotheses to account for the observed patterns of sharing, finding support for 'sharing under pressure' (buying-off supplicants) and more limited evidence for kin selection. We learn that bonobos, unlike most chimpanzees, share not only vertebrate flesh, but also the beach-ball-sized fruits of *Treculia africana*. Their description of a bonobo climbing a tree to secure one of these fruits (which can weigh, on average, 28 kg), stamping on the fruit to loosen it, and then descending to the forest floor to be surrounded by others eager to have a portion, is an arresting image. It reminds us that sharing in hominid evolution need not have focused solely on meat.

Another departure from the more familiar pattern known for the chimpanzee is *female* ownership of the carcass or fruit. Lomako bonobos prefer to target duikers (*Cephalophus* spp.) instead of other primates (*Procolobus badius*), as the chimpanzee typically does. Hidden duikers are opportunistically hunted by individual bonobo females at Lomako; upon finding the duikers, the female bonobos then engage in sharing episodes that may last more than 3 hours.

The hunting activities of chimpanzees are well documented for many field sites. Field studies in the 1990s rekindled our interest in the importance of hunting as a political and sexual strategy used by males to secure allies and reproductive partners (Nishida & Hosaka 1996; Stanford 1998). The reassertion of hunting as a significant factor in explaining early hominid foraging has also been explored in light of these studies (e.g. Stanford 1996, 1999). Meanwhile, virtually no evidence of scavenging by apes has emerged (cf. Muller *et al.* 1995). However, some paleoarchaeologists continue to focus more on scavenging as the primary adaptive strategy for hominids, or at least as a significant adjunct to hunting (e.g. Blumenschine & Cavallo 1992; Bunn 2001).

The final chapter in this section, written by Watts &

Mitani (Chapter 18), is a potent analysis of hunting and meat sharing by the chimpanzees of Ngogo in the Kibale National Park, Uganda. The 24 males of this large community of 140–150 members provide the kind of empirical data that have previously been unavailable. Watts & Mitani supply compelling evidence of how canopy structure affects hunting success or failure. These chimpanzees more often hunt in regenerating forest and other similar habitats with broken canopies and, in such habitats, they are rewarded with a 92% success rate. Undeniably, these apes are successful hunters. Their overall success rate is 84%, they score 84% in multiple kills (as one episode had 13 prey!), and they have a mean of 3.74 individuals killed per hunt. These figures exceed those from any other long-term chimpanzee study site.

Using phenological data on ripe fruit, Watts & Mitani show that at Ngogo, energy surpluses promote increases in hunting frequency and success. Like chimpanzees elsewhere, these apes go on hunting binges. In 1998 at Ngogo, this took the form of a 57-day rampage that yielded 69 red colobus, one mangabey (*Lophocebus albigena*), and one red duiker (*Cephalophus monticola*). These Ngogo males also go on hunting 'patrols' that may last several hours, a pattern not reported as such from other sites. These 'patrols' account for almost half of the successful monkey hunts.

Given the abundance of males and meat at Ngogo, Watts & Mitani examine the extent to which meat sharing can be explained by 'male bonding' versus 'meat for sex.' They use direct and indirect measures such as meat transfer, copulation, grooming, and coalitionary support, concluding that male chimpanzees at Ngogo typically use meat to 'lubricate' male social relations.

Taken together, these three chapters add significantly to our understanding of the hunting and sharing patterns of *Pan*. We can also now better appreciate the uses and power of meat or other highly prized food items in the lives of chimpanzees and bonobos. These chapters make a persuasive case not only for the value of long-term monitoring of prey species, but also for the value of across-site comparisons.

REFERENCES

Blumenschine, R. J. & Cavallo, J. A. (1992). Scavenging and human evolution. *Scientific American*, 267 (4), 90–6.

Boesch, C. (1994). Chimpanzee–red colobus: a predator–prey system. *Animal Behaviour*, 47, 1135–48.

Bunn, H. T. (2001). Hunting, power scavenging, and butchering by Hadza foragers and by Plio-Pleistocene *Homo*. In *Meat-Eating and Human Evolution*. ed. C. B. Stanford & H. T. Bunn, pp. 199–218. New York: Oxford University Press.

Dart, R. A. (1953). The predatory transition from ape to human. *International Anthropological and Linguistic Review*, 1, 201–18.

Goodall, J. (1963). Feeding behaviour of wild chimpanzees. *Symposia of the Zoological Society of London*, 10, 39–48.

Hohmann, G. & Fruth, B. (1993). Field observations on meat sharing among bonobos (*Pan paniscus*). *Folia Primatologica*, 60, 225–9.

Ihobe, H. (1992). Observations on the meat-eating behavior of wild bonobos (*Pan paniscus*) at Wamba, Republic of Zaïre. *Primates*, 33, 247–50.

Isaac, G. L. (1978). The food-sharing behavior of protohuman hominids. *Scientific American*, 238 (4), 90–108.

Lovejoy, C. O. (1981). The origin of man. *Science*, 211, 341–50.

Marchant, L. F. (2002). The limits of chimpanzee charity: Strategies of meat sharing in communities of wild apes. In *Welfare, Ethnicity, and Altruism: Bringing in Evolutionary Theory*, ed. F. Salter, London: Frank Cass Publishers (in press).

Muller, M., Mpongo, E., Stanford, C. B. & Boehm, C. (1995). A note on the scavenging behavior of wild chimpanzees. *Folia Primatologica*, 75, 43–7.

Nishida, T. & Hosaka, K. (1996). Coalition strategies among adult chimpanzees of the Mahale Mountains, Tanzania. In *Great Ape Societies*, ed. W. C. McGrew, L. F. Marchant & T. Nishida, pp. 114–34. Cambridge: Cambridge University Press.

Stanford, C. B. (1996). The hunting ecology of wild chimpanzees: implications for the evolutionary ecology of Pliocene hominids. *American Anthropologist*, 98, 96–113.

Stanford, C. B. (1998). *Chimpanzee and Red Colobus: The Ecology of Predator and Prey*. Cambridge, MA: Harvard University Press.

Stanford, C. B. (1999). *The Hunting Apes*. Princeton: Princeton University Press.

Stanford, C. B. & Bunn, H. T. (eds.) (2001). *Meat-Eating and Human Evolution*. New York: Oxford University Press.

Uehara, S., Nishida, T., Hamai, M., Hasegawa, T., Hayaki, H., Huffman, M., Kawanaka, K., Kobayashi, S., Mitani, J. C., Takahata, Y., Takasaki, H. & Tsukahara, T. (1992). Characteristics of predation by the chimpanzees in the Mahale Mountains National Park, Tanzania. In *Topics in Primatology. Volume 1, Human Origins*. ed. T. Nishida, W. C. McGrew, P. Marler, M. Pickford & F. B. M. de Waal, pp. 143–58. Tokyo: University of Tokyo Press.

16 • Variations in chimpanzee–red colobus interactions

CHRISTOPHE BOESCH, SHIGEO UEHARA & HIROSHI IHOBE

INTRODUCTION

Predator–prey interactions are expected to be dynamic, as the outcome of each interaction depends upon the reactions of the two species and the contextual circumstances. Comparisons of the hunting behaviour of lions living in different environments revealed that selection of prey and hunting success was affected not only by the density of available cover but also by the density of potential prey (Van Orsdol 1984; Packer *et al.* 1990; Stander 1992). Studies on wild dogs have shown that hunting behaviour is affected by the ecological conditions under which the hunt occurred (Creel & Creel 1995). Similarly, fish modified their hunting behavior depending upon their hunger level as well as the available cover (Milinski 1975). Accordingly, empirical evidence supports the expectations that predator–prey interactions vary under the influence of different factors, including the behaviour of both opponents, as well as the prevailing spatial conditions. These interactions may also vary over time since species density as well as other environmental factors may vary.

The chimpanzee (*Pan troglodytes*) is an ideal species in which to examine these interactions, as all well-studied populations have been reported as hunting mammals for meat (Boesch & Boesch-Achermann 2000; Goodall 1986; Uehara *et al.* 1992; Uehara 1997; Mitani & Watts 1999). In addition, in every long-term study site where red colobus monkeys (*Colobus badius*) are sympatric with chimpanzees, that is, Mahale and Gombe in Tanzania, Kanyawara and Ngogo of Kibale in Uganda (all in East Africa), and Taï in Côte d'Ivoire (West Africa) this species is the most common prey. Comparison between populations reveals that the prey profile of chimpanzees sometimes differs greatly between sites even if the sympatric mammalian fauna seems similar to each other. At Mahale, at least 17, possibly 19, species from six mammalian orders have been confirmed as prey items of the chimpanzee. Of these, M Group chimpanzees have been known to prey upon at least 14 species. In contrast, the Taï chimpanzees selectively killed only 5 monkey species out of 15 possible mammalian species (Boesch & Boesch 1989). The impact of hunting pressure on various prey populations, that is, evaluation of predation rate or prey offtake by the chimpanzees, has only occasionally been estimated (Boesch & Boesch-Achermann 2000; Wrangham & Bergman-Riss 1990; Ihobe & Uehara 1999), except for that of red colobus at Gombe and Taï (Busse 1977; Stanford 1998; Boesch & Boesch-Achermann 2000). This may be because, in general, there are few data on density estimates of sympatric mammals. Here we investigate how temporal changes in abundance and how distribution of prey mammals affects the hunting behaviour of the chimpanzee.

However, presence or absence of a prey species is only one aspect that might affect predator–prey interactions. When prey are present, the outcomes of the predator–prey interactions may vary. How important are such variations in chimpanzees? Previous comparisons of the interactions between chimpanzees and red colobus monkeys in Gombe and Taï have shown striking differences in the way each species reacts to the other (Boesch 1994; Stanford 1998). Forest structure is the primary explanation for the observed differences: red colobus monkeys behave more aggressively toward chimpanzees in low interrupted forest, such as Gombe Stream National Park, than in high continuous forest, such as that of Taï National Park. At Gombe, as long as the chimpanzees could overcome the aggressive defense of the monkeys, their red colobus hunting was highly successful, even without working as a team (Boesch 1994). A similar effect has been suggested in the Ngogo forest, where chimpanzees are more successful in their hunting efforts in interrupted forest areas (Watts & Mitani, Chapter 18). Given that the Mahale Mountains National Park is more similar to the Gombe forest structure than to the forest of Taï National Park, we searched here for another factor, beside forest structure, that might help to explain the variability of the predator–prey relationships between these two primates.

METHODS

Study populations: our description is limited to the aspect that directly affects our comparison (for a general description of the study area see Nishida 1990; Boesch & Boesch-Achermann 2000; Goodall 1986).

Mahale Mountains National Park: the study community, M Group, uses a territory of about 30 km^2, which is a thin strip of area stretching from north to south along Lake Tanganyika. The area is covered by a semi-deciduous or semi-evergreen gallery forest, which is present from 780 m to 1300 m of altitude. The main forest types are the gallery forests that are along the rivers and that include high trees, the *Brachystegia* woodland that occupies the major part of the forested slopes in the Mahale area with short trees for a regularly interrupted canopy, and a limited number of high altitude montane savanna characterized by scattered trees (Nishida 1990). Chimpanzees have been fully habituated to human observers for over 35 years and target males were followed by sight during their daily forays. Due to the thick undergrowth, observations can be difficult and bushknifes are routinely used to cut through the vegetation. As at Gombe, tourists come to see the chimpanzees during the dry months of the year. When C. Boesch visited the community in summer 1999, M Group included 9 adult males, 19 adult females and 22 subadults.

Taï National Park: the forest in this area is a homogeneous moist tropical rainforest with only swamp forest along the rivulet less densely forested. Canopy trees average 30 m in height and are dominated by the emergent trees, 40–60 m high. Chimpanzees are fully habituated to human observers and target males were followed during day follows from dawn to dusk. When comparative data were collected in autumn 1990, the study community included 7 adult males, 15 adult females and 36 subadults.

Observation methods

C. B.'s observations of the chimpanzees in Mahale National Park, Tanzania, were performed in the same way as in Gombe National Park, Tanzania, and in Taï National Park, Côte d'Ivoire. Individual targets were followed on foot during the whole day and their behaviour recorded continuously. Habituation level was very comparable in the three populations. To compare the hunting strategies, adult males were followed daily and all auditory and visual encounters with potential prey were recorded.

Hunting rates

Based on the 1995–96 census data on prey mammals (Uehara & Ihobe 1998) and the frequency of hunting by the chimpanzees in 1991–95 (Hosaka *et al.* 2001), we estimated the proportion of prey killed annually for each species. Annual hunting frequency in the 1980s was estimated as well, but no comparative census data before 1995 are available. For convenience sake, we also approximated the yearly prey offtake in the 1980s using the same census data (1995–96) solely to suggest the order of magnitude. For gregarious species, we calculated the population density by multiplying group density and mean group size. We divided the estimated number of kills made per year by the product of the territory of M Group (24 km^2 of forest plus 6 km^2 of woodland) and population densities of prey mammals to derive a crude estimate of chimpanzee hunting rates (Ihobe & Uehara 1999).

Census methods

In 1969, Nishida (1972) estimated population density of red colobus within 3 km^2 of a narrow strip of relatively undisturbed forest and found six to eight groups and average group size of 40 individuals. In 1979, Takahata (1981) reported population densities from 15 groups of redtailed monkeys and seven groups of red colobus, and found an average group size of 8.4 for redtailed monkeys and 40 for red colobus within the study area of 4 km^2, although the figure for the latter species must have been underestimated because of the overestimation of home range areas (for more detail, see Uehara & Ihobe 1998). Between 1995 and 1996, S. U. and H. I. censused eight species of medium- and large-sized diurnal mammals along observational paths in the Kasoje area of the Mahale Mountains (Uehara & Ihobe 1998). The study area covered about 9 km^2 within the 30 km^2 territory of M Group chimpanzees, and chimpanzees had been observed to have eaten all eight species at least once. The ratio of forest: woodland in length along the census routes was 6.5 : 3.5, which coincides well with an earlier estimate of approximately 80% of forest vegetation, with the remaining 20% being woodland vegetation (Nishida *et al.* 1979). Densities of respective species were estimated independently in forest and in woodland, and data in 1995 and 1996 were treated separately due to the difference in sample width adopted in each year (for more details on method, see Uehara & Ihobe 1998).

Detectability of prey is the hourly rate of noticing a group of a given potential prey species, while moving in the home range of the chimpanzees. A group was detected either because it was seen by the observer, or, more frequently, when a call was heard. To decide whether calls belonged to the same group of a given species, we categorized all calls or sightings within the same 60° angle as belonging to the same

Table 16.1. *Density estimates of eight species of mammals at Mahale (per km²). Group densities for gregarious species are given in parentheses*

Species		1969[a]	1979[b]	1995–96[c]
Red colobus	Forest	80–100 (2–2.7)	68 (1.7)	99–126 (3.3–4.2)
	Woodland	— (—)	— (—)	27–29.4 (0.9–1.0)
Redtailed monkey	Forest	— (—)	40 (3.8)	50.4–82.8 (4.2–6.9)
	Woodland	— (—)	— (—)	33.6–45.6 (2.8–3.8)
Blue monkey	Forest	— (—)	— (—)	9 (0.9)
	Woodland	0 (0)	0 (0)	0 (0)
Yellow baboon[d]	Forest	0 (0)	0 (0)	6.2 (0.2)
	Woodland	— (—)	— (—)	60–70 (1.5–1.8)
Warthog[d]	Forest	0 (0)	0 (0)	— (—)
	Woodland	0 (0)	— (—)	9.1 (4.6)
Blue duiker	Forest	—	—	19.5
	Woodland	—	—	7.8
Bushbuck	Forest	—	—	—
	Woodland	—	—	7.2
Forest squirrel	Forest	—	—	4.5
	Woodland	—	—	—

Notes:
—, Present but no data available.
[a] After Nishida (1972).
[b] After Takahata (1981).
[c] After Uehara & Ihobe (1998).
[d] Immigrant species.

group (see Boesch 1994). Results of an initial analysis comparing detectability at the two sites, Taï and Gombe, has been published previously (Boesch 1994). Here we complement this comparison by adding C. B.'s observations of Mahale chimpanzees from the fall of 1999.

RESULTS

Temporal variation in chimpanzee–red colobus interactions

DENSITY AND POPULATION SIZE OF PREY MAMMALS IN 1995–1996, AND THEIR TEMPORAL CHANGES

Individual densities were estimated for non-gregarious mammals, while both group and individual densities were calculated for gregarious species (Table 16.1). It should be noted that, for several reasons (as explained in Uehara & Ihobe 1998), these results were conservative estimates (e.g. solitary monkeys and small dependent ungulates were excluded from the estimation).

Among the non-gregarious species, blue duikers (*Cephalophus monticola*) were the most abundant ungulates, followed by bushbucks (*Tragelaphus scriptus*). Among the gregarious mammals, red colobus (*Colobus badius*) were the most abundant species, followed by redtailed monkeys (*Cercopithecus ascanius*) (Table 16.1). Although no longer rare, yellow baboons (*Papio cynocephalus*) and warthogs (*Phacochoerus aethiopicus*) are immigrant mammals, arriving in the study area after the 1960s. Regrettably, we have only anecdotal evidence for the historical changes in their distribution and abundance. Nevertheless, we consider these other species to be residents.

Some evidence indicates temporal changes in the abundance of two resident species. Table 16.1, for example,

Table 16.2. *Estimated average number of mammalian prey killed by the chimpanzees of M Group in Mahale (per year)*[a]. *The number of individual prey killed per year, which equals the estimated number of individuals for each prey species killed per year/estimated total number of individuals for each species in M group's range, based on 1995–96 census data (× 100), is given in parentheses*

		Prey species					
Study period	Average no. of prey	Red colobus	Redtailed & Blue monkey[d]	Bushbuck	Blue duiker	Bushpig	Warthog
1981–1990[b]	61.0	32.3 (1.1–1.3)	3.7 (0.1–0.2)	4.3 (10.0)	12.2 (2.4)	4.3 (?)	1.0 (1.8)
1991–1995[c]	116.8	97.3 (3.0–3.8)	2.0 (0.1)	4.0 (9.3)	8.7 (1.7)	2.0 (?)	0 (0)

Notes:

[a] Modified from Ihobe & Uehara (1999).

[b] After Uehara *et al.* (1992).

[c] After Hosaka *et al.* (2001).

[d] Two species combined because of the low predation frequency.

suggests an increase of red colobus and redtailed monkey populations from 1979 onward, although different methods have been employed in arriving at these estimates (see Census methods). In 1985, Mahale became Tanzania's eleventh national park. However, as early as 1974, farming and burning were banned and hunting was prohibited. With the exception of vervet monkeys (*Cercopithecus aethiops*), the wild mammal population in general appears to have gradually increased from that point on (Table 16.1).

HUNTING RATES AND IMPACT OF CHIMPANZEE PREDATION ON PREY POPULATIONS AT MAHALE

Table 16.2 shows the estimated number of mammalian prey killed annually by the chimpanzees of M group in 1981–90 and 1991–95, respectively (modified from Ihobe & Uehara 1999; Hosaka *et al.* 2001). Unfortunately, the density of nocturnal bushpigs (*Potamochoerus porcus*) has not been estimated. Predation frequencies for red colobus monkeys appear to correspond to their relative abundance, although their density jumped from the 1981–90 period to the 1991–95 period. In addition, as they expanded their distribution into M Group's range, the two immigrant species began falling victim to chimpanzee predation after the 1970s. This does not, however, apply to the redtailed monkeys (hunting frequencies decreased when population size increased), nor for the baboons, who are not hunted despite their common occurrence within chimpanzee territory. Thus, with respect to the frequency of their hunting, Mahale chimpanzees' behavior is rather selective.

The impact of chimpanzee predation shown in Table 16.2 is only a conservative approximation and the figures should not be taken as extremely precise. Chimpanzee predation at Mahale does not seem to have had any conspicuous impact on populations in any of the years for which data are presented. Rather, predation coincides with the estimates of temporal changes in abundance and distribution of these mammals (Uehara & Ihobe, 1998). Caution seems required when interpreting bushbuck mortality due to chimpanzee hunting because we do not believe that this bovid is decreasing in number. At the moment, we cannot tell exactly why the chimpanzees rarely hunt for the abundant redtailed monkeys, although some evidence indicates that interactions between redtailed monkeys and chimpanzees differ from those between red colobus and chimpanzees. Red colobus at Mahale adopted a 'stand and defend strategy' as did the red colobus at Gombe (Stanford 1998), whereas redtailed monkeys immediately moved away when they heard chimpanzee voices.

TEMPORAL VARIATION OF CHIMPANZEE HUNTING AT MAHALE

In contrast with predation rates for other prey species, the frequency of red colobus hunting by M group chimpanzees increased markedly from the 1981–90 period to that of

Table 16.3. *Detectability of the prey in three populations of chimpanzees (in number of calls per hour). Only species common to more than one site are presented. Chimpanzees were considered to be present when some of them were being followed, and absent when none were heard or seen for more than one hour*

	Gombe chimpanzees		Mahale chimpanzees		Taï chimpanzees	
Monkey species	present	absent	present	absent	present	absent
Colobus badius	0.14	0.18	0.29	0.67	0.39	1.08
Papio spp[a]	0.73	1.43	0.01	0.02	—	—
Cercopithecus ascanius	0.10	0.06	0.08	0.8	—	—

Note:
[a] *Papio anubis* are present at Gombe and *Papio cynocephalus* are present at Mahale.

1991–95 (Table 16.2). The three-fold increase in hunting frequency may be explained in part by the increase in the red colobus monkey population during the same period. This seems to support the opportunistic nature of Mahale chimpanzee hunting. However, observations suggest that prey demography provides only a partial explanation. Although we cannot present quantitative data, the frequency of red colobus hunting dropped again after 1995, as the size of M Group decreased dramatically (Nishida *et al.*, unpublished data).

Regrettably, with the exception of red colobus hunting in the 1991–95 period, actual encounters between the chimpanzees and their potential prey mammals have not yet been intensively studied at Mahale. In 1999, however, C. B. systematically observed interactions between chimpanzees and red colobus from the perspective of cross-population comparison (100 chimpanzee–red colobus interactions in 328 hours of observations; see sections on chimpanzee–red colobus interactions that follow).

Spatial variations in chimpanzee–red colobus interactions

How do chimpanzee and red colobus interactions vary in different populations? In a previous report on the Gombe and Taï chimpanzee populations, Boesch (1994) showed that the forest structure was an important factor in explaining the observed differences. To ensure methodological and observational consistency, C. B. observed chimpanzee–red colobus interactions at three sites: Taï National Park (Côte d'Ivoire, West Africa) in 1990 and 1991, Gombe National Park (Tanzania, East Africa) in 1990 and 1992, and Mahale Mountains National Park (Tanzania, East Africa), in 1999.

FINDING THE PREY

Throughout their range of distribution, monkeys are the main prey of chimpanzees. Clearly, finding prey is related to its abundance. Yet more important is how easily the chimpanzees can detect that prey. For this, C. B. measured detectability according to how frequently a prey species could be seen/heard by the chimpanzees within the forest (see Methods). In addition, detectability also included those instances when no chimpanzee was heard for a minimum of 1 hour. This provided a control condition for prey behaviour in the presence or absence of chimpanzees within the area.

Table 16.3 indicates that Mahale and Taï red colobus monkeys become more silent when chimpanzees are nearby, while Gombe red colobus continue making noise but switch to alarm calling as the chimpanzees come closer. However, note that red colobus monkeys at Gombe tend to be more silent than their Taï or Mahale counterparts. Thus, the reaction of the red colobus to the presence of chimpanzee differs among populations. From the hunters' point of view, red colobus monkeys are the most difficult to detect in Gombe, not only because they are less abundant but also because they tend to be more silent. Absolute measures of red colobus population density suggest that Mahale has the highest density (average between 1995 and 1996: 112 ind km^{-2} or 3.8 group km^{-2}), followed by the Taï forest (66 ind km^{-2} or 1.65 group km^{-2}) (Galat & Galat-Luong 1985), and Gombe (42 ind km^{-2} or 1.82 group km^{-2}) (Stanford 1998).

In none of the three populations does detectability or prey density within the territory predict the hunting frequency. In addition, in the three observed chimpanzee populations, the density of other monkey species indicates that potential prey are abundant but are neglected by the chimpanzees (for Taï see Boesch 1994). Thus, in these

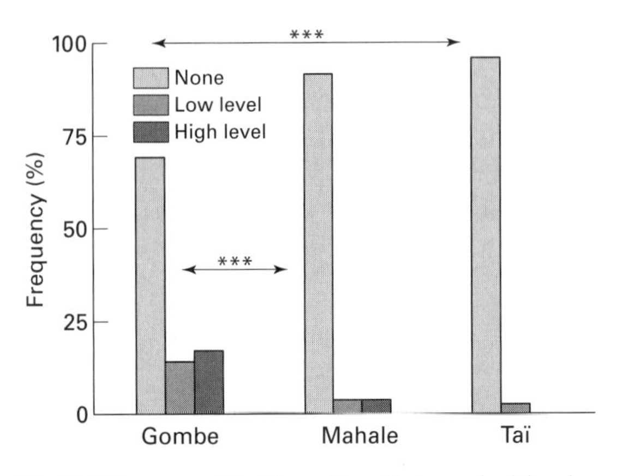

Fig. 16.1. Frequency of double predator effect on red colobus due to humans following chimpanzees in three populations. A 'low' double predator effect was recorded when the monkey moved away within the same tree, and 'high' when they moved away into adjacent trees. In all three figures, χ^2 tests were performed with $* = p < 0.05$, $*** = p < 0.001$.

populations, chimpanzees have a strong preference for red colobus monkeys.

COLOBUS REACTION TO CHIMPANZEES

The human influence

The difference in visibility among the three sites is striking, the lowest visibility being in the tropical rainforest of Taï National Park with its continuous forest cover and rather flat topography. At Gombe and Mahale, the interruption of the tree canopy and the steep slope dramatically increase the ability to see arboreal monkeys from the ground. In addition, at both Gombe and Mahale, tourists are regular visitors, and in Gombe this sometimes dramatically increases the number of human observers near a party of chimpanzees. This has also become typical at Mahale since the establishment of a tourist camp in 1989 within the territory of the M Group. Furthermore, during most of C. B.'s stay in Mahale, a film crew was following the chimpanzees, and there were several instances when the number of people with the chimpanzees exceeded a dozen. We might predict that the more humans are with the chimpanzees, the more fearful the colobus monkeys will be. Thus, the double predation effect, occurring when a prey is simultaneously threatened by two different types of predators (in this case, human and chimp), should decrease from Mahale, and from Gombe to Taï.

However, this prediction may be affected by the red colobus reactions when they face humans. Figure 16.1 shows that, contrary to expectation, the Mahale colobus population, with a high number of human observers present, displayed a level of double predator effect similar to that shown by the Taï colobus population, which has the least number of human observers. The Gombe colobus were the most disturbed by human observers (see also Boesch 1994; Goodall 1986). The difference between the Gombe and the Mahale colobus is intriguing since both have been followed for decades by human observers and live in a similarly structured habitat. One possible explanation may be the different observational methods used at the two sites. In Mahale, human observers use a bushknife to cut their way through the undergrowth while following the chimpanzees in their daily forays. With this technique, there is simply no way a human observer can surprise an animal, and the red colobus monkeys in Mahale always know well ahead who is approaching them.

COLOBUS FACING CHIMPANZEES

In all chimpanzee populations where interactions with red colobus monkeys have been closely observed, the chimpanzees were described as being afraid of the colobus (Boesch 1994). This is also true for the Mahale chimpanzees. Colobus monkeys responded in different ways to the chimpanzees. The East African red colobus were clearly more aggressive than their West African counterpart (Figure 16.2): the Mahale and Gombe populations regularly threatened and mobbed the chimpanzees. The tall and continuous forest structure in the Taï forest affords the colobus time to retreat and provides escape paths in all possible directions. When retreat or escape is not possible for the prey, attacking the hunter is the next possible solution. Both East African populations physically attacked the chimpanzee hunters by jumping on their backs and trying to bite them. They were also seen pursuing chimpanzee hunters on the ground whenever the latter were pressing some of the colobus too closely or when the colobus were cornered in isolated trees (Mahale, $n = 1$; Gombe, $n = 3$). In both populations, the chimpanzees screamed and reacted fearfully to the monkeys.

However, important differences were also visible within the East African populations of red colobus monkeys. The Mahale colobus monkeys appear the least affected by the double predator effect, as they are most aggressive – they threatened and attacked chimpanzees whenever they were in close proximity (Figure 16.2). It was common to see colobus males coming close to chimpanzees who were looking at them. The colobus would regularly attack the chimpanzees

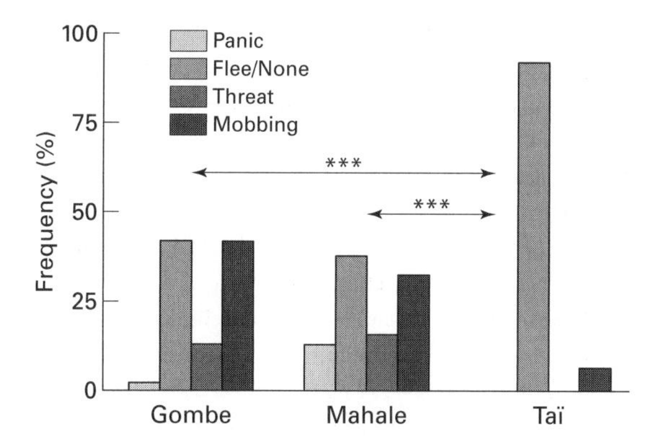

Fig. 16.2. Reactions of red colobus monkeys to the visual presence of chimpanzees. 'Panic' was recorded when most monkeys rapidly left their trees and ran away from the chimpanzees; 'flee' was recorded when the monkeys avoided direct visual contact with the chimpanzees by changing position in the same tree; 'threat' was when the monkeys faced the chimpanzees and threatened them; and 'mobbing' was when more than one monkey moved toward chimpanzees, threatening or attacking them.

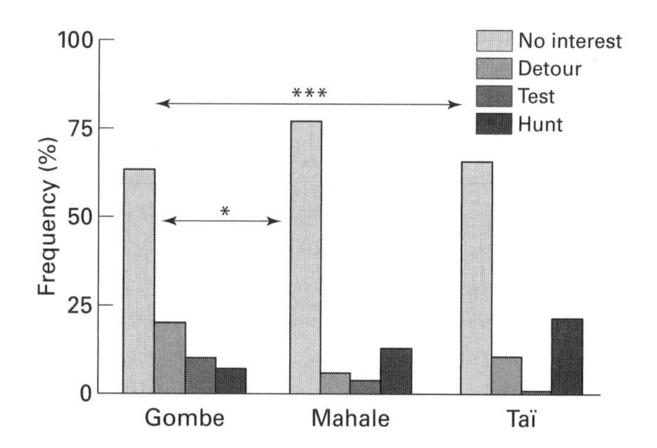

Fig. 16.3. Chimpanzee reactions to the red colobus monkeys. 'No interest' was recorded when the chimpanzees showed no signs of being interested in the monkeys; a 'detour' was recorded when the chimpanzees changed direction in order to approach the monkeys; a 'test' was recorded when chimpanzees who were in visible contact with the monkeys displayed or climbed toward them and looked to see if they reacted; a 'hunt' was recorded when the chimpanzees climbed high enough to be able to make a capture. Note that in most cases, chimpanzees needed to make a detour in order to arrive under the monkeys prior to 'testing' or 'hunting' them, but 'test' and 'hunt' were not recorded as 'detour'.

even before they started to move towards them. They were more prone to do so when in the more open parts of the home range, where the woodlands intermingled with savannah. In those areas, the Mahale colobus monkeys would even threaten a lone human observer by shaking branches ($n=4$ out of 14 encounters). These observations were possible because the Mahale colobus are less afraid of humans: our presence did not prevent the colobus from threatening the chimpanzees, and they were still willing to charge them even when the chimpanzees were within 5 m of the observer(s).

CHIMPANZEE BEHAVIOUR

In all three populations, chimpanzees hear or encounter colobus much more regularly than they hunt them. Thus chimpanzees can be characterized as opportunistic in that they most often start to hunt when they have heard red colobus in the vicinity (see above). How do chimpanzees behave when they hear red colobus monkeys? Figure 16.3 shows that chimpanzees regularly take detours leading to the prey but do not hunt, giving the impression that they are evaluating the prospects for a hunt. Only in the Taï forest have researchers observed chimpanzees searching intensively for prey before starting to hunt (Boesch & Boesch 1989).

Testing colobus' reaction is a strategy rarely used by Taï chimpanzees, but one that is regularly employed by the East African (Gombe and Mahale) chimpanzees. This testing on the part of the Gombe and Mahale chimps may be because they have specialized in the capture of infant colobus. As a result, they want to ensure that their hunting targets will be groups that contain infants, and in which they are less likely to have to face adult males. The East African chimpanzees have two different ways of testing the colobus monkeys. Gombe chimpanzees first display on the ground or in trees beneath or near the group of colobus and watch their reaction. Whenever the colobus face or threaten them, the chimpanzees retreat or test another part of the group.

Mahale chimpanzees use a technique in which they place themselves beneath a group of red colobus and carefully observe each of their movements. Most of the chimpanzees sit or lay under the group, watching intently and positioning themselves in order to have the best view of the monkeys' behaviour. The chimpanzees remain silent and may reposition themselves at some trees if the colobus appear to move lower in the trees or become more active. Only rarely will one do a silent display on the ground. After one or two hours, if nobody starts a hunt they will move farther away. At both Mahale and Gombe, the chimpanzees may seek a group of infant/juvenile colobus monkeys, their preferred prey, but do not dare to search for them actively by climbing in the trees for fear of being attacked by the large adult colobus males.

This fear that the East African chimpanzees have for the adult colobus seems to affect them in two additional ways. Gombe chimpanzees regularly neglect adult colobus that they have killed in the heat of the hunt (Boesch 1994; Goodall 1986). This seems to be the case with hunters that continue to hunt in order to capture younger prey. This puzzling behaviour could be explained because either large prey are more difficult for the owner to monopolize (although this would not apply to the alpha male, who also neglected the dead adult prey), or because they have a taste for the meat of young colobus (Stanford 1998). This behavior has never been seen in Taï nor in Mahale chimpanzees (Nishida, personal communication).

In contrast, Mahale chimpanzees have been observed refraining from attacking and harming adult colobus when they have trapped them on the ground. On 7 October, 1999, at 9:30, three male chimpanzees trapped an adult female colobus in such a way that she missed her escape jump and fell directly to the ground. Pim, an adolescent male, pursued her and held her on the ground for a few seconds. The exhausted female remained on the ground for 10 minutes surrounded by as many as seven chimpanzees, including four adult males. Having recovered her breath, the female colobus slowly climbed 20 m up into a tree, followed by at least 11 chimpanzees. After 40 minutes, the chimpanzees moved away, leaving the colobus unharmed. Although not systematic, this behaviour has regularly been observed among Mahale chimpanzees whenever they capture an adult monkey on the ground without an infant (Nishida, personal communication). This behavior has never been observed in Gombe or in Taï chimpanzees.

In contrast to the East African chimps, Taï chimpanzees hunt adult red colobus monkeys directly (Boesch & Boesch-Achermann 2000).

DISCUSSION

The present analysis suggests that two factors affect the predator–prey relationship: demography and habitat composition, which seem to explain many of the differences that have been observed to date in the hunting behaviour of chimpanzees in different populations. Mahale chimpanzees changed their hunting frequency in response to demographic changes in the prey. The more abundant a prey species, the more frequently it was hunted. Similar observations have been reported for many predator–prey systems in fish, snails and insects, and help explain predators switching preference according to the density of prey (Begon *et al.* 1990). In addition, demographic parameters *within* the hunter population affect hunting frequency. In the Taï forest, the hunting frequency decreased after the population of adult males in the community fell to less than six (Boesch & Boesch-Achermann 2000). A similar effect was reported to have occurred in M Group when the number of males dramatically decreased after 1995. Thus, in this system the demography of both prey and predator affect the predator–prey relationship, making it unstable over time.

The effect of individuals in the hunter populations can be more subtle. Given that most chimpanzees fear red colobus we can often see the impact on hunting frequency of brave hunters who do not fear the prey. Frodo in Gombe, Ulysee in Taï (Boesch 1994), and Toshibo in Mahale (Nishida, personal communication) had a large impact on hunting frequency because they had overcome their fear of the colobus and initiated hunts much more often than other group members. In addition, Hosaka *et al.* (2001) suggested that some subtle changes might explain the decrease in hunting frequency against red colobus after 1995; either larger parties of chimpanzees may have become more common as secondary succession gradually progressed and fruit availability increased, or extraordinarily long-lasting social conflicts among adult male chimpanzees of M Group, which may have been responsible for bringing more males together from time to time during the 1991–95 period, stopped after many males had disappeared (Nishida, personal communication). The latter explanation (decrease in number of adult males) appears the most likely in terms of the short-term drastic variation in the red colobus–chimpanzee interactions at Mahale. The larger the number of males, the more likelihood that bold individuals will be present. This might explain why among the 24 adult males in the Ngogo community in the Kibale Forest, Uganda, there is such a high hunting frequency and level of success (Mitani & Watts 1999).

Why do East African colobus harass their potential predators?

Why do East African colobus monkeys harass their potential predators more often than do Taï colobus monkeys? Boesch (1994) proposed the forest structure as a plausible explanation – in low and interrupted tree cover, colobus tend to be more aggressive. This factor may also explain part of the differences observed in the behaviour of the Ngogo chimpanzees. A detailed analysis by Watts & Mitani (Chapter 18) shows that the high level of success of Ngogo chimpanzees occurred most often in hunts in areas with broken canopies. When Ngogo chimpanzees hunted in continuous and uniform forest areas, their success rate dropped and was very similar to that observed in the West African Taï forest, although the latter had far fewer males. This illustrates how two factors, demography and forest structure, may interact; within similar forest structures, demography strongly affects hunting rate and success, but the effect of forest structure overcomes the demographic effect. Thus, for chimpanzees, as with some other social carnivores, the conditions prevailing during the hunt play an important role in the outcome. The more difficult the conditions, the greater the need for the hunters to cooperate with each other (Boesch & Boesch-Achermann 2000; Creel & Creel 1995; Packer & Ruttan 1988).

The observations from Mahale suggest that an additional factor, namely feeding competition, strongly influences colobus–chimpanzee interactions. In both Mahale and Gombe, overlap of different fruits in the diet of the two species has been documented (Nishida 1972; Stanford 1998). A preliminary analysis showed that the overlap at Mahale did not differ from that at Gombe (Ihobe, unpublished data). Boesch saw the two species overtly fight for food three times in the Mahale Mountains, and chimpanzees in Taï were seen passively driving colobus out of trees several times; this was never seen at Gombe. During the Mahale Mountains study period, the two species were seen actively competing for the fruits of *Pseudospondias microcarpa*, *Pycnanthus angolensis*, and the abundantly eaten fruits of *Garcinia hulliensis*. In such situations, the chimpanzees would try to chase the colobus out the trees and the colobus would face-off with the intruders. Invariably, chimpanzees who were attacked would scream and might obtain help of other group members, but the respect that the chimpanzees had for the adult colobus made such situations very tense and the outcomes uncertain.

In one very spectacular instance, on 15 September, 1999, at 16:50, Boesch joined a group of 12 adult chimpanzees, including the alpha and the gamma males of the community, that had been screaming and aggressively barking for 20 minutes at a group of red colobus that was eating in a very isolated *Pseudospondias microcarpa* tree. The chimpanzees did not attempt to enter the tree as long as the colobus were present. Due to the terrain, Boesch was unable to approach without being spotted by the monkeys, and the last two adults left as Boesch neared. Only then did the chimpanzees enter the tree and eat the fruit. In another instance, it took 20 minutes for six chimpanzees, including three adult males, to chase two adult colobus from a small *Garcinia* tree. During and after these conflicts with colobus monkeys over food, chimpanzees never showed any signs of hunting. These scenarios illustrate how feeding competition might play out, but more systematic data is needed in order to confirm the importance of the proposed feeding competition factor.

Interspecific food competition might explain why red colobus monkeys are so much more willing to face chimpanzees in Mahale than has been observed in Gombe. This potential effect amplifies the effect of discontinuous and lower forest structure that is common to both Gombe and Mahale, and explains why both East African red colobus populations are generally more aggressive against chimpanzees than are Taï colobus monkeys. Red colobus monkeys in Mahale tend to be more aggressive than those at Gombe, standing their ground against chimpanzees when competing for fruit. The more aggressive behaviour of the colobus monkeys leads to the chimpanzees specializing in capturing infant prey, and either neglecting captured adults (Mahale) or neglecting the meat of adults they have killed (Gombe). Further observations of actual encounters between the two species is necessary to clarify the effect of feeding competition on predator–prey interactions.

In conclusion, we see that demographic factors, such as food competition and forest structure, might all affect predator–prey interactions and help explain why the interactions between chimpanzee and red colobus differ in the three study populations. Additionally, these factors may also vary temporally within the same site, resulting in changes in the interactions between the two species.

REFERENCES

Begon, M., Harper, J. L. & C. R. Townsend (1990). *Ecology: Individuals, Populations and Communities* (2nd Edition). Oxford: Blackwell Scientific Publications.

Boesch, C. (1994). Chimpanzees–red colobus: a predator–prey system. *Animal Behaviour*, 47, 1135–48.

Boesch, C. & Boesch, H. (1989). Hunting behavior of wild chimpanzees in the Taï National Park. *American Journal of Physical Anthropology*, 78, 547–73.

Boesch, C. & Boesch-Achermann, H. (2000). *The Chimpanzees of the Taï Forest: Behavioural Ecology and Evolution*. Oxford: Oxford University Press.

Busse, C. D. (1977). Chimpanzee predation as a possible factor in the evolution of red colobus monkeys social organization. *Evolution*, 31, 907–11.

Creel, S. & Creel, N. M. (1995). Communal hunting and pack size in african wild dogs, *Lycaon pictus*. *Animal Behaviour*, 50, 1325–39.

Galat, G. & Galat-Luong, A. (1985). La communauté de primates diurnes de la forêt de Taï, Côte d'Ivoire. *Revue d'Ecologie (Terre Vie)*, 40, 3–32.

Goodall, J. (1986). *The Chimpanzees of Gombe: Patterns of Behavior*. Cambridge: The Belknap Press of Harvard University Press.

Hosaka, K., Nishida, T., Hamai, M., Matsumoto-Oda, A. & Uehara, S. (2001). Predation of mammals by chimpanzees of the Mahale Mountains, Tanzania. In *All Apes Great and Small, Vol. 1: African Apes*, ed. B. Galdikas, N. Briggs, L. Sheeran & G. Shapiro, pp. 105–28. New York: Kluwer Academic Publishers.

Ihobe, H. & Uehara, S. (1999). A preliminary report on the impact of chimpanzee hunting on mammal populations at Mahale, Tanzania. *Primate Research*, 15, 163–9 (in Japanese with English summary).

Milinski, M. (1975). Experiments on the selection by predators against spatial oddity of their prey. *Zeitschrift für Tierpsychologie*, 43, 311–25.

Mitani, J. & Watts, D. (1999). Demographic influences on the hunting behavior of chimpanzees. *American Journal of Physical Anthropology*, 109, 439–54.

Nishida, T. (1972). A note on the ecology of the red colobus monkey (*C. badius tephrosceles*) living in the Mahali Mountains. *Primates*, 13(1), 57–64.

Nishida, T. (1990). *The Chimpanzees of the Mahale Mountains: Sexual and Life History Strategies*. Tokyo: University of Tokyo Press.

Nishida, T., Uehara, S. & Ramadhani, N. (1979). Predatory behavior among chimpanzees of the Mahale Mountains. *Primates*, 20, 1–20.

Packer, C. & Ruttan, L. (1988). The evolution of cooperative hunting. *American Naturalist*, 132(2), 159–98.

Packer, C., Scheel, D. & Pusey, A. E. (1990). Why lions form groups: food is not enough. *American Naturalist*, 136, 1–19.

Stander, P. E. (1992). Foraging dynamics of lions in a semi-arid environment. *Canadian Journal of Zoology*, 70, 8–21.

Stanford, C. (1998). *Chimpanzee and Red Colobus: The Ecology of Predator and Prey*. Cambridge, MA: Harvard University Press.

Takahata, Y. (1981). A preliminary report on distribution, habitat preference, and food habit of sympatric primates in the Mahale Mountains. *Mahale Mountains Chimpanzee Research Project*, Ecological Report 14a.

Uehara, S. (1997). Predation on mammals by the chimpanzee (*Pan troglodytes*). *Primates*, 38, 193–214.

Uehara, S. & Ihobe, H. (1998). Distribution and abundance of diurnal mammals, especially monkeys, at Kasoje, Mahale Mountains, Tanzania. *Anthropological Science*, 106, 349–69.

Uehara, S., Nishida, T., Hamai, M., Hasegawa, T., Hayaki, H., Huffman, M., Kawanaka, K., Kobayashi, S., Mitani, J., Takahata, Y., Takasaki, H. & Tsukahara, T. (1992). Characteristics of predation by the chimpanzees in the Mahale Mountains National Park, Tanzania. In: *Topics in Primatology, Vol: 1: Human Origins*, ed. T. Nishida, W. C. McGrew, P. Marler, M. Pickford & F. de Waal, pp. 143–158. Tokyo: University of Tokyo Press.

Van Orsdol, K. G. (1984). Foraging behavior and hunting success of lions in Queen Elizabeth National Park, Uganda. *African Journal of Ecology*, 22, 79–99.

Wrangham, R. W. & van Bergmann-Riss, E. (1990). Rates of predation on mammals by Gombe chimpanzees. *Primates*, 31, 157–170.

17 • How bonobos handle hunts and harvests: why share food?

BARBARA FRUTH & GOTTFRIED HOHMANN

INTRODUCTION

In nature, access to food sources is constrained and, depending on their diet, individuals compete directly or indirectly to meet their nutritional needs. Sometimes, however, food resources can be divided. Division of food has been seen in vertebrates as diverse as the raven (*Corvus corax*, Heinrich *et al.* 1995), vampire bat (*Desmodus rotundus*, Wilkinson 1984), and chimpanzee (*Pan troglodytes,* Goodall 1968). For mammalian species, reports on food sharing are mostly for primates and, within this order, *Pan* dominates the other genera.

In chimpanzees, food sharing occurs most often between mother and dependent offspring (McGrew 1975; Nishida & Turner 1996) and involves plant food. Although food sharing between mature individuals sometimes involves plant food (Lanjouw, Chapter 3; Watts & Mitani, Chapter 18), it is more common in the context of hunting and meat consumption. Hunting of mammalian prey has been reported from all sites where chimpanzees have been studied long term: Assirik (Hunt & McGrew, Chapter 2), Budongo (Suzuki 1975), Gombe (Goodall 1968; Stanford 1995), Kahuzi (Basabose & Yamagiwa 1997), Lopé (Tutin & Fernandez 1993), Mahale (Nishida *et al.* 1979), Ngogo (Mitani & Watts 1999), Kanyawara (Wrangham, personal communication), Ndoki (Kuroda *et al.* 1996), Taï (Boesch & Boesch 1989), Tongo (Lanjouw, Chapter 3), Sapo (Anderson *et al.* 1983), and Semliki (Hunt & McGrew, Chapter 2).

Frequency and style of hunting vary both within and across sites, and appears to be affected by forest structure, prey species and group composition (Lenglet 1987; Boesch 1994; Stanford 1998a; Boesch *et al.*, Chapter 16). Accordingly, food sharing is also likely to vary across different sites or populations. Nevertheless, several patterns recur across all sites: capture of prey and ownership of the carcass are male-biased, males share more often with other males than with females, and amount of meat received by females varies with reproductive status and social dominance (but see Boesch & Boesch 1989). While these patterns may be typical for hunting in mixed-sex parties, a word of caution is needed since few studies have ever focused on hunting behavior and meat sharing by female chimpanzees, which is considered to be largely underestimated (Takahata *et al.* 1984). At Gombe, however, where focal subject sampling from nest to nest gives true frequencies by sex, females rarely hunted, and the main hunter was the masculinised female Gigi (Goodall 1986, p. 304 ff.).

In most populations hunting is frequent and targeted mostly at other primates, but also at ungulates and a few species of small mammals (Nishida *et al.* 1979; Nishida & Uehara 1983). Among primates, red colobus monkeys (*Colobus badius*) make up most of the chimpanzees' prey. Hunting of arboreal monkeys is thought to be costly in that the risk for each hunter is high. Hunting success increases with the number of active participants, and cooperation takes place whenever the per capita gain achieved by multiple hunters surpasses that of a single one (Boesch 1994). Chimpanzees may share meat in order to reinforce collaborators or allies, or to retaliate against cheats. Other potential causes for the observed bias for meat rather than plant food sharing in chimpanzees are prey size and nutritional value. Most fruits eaten by primates are relatively small (Mack 1993), and the costs of obtaining an equivalent amount of calories or essential micronutrients from plant food may exceed those that derive from faunivory (Boesch 1994, but see Stanford 1998b for a critical discussion).

In bonobos (*Pan paniscus*), food sharing between mature individuals is common. Hunting of mammalian prey has been reported from various sites and involves black-and-white colobus (*Colobus angolensis*) and redtailed monkey (*Cercopithecus ascanius*) (Sabater Pi *et al.* 1993), flying squirrels (species unknown, Kano & Mulavwa 1984; Ihobe 1992) and forest antelopes (*Cephalophus* spp., Hohmann & Fruth 1993). Carnivory is known from Wamba (Ihobe 1992) and Lomako (Badrian & Malenky 1984), but only at Lomako is the consumption of meat accompanied by food sharing. While hunting and meat consumption are relatively rare, bonobos often divide large-sized fruits such as *Treculia africana* and *Anonidium mannii* (Kuroda 1984, Hohmann & Fruth 1996). Previous studies at Lomako suggest that divisible food is controlled by females rather than by males (Hohmann & Fruth 1993, 1996; Fruth 1998).

Here, we present data on the division of meat and plant foods, data that were collected over a period of 8 years at Lomako (Democratic Republic of Congo). We describe hunting techniques used to obtain mammalian prey, and investigate the relationship between owner and participant by considering kinship, rank and role allocation. The data are used to test hypotheses on the forces that promote food sharing in bonobos.

1. *Reciprocity*: this hypothesis predicts that an individual's actions depend on the behavior of its partner during the preceding interaction (Axelrodt & Hamilton 1981). The major principle is tit-for-tat, with responses of either cooperation, retaliation or forgiveness. Reciprocity is seen as a driving force to explain the evolution of cooperation. Evidence for reciprocity in the context of food sharing comes from studies on humans (Bird & Bird 1997), primates (de Waal 1989, 2000), and bats (Wilkinson 1987).
2. *Kin selection*: this hypothesis implies that altruism or cooperation is most likely to evolve among relatives since any asymmetry among kin could be explained by inclusive fitness. The division of food is not randomly distributed among companions, but preferably directed to close kin (Hamilton 1963; Gadakar 1990). Kinship promotes baby-sitting in mongooses (*Helagale undulata*, Rasa 1977), grooming in primates (Gouzoules 1984), agonistic aid in vervet monkeys (*Cercopithecus aethiops*, Cheney & Seyfarth 1980), and blood sharing among vampire bats (Wilkinson 1984).
3. *By-product mutualism*: this term refers to events of cooperative behavior that occur as an 'incidential' consequence of selfishness (Brown 1983, 1987). It may be the simplest type of cooperation. Neither kinship nor complex cognitive mechanisms are required for score-keeping in reciprocal relations. Instead, both individuals gain the highest pay-off from mutual co-action rather than from mutual defection. The concept of by-product mutualism has been used to explain cooperative foraging in fish (Foster 1985), food sharing in birds (Roell 1978; Heinrich & Marzluff 1991), hunting in lions, *Panthera leo* (Packer *et al.* 1991), and allomothering in vervet monkeys (Fairbanks & McGuire 1984).
4. *Buy-off*: this occurs when individuals give away part of a resource in order to avoid disputes or aggression from supplicants (Wrangham 1975; Moore 1984).

We contrast our results with current explanations given for chimpanzee and bonobo food sharing and seek the most parsimonious interpretation for this complex behavior seen in both species.

METHODS

Data collection

We collected data during nine field seasons ranging from 2 to 9 months (1990–98; total: 46 months). We studied members of the Eyengo community living in the eastern part of the Lomako study site, Equateur, Democratic Republic of Congo (20°40′–21°40′E, 00°39′–01°12′N). Community members were identified by physical traits (disfigured limbs, sexual swellings, pigmentation) and, if born before 1991, were assigned ages based on traits such as body size, body proportions, condition of teeth, or length of mammillae. The number of adult and adolescent females ranged from 11 to 15 and of mature males from four to eight. Details on demography are published elsewhere (Fruth 1995; Hohmann *et al.* 1999). All data given here refer to 31 mature individuals. Since not all of them lived simultaneously in the community (some had vanished when others immigrated or matured), only 404 of theoretically 465 dyads are considered here.

We collected data on food sharing *ad libitum*, by online recording as spoken protocols using either a dictaphone, an audio cassette recorder or a video-camcorder. We regularly visited *Treculia* trees and measured *weight* and size of fruit on the ground, using a portable balance and a tape measure. From these data, we draw a linear relation of both weight and diameter (Pearson: $R^2 = 0.98$, $p < 0.001$). By estimating the fruit diameter, this calculation served to infer the weight of fruits consumed directly after harvesting.

Definitions

We considered a *sharing session* to be consumption and distribution of one food-item from beginning to end. An *owner* was an individual that monopolised an item. *Monopolisation* was expressed by the owner having the item on his/her lap, holding both hands/arms on the item, or carrying the item. An individual became a *supplicant* as soon as it showed interest in the item, seeking a share. When a supplicant obtained a share, it became a *participant*. A sharing session was split into *owner episodes*, and the duration of each episode was the *owner's tenure*, as defined by the time an owner was able to monopolise the item. During an owner's tenure, an owner might have several supplicants and participants. Each supplicant was linked in a dyad with the owner; the length of time any one supplicant spent with the owner defined a *supplicant–owner episode*. The episode started with the supplicant indicating interest in a share and ended when either the owner or the supplicant left. The time they were together was named *dyadic time*. Owners were considered to be without supplicants if, even when others were present, no individual showed interest in getting a share.

In any supplicant–owner episode, a dyad was either uni- or bidirectional. In a *unidirectional* dyad, *A* shared with *B*, but whenever *B* owned food, *A* either was not interested or was not in the food sharing party. In a *bidirectional* dyad, *A* shared with *B* and *B* had the opportunity to reciprocate.

Assessment of *rank* was based on the outcome of agonistic interactions and non-agonistic displacements (approach–retreat). Only dyadic relations were considered here. An individual winning more often than losing against another was considered to be dominant within their dyad. From these data, three classes emerged: high-, medium- and low-ranking individuals.

Reciprocity: A dyad was considered to reciprocate when *A* gave to *B* and *B* gave to *A*. This was independent of the quantity, frequency and time of hand-outs. In order to test reciprocity with another currency, all episodes of *social grooming* from 1993 to 1998 were analysed. For each season, we compared directionality of food sharing with that of grooming for the same dyads. Analysis was restricted to dyads that had at least one score for each of the two types of behavior.

Kinship was assessed by DNA fingerprinting (Gerloff *et al.* 1999). Here, only close kin ($r_s \geq 0.25$) were considered.

Statistics

Data were analysed with SPSS 8.0 or SsS (Engel 1998). All analyses done were non-parametric and two-tailed.

RESULTS

Meat

PREY

Bonobos were seen to eat meat nine times. In each case, the meat was divided among adult community members. When the prey was identifiable ($n=7$) bonobos ate the meat of duikers (*Cephalophus* spp.) of large size (estimated weight 9–15 kg). Twice, they consumed unidentified mammals of smaller size.

HUNTING

Encounters between bonobos and duikers occurred daily, usually with no obvious interaction between the species. Occasionally, a single bonobo lunged at a duiker but did not pursue the target when it ran away. More often, bonobos inspected resting places of duikers, such as hollow trees or corners between buttresses of large trees. In two of these stalks, we saw a bonobo catch a duiker from such a site. All attempts by bonobos to grab a duiker were made by lone individuals.

CONSUMPTION

Duikers were not killed before being eaten, but were opened ventrally at the abdomen while still alive. Consumers stuck their fingers into the opening and licked them. Intestines were eaten first. In the first half hour, the hindlegs were usually ripped off and carried away by other individuals. After the ingestion of intestine, muscle and bone, the pelt was turned inside out and the remaining meat was scraped down to the skin with the incisors. In all cases the skull was consumed last. The owner inverted the skin around the head and held it in one fist. Then, by poking with the index finger, it extracted brain matter through the *foramen magnum*. Eventually the skull was cracked open with the teeth and emptied completely. Tool use was never seen during meat consumption. Few remnants remained, but we once found parts of a mandible. Occasionally, individuals carried the skin of a duiker the next day.

Of the nine kills, eight meat-sharing sessions were well-documented; their average duration was almost 3 hours (Table 17.1).

OWNERS

Eleven mature community members, ten females and one male, owned a kill. They were in 23 owner episodes, of which 91% were by females, and the rest by the male. A kill was owned by several individuals in sequence (average number of owners/kill: 2.75; SD = 1.91; range = 1–7, $n=8$ kills). Owners' tenure lasted on average for more than an hour (Table 17.1), and first owners monopolised the kill longer than did successive owners (78.8% versus 16.9%, total owner time: U-test: $z=-2.77$, $p=0.006$)

We could rank 10 of the 11 owners. Three of them were of high-, four of middle-, and three of low-rank.

Table 17.1. *Frequencies in sharing episodes*

	Meat-sharing ($n=9$ animals)	*Treculia*-sharing ($n=80$ fruits)	SSD
Duration of session (min; $\bar{x}\pm$SD)	171 ± 104	86 ± 63	*
Female owner	91%	88%	ns
Male owner	9%	12%	ns
Owner/item ($\bar{x}\pm$SD)	2.75 ± 1.91	1.68 ± 1.03	*
Bystander/item ($\bar{x}\pm$SD)	6.38 ± 4.31	3.08 ± 2.24	*
Participants/owner ($\bar{x}\pm$SD)	3.36 ± 3.43	1.52 ± 1.94	**
Owners' tenure ($\bar{x}\pm$SD)	68.74 ± 62.75	40.06 ± 35.75	ns
Participants' tenure ($\bar{x}\pm$SD)	33.52 ± 45.82	27.12 ± 38.26	ns

Notes:
$\bar{x}\pm$SD: mean ± standard deviation; SSD: statistically significant difference between meat and *Treculia* (Mann–Whitney U-Test); ns = not significant; *: $p<0.05$; **: $p<0.01$; n = sample size.

Considering the episodes ($n=18$) of all ranked owners, high-ranking individuals monopolised prey most often, followed by middle- and low-ranking individuals. These frequencies did not deviate from the expected ones (rank: expected versus observed; high: 5.4 versus 8, middle: 7.2 versus 6, low: 5.4 versus 4; $\chi^2=1.33$, df=2, $p=0.51$; ns). First owners ($N=8$) were either high- (63%) or middle-ranking (37%), but low-ranking individuals were never initial owners of a kill.

SUPPLICANTS AND PARTICIPANTS

Twenty-three (15 females, 8 males) mature individuals showed interest in obtaining a share of meat. On average, six waited at a kill seeking food, but only four succeeded, the rest waited in vain (Table 17.1). Females were more successful than were males. Considering the 87 episodes involving the 23 individuals, all 15 females got a share at least once, while four of the eight males never succeeded. In 32 of these episodes, individuals begged in vain until the very end of an owner's tenure. In seven of these episodes, however, individuals got the kill's remnants when the owner left, while the rest obtained nothing. On average, individuals seeking a share had to wait one-third (35% ± 38%) of their dyadic time before they got their first share. Pooling both successful and unsuccessful episodes, the average time for eating the kill was, at 18.7 min, relatively short.

Of the 22 participants that were ranked, five were high-, six middle-, and eleven low-ranking. They were involved in 61 owner-participant dyads, whereby low-ranking individuals showed their interest most often, followed by middle-ranking and high-ranking individuals. This was as expected by the ranks' distribution (rank: expected versus observed; low: 30.5 versus 25, middle: 16.5 versus 21, high: 14 versus 15; $\chi^2=2.49$, df=2, $p=0.29$; ns). However, when high-ranking individuals were present to seek a share, they got one more often than did low-rankers (U-test; $z=-2.23$, $p=0.025$).

Plant food

PREY

Occasionally, bonobos shared medium-sized fruits such as *Gambeya* spp., *Mammea africana* and *Garcinia cola*. The large-sized fruits of *Treculia africana* and *Anonidium mannii*, however, made up most of the shared plant food. Of 165 of these large-sized fruits eaten by bonobos (*Treculia*: 93%; *Anonidium*: 7%), about half ($n=85$) were shared. Bonobos spent 111 hours eating these shared plant items, 97% of the time on *Treculia* and 3% on *Anonidium*. Therefore, *Treculia* played the key role in food sharing.

As shown in Table 17.2, *Treculia* is a large tree with an average dbh (diameter at breast height) of almost 1 m, but usually bearing only a small number of fruits at a time. The fruits' diameters measured 10–45 cm and the median weight was 15 kg. The fruits consisted of fibre, seeds and a stalk that was never consumed.

Table 17.2. *Dimensions of* Treculia africana *and its consumption*

	Mean	SD	range	n
Tree dbh (cm)	92	6.00	—	2
No. of fruits per tree	—	—	1–50*	20
Fruit size (∅ cm)	23.51	7.08	10–45	40
Fruit weight (kg)	28.42	49.51	0.20–30.00	75
Fruits eaten per day	0.49	0.91	0–8	152
Feeding time per fruit (min)	66.65	57.49	1–348	127

Notes:
SD = standard deviation; n = sample size; ∅ = diameter; * = estimation from regularly used feeding trees.

Treculia is a seasonal fruit. In 9 years of observation, the peak of *Treculia* consumption (57% of all items, $n=152$) was in June and July, a relatively dry season of the year. Fruit consumption was observed first in March and last in October.

COLLECTION

Bonobos sometimes climbed a tree to harvest a fruit. Clutching an overhead branch with both hands for support, individuals stood on the fruit and actively pushed or stamped in order to detach it. At other times, bonobos used their hands and teeth to remove fruits from a tree.

CONSUMPTION

The harvested fruits were eaten directly after they hit the ground. However, we often found unripe fruits on the ground, and these sometimes stayed there for weeks. Bonobos occasionally visited and tested (by smell and pressure) the fruits, before they eventually consumed them.

On days when we spent at least half the day with the bonobos, one fruit was consumed every other day. The time to consume a fruit was on average about an hour (Table 17.2).

OWNERS

Thirty individuals owned *Treculia africana* fruits, 20 females and 10 males. The owners were involved in 219 owner episodes, dealing with 0–12 supplicants. Females possessed fruit in 80% and males in 20% of cases. When only the episodes with supplicants were considered, the number dropped to 136; 88% were controlled by females, 12% by males. In 102 of these episodes, the owners shared with the supplicants, and females owners did so more often than males (expected versus observed: females: 78.5 versus 97; males: 23.5 versus 5; $\chi^2=81.19$, df=1, $p<0.001$). In the other 34 episodes, the owners refused to share with their supplicants, and both females and males did so as expected by chance (expected versus observed: females: 26 versus 23, males: 8 versus 11; $\chi^2=3.56$, df=1, $p=0.06$; ns).

A fruit was often owned by several individuals in sequence (mean number of owners/fruit: 1.68; SD=1.03; range=1–6, $n=119$). An owner's tenure lasted on average 40 min (Table 17.1). When the relative time an owner was in possession of an item was considered, first owners monopolised a fruit longer than later owners (73% versus 42%: U-test: $z=-5.62$, $p<0.001$).

Treculia was owned by individuals of all ranks. When only owners with supplicants were counted, 6 of 26 were high-, 6 were middle- and 14 were low-ranking. High-ranking individuals monopolised *Treculia* more often than low-ranking ones (rank: expected versus observed; high: 29 versus 47, middle: 29 versus 29, low: 70 versus 52, $\chi^2=6.86$, df=2, $p=0.03$; $n=128$ dyads).

SUPPLICANTS AND PARTICIPANTS

Thirty-two mature individuals showed interest in getting a share, and they were involved in 432 supplicant–owner episodes. The 21 females were supplicants in 312 episodes, the 11 males in 120 episodes. Each of them got a share at least once. On average, however, one out of three seeking a share came away empty-handed. Males were unsuccessful more often than females: while females begged in vain in only 21% of their episodes, males did so in 46%. Supplicants had to wait about one-third of their invested time before obtaining their first share. When they were successful, they spent, on average, 35 min at the fruit.

Of all ranked supplicants and participants ($n=27$), 6 were high-, 6 were middle- and 15 were low-ranking. Low- and middle-ranking individuals showed less interest than did high-ranking individuals (rank: expected versus observed; low: 176 versus 161, middle: 70 versus 64, high: 70 versus 91; $\chi^2=47.59$, df$=2$, $p<0.001$; $n=316$ dyads)

COMPARISON OF MEAT- AND PLANT-FOOD SHARING

As shown in Table 17.1, meat-sharing sessions lasted longer, had more owners and participants and more participants per owner than *Treculia*-sharing sessions. Ownership was clearly biased to females for both foods. Owners' and participants' tenure did not differ.

Begging during sharing sessions was common, and included the entire repertoire from simple peering with hand outstretched to intense display of infantile behavior, including body rocking, pout face and whimpering. When there was more than one male among the supplicants, male displays became rough and included branch-dragging, continuous screams of increasing intensity, and jumps leading to physical contact with the individuals close by the kill or fruit. However, these displays did not result in the receipt of food.

Despite the absence of any collaborative hunting or collection technique, bonobos shared food. So, if it is not collaboration that needs to be reinforced by means such as food sharing, then why do bonobos share food?

HYPOTHESES TESTED

Reciprocity

If food sharing is based on reciprocity, we predict that (1) individuals who receive food will later share with the donor, and (2) individuals who do not get a share will 'retaliate' later when they own food. Alternatively, reciprocity may involve a different 'currency', that individuals receiving now may reciprocate later, not with food but with, for example grooming.

RECIPROCITY AND RETALIATION WITH FOOD

Of all 31 mature community members, 25 were both owner and recipient of divisible food. However, time of possession or participation across individuals, was unequal. Of the 159 hours that individuals were involved in food sharing, possession across owners ranged from 1 minute to 40 hours, and time of consumption across participants varied from 18 minutes to 8 hours.

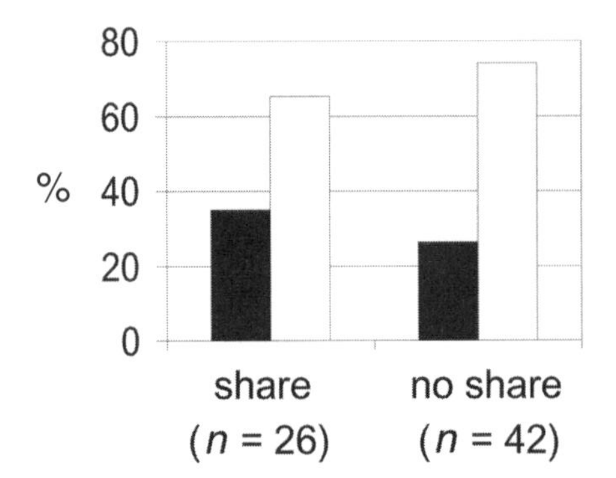

Fig. 17.1. Reciprocity of food sharing with grooming: relative frequency of grooming for all unidirectional dyads ($n=68$). Dark bars represent grooming dyads, clear bars represent non-grooming dyads.

Of the 404 potential dyads, 164 were seen to share food. Of these, 39 dyads were involved in bidirectional actions, 125 in unidirectional actions. Of the bidirectional dyads 82% reciprocated, 10% did not and 8% retaliated.

GROOMING AS CURRENCY

To see if currencies other than food could be used to reciprocate, we investigated the relation between food sharing and grooming. We looked at the 125 unidirectional dyads. In 66% of the dyads, *A* shared with *B*; in 34% *A* refused to share with *B*. If grooming is a currency for reciprocation, the dyads where *A* shared with *B* would be expected to be more involved in grooming than those where *A* refused to share with *B*. Figure 17.1 shows no difference between sharing dyads ($n=26$) versus non-sharing dyads ($n=42$) (35% versus 26%: $\chi^2=0.22$, $p=0.64$; ns).

Kin selection

If food sharing is based on kinship, we predict that owners will respond differently to supplicants who are closely related to them than they will to others. Preferences may be expressed in various ways: (1) owners may share more often with kin than with non-kin; (2) kin may receive a share earlier than non-kin; or (3) kin may receive larger shares than non-kin. Moreover, (4) if kinship is the driving force of food sharing, reciprocity is less likely to occur.

Of 404 possible dyads of the community, only 3.2% were

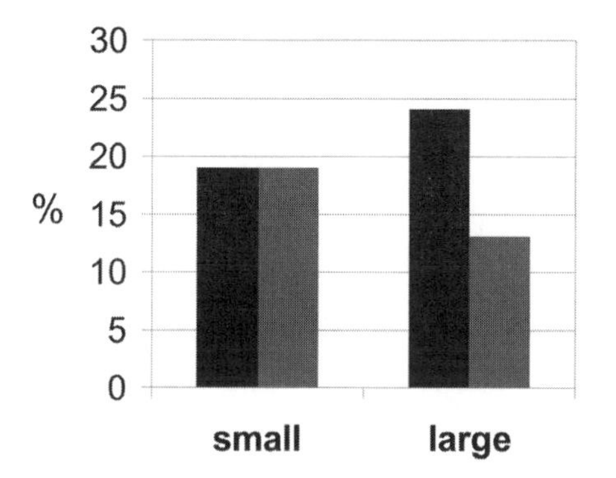

Fig. 17.2. Quantity of food given away in relation to relatedness. Dark bars represent events among kin ($n=21$), light bars refer to events among non-kin ($n=344$).

kin. Of the 164 dyads that shared food, the proportion of related dyads was the same. These six closely related, owner–participant dyads met in 22 episodes, in 68% of which they shared, and the 158 unrelated dyads met in 361 episodes, in 71% of which they shared. Kin therefore, shared no more often than did non-kin ($\chi^2=0.0$, df=1, $p=1.0$; ns).

Considering the relative time an individual was kept waiting until he or she got a first share as a measure of the owner's reluctance to share, kin waited almost 90% of their participation time, while non-kin waited only 50% of this time. These delays were not significantly different (median ± interquartile range: 89% ± 100% versus 50% ± 100%; U-Test: $z=-1.129$, $p=0.26$; ns).

We estimated food quantity by distinguishing between small shares (ranging from small bundles of fibres to hand size) and large shares (larger than the size of hand). Kin received large shares in 24% of the cases ($n=21$), non-kin did so in only 13% of the cases ($n=344$). This was not significantly different (χ^2-Test/z-Test: $\chi^2=1.21$, df=1, $z=1.099$, $p=0.27$; ns) (Figure 17.2).

In order to test whether or not asymmetry in sharing among kin could be explained by inclusive fitness, we looked at all dyads in which *A* gave to *B* and *B* reciprocated ($n=32$) compared to dyads in which *A* gave to *B* without reciprocation ($n=83$). We expected a larger proportion of kin to be among the unreciprocal dyads rather than the reciprocal ones. While 7% of the asymmetric dyads involved closely related individuals, only 3% of the reciprocating dyads did so, but this was not significantly different (dyad: kin versus non-kin: unidirectional: 6 versus 77, bidirectional: 1 versus 31; Extension of Fisher-test: $p=0.276$; ns).

Mutualism

If food sharing is based on mutualism, we predicted that both owners and supplicants would gain by working together. Participants always get a pay-off from the food obtained. They could, however, also improve their gain by extending their feeding time through helping owners to defend the item. Similarly, owners may gain from participants assisting them in defending their possession against others.

Participants who prevented others from joining the food-sharing cluster tended to obtain an extended feeding time, while those whose loyalty was not openly demonstrated had a shorter feeding time. This tendency, however was not statistically substantiated (time eating on share (min): $n=38$ versus $n=336$ supplicant–owner episodes (median ± interquartile range): 8 ± 34 versus 16 ± 71; U-Test: $z=-1.848$, $p=0.065$; ns).

In order to evaluate the gain an owner realized by sharing with supplicants, we looked at the absolute time the owner was in possession of an item. Owners with supplicants were in possession of an item longer than those without supplicants ($n=159$ versus $n=46$ supplicant–owner episodes (median ± interquartile range): 20 min ± 28 versus 34 min ± 46; U-Test: $z=-2.922$, $p=0.003$). When owners had supplicants, those who shared were in possession longer than those who did not share ($n=124$ versus $n=35$ supplicant–owner episodes (median ± interquartile range): 21 ± 32 versus 42 ± 52; U-Test: $z=-2.626$, $p=0.009$). Figure 17.3 shows the benefit of owners with helpers: those with supplicants that prevented others from joining, were in possession of the item longer than those without helpers ($n=21$ versus $n=103$ supplicant-owner episodes (median ± interquartile range): 35 ± 44 versus 65 ± 99; U-Test: $z=-2.828$, $p=0.005$).

Buy-off

Distribution of food may result from pressure exerted by supplicants. We predicted that (1) sharing is passive (tolerated theft) rather than active (donation of food), and (2) sharing increases with the number of supplicants.

To determine the mode of food sharing, we used a rough indicator of the occurrence of both passive and active transfer within each owner–participant dyad. Of 380 scores, 66% were passive, while 34% were active. We cannot say if the active handing over was voluntary or not.

Figure 17.4 shows a positive correlation between the

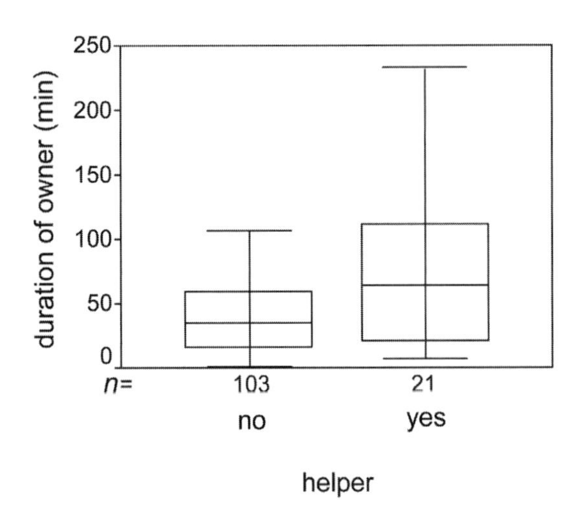

Fig. 17.3. Benefit for the owner in relation to presence of helpers. Duration of ownership (in minutes) is the time that an owner is in possession of the item. Helper is a supplicant that prevents other supplicants from joining. Horizontal bars in boxes indicate median with 75th percentile above and 25th percentile below. Bars outside boxes indicate range of observed values (excluding outliers); *n* refers to sample size for owner–supplicant episodes.

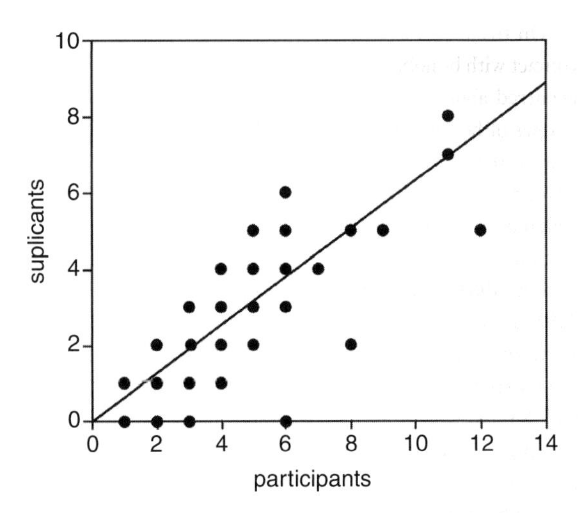

Fig. 17.4. Correlation between the number of supplicants and the number of participants. (ANOVA: $r^2 = 0.57$, $F = 180.36$, $p < 0.001$).

number of supplicants and the number of participants. The more who beg, the more who get (analysis of variance (ANOVA): $r^2 = 0.57$, $F = 180.36$, $p < 0.001$). The tendency also emerged that the more supplicants present, the shorter their wait to get a first share, but this was not significant.

We expected that owners should be more inclined to share with supplicants when the owners were the target of agonistic behavior. That is, an owner may occasionally have to be pushed to share. A sub-sample of 387 supplicant–owner dyads had scores for either presence or absence of agonism. Agonism between owner and supplicant was a rare event. In those few cases in which it occurred (4%), it did not increase the owners' willingness to share. On the contrary, when a supplicant became aggressive, he had to wait significantly longer to get a first share (relative time waiting (median ± interquartile range) 50% ± 95 versus 96% ± 54: U-Test: $z = -2.03$, $p = 0.04$). In 65% of episodes, the owners shared nothing with aggressive supplicants.

SUMMARY

For the *reciprocity hypothesis*, we conclude that reciprocity explains only a few food transfers between owner and recipient, and retaliation was of no significance. There was no evidence that grooming was a currency of reciprocation.

For the *kin selection* hypothesis, we found no difference in sharing with kin versus non-kin.

The *mutualism hypothesis* seems likely since both participants and owners benefited.

The *buy-off hypothesis* seems likely since most sharing was passive, and the more supplicants an owner had, the more individuals the owner shared with. However, the degree of pressure had its limits: when owners were threatened they were reluctant to share with the aggressors.

DISCUSSION

Until recently, hunting, consumption of mammalian prey and meat sharing were thought to be characteristic traits separating the two *Pan* species; all traits were thought to be common in chimpanzees but absent or rare in bonobos. In a review on the behavioral ecology of both *Pan* species, Stanford (1998a) speculated that the lack of information on hunting and meat sharing in bonobos might be the result of not enough data from different populations, incomplete habituation of some study groups, or from human interference. While these explanations are plausible, the conventional view of bonobos as a non-hunting, vegetarian *Pan* species still persists (e.g. Doran *et al.*, Chapter 1). Thus, the search for empirical data that can illuminate this topic remains a challenge.

On the 454 days when observers at Lomako had visual contact with bonobos for 1 hour or more, meat consumption occurred about once every 7 weeks. In contrast to earlier studies of bonobos (Badrian & Malenky 1984; Ihobe 1992), prey were not restricted to small mammals such as squirrels or infant duikers, but included adult mammals estimated to weigh more than 10 kg. Differences across sites are striking: although duikers are present at all sites (Lomako: Badrian & Badrian 1984; Wamba: Kano 1992, Lilungu: Sabater Pi *et al.* 1993; Lukuru: Myers-Thompson 1997), only bonobos at Lomako were seen to hunt them. Flying squirrels, also ubiquitous, seem to be preyed upon only at Wamba (Ihobe 1992), and for arboreal monkeys, present in all sites, the only hunting record is from Lilungu. There, both redtailed monkeys and black-and-white colobus were captured, although consumption was not observed (Sabater Pi *et al.* 1993). As these examples show, prey preference may be a good candidate for cultural variation in the genus *Pan*.

The monkey species on which chimpanzees prey most frequently is the red colobus (Nishida *et al.* 1983; Boesch 1994; Stanford *et al.* 1994). So far, hunting of red colobus by bonobos has neither been seen directly nor confirmed indirectly through faecal analysis. Red colobus monkeys are absent at Lomako, but present at Wamba, where peaceful encounters with interspecific grooming were known (Ihobe 1990).

The data from Lomako show that meat consumption in bonobos is not as high as in some populations of chimpanzees, but seems to fall within the range of *Pan troglodytes*. While data on hunting frequencies are still scarce, indirect evidence obtained by faecal analyses suggests that in a number of populations, for example those at Mt. Assirik (McGrew *et al.* 1988), Bossou (Sugiyama & Koman 1987), Lopé (Tutin & Fernandez 1993) and Ndoki (Kuroda *et al.* 1996), meat consumption is relatively rare and does not exceed what we found in bonobos.

We investigated 88 food items shared among bonobos, most of which were fruit. Although meat-sharing sessions stood out in terms of length of session, number of owners and participants and a pronounced bias to female owners, the patterns of sharing behavior remained consistent across prey items. The procedure of begging and hand-outs, the excitement of supplicants and participants during the sharing sessions, the fact of sharing itself, coincided with what is known from chimpanzees. Nevertheless, there are many aspects that do not match with patterns described for chimpanzees, such as prey species that are solitary rather than group-living mammals, a predominance of fruit rather than meat among shared food items, and female- rather than male-biased sex of owner and recipient.

Chimpanzees mostly hunt monkeys. Depending on prey population, risk of predation, and habitat density, social hunts are more successful than solitary ones, and sharing is thought to be the motor that drives the system. In contrast to arboreal monkeys, duikers are a cheap prey: investment of time, effort and risk is low, due to their terrestriality and defencelessness. Therefore, solitary hunting is profitable for bonobos. So why do bonobos share, when neither the hunting technique nor the process of fruit harvesting or consumption requires cooperation?

Only when bidirectional dyads were considered did most reciprocate, but these accounted for merely one quarter of all dyads that shared food. This result can be interpreted in different ways. It may be that reciprocity in food exchange exists but was not detected because too many food sharing episodes were missed. Reciprocity may involve other 'currencies' that we did not identify. Alternatively, food transfer in bonobos may not be based on reciprocity.

We know that some sharing episodes were missed. However, the finding that ownership of divisible food was biased towards a few individuals makes reciprocity an unlikely explanation. Moreover, we expected a dyad to act reciprocally when its individuals had exchanged food once in a season, independent of the quantity or frequency of the exchanged goods. Without this generous time frame, the number of reciprocal dyads would have been even smaller, and the likelihood that these few cases of reciprocity promoted sharing is, indeed, low.

In a group of captive chimpanzees, recipients of food reciprocated by grooming the owner of divisible food (de Waal 1997). However, the correlation between the two acts was found when food sharing was compared with grooming 2 hours *before* the food transfer took place. Our data from Lomako did not allow us to apply the same procedure. In another study of chimpanzees, no such trade of social services like food for grooming was found (Hemelrijk *et al.* 1999).

This suggests that the mechanism that promotes food sharing varies between groups, populations and, perhaps, different periods.

Variation in resource value may affect the mode of food transfer (Bird & Bird 1997; Hakoyama & Iguchi 1997), suggesting that the more valuable a resource, the more complex will be its division. The proportion of fat and protein is high in both meat and *Treculia*. *Treculia* seeds are not only rich in fat (12%) but also in protein (23%) (Foma & Abdala 1985).

A resource of high value spurs interest and prompts competition, which implies additional advantages for owners who value their item as currency and use them for political purposes.

While the predictions that related food sharing to kinship were not met, both the mutualism and the buy-off hypothesis seem to provide explanations. The benefit for participants is evident. However, our data showed that owners may also benefit from sharing with supplicants: owners with supplicants were in possession of an item longer than those without supplicants. Of course, individuals with small or low-valued items, such as fibres without seeds, or skin, may have few or no supplicants, and thus may be able to finish off their part comparatively quickly. Eating without supplicants allows efficient intake without the costs that derive from efforts to prevent others from participation. The extended time of possession for owners with supplicants, however, was not due to their interruptions of the meal. Owners who shared with supplicants indeed ate longer. Monopolisation of highly valued food that was of interest to supplicants required exertion of the owners' social dominance in order to be able to share food without losing it. While low-ranking individuals quickly lost an item to individuals who badgered, high-ranking individuals were able to gain feeding time by allowing others access to their food. This was an even smarter move in situations in which partners who displaced others from the food source were also among the owners' supplicants. The tendency to allow these supplicants an extended participation period may indicate the owners' reinforcement of this support.

While the mutual benefit for owners and helpers may develop during sharing, pressure is a good candidate for the motor that incites sharing. The more supplicants badgered an owner, the more received a share. Individuals who ate alone had a higher intake than individuals who had supplicants, whether or not they shared. So why did individuals who were alone with a fruit call and attract others within earshot, which then led to the pressure to share?

It may be that providing others with a large divisible resource increases status. De Waal (1982) and Moore (1984) proposed this, but it is hard to test, as status covaries with age, reproductive state, number and sex of offspring, allies, group composition, and so on. It is also contradicted by the phenomenon of low-ranking owners more often refusing to share, thereby missing a chance to promote their popularity.

It may be that an owner who eats alone but is found by group members risks punishment, or at least a loss of status.

Alternatively, it may be safety in numbers. Studies on fish, birds and mammals show the advantages of group foraging, including the combined effects of dilution (Hamilton 1971), improved detection (Kenward 1978), predator confusion (Edmunds 1974), and predator deterrence (Curio 1978). The risk of encounters with leopards may enhance aggregations, but the data to test the impact of predators on group formation in bonobos are not yet available. When bonobos ate in trees rather than on the ground, monopolisation and consumption often became mutually exclusive since much effort was needed to position the food (> 10 kg), maintain balance, remove bite-sized pieces, and deal with beggars simultaneously. Therefore, food was usually eaten on the forest floor in larger parties that may have been beneficial to the owner.

Why did individuals who were left waiting stay instead of leave?

First, it is likely that ripe *Treculia* fruits are scarce, so bonobos may compete for access to them. Moving on to other food patches may be more costly than waiting to get a share. This, however, is contradicted by the observation that bonobos sometimes consumed nearby fruits in sequence rather than simultaneously.

Second, female bonobos are gregarious and associate even when food is scarce. Females are likely to derive benefits from travelling together with other community members instead of foraging alone. Parish (1993) proposed that female bonobos associate in order to resist male aggression and to control food sources. Males often acted aggressively in order to get a share, but willingness (by the females) to give a share did not increase with the intensity of such agonism. On the contrary, when pestering turned into aggression, owners became reluctant to share with the aggressors.

All in all, of the four hypotheses presented here, the buy-off hypothesis and the mutualism hypothesis seem to provide reasonable explanations for why bonobos share food: both the pressure by the number of interested individuals as well as the mutual benefit of sharing may work together. In addition there may be a number of other factors that we did not investigate here but that act on the owner in each sharing episode, such as availability of food, state of nutrition, time of day, identity of individuals present, place of division, and on on. The fact is that members of a bonobo community who usually have a high degree of individual independence gather under conditions that require high levels of social tolerance, which may then provide them with opportunities to develop cohesiveness, as well as consolidate friendships and status.

ACKNOWLEDGEMENTS

We thank the Ministère de l'Education Nationale, République Démocratique du Congo for permission to conduct fieldwork at Lomako. This research was financially supported by the Max-Planck-Society (MPG), the German Research Council (DFG), and the University of Munich (LMU). We thank the external reviewer, W.C. McGrew, as well as the editors C.B. and L.F.M. for helpful comments on the manuscript. The results were presented at the conference on 'Behavioural Diversity in Chimpanzees and Bonobos' that was supported by the MPG, the DFG, the Wenner-Gren Foundation for Anthropological Research, and the Ministry for Science, Research & Art of the State of Bavaria.

REFERENCES

Anderson, J. R., Williamson, E. A. & Carter J. (1983). Chimpanzees of Sapo forest, Liberia: density, nests, tools and meat-eating. *Primates*, 24, 594–601.

Axelrodt, R. & Hamilton, W. D. (1981). The evolution of cooperation. *Science*, 211, 1390–6.

Badrian, N. & Badrian, A. (1984). Social organization of *Pan paniscus* in the Lomako forest, Zaïre. In *The Pygmy Chimpanzee*, ed. R. L. Susman, pp. 325–46. New York: Plenum Press.

Badrian, N. & Malenky, R. K. (1984). Feeding ecology of *Pan paniscus* in the Lomako Forest, Zaïre. In *The Pygmy Chimpanzee*, ed. R.L. Susman, pp. 275–99. New York: Plenum Press.

Basabose, K. & Yamagiwa, J. (1997). Predation on mammals by chimpanzees in the montane forest of Kahuzi, Zaïre. *Primates*, 38, 45–55.

Bird, R. L. B. & Bird, D. W. (1997). Delayed reciprocity and tolerated theft. The behavioral ecology of food-sharing strategies. *Current Anthropology*, 38, 49–78.

Boesch, C. (1994). Cooperative hunting in wild chimpanzees. *Animal Behaviour*, 48, 653–67.

Boesch, C. & Boesch, H. (1989). Hunting behavior of wild chimpanzees in the Taï National Park. *American Journal of Physical Anthropology*, 78, 547–73.

Brown, J. L. (1983). Cooperation – a biologist's dilemma. In *Advances in the Study of Behavior*, ed. J. S. Rosenblatt, pp. 1–37. New York: Academic Press.

Brown, J. L. (1987). *Helping and Communal Breeding in Birds*. Princeton: Princeton University Press.

Cheney, D. & Seyfarth, R. (1980). Vocal recognition in free-ranging vervet monkeys. *Animal Behaviour*, 28, 362–67.

Curio, E. (1978). The adaptive significance of avian mobbing. I. Telenomic hypothesis and predictions. *Zeitschrift für Tierpsychologie*, 48, 175–83.

de Waal, F. B. M. (1982). Appeasement, celebration, and food sharing in the two *Pan* species. In *Topics in Primatology*, Vol. 1, ed. T. Nishida, W. C. McGrew, P. Marler, M. Pickford & F. B. M. de Waal, pp. 37–50, Tokyo: Tokyo University Press.

de Waal, F. B. M. (1989). *Peacemaking among Primates*. Cambridge, MA: Harvard University Press.

de Waal, F. B. M. (1997). The chimpanzees service economy – food for grooming. *Evolution and Human Behavior*, 18, 375–86.

de Waal, F. B. M. (2000). Attitudinal reciprocity in food sharing among brown capuchin monkeys. *Animal Behaviour*, 60, 253–61.

Edmunds, M. (1974). *Defence in Animals: A Survey of Anti-Predator Defences*. London: Longmans.

Engel, J. (1998). *SsS Benutzerhandbuch*. Rubisoft Software.

Fairbanks, L. A. & McGuire, M. M. (1984). Determinants of fecundity and reproductive success in captive vervet monkeys. *American Journal of Primatology*, 7, 27–38.

Foma, M. & Abdala, T. (1985). Kernel Oils of Seven Plant Species of Zaïre. *JAOCS: Journal of the American Oil Chemist's Society*, 62, 910–1.

Foster, S. A. (1985). Group foraging by a coral reef fish: a mechanism for gaining access to defended resources. *Animal Behaviour*, 33, 3782–92.

Fruth, B. (1995). *Nests and Nest Groups in Wild Bonobos* (*Pan paniscus*): Ecological and Behavioural Correlates. Aachen: Shaker Verlag.

Fruth, B. (1998). Comment. *Current Anthropology*, 39, 408–9.

Gadakar, R. (1990). Evolution of eusociality: the advantages of assured fitness returns. *Philosophical Transactions of the Royal Society of London*, 329, 17 25.

Gerloff, U., Hartung, B., Fruth B., Hohmann, G., & D. Tautz. (1999). Intra-community relationships, dispersal pattern and control of paternity in a wild living community of bonobos (*Pan paniscus*) determined from DNA analyses of faecal samples. *Proceedings of the Royal Society of London*, 266, 1189–95.

Goodall, J., v. L. (1968). The behavior of free-living chimpanzees in the Gombe Stream Reserve. *Animal Behaviour Monographs*, 1, 163–311.

Goodall, J. (1986). *The Chimpanzees of Gombe: Patterns of Behavior.* Cambridge, MA: Belknap Press of Harvard Univiversity Press.

Gouzoules, S. (1984). Primate mating systems, kin associations and cooperative behavior: evidence for kin recognition. *Yearbook of Physical Anthropology*, 27, 99–134.

Hakoyama, H. & Iguchi, K. (1997). Why is competition more intense if food is supplied more slowly? *Behavioral Ecology & Sociobiology*, 40, 159–68.

Hamilton, W. D. (1963). The evolution of altruistic behavior. *American Naturalist*, 97, 354–56.

Hamilton, W. D. (1971). Geometry of the selfish herd. *Journal of Theoretical Biology*, 31, 295–311.

Heinrich, B. & Marzluff, J. M. (1991). Do common ravens yell because they want to attract others. *Behavioral Ecology & Sociobiology*, 28, 13–21.

Heinrich, B., Marzluff, J. & Adams, W. (1995). Fear and food recognition in naive common ravens. *Auk*, 112, 499–503.

Hemelrijk, C. K., Meier, C. & Martin, R. D. (1999). 'Friendship' for fitness in chimpanzees? *Animal Behaviour*, 58, 1223–9.

Hohmann, G. & Fruth, B. (1993). Field observations on meat sharing among bonobos (*Pan paniscus*). *Folia Primatologica*, 60, 225–9.

Hohmann, G. & Fruth, B. (1996). Food sharing and status in unprovisioned bonobos (*Pan paniscus*): preliminary results. In *Food and the Status Quest*. ed. P. Wiessner & W. Schiefenhövel, pp. 47–67, Providence: Marion Berghahn Press.

Hohmann, G., Gerloff, U., Tautz, D. & Fruth, B. (1999). Social bonds and genetic ties: kinship, association and affiliation in a community of bonobos (*Pan paniscus*). *Behaviour*, 136, 1219–35.

Ihobe, H. (1990). Interspecific interactions between wild pygmy chimpanzees (*Pan paniscus*) and red colobus (*Colobus badius*). *Primates*, 31, 109–12.

Ihobe, H. (1992). Observations on the meat-eating behavior of wild bonobos (*Pan paniscus*) at Wamba, Republic of Zaïre. *Primates*, 33, 247–50.

Kano, T. (1992). *The Last Ape*. Stanford, CA: Stanford University Press.

Kano, T. & Mulavwa, M. (1984). Feeding ecology of the pygmy chimpanzee (*Pan paniscus*) of Wamba. In *The Pygmy Chimpanzee*, ed. R. L. Susman, pp. 233–74. New York: Plenum Press.

Kenward, R. E. (1978). Hawks and doves: factors affecting success and selection in goshawk attacks on woodpigeon. *Journal of Animal Ecology*, 47, 449–60.

Kuroda, S., Suzuki, S. & Nishihara, T. (1996). Preliminary report on predatory behaviour and meat sharing in tschego chimpanzees (*Pan troglodytes troglodytes*) in the Ndoki Forest, Northern Congo. *Primates*, 37, 253–59.

Kuroda, S. (1984). Interactions over food among pygmy chimpanzees, *Pan paniscus*. In *The Pygmy Chimpanzee*, ed. R. L. Susman, pp. 65–87. New York: Plenum Press.

Lenglet, G. L. (1987). Animals eaten by chimpanzees: a bibliographical review. *Revue Zoologique Africaine*, 101, 197–219.

Mack, A. L. (1993). The sizes of vertebrate-dispersed fruits: a neotropical-paleotropical comparison. *American Naturalist*, 142, 840–56.

McGrew, W. C. (1975). Patterns of plant food sharing by wild chimpanzees. In *Contemporary Primatology: Proceedings of the Fifth Congress of the International Congress of Primatology, Nagoya, 21–24 August, 1974*, ed. S. Kuodo, pp. 304–9, Basel: Karger.

McGrew, W. C., Baldwin, P. J. & Tutin, C. E. G. (1988). Diet of wild chimpanzees (*Pan troglodytes verus*) at Mt. Assirik, Senegal: I. composition. *American Journal of Primatology*, 16, 213–26.

Mitani, J. & Watts, D. (1999). Demographic influences on the hunting behavior of chimpanzees. *American Journal of Physical Anthropology*, 109, 439–54.

Moore, J. (1984). The evolution of reciprocal sharing. *Ethology and Sociobiology*, 5, 5–14.

Myers-Thompson, J. A. (1997). The history, taxonomy and ecology of the bonobo (*Pan paniscus*, Schwarz 1929) with a first description of a wild population living in a forest/savanna mosaic habitat. PhD dissertation. University of Oxford, UK.

Nishida, T. & Uehara, S. (1983). Natural diet of chimpanzees (*Pan troglodytes schweinfurthii*): long-term record from the Mahale Mountains, Tanzania. *African Study Monographs*, 3, 109–30.

Nishida, T. & Turner, L. A. (1996). Food transfer between mother and infant chimpanzees of the Mahale Mountains National Park, Tanzania. *International Journal of Primatology*, 17, 947–68.

Nishida, T., Uehara, S. & Ramadhani, N. (1979). Predatory behavior among wild chimpanzees of the Mahale Mountains. *Primates*, 20, 1–20.

Nishida, T., Wrangham, R. W., Goodall, J. & Uehara, S. (1983). Local differences in plant-feeding habits of chimpanzees between Mahale Mountains and Gombe National Park. *Journal of Human Evolution*, 12, 467–80.

Packer, C., Geilbert, D., Pusey, A. E. & O'Brien, S. (1991). A molecular genetic analysis of kinship and cooperation in African lions. *Nature*, 351, 562–5.

Parish, A. R. (1993). Sex and food control in the 'uncommon chimpanzee': How bonobo females overcome a phylogenetic legacy of male dominance. *Ethology and Sociobiology*, 15, 157–79.

Rasa, O. A. (1977). The ethology and sociology of the dwarf

mongoose (*Helagale undulata rufala*). *Zeitschrift für Tierpsychologie*, 43, 160–2.

Roell, A. (1978). Social behaviour of the jackdaw, *Corvus monedula*, in relation to its niche. *Behaviour*, 64, 1–124.

Sabater Pi, J., Bermejo, M., Illera, G. & Véa, J. J. (1993). Behavior of bonobos (*Pan paniscus*) following their capture of monkeys in Zaïre. *International Journal of Primatology*, 14, 797–804.

Stanford, C. B. (1995). Chimpanzee hunting behaviour and human evolution. *American Scientist*, 83, 256–61.

Stanford, C. B. (1998a). The social behavior of chimpanzees and bonobos: Empirical evidence and shifting assumptions. *Current Anthropology*, 39, 399–420.

Stanford, C. B. (1998b). *Chimpanzee and Red Colobus*. Cambridge, MA: Harvard University Press.

Stanford, C. B., Wallis, E., Mpongo, E. & Goodall, J. (1994). Hunting decisions in wild chimpanzees. *Behaviour*, 131, 1–20.

Sugiyama, Y. & Koman, J. (1987). A preliminary list of chimpanzees' alimentation at Bossou, Guinea. *Primates*, 28, 133–47.

Suzuki, A. (1975). The origin of hominid hunting: a primatological perspective. In *Socioecology and Psychology of Primates*, ed. R. H. Tuttle, pp. 259–78. Mouton: The Hague.

Takahata, Y., Hasegawa, T. & Nishida, T. (1984). Chimpanzee predation in the Mahale Mountains from August 1979 to May 1982. *International Journal of Primatology*, 5, 213–33.

Tutin, C. E. G. & Fernandez, M. (1993). Composition of the diet of chimpanzees and comparisons with that of sympatric lowland gorillas in the Lopé Reserve, Gabon. *American Journal of Primatology*, 28, 195–211.

Wilkinson, G. S. (1984). Reciprocal food sharing in the vampire bat. *Nature*, 308, 181–4.

Wilkinson, G. S. (1987). Altruism and co-operation in bats. In *Recent Advances in the Study of Bats*, ed. M. B. Fenton, P. Racey & J. M. V. Rayner, pp. 299–323. Cambridge: Cambridge University Press.

Wrangham, R. W. (1975). The behavioral ecology of chimpanzees in Gombe National Park, Tanzania. PhD thesis: University of Cambridge, UK.

18 • Hunting and meat sharing by chimpanzees at Ngogo, Kibale National Park, Uganda

DAVID P. WATTS & JOHN C. MITANI

INTRODUCTION

Chimpanzees (*Pan troglodytes*) in all well-studied wild populations hunt a variety of vertebrates and share meat extensively (reviews: Wrangham & Bergmann-Riss 1990; Uehara 1997; Stanford 1998). Meat is a highly valued resource and nutritional reasons presumably explain the origin of hunting, although its exact nutritional significance is unclear. Hunting and meat sharing also have considerable social importance. Two major similarities across chimpanzee populations are that red colobus monkeys (*Procolobus badius*) are the primary prey species where the two species are sympatric (Goodall 1986; Boesch & Boesch 1989; Uehara 1997; Ihobe & Uehara 1999; Mitani & Watts 1999; Boesch & Boesch-Achermann 2000; Hosaka *et al.* 2001), and that predation rates vary over time (Mahale: Takahata *et al.* 1984; Uehara 1997; Hosaka *et al.* 2001; Gombe: Goodall 1986; Wrangham & Bergmann-Riss 1990; Stanford *et al.* 1994a; Stanford 1998; Taï: Boesch & Boesch 1989; Kahuzi-Biega: Basabose & Yamagiwa 1997; Kibale: Mitani & Watts 1999). Other aspects of hunting vary considerably across and sometimes within populations. In this chapter, we summarize data on hunting and meat sharing by chimpanzees at Ngogo, in Kibale National Park, Uganda. We focus mostly on hunts of red colobus, by far the most important prey species. We compare our data to those from other sites to highlight questions about hunting and meat sharing that are partly resolved and those that remain open. We summarize results of our earlier analyses of hunting and meat sharing at Ngogo (Mitani & Watts 1999, 2001; Watts & Mitani 2002), and use data collected more recently to update analyses of hunting success, cooperation, and the relationship of meat sharing to mating behavior.

We focus on four major issues:

(1) What determines whether chimpanzees hunt red colobus on encounter? Successful hunters often share their captures, and this question is inseparable from hypotheses about how meat is distributed after hunts. Ecological, social, and reproductive factors can influence hunting and meat sharing decisions. Researchers have often hypothesized that forest structure influences hunting decisions, and that chimpanzees are more likely to hunt in areas where the canopy is low and broken than where it is tall and continuous (e.g. Teleki 1973; Wrangham 1975; Boesch 1994b). However, quantitative analyses of this hypothesis have not been performed. Various researchers have proposed that meat procurement and sharing may be most important for the maintenance of cooperative social relationships between males (Nishida *et al.* 1992; Stanford *et al.* 1994b; Stanford 1998); we call this the 'male-bonding' hypothesis. Also, or alternatively, males may increase their mating success by sharing meat with estrous females (Teleki 1973; McGrew 1992; Stanford *et al.* 1994a, b; Stanford 1996, 1998); we call this the 'meat for sex' hypothesis. The male-bonding hypothesis predicts that the number of males present significantly influences decisions to hunt red colobus, that males share meat mostly with each other, that they show reciprocity in meat sharing, and that they share preferentially with allies and important grooming partners. The meat for sex hypothesis predicts that the presence of estrous females significantly influences decisions to hunt on encounter, that males share meat preferentially with estrous females, and that meat sharing increases male mating success (Mitani & Watts 2001).

(2) What factors influence hunting success? Note that these include ecological and demographic variables. Forest structure is again a candidate: areas with tall and continuous canopies give red colobus multiple escape routes, and hunts are presumably more difficult in such areas than in others where the canopy is shorter and broken (Boesch 1994b). Variation in hunting party size influences hunting success within communities (below), and variation in community size and composition should influence variation in hunting success across communities. Ngogo provides excellent oppor-

tunities for investigating the effects of hunting party size because the chimpanzee community there is unusually large.

(3) What leads to temporal variation in hunting frequency and predation intensity? If chimpanzees hunt to compensate for energy shortfalls, they should hunt most often when fruit, their main source of easily-assimilable energy, is in short supply (Mitani & Watts 2001; Watts & Mitani 2002). Hunting was more common during rainy seasons at Taï (Boesch 1994a), but at Mahale it occurred more often during dry seasons and at the start of subsequent rainy seasons (Takahata *et al.* 1984; Uehara 1997; Hosaka *et al.* 2001). Predation pressure on red colobus (kills per month) was higher during dry seasons in Stanford *et al.*'s (1994a) Gombe sample. Various researchers have speculated about the role of energetics in explaining this variation. Fruit abundance is usually high during rainy seasons at Taï (Boesch & Boesch-Achermann 2000; but see Boesch 1996), and hunting peaks roughly correspond to fruiting peaks at Mahale (Takahata *et al.* 1984; Uehara 1997). By implication, energy surpluses, not shortfalls, promote hunting at these sites, although Takahata *et al.* (1984) speculated that dry season protein shortfalls might have spurred hunting at Mahale. Stanford (1996) raised the possibility that dry season hunting compensated for energy shortfalls at Gombe, but also noted that average party size was higher during the dry season, contrary to expectations if dry season fruit availability was low. He suggested that seasonal variation in hunting depended more on seasonal variation in the number of cycling females (ibid.; Stanford *et al.* 1994a). However, quantitative data on food availability with which to evaluate these arguments have not been collected. Data from Kanyawara, in Kibale National Park, Uganda, show that fruit production by tree species important in the chimpanzees' diet varies unpredictably (Wrangham *et al.* 1996; Chapman *et al.* 1997); this cautions against use of rainfall as a universal proxy for fruit availability.

(4) To what extent do chimpanzees hunt cooperatively? No consensus exists on this issue, partly because cooperation can be defined in terms of outcomes or in terms of behavior during hunts, and partly because of variation among communities and populations. Packer & Ruttan (1988) proposed one outcome-based criterion: cooperation occurs when some measure of success – probability of at least one kill, total number of kills, total amount of meat captured – increases as the number of hunters increases. Results of some chimpanzee studies (Gombe: Packer & Ruttan 1988; Stanford 1996; Taï: Boesch 1994b) meet this criterion, but those of others do not (Gombe: Busse 1978; Boesch 1994b). A stricter outcome-based criterion is that meat available per individual increases with the number of hunters. Stanford (1996) concluded that despite increases in hunting success and in the number of kills per hunt with increases in hunting party size, hunting was not cooperative at Gombe because per capita meat intake did not also increase with party size. However, he later noted that the amount of meat available per capita was higher when seven or more males were present than when fewer than seven were present (Stanford 1998). Ideally, though, we should measure individual net energy gain from hunting. Hunting is cooperative when this increases with hunting group size, or is at least higher for individuals in certain-sized groups than for solitary hunters (Boesch 1994b; Creel & Creel 1995). Boesch (1994b) showed that per capita energy intake increased with hunting party size at Taï, but at Gombe this held for only four or fewer hunters. We can also define cooperation in social terms: cooperation occurs when individuals coordinate actions in time and space in ways that increase hunting success (Boesch & Boesch 1989). Behavioral coordination could contribute to any positive relationship between party size and hunting success or net energy gain. By this definition, cooperation is common at Taï, but rare at Gombe (ibid.; Boesch 1994b) and Mahale (Hosaka *et al.* 2001). Presumably the taller, more continuous canopy at Taï and the larger size of red colobus there means that Taï chimpanzees must coordinate their behavior more effectively to achieve levels of success comparable to those at Gombe and Mahale (Boesch 1994a; Stanford *et al.* 1994a).

STUDY SITE AND METHODS

The Ngogo study area covers about 35 km^2 of mixed mature and regenerating forest, transitional between lowland and montane evergreen forest and other minor vegetation types (Butynski 1990; Struhsaker 1997). We maintain a permanent research presence there in collaboration with J. Lwanga and four Ugandan field assistants, but behavioral data reported here come from observations made while one or both of us were engaged in fieldwork during 37 months

Table 18.1. *Frequency of predation episodes at Ngogo, by prey species*

Species	H	S	ME	CC	Total	Kills	Percentage of kills
Red colobus, *Procolobus badius*	98	82	6	4	108	317	87.8
Black and white colobus, *Colobus guereza*	17	10			17	17	4.7
Redtail monkeys, *Cercopithecus ascanius*	9	5	1	1	11	12	3.3
Mangabeys, *Lophocebus albigena*	3	1			3	1	0.3
Blue monkeys, *Cercopithecus mitis*	1	1			1	1	0.3
Baboon, *Papio cynocephalus*			1		1	2	0.6
Red duiker, *Cephalophus monticola*	2	2	5		7	7	1.9
Blue duiker, *Cephalophus callipyga*			2		2	2	0.6
Bushpig, *Potomochoerus porcus*	1	0	1		1	1	0.3
Guinea fowl, *Guttera pucherani*			1		1	1	0.3
Total	131	101	17	5	152	361	

Notes:
H, hunts; ME, meat eating; CC, carcass carrying; S, number of successful hunts. Total is the total number of hunts, meat eating episodes, and carcass carrying episodes. Kills are the total number of observed kills.

between 1995 and 2000. Field assistants collect phenology data monthly. Most chimpanzees in the Ngogo community, including all males, are well habituated, but its exact composition is still uncertain. However, it is the largest yet documented (Watts, 1998, Mitani & Watts 1999, 2001; Mitani *et al.* 1999, 2000; Pepper *et al.* 1999; Watts & Mitani 2001, 2002). It has had about 140–150 members, including 23–24 adult males and 15–16 adolescent males, during the periods considered here.

Our methods follow those in Mitani & Watts (1999, 2001) and in Watts & Mitani (2002). We collected data ad libitum on three kinds of predation episodes (Table 18.1). Occasionally we found chimpanzees eating meat just after a hunt (meat eating episodes) or found individuals carrying carcasses (carcass carrying episodes). Most episodes were hunts, during which chimpanzees rushed at prey (Table 18.1). Hunts were successful when the chimpanzees captured one or more prey. During hunts, we recorded the prey species, the time until the first kill, the number of kills, and the age–sex class of the prey. We identified all chimpanzees present, those who captured prey, and those who got meat. We collected data on meat sharing and theft ad libitum; sometimes we recorded all occurrences, but when the chimpanzees made multiple kills and dispersed widely to eat meat, we undoubtedly missed some sharing.

Boesch & Boesch (1989; cf. Boesch 1994b; Boesch & Boesch-Achermann 2000) rightly stress the importance of accurate data on the activities of individuals during hunts for analyses of whether hunting is cooperative. At most hunts at Ngogo, however, the large number of chimpanzees present, the wide area over which they spread, the swiftness of events during rushes at prey, and the density of the foliage prevented us from monitoring all individuals simultaneously. Our measures of hunting party size are thus not as accurate as those from Taï (ibid.), and we lack systematic data on the behavior of each potential hunter present. Nevertheless, we consider the number of adult males present to be a reasonable estimate of hunting party size, for several reasons. First, adult males made nearly all (311/336, or 93%) kills during group hunts of monkeys, including 93% (284/307) of red colobus kills. Adolescent males made most other kills (21/311, or 7%, and 19/307 red colobus kills, or 6%). Including them in counts of hunting party size changes none of the results of analyses below. Second, bystanders (sensu Boesch & Boesch 1989) could, and often did, switch to hunting during the course of a hunt. Finally, if success or per capita meat availability initially increases with group size, but then remains steady or declines, this could indicate that individuals cooperate in small groups, but become more likely to cheat as group size increases

(Packer & Ruttan 1988). In their analysis of hunting by African wild dogs (*Lycaon pictus*), Creel & Creel (1995) describe similar problems due to observational constraints, but justify their use of the number of adults per group as a measure of hunting group size by noting that bystanders can affect prey behavior and that when hunts ended, they often discovered that group members they had been unable to monitor continuously had made kills.

For successful hunts of red colobus, we estimated the biomass of prey harvested per hunt by multiplying the number of kills for a given age–sex class by the estimated body mass of a member of that class, and then summing across classes. Body masses were 11 kg for adult males, 6 kg for adult females and for subadults, 3 kg for juveniles, and 1 kg for infants (Struhsaker 1975; Struhsaker & Leland 1987; Stanford 1996).

During fieldwork in 1998–99, we noted all visual encounters between chimpanzees and red colobus. We assigned the forest structure at encounter sites to one of three categories, and did the same for hunts prior to 1998; this yielded 192 encounters, 82 of which led to hunts. Categories were: (1) primary forest with a tall, continuous canopy; (2) regenerating forest or swamp forest with a broken, mostly low, canopy; and (3) forest with a tall canopy, but within about 100 m of regenerating forest, open swamp forest, or grassland or bush with no tall trees. Forest in the second and third categories offered the monkeys fewer escape routes than tall primary forest. We used logistic regression to examine the effect of forest structure on decisions to hunt, represented by the categorical variable 'hunt/no hunt.' We did a similar analysis for hunts to examine the effects of forest structure on hunting success, represented by the categorical variable 'success/no success.' We pooled data from forest structure categories two and three in the analysis of hunting success because so few hunts in these areas were unsuccessful.

We performed similar logistic regressions on a subset of these data (164 encounters with 64 hunts) to examine the effect of male party size and that of the presence of estrous females on hunting decisions. The dependent variable was again 'hunt/no hunt,' while independent variables were the number of adult males present (a continuous variable) and the presence or absence of at least one estrous female (a dichotomous variable).

We used data on phenology, tree size, and tree densities of the top 20 fruit species in the chimpanzees diet, as measured in 263 plots, each 5 m × 50 m, to compute an index of ripe fruit availability ('ripe fruit score,' or RFS) each month. This was:

$$RFS = \sum_{i=1}^{20} p_i \cdot d_i \cdot s_i \qquad (18.1)$$

where p_i is the percentage of the ith tree species possessing ripe fruit; d_i is the density of the ith tree species (trees ha^{-1}), and s_i is the mean size of the ith tree species in cm dbh (diameter at breast height).

Phenology data were collected on the first 5 days of a month, and we centered months around the midpoint of each sample for analytical purposes, under the assumption that RFS values represented mean fruit availability during the resulting interval.

We measured the monthly frequency of hunts and of hunting patrols (see below) as the number of hunts or patrols per day per month, with months defined as in the phenology sample. We only included days on which we observed the chimpanzees for at least 6 hours, so that we could be reasonably sure we did not miss hunts (cf. Boesch & Boesch-Achermann 2000). We then regressed the number of red colobus hunts, the number of hunting patrols, and the number of kills per month on RFS values to examine the relationship between hunting and fruit availability. Data on hunting and phenology were available for 16 months in 1998–99.

We used Hemelrijk's (1990) MATSQUAR matrix permutation program to test for reciprocity of meat sharing, for interchange of meat for grooming, and for agonistic support between males. We give results of K_r tests for relative reciprocity and interchange. That is, we tested the hypotheses that positive correlations held between: (1) the rank order with which males shared meat with others and the rank order with which they received meat from others; (2) the rank order with which males shared meat with others and the rank order with which they were groomed by others; (3) the rank order with which males received meat from others and the rank order with which they groomed others; (4) the rank order with which males shared meat with others and the rank order with which they received agonistic support from others; and (5) the rank order with which males received meat from others and the rank order with which they gave agonistic support to others. MATSQUAR tests for reciprocity and interchange at the group level (ibid.); we plan to analyze reciprocity and interchange within dyads in more detail in the future. Grooming values come from data collected during focal samples of adult males in 1998–99 and were recorded as total grooming duration per dyad (cf. Watts 2000a,b). A 'coalition' occurred when two males jointly displayed at one or more other adult males, or when a male intervened in an ongoing agonistic interaction to support one of the opponents.

We used three methods to test the 'meat for sex hypothesis' (Mitani & Watts 2001). First, we used a re-sampling technique to test the null hypothesis that males simply shared meat with estrous and anestrous females in direct proportion to the number of females in each class present at hunts (Mitani & Watts 2001). To test the prediction that meat sharing led directly to mating, we used a binomial test to compare the observed proportion of times that males copulated with estrous females after sharing meat with them to an expected value of 50% (ibid.) We used a Wilcoxon matched-pairs test to evaluate the prediction that males obtained higher shares of matings with estrous females for cycles during which they shared meat with those females than for cycles during which no sharing occurred ($n = 15$ male–female dyads).

RESULTS

Prey species and pursuit frequencies

The total Ngogo sample is 131 predation episodes, mostly hunts, involving ten prey species (Table 18.1). Most hunts and most kills were of red colobus (Table 18.1). Black-and-white colobus monkeys (*Colobus guereza*) are the second-most frequently targeted primate prey, followed by redtails (*Cercopithecus ascanius*), mangabeys (*Lophocebus albigena*), and blue monkeys (*Cercopithecus mitis*; Table 18.1). We have probably underestimated the frequency of redtail and black-and-white colobus hunts because these are opportunistic and quick (Watts & Mitani 2002). Red duiker (*Cephalophus monticola*) are the most common non-primate prey. As also reported for bonobos (*Pan paniscus*), the chimpanzees opportunistically hunt both this species and blue duiker (*Cephalophus callipyga*).

Influences on hunting decisions

The chimpanzees were significantly less likely to hunt red colobus when they encountered them in primary forest, far from areas with broken canopy or no canopy, than when encounters took place in or near areas with broken and mostly low canopies ($\chi^2 = 38.32$, df$=2$, $p < 0.0001$; Watts & Mitani 2002). When they encountered red colobus in regenerating or swamp forest they hunted in 68% of the encounters (32/47). When they encountered red colobus in mature forest, but near regenerating or swamp forest or near grassland, they hunted in 63% (28/44) of those encounters. They hunted in primary forest in only 21.8% of encounters (22/101).

Data on hunting decisions support the male bonding hypothesis, but not the meat for sex hypothesis. Logistic regressions showed that both the number of males and the number of estrous females per party were significantly higher when the chimpanzees hunted than when they did not hunt (males: Wald statistic$=49.76$, 1 df, $r = 0.47$, $p < 0.001$; females: Wald statistic$=17.88$, 1 df, $r = 0.27$, $p < 0.001$; Mitani & Watts 2001). However, numbers of adult males and of estrous females were significantly correlated with each other, and only the number of males had a significant effect in a multiple regression that included both of these variables (Wald statistic$=42.69$, 1 df, partial $r = 0.47$, $p < 0.001$; ibid.).

Meat sharing decisions

Our results provide only limited support for the meat for sex hypothesis. Data on sharing between 22 pairs of males and estrous females and 56 pairs of males and anestrous females showed that males shared with estrous females more often than expected by chance ($p < 0.02$) and with anestrous females less often than expected ($p < 0.01$; Mitani & Watts 2001). However, estrous females were successful in only 45% (37/82) of their begging attempts, and males copulated with them after only 43% of sharing episodes, a proportion not significantly different from 50% (binomial $p > 0.40$). In comparison, anestrous females succeeded in 48% (66/139) of their begging attempts. Males did not achieve greater shares of matings with females during cycles when they shared meat than during cycles with no sharing (Wilcoxon $T^+ = 68.5$, $n = 15$, $p > 0.30$). Results of these updated analyses of mating behavior corroborate our earlier findings (Mitani & Watts 2001).

In contrast, we find strong support for the male bonding hypothesis (cf. Mitani & Watts 2001): Males shared meat reciprocally ($K_r = 1619$, $p < 0.001$; 10000 matrix permutations). The number of times that males shared meat with others was significantly correlated with the number of times that they received agonistic support from others ($K_r = 954$, $p < 0.001$; 10000 matrix permutations; Figure 18.1).

Correspondingly, the number of times that males re-

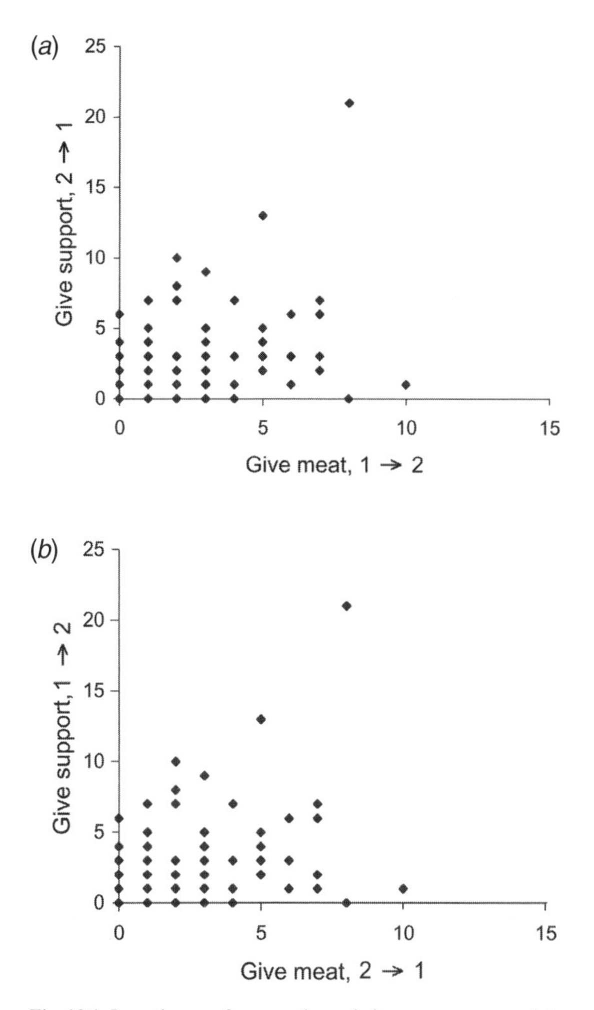

Fig. 18.1. Interchange of meat and agonistic support among adult males at Ngogo. (*a*) Meat given for support received, from the perspective of male 1: for a given pair (1, 2), the *x*-value is the number of times that male 1 gave meat to male 2 and the *y*-value is the number of times that male 2 gave coalitionary support to male 1. (*b*) Meat received for support given, from the perspective of male 1: for a given pair (1, 2), the *x*-value is the number of times that male 2 gave meat to male 1, and the *y*-value is the number of times that male 1 gave coalitionary support to male 2. Some points represent multiple dyads.

ceived meat from others was significantly correlated with the number of times they gave agonistic support to others ($K_r = 1082$, $p < 0.001$; 10000 matrix permutations). Significant positive correlations also held between the number of times males who possessed meat shared it with others and the amount of grooming that they received from others ($K_r = 1370$, $p < 0.001$; 10000 matrix permutations), and between meat received and grooming given ($K_r = 1235$, $p < 0.001$; 10000 matrix permutations).

Hunting success and kills per red colobus hunt

Eighty-four percent of red colobus hunts were successful (Table 18.2). This is a far higher proportion than at Taï, Mahale, and Gombe (Table 18.2). Success rates for hunts of black-and-white colobus and of redtails were lower, but still over 50% (Table 18.1), comparable to values documented elsewhere for red colobus hunts. We have seen too few hunts of other species to estimate success rates.

The chimpanzees made multiple kills in at least 84% (68/81) of red colobus hunts for which we were reasonably sure that we missed no kills (Table 18.2). The mean number of kills per hunt was 3.74 (SD = 2.33, range = 1–13). These values were also far higher in value than those from other sites (Table 18.2; cf. Mitani & Watts, 1999; Watts & Mitani 2002), even taking into account variation over time at those sites (e.g. Gombe: Stanford *et al.* 1994a).

Canopy structure influenced hunting success. Only 62% (16/26) of red colobus hunts in primary forest succeeded, while 92% (66/72) of those in or near regenerating forest, swamp forest, grassland, or bush succeeded ($\chi^2 = 11.39$, df = 1, $p < 0.01$). The five hunts with the most kills ($n = 13$, 10, 9, 9, and 8) started in areas with some tall trees, but these were in narrow strips on shallow slopes bounded by open swamp forest and grassland or low regenerating forest. The chimpanzees isolated the monkeys in one or a few tall trees or drove them up or downhill into small trees, where they easily captured many of them.

Male party size also seemed to influence whether the chimpanzees succeeded (see also below). On average, significantly more males were present for successful hunts ($\bar{x} = 15.1$, SD = 4.7) than for unsuccessful hunts ($\bar{x} = 10.5$, SD = 5.0; unpaired t-test, $t = 3.47$, df = 96, $p < 0.001$). Parties with five or fewer males hunted red colobus only five times and made captures in three of these hunts. The likelihood of success increased with the number of adult males, and all hunts at which 20 or more adults were present were successful (Watts & Mitani in press).

Hunting frequency, hunting patrols, and variation in fruit availability

The frequency of red colobus hunts and the intensity of predation increased as the amount of ripe fruit at Ngogo increased (Mitani & Watts 2001; Watts & Mitani 2002). The

Table 18.2. *Variation across research sites in hunting success and in multiple kill hunts for hunts of red colobus monkeys*

	Ngogo ($n=98$)	Gombe ($n=414$)	Taï ($n=95$)	Mahale ($n=285$)
Percentage of hunts successful	84	52	54	60
Percentage of hunts with multiple kills	84	30	25	23.1
Mean number of kills per successful hunt	3.74	1.62	1.15	1.4

Notes:

Sources: Gombe: Stanford et al. 1994a; Stanford 1998; Tai: Boesch & Boesch 1989; Mahale: Hosaka et al. 2001.

RFS was positively and significantly correlated with both the number of hunts per day ($F_{1,14}=9.22$, $r^2=0.38$, $p=0.008$) and the number of kills per month ($F_{1,14}=7.30$, $r^2=0.30$, $p=0.017$).

Chimpanzees at Ngogo often go on hunting 'patrols,' during which they move quietly and in single file, sometimes for hours, while deliberately searching the canopy for monkeys (Mitani & Watts 1999; Watts & Mitani 2002). In all, 48% (62/129) of the monkey hunts came during patrols. This included 55% (53/98) of the red colobus hunts. We have also documented seven patrols during which the chimpanzees did not encounter monkeys, and one during which they encountered, but did not hunt, a red colobus group. The number of patrols per observation day also increased significantly with the RFS ($F_{1,14}=11.07$, $r^2=0.40$, $p=0.005$; Watts & Mitani 2002).

The chimpanzees went on a 57-day binge from October until December, 1998, during which they killed 69 red colobus in 17 hunts (15 successful), one mangabey in two hunts, and one red duiker. They also unsuccessfully hunted black-and-white colobus twice. Fifteen of these hunts occurred during hunting patrols, and the chimpanzees made three other patrols during which they did not encounter monkeys. This binge coincided with a major crop of *Uvariopsis congensis* fruit. We saw 17 hunts of monkeys, 16 of them successful, and one carcass-carrying episode during a 2-month period in 2000 that coincided with another major *U. congensis* crop and a subsequent major crop of *Chrysophyllum albidum*. The chimpanzees killed 54 red colobus, six black-and-white colobus, two redtails, and two baboons during these hunts. Another *U. congensis* fruiting peak earlier in 1998 also saw a hunting binge (Mitani & Watts 1999). Binges in 1995 and 1996 had also coincided with major fruit crops (ibid.).

Cooperation

Both the number of kills per red colobus hunt and estimated prey offtake in kilograms increased in association with the number of adult males present (kills: $F_{1,95}=61.98$, $r^2=0.63$, $p<0.0001$; Figure 18.2; offtake: $F_{1,95}=41.24$, $r^2=0.30$, $n=97$ hunts, $p<0.0001$; Figure 18.3). The number of males who obtained meat increased with the number of males present ($F_{1,65}=36.85$, $r^2=0.36$, $n=67$ hunts, $p<0.001$) and with the estimated amount of meat captured ($F_{1,65}=77.75$, $r^2=0.55$, $n=67$, $p<0.001$; Watts & Mitani 2002). However, meat per capita was independent of the number of males present ($F_{1,95}=1.76$, $p=0.19$, $n=97$ hunts; Figure 18.4). That is, individual males did not get more meat in large hunting parties than in small ones, despite greater overall capture success in large parties. Without data on energy expended per hunt, we cannot estimate per capita net energy intake. However, the absence of a significant positive relationship between per capita meat acquisition and party size also suggests that net energy intake does not increase with the number of potential hunters. Our results suggest that males in large hunting parties commonly cheat. Cheating could take two forms. First, the probability that individuals refrain from hunting could increase with the number of potential hunters present – that is, males may fail to solve a collective action problem. As noted above, collecting accurate quantitative data with which to test this prediction is difficult. Second, the tendency of individuals to end pursuits, and instead, to beg for meat from those who have made captures may be particularly high in large parties, precisely because hunters often quickly make several captures soon after such hunts have begun.

Social cooperation sometimes occurs, but seems to be less common than at Taï. Members of hunting parties

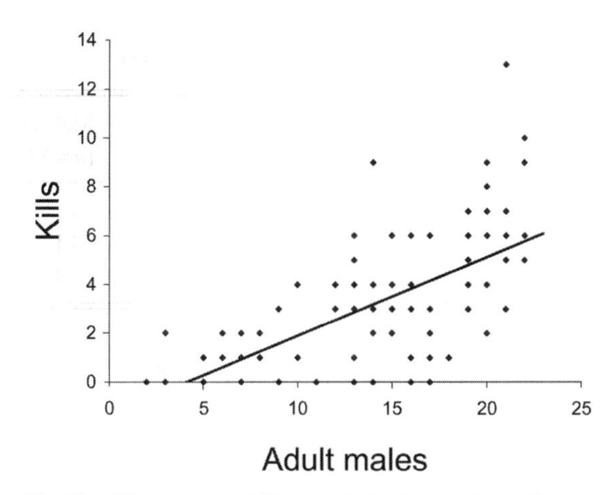

Fig. 18.2. The number of kills made during hunts of red colobus, in relation to the number of adult male chimpanzees present at hunts.

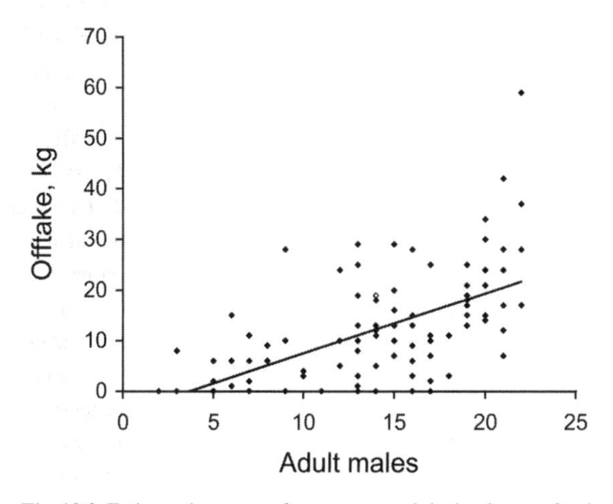

Fig. 18.3. Estimated amount of meat procured during hunts of red colobus, in relation to the number of adult male chimpanzees present at hunts.

spread out widely, both in trees and on the ground, during hunts. Males often position themselves in trees surrounding those where red colobus have taken refuge. During one hunt, for example, male PA climbed into a tree as a number of red colobus in a neighboring tree, under pressure from several other male chimpanzees, edged towards the periphery of its canopy. When monkeys started to leap into the tree where PA was stationed, he quickly caught a juvenile. Surrounding the monkeys also allows pursuit from several directions, sometimes of single prey, and facilitates capture

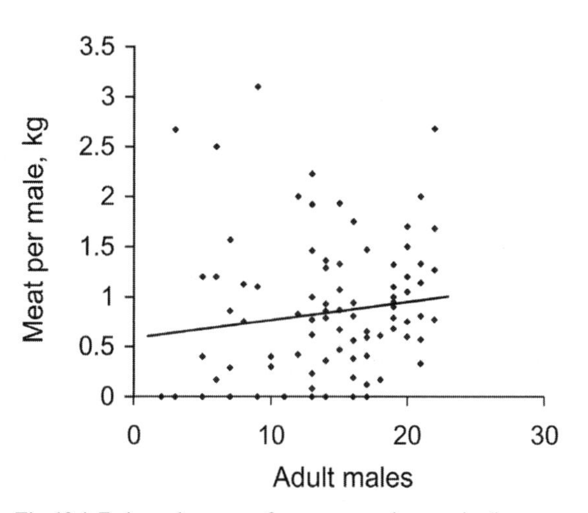

Fig. 18.4. Estimated amount of meat procured per capita (i.e. per adult male chimpanzee) during hunts of red colobus, in relation to the number of adult male chimpanzees present at hunts.

of fleeing monkeys. For example, male CO chased a subadult red colobus that fled along a large bough during one hunt; then leapt on to a smaller branch that came off of the bough. While CO lunged at the monkey, who tried to fend him off, male PI climbed into the tree from an adjoining one, ran along the bough from the opposite direction, and grabbed it. PI and CO then divided the carcass. In other cases, the chimpanzees' behavior also fits descriptions of synchrony or collaboration (Boesch & Boesch 1989) and, in general, individuals seem good at predicting the effects that others' behavior will have on the prey. However, they often just seem to be responding to these effects, rather than anticipating them.

Chimpanzees on the ground are sometimes bystanders, but sometimes also rush ahead when red colobus flee the chimpanzee pursuers operating in the canopy, and who climb to advantageous positions along the monkeys' route. We have seen 'bystanders' make captures by anticipating (Boesch & Boesch-Achermann 2000) prey movements in this way. For example, during a hunt in which the chimpanzees made ten kills, most of a red colobus group was trapped in two tall but isolated, partly-dead trees on the edge of an area of low, regenerating forest by several male chimpanzees positioned in neighboring smaller trees. Male RU stayed on the ground monitoring the monkeys until, after about 15 minutes, they panicked and tried to flee. RU then quickly ran to a tree that was in the path of an isolated, fleeing adult female, ascended as she entered it, and caught her easily.

Chimpanzees already in the canopy and others who had been on the ground quickly caught six more monkeys. Chimpanzees in the canopy sometimes also cause monkeys to fall by hitting them, pulling their tails, or shaking branches to which they are clinging. In one hunt, for example, males on the ground caught two adult male red colobus that others in the canopy had knocked to the ground. We have seen 21 red colobus kills on the ground, or about 7% of all captures, including at least 16 of 21 adult males in the prey sample. Seeking terrestrial positions that may afford opportunities for such captures is a common hunting tactic. Thus, no clear division between bystanders and hunters exists in many hunts at Ngogo.

Captures on the ground, by bystanders who switch to hunting, are also common at Mahale (Hosaka *et al.* 2001). Hosaka *et al.* (2001) call this the 'wait for monkey to fall' tactic. Successful use of this tactic depends on close monitoring of the action in the canopy, as well as on the behavior of the other chimpanzees stationed there. We have seen three adult male red colobus escape back into the trees after being knocked or thrown to the ground at a spot where no adult male chimpanzee was close enough to attack them. We cannot yet assess the extent to which different individuals routinely pursue different roles, as can happen during cooperative group hunts by lions (Stander 1991), although certain males seem especially likely to use the 'wait for monkey to fall' tactic at Ngogo. However, males who make ground captures may refuse to share meat with those males who sent the monkeys to the ground, unlike in the lion case. Our impression is that switches from bystander to arboreal or terrestrial pursuer are mostly opportunistic responses to changes in prey vulnerability, rather than systematic coordination of hunting tactics. Hosaka *et al.* (2001) similarly conclude that systematic coordination is uncommon at Mahale.

DISCUSSION

Chimpanzees at Ngogo are most likely to undertake group hunts of red colobus when many males encounter the monkeys in areas where the forest canopy is low and/or broken, and at times when ripe fruit is abundant in their range. The likelihood that the chimpanzees kill some monkeys increases with party size. Packer & Ruttan (1988) obtained a similar result using some data from Gombe, and argued that this showed that chimpanzees there hunted cooperatively. The number of kills and the amount of meat available also increases significantly with the number of males present at Ngogo. However, these results do not necessarily mean that individuals do better by hunting in large parties than in smaller ones (Boesch 1994b; Creel & Creel 1995). Although males at Ngogo show some synchrony and collaboration, the absence of a significant relationship between per capita meat availability and the number of males present implies that hunting is generally not cooperative. Males share meat reciprocally, and especially share meat with other males with whom they cooperate in within-community competition and with whom they often groom. They do not routinely use meat as a tool to increase their mating opportunities. Multiple factors, starting with nutritional ones, probably help to explain why chimpanzees hunt and why they share meat (Stanford 1998), but our data support the hypothesis that hunting and meat sharing are important contributions to social bonding between males, and also help maintain cooperation in other contexts.

Quantitative data from other sites would probably show similar influences of variation in canopy structure on hunting decisions and hunting success. The overall success rate for red colobus hunts at Ngogo is much higher than at Taï, but this is at least partly because most Ngogo hunts occur in areas where the canopy is broken. The proportion of hunts in primary forest that succeed is similar to the overall proportion of successful hunts at Taï, where the canopy is uniformly tall and continuous, even though hunting party sizes are typically larger at Ngogo. This supports the argument that behavioral cooperation, especially coordination, can be crucial for consistent success in tall primary forest habitats (Boesch & Boesch 1989; Boesch & Boesch-Achermann 2000). Nevertheless, when we restrict the analysis to hunts in areas with broken canopies, we still find significant positive relationships of hunting party size to the number of kills per hunt and to total meat availability. This supports the argument that demographic variation – especially variation in the number of males per community – contributes to variation in hunting success across chimpanzee communities (Mitani & Watts 1999). Success also increases with the number of males per hunting party at Taï (Boesch & Boesch 1989; Boesch & Boesch-Achermann 2000) and Gombe (Stanford 1996). One reason for the effect of the number of males is that when enough male chimpanzees pursue a red colobus group, they can overwhelm the defenses of the male red colobus, as noted by Stanford (1996). The greater ease with which males at Ngogo overwhelm the monkeys' defenses accounts for some of the enormous difference in degrees of success between Ngogo and elsewhere. How much, and how generally, chimpanzees in lowland primary forest can compensate somewhat for the difficulties of hunting in tall canopy by cooperating more than we see at Ngogo, are still open questions.

Ripe fruit production is independent of rainfall at Ngogo, but positively correlated with party size (Mitani & Watts 2001; Mitani *et al.*, Chapter 7). Also, the frequency of hunts and hunting patrols per month and the number of kills per month are all positively related to the abundance of ripe fruit. Together, these results indicate that the chimpanzees hunt more when they can easily meet their energetic needs from fruit. This is contrary to the prediction of the energy shortfall hypothesis. However, it makes sense given that: hunting is energetically expensive (Boesch 1994b); hunting patrols impose foraging opportunity costs and energy costs; hunts are not necessarily successful; and not all members of successful hunting parties receive appreciable amounts of meat (Mitani & Watts 2001; Watts & Mitani 2002). Large parties contain many males and are highly likely to succeed when they hunt. Members of large parties thus have some chance to accrue net energy gains by hunting, but most hunts occur at times when they can afford net energy losses. High fruit abundance and good prospects for hunting success give males opportunities to reinforce bonds by sharing meat, and resulting social gains may outweigh any net energy costs. Our results do not preclude the possibility that other nutritional factors motivate hunting. However, together with data on hunting decisions and meat sharing, they indicate that male social strategies strongly influence hunting behavior, particularly at times of energy surpluses, not shortfalls. Qualitative observations that hunting was more common during fruiting peaks at other sites (e.g. Mahale: Uehara 1997; Hosaka *et al.* 2001) suggest that the same relationship of hunting frequency to energy availability holds widely, although we need comparable tests using quantitative data on food availability.

Gombe may be an exception. Long-term body mass data show that the dry season is a time of energy stress (Williams *et al.*, Chapter 14). Yet predation intensity is highest during dry season months (Stanford *et al.* 1994a; Stanford 1996, 1998): hunts are not more common, but success rates and the percentage of multiple kills are higher than at other times because average hunting party size is higher. Still, whether these data support the energy shortfall hypothesis depends on the net energy returns from hunting. Support seems unlikely given Boesch's (1994b) estimates of these returns and Stanford's (1996) estimates that per capita meat availability does not increase significantly in relation to hunting party size.

Our analyses of reciprocity in meat sharing and of interchange of meat for grooming and for agonistic support show that these occur at the group level (Hemelrijk, 1990). Males who possess meat share it relatively often with others from whom they receive meat relatively often, and they rarely share meat with those from whom they rarely receive it. They obtain meat relatively often from others whom they groom relatively often and to whom they give agonistic support relatively often; they give meat relatively often to those who often groom them and who often give them agonistic support. Meat sharing is relatively uncommon in dyads that groom little and give each other little or no support. These results all support the hypothesis that males use meat as an important commodity in negotiating and maintaining cooperative social relationships. Nishida *et al.* (1992) proposed that alpha males, in particular, can share meat selectively with other males to gain coalitionary support. Based on a small sample of meat sharing episodes at Mahale, Nishida & Hosaka (1996) later qualified this argument by noting that alpha males may share meat as part of a strategy to maintain alliances with loyal partners, but may not share with potential allies who are 'untrustworthy.' Stanford (1998) also notes the importance of meat sharing as a male political tool, and stresses that political considerations are one of multiple factors that can influence both hunting decisions and meat distribution patterns.

Also at Ngogo, the frequency with which males hunt together is correlated with the frequency with which they go on boundary patrols together. In turn, the frequency of joint patrolling is also positively correlated with the amount of grooming between males and the frequency with which they form coalitions (Watts & Mitani 2001). Hunting success varies considerably among males, as at other sites (e.g. Gombe: Stanford *et al.* 1994b, Stanford 1998), and good hunters go on boundary patrols relatively often (Watts & Mitani 2001). Hunts may give males opportunities to demonstrate their willingness to take risks in general; combined with information on others' willingness to provide agonistic support and their investment in grooming, this could allow individuals to assess how reliable others will be in the face of risks associated with patrolling (ibid.).

We have not yet analyzed within-dyad reciprocity, but have the impression that sharing is not balanced within all dyads. In part, this is because some sharing seems to take place under pressure from males who outrank meat possessors. Also, while meat theft is uncommon among adult males (Mitani & Watts 1999), most cases involve high-ranking males who take whole or partial carcasses from males whom they outrank. We have also seen coalitions of two males steal meat from single males who outrank them, however.

Boesch (1994b) used data on individual variation in prey pursuit at Taï to investigate another hypothesis concerning meat sharing: that successful hunters share meat mostly with others who also hunted, rather than with bystanders or with latecomers who arrived after hunts had ended. He found that all males who hunted were also sometimes bystanders or latecomers, but that they obtained the most meat when they hunted, and concluded that whether males pursued prey was the most important factor that influenced access to meat. Males who had meat presumably discriminated against non-hunters in distributing it (although this did not apply to adult females, with whom males shared meat non-reciprocally), which should make cooperation among males more stable (ibid.). Successful hunters at Ngogo and others who get large shares of their kills do not necessarily discriminate against bystanders, including adult male bystanders. The case of MZ, the oldest male in our sample, illustrates this clearly: he made only one known kill and did not pursue monkeys during most hunts for which he was present, but he received meat from 11 other males, more than all but one other male in the community.

We cannot test this 'meat for hunting' hypothesis at Ngogo without more detailed quantitative data on the behavior of individuals during hunts. We also need such data to determine the extent to which males share freely versus under pressure, the extent to which reciprocity is balanced within dyads, and to address questions about variation in cooperation across populations. Clearly such analyses are important if we are to have a better understanding of the social importance of hunting and meat sharing.

REFERENCES

Basabose, K. & Yamagiwa, J. (1997). Predation on mammals by chimpanzees in the montane forest of Kahuzi, Zaire. *Primates*, 38, 45–55.

Boesch, C. & Boesch, H. (1989). Hunting behavior of wild chimpanzees in the Taï National Park. *American Journal of Physical Anthropology*, 78, 547–73.

Boesch, C. (1994a). Chimpanzees and red colobus monkeys: a predator prey system. *Animal Behaviour*, 47, 1135–48.

Boesch, C. (1994b). Cooperative hunting in wild chimpanzees. *Animal Behavior*, 48, 653–67.

Boesch, C. (1996). Social grouping in Tai chimpanzees. In *Great Ape Societies*, ed. W. C. McGrew, L. F. Marchant & T. Nishida, pp. 101–13. Cambridge: Cambridge University Press.

Boesch, C. & Boesch-Achermann, H. (2000). *The Chimpanzees of the Taï Forest*. Oxford: Oxford University Press.

Busse, C. (1978). Do chimpanzees hunt cooperatively? *American Naturalist*, 112, 767–70.

Butynski, T. (1990). Comparative ecology of blue monkeys (*Cercopithecus mitis*) in high- and low-density subpopulations. *Ecological Monograph*, 60, 1–26.

Chapman, C. A., Chapman, L. J., Wrangham, R., Isibirye-Basuta, G. & Ben-David, K. (1997). Spatial and temporal variability in the structure of a tropical forest. *African Journal of Ecology*, 35, 287–302.

Creel, S. & Creel, N. M. (1995). Communal hunting and pack size in African wild dogs. *Animal Behaviour*, 50, 1325–39.

Goodall, J. (1986). *The Chimpanzees of Gombe*. Cambridge, MA: Belknap Press of Harvard University Press.

Hemelrijk, C. K. (1990). Models of and tests for reciprocity, unidirectionality and other social interaction patterns at a group level. *Animal Behaviour*, 39, 1013–29.

Hosaka, K., Nishida, T., Hamai, M., Matsumoto-Oda, A. & Uehara, S. (2001). Predation of mammals by the chimpanzees of the Mahale Mountains, Tanzania. In *All Apes Great and Small*, ed. B. M. F. Galdikas, N. E. Briggs, L. Sheeran & G. Shapiro, pp. 105–28. New York: Kluwer.

Ihobe, H. & Uehara, S. (1999). A preliminary report on the impact of chimpanzee hunting on mammal populations at Mahale, Tanzania. *Primate Research*, 15, 163–9.

McGrew, W. C. 1992. *Chimpanzee Material Culture*. Cambridge: Cambridge University Press.

Mitani, J. & Watts, D. P. (1999). Demographic influences on the hunting behavior of chimpanzees. *American Journal of Physical Anthropology*, 109, 439–54.

Mitani, J. C. & Watts, D. P. (2001). Why do chimpanzees hunt and share meat? *Animal Behavior*, 61, 915–24.

Mitani, J. C., Hunley, K. L. & Murdoch, M. E. (1999). Geographic variation in the calls of wild chimpanzees: a reassessment. *American Journal of Primatology*, 47, 133–51.

Mitani, J. C., Merriwether, D. A. & Zhang, C. (2000). Male affiliation, cooperation, and kinship in wild chimpanzees. *Animal Behaviour*, 59, 885–93.

Nishida, T. & Hosaka, K. (1996). Coalition strategies among adult male chimpanzees of the Mahale Mountains, Tanzania. In: *Great Ape Societies*, ed. W. C. McGrew, L. F. Marchant & T. Nishida, pp. 114–34. Cambridge: Cambridge University Press.

Nishida, T., Hasegawa, T., Hayaki, H., Takahata, Y. & Uehara, S. (1992). Meat sharing as a coalition strategy by an alpha male chimpanzee? In *Topics in Primatology, Vol. 1, Human Origins*, ed. T. Nishida, W. C. McGrew, P. Marler, M. Pickford & F. B. M. de Waal, pp. 159–74. Tokyo: University of Tokyo Press.

Packer, C. & Ruttan, L. (1988). The evolution of cooperative hunting. *American Naturalist*, 132, 159–98.

Pepper, J., Mitani, J. C. & Watts, D. P. (1999). General gregariousness and specific partner preference among wild chimpanzees. *International Journal of Primatology*, 20, 613–32.

Stander, P. E. (1991). Cooperative hunting in lions: the role of the individual. *Behavioral Ecology and Sociobiology*, 29, 445–54.

Stanford, C. B. (1996). The hunting ecology of wild chimpanzees: implications for the evolutionary ecology of Pliocene hominoids. *American Anthropologist*, 98, 96–113.

Stanford, C. B. (1998). *Chimpanzee and Red Colobus: The Ecology of Predator and Prey*. Cambridge, MA: Harvard University Press.

Stanford, C. B., Wallis, J., Matama, H. & Goodall, J. (1994a). Patterns of predation by chimpanzees on red colobus monkeys in Gombe National Park, 1982–1991. *American Journal of Physical Anthropology*, 94, 213–28.

Stanford, C. B., Wallis, J., Mpongo, E. & Goodall, J. (1994b). Hunting decisions in wild chimpanzees. *Behaviour*, 131, 1–20.

Struhsaker, T. T. (1975). *The Red Colobus Monkey*. Chicago: University of Chicago Press.

Struhsaker, T. T. & Leland, L. (1987). Colobines: infanticide by adult males. In *Primate Societies*, ed. B. B. Smuts, D. L. Cheney, R. M. Seyfarth, R. W. Wrangham & T. T. Struhsaker, pp. 83–97. Chicago: University of Chicago Press.

Struhsaker, T. T. (1997). *The Ecology of an African Rainforest.* Gainesville: University Presses of Florida.

Takahata, Y., Hasagawa, T. & Nishida, T. (1984). Chimpanzee predation in the Mahale Mountains from August 1979 to May 1982. *International Journal of Primatology*, 5, 213–33.

Teleki, G. (1973). *The Predatory Behavior of Wild Chimpanzees.* Lewisburg, PA: Bucknell University Press.

Tutin, C. E. G., Fernandez, M., Rogers, M. E., Williamson, E. A. & McGrew, W. C. (1991). Foraging profiles of sympatric lowland gorillas and chimpanzees in the Lope Reserve, Gabon. In *Foraging Strategies and Natural Diet of Monkeys, Apes, and Humans*, ed. A. Whiten & E. M. Widdowson, pp. 19–26. Oxford: Oxford University Press.

Uehara, S. (1997). Predation on mammals by the chimpanzee (*Pan troglodytes*). *Primates*, 38, 198–214.

Watts, D. P. (1998). Coalitionary mate guarding by male chimpanzees at Ngogo, Kibale National Park, Uganda. *Behavioral Ecology and Sociobiology*, 44, 43–55.

Watts, D. P. (2000a). Grooming between male chimpanzees at Ngogo, Kibale National Park, Uganda. I. Partner number and diversity and reciprocity. *International Journal of Primatology*, 21, 189–210.

Watts, D. P. (2000b). Grooming between male chimpanzees at Ngogo, Kibale National Park, Uganda. II. Male rank and priority of access to partners. *International Journal of Primatology*, 21, 210–38.

Watts, D. P. & Mitani, J. C. (2000). Infanticide and cannibalism by male chimpanzees at Ngogo, Kibale National Park, Uganda. *Primates*, 41, 357–64.

Watts, D. P. & Mitani, J. C. (2001). Boundary patrols and intergroup encounters in wild chimpanzees. *Behaviour*, 138, 299–327.

Watts, D. P. & Mitani, J. C. (2002). Hunting behavior of chimpanzees at Ngogo, Kibale National Park, Uganda. *International Journal of Primatology*, 23, 1–28.

Wrangham, R. W. (1975). The behavioral ecology of chimpanzees in Gombe National Park, Tanzania. PhD thesis, Cambridge University.

Wrangham, R. W. & Bergmann-Riss, E. L. (1990). Rates of predation on mammals by Gombe chimpanzees, 1972–1975. *Primates*, 38, 157–70.

Wrangham, R. W., Chapman, C. A., Clark-Arcadi, A. P. & Isabirye-Basuta, G. (1996). Social ecology of Kanyawara chimpanzees: implications for understanding the costs of great ape groups. In *Great Ape Societies*, ed. W. C. McGrew, L. F. Marchant & T. Nishida, pp. 45–77. Cambridge: Cambridge University Press.

Part V
Genetic diversity

19 • The evolutionary genetics and molecular ecology of chimpanzees and bonobos

BRENDA J. BRADLEY & LINDA VIGILANT

INTRODUCTION

In the past few decades, developments in molecular biology and genetics have contributed a new dimension to the study of evolutionary systematics and socioecology. This has led to the creation of several new fields of research, including 'molecular systematics' and 'molecular ecology.' The study of molecular systematics applies methods of genetic analysis to such problems as: examining taxonomic relationships on a molecular level; identifying molecular phylogenies among taxa; and estimating the time of these taxa's most recent common ancestor. Molecular data can also be compared with morphological and behavioral data to gain a more comprehensive understanding of evolutionary genetics and phylogenetic relationships. The relatively new field of molecular ecology utilizes methods of DNA analysis to address questions about behavioral ecology, evolution, and conservation through direct measures of relatedness and genetic variability. Genetic analysis is now commonly used to describe social structure and dispersal patterns, to verify mating systems, and to identify and census individuals in a population (Sunnucks 2000). Similarly, behavioral ecologists can better understand aspects of social dynamics, such as the evolution of altruism through kin selection (Hamilton 1964), by combining direct observational data from the field with DNA analysis of relatedness in the laboratory.

The earliest molecular studies of chimpanzees sought to understand the degree of similarity and difference among humans and apes. Goodman (1962) examined the immunological properties of the albumin protein in apes and found that chimpanzees, gorillas, and humans showed a strong degree of similarity to the exclusion of orangutans and gibbons. The results indicated that chimpanzees and gorillas are as similar to humans as they are to each other, which was considered quite surprising at the time, given the morphological similarities between chimpanzees and gorillas, and, more importantly, the perceived 'uniqueness' of humans. The next landmark in the development of molecular systematics was Sarich & Wilson's (1967) use of the molecular clock to assign a date to the divergence of chimpanzees, gorillas, and humans. Also using immunological properties to measure differences in protein structure between the three species, they estimated a most recent common ancestor at 5 million years ago. However, the power of resolution was insufficient to determine which two species among the three share a more recent common ancestor, and the 'trichotomy' at that point remained unresolved. Over the next 10 years, further comparisons were made of numerous other proteins, and in 1975 King & Wilson provided a comprehensive review of protein and DNA hybridization studies that revealed that the average human protein (polypeptide) is more than 99% identical to its chimpanzee homologue. This was particularly remarkable in that this degree of genetic similarity is equivalent to that observed between sibling species of fruit flies or rodents, which are virtually identical in their morphology. King and Wilson concluded by postulating that the considerable phenotypic differences between chimpanzees and humans are potentially the result of small changes in the regulatory systems controlling gene expression. Although the phenotypic effects resulting from differential gene expression have been rather well studied in *Drosophila* species and *Caenorhabditis elegans* (e.g. Matova & Cooley 2001), this topic is only now, with the advent of the complete human genome sequence and microarray technology, amenable to investigation in primates (Normile 2001b; Enard *et al.*, in press).

In the 1980s and 1990s, restriction analysis and, later, direct sequencing of mitochondrial DNA (mtDNA) segments (see below) became the preferred tool for looking at genetic distances and phylogenies within and among primate taxa (Brown *et al.* 1982; Cann *et al.* 1987; Vigilant *et al.* 1989; Kocher & Wilson 1991). Ruvulo *et al.* (1991, 1994) provided the first clear resolution of the human–chimp–gorilla trichotomy, using sequences from the mitochondrial cytochrome oxidase subunit II gene from several humans, chimpanzees, bonobos, orangutans, and siamangs. Phylogenetic analyses of these sequences, using both parsimony and distance methods, supported a scenario in which humans and chimpanzees share a more recent common ancestor than do chimpanzees and gorillas. Although a few researchers examining other loci found conflicting results at the time (e.g. nuclear protamine sequences supported a *Pan–Gorilla* clade, Retief & Dixon 1993), the *Homo–Pan* clade has been further supported by the vast majority of studies, and likelihood-based statistical tests favor the *Homo–Pan* clade with a high level of significance (Ruvolo 1997; Satta *et al.* 2000).

Early research on ape molecular systematics examined relatively few samples from captive individuals, typically of unknown origin. Although broad molecular studies of systematics and social structure in wild populations of the genus *Pan* only began in the mid-1990s (Sugiyama *et al.* 1993; Morin *et al.* 1994a), numerous independent studies covering a wide geographic sampling distribution have already been conducted (*P. troglodytes*: Sugiyama *et al.* 1993; Morin *et al.* 1994a,b; Deinard 1997; Goldberg & Ruvolo 1997; Gagneux *et al.* 1999; Gonder 2000; Constable *et al.* 2001; *P. paniscus*: Hashimoto *et al.* 1996; Gerloff *et al.* 1999; Reinartz *et al.* 2000). The combined results of this research now provide an interesting opportunity to evaluate the genetic diversity of chimpanzees and bonobos at numerous levels (species, subspecies, population, community) using a variety of genetic markers, both mitochondrial and nuclear. This chapter reviews our current understanding of chimpanzee and bonobo genetics in terms of phylogeny, systematics, and socio-ecology, and includes a discussion of the general methods used in molecular analysis. In addition, this chapter points out the current limitations on the types of information molecular studies can provide and describes potential areas for future research on *Pan* genetic diversity.

METHODS OF GENETIC ANALYSIS

The genetic similarity between humans and apes has allowed investigators to apply to molecular studies of *Pan* many of the methods and markers that were first developed to study human populations. A general overview of these methods will be briefly given, followed by a review of the results from molecular research, employing these methods in order to examine the systematics and molecular ecology of chimpanzees and bonobos.

PCR and noninvasive sampling

The polymerase chain reaction (PCR), first described in the late 1980s, is an enzymatic method of replicating a targeted region of DNA in vitro, thus producing several million copies of the specified fragment (Saiki *et al.* 1985; Mullis & Faloona 1987). Conventional PCR requires the use of sequence-specific oligonucleotide primers (short synthetic fragments of DNA), usually about 20 base pairs in length, to anneal with complementary sequences flanking each side of the target DNA segment and to provide the starting points for DNA synthesis. Therefore, design of primers requires some prior knowledge of the sequence segment to be amplified but, because of the high level of genetic similarity, primers based on human sequences can often be utilized when amplifying ape DNA.

Because PCR does not require large amounts of starting template DNA, it has facilitated the genetic analysis of the small amounts of DNA obtained from noninvasive samples, such as hair and feces (Woodruff 1993). This has led to new opportunities to study wild populations from which obtaining blood or tissue samples is rarely feasible, as in the case of apes. Several of the early genetic studies using noninvasive samples from wild individuals were studies of chimpanzees and bonobos, including successful extraction and amplification of DNA from chimpanzee hair (Sugiyama *et al.* 1993; Morin *et al.* 1994a), bonobo feces (Gerloff *et al.* 1995), and food wadges (Hashimoto *et al.* 1996). Although the DNA extracted from these samples is invariably in extremely small quantities (picogram range), it can typically be amplified by PCR and the PCR products subsequently used for applications such as DNA sequencing or genotyping. In addition, noninvasive sampling can be done in combination with behavioral data collection or, in cases where the target animals are not habituated, samples can often be collected from abandoned nesting or sleeping sites without the necessity of contacting the individuals.

Although the low quantity of DNA extracted from noninvasive samples can typically be amplified by PCR, the use of such samples in genetic studies remains limited, particularly when nuclear DNA rather than mitochondrial DNA (see below) is of interest. Similar to DNA extracted from ancient samples, the molecules of DNA recovered from hair and feces samples are often degraded into small pieces less than 300–600 base pairs long (Hofreiter *et al.* 2001). Therefore, although PCR has enabled the use of noninvasive samples for some types of genetic analysis, studies that require high quality, high molecular weight DNA (such as sequencing several kilobases of DNA; Kaessmann *et al.* 2001) still must rely upon blood, tissue, or cell line samples.

Genetic studies of wild chimpanzees and bonobos have used a wide variety of markers to achieve various goals.

These genetic markers can be broadly classified into three types: mitochondrial DNA, autosomal or X-chromosome nuclear DNA, and Y-chromosome DNA (Table 19.1).

Mitochondrial DNA

Mitochondrial (mt) DNA studies have largely dominated evolutionary genetics due to several unique characteristics of mtDNA: (1) strictly maternal inheritance; (2) no recombination (recent claims to the contrary (Eyre-Walker *et al.* 1999) have been refuted (e.g. Stoneking 2000)); (3) high variability relative to nuclear DNA; and (4) high copy number in most cells.

Since mtDNA is inherited only from the mother and there is no recombination to cause variation from one generation to the next, an evaluation of mtDNA sequence types (referred to as haplotypes) in a population can potentially allow inferences of maternal relationships, genetic diversity, and estimates of female dispersal rates to be made. Furthermore, by comparing mtDNA sequences, one can reconstruct phylogenetic relationships among various taxonomic levels (Brown *et al.* 1979; Kocher & Wilson 1991; Vigilant *et al.* 1991). The majority of studies on molecular phylogenies of mtDNA involve sequence analysis of the rapidly evolving, first hypervariable segment of the mtDNA control region (referred to as Hypervariable Region 1 or HVR1; sometimes referred to as D-loop). Since HVR1 is noncoding, mutational differences are thought to be less affected by selection than other, coding regions. Therefore, one can examine and compare polymorphisms between multiple species or subspecies and determine the genetic distance between taxa. The inferred mutation rate can then be used to calculate the time since the divergence, or the number of generations to the most recent common ancestor. However, it is clear that mutations are not randomly distributed even across the noncoding control region, and that estimates of mutation rates and times of divergence should take into account this rate of heterogeneity (Meyer *et al.* 1999).

Compared to nuclear DNA, mtDNA shows more variation across the same taxonomic levels due to its relatively high mutation rate, which is five to ten times the mutation rate of nuclear DNA (Brown *et al.* 1979). Therefore, even using relatively short segments of DNA, one can observe enough sequence variation (nucleotide differences) for comparative analysis. In addition, the mitochondrial genome is present in many more copies (100s-1000s of copies per cell, Robin & Wong 1988) than the nuclear genome (only two copies/cell for autosomes, one from each parent; only one copy each of X and Y in males). As a result, the quantity of mtDNA that can be obtained from genetic samples – especially noninvasive samples that contain few cells to begin with – is typically much higher than that from nuclear DNA. It is also worth clarifying that although the copy number of mitochondrial DNA is much higher than nuclear DNA in a typical cell, the mitochondrial genome is haploid and, therefore, its effective population size is one-fourth that of nuclear autosomes. This fact results in mtDNA having a higher probability of accurately reflecting the true species tree in some cases (see modeling and review in Moore 1995).

Although mtDNA analysis provides important insights into phylogenetic relationships and maternal relationships, it is, nonetheless, information about the evolution of a single, maternally inherited, genetic locus. The mitochondrial genome is a small fraction, only 0.00006%, of the total human genome (Stoneking 1994), and many questions of interest to molecular ecology and systematics, such as paternal relatedness, cannot be answered by analysis of mtDNA alone.

Nuclear DNA (sequencing, microsatellite analysis)

Each nuclear DNA segment in the genome has its own history and studies of nuclear DNA sequences have lagged behind those of mtDNA for several reasons. First, since much of the nuclear genome undergoes recombination, tracing DNA lineages from generation to generation is typically not possible with nuclear DNA (except for low-recombination regions, see below). Second, the slower mutation rate and relatively low level of sequence variability means that much longer stretches of DNA (on the order of kilobases), or numerous smaller loci, must be sequenced before one can observe enough informative differences among taxa or individuals for comparison. As mentioned above, since noninvasive samples only provide DNA in small fragments, they cannot be used in large sequencing studies and therefore, blood, tissue, or cell line samples are required. Once enough data are obtained, preferably from noncoding areas of the genome with low recombination, phylogenetic comparisons can be made as described for mtDNA above.

Microsatellite analysis differs from nuclear sequencing in that allelic genotypes, rather than nucleotide sequences, are determined. Microsatellites (Tautz 1989; Weber & May 1989), also known as short tandem repeats (STRs) or simple sequence repeats (SSRs), are polymorphic nuclear loci containing units (one to six bases in length) of nucleotide

Table 19.1. *Characteristics of molecular markers utilized in genetic studies of* Pan

Markers	Methods of analysis	Benefits	Limitations	Studies in *Pan*
mtDNA	Sequencing (typically ~350 bp HVR1)	More abundant than nuclear DNA Relatively high mutation rate; more variability than nuclear DNA Maternally inherited; no recombination High copy number; can use noninvasive samples	Only a single locus Strictly maternally inherited Typically not useful for individual identification	Community-level: 1,3,5,7,15 Species-level: 1,6,8,10,14,19
Nuclear DNA (autosomal or X-chromosome)	Large scale sequencing	Mutations likely to represent unique events in evolutionary time Possible to evaluate multiple loci independently Once variants are identified, SNP typing reduces effort of population screening	Recombination can complicate analysis Requires high quality DNA from blood or tissue Not useful for individual identification	Species-level: 4,12,13
	Microsatellite genotyping	Highly variable; allowing for relatedness determination and identification of individuals Typically <350bp in size; can use noninvasive samples	Limited information on higher-order relationships; shared alleles may not be due to identity by descent	Community-level: 2,5,9,11,18,21,22 Species-level: 14,16
Y-chromosome	Large scale sequencing	Paternally inherited; little recombination Once variants are identified, SNP typing reduces effort of population screening	Requires high quality DNA from blood or tissue Large segments need to be examined due to low level of variability	Species-level: 20
	Microsatellite genotyping	Typically <350bp in size; can use noninvasive samples	Information limited to males/paternal lineages Low variability within populations may limit analysis Limited information on higher-order relationships; shared alleles may not be due to identity by descent	Species-level: 17

Notes:
(1) Morin *et al.* 1994a; (2) Morin *et al.* 1994b; (3) Hashimoto *et al.* 1996; (4) Deinard 1997; (5) Gerloff *et al.* 1999; (6) Goldberg & Ruvolo 1997; (7) Goldberg & Wrangham 1997; (8) Gonder *et al.* 1997; (9) Ely *et al.* 1998; (10) Gagneux *et al.* 1999; (11) Gagneux *et al.* 1997; (12) Kaessmann *et al.* 1999; (13) Deinard & Kidd 2000; (14) Gonder 2000; (15) Mitani *et al.* 2000; (16) Reinartz *et al.* 2000; (17) Benn *et al.* 2001; (18) Constable *et al.* 2001; (19) Gagneux *et al.* 2001; (20) Stone *et al.* 2001; (21) Vigilant *et al.* 2001; (22) Sugiyama *et al.* 1993.

sequences sequentially repeated some 10–30 times (Beckman & Weber 1993). STRs are abundant throughout the nuclear genomes of eukaryotes, and the number of unit repeats at each locus varies greatly among individuals within a population (Tautz & Renz 1984; Edwards *et al.* 1992). Alleles at microsatellite loci generally differ in the number of repeats rather than in overall nucleotide sequences, therefore genotypes at microsatellite loci can be determined by separating alleles by size through gel electrophoresis followed by fluorescent detection (Dowling *et al.* 1996). Most importantly, because microsatellite repeats are typically less than 350 base pairs in length (e.g. Bradley *et al.* 2001), specific loci can be amplified using PCR from the small degraded DNA fragments found in noninvasive samples. More than 10000 human microsatellites are now described in the publicly available databases. Thus, studies of apes can utilize these microsatellite markers previously described for humans, provided that meticulous precautions are taken against contamination, and numerous proper controls are included at each step of analysis so that contamination, if present, will be detected.

Microsatellites are ideal markers for assessing individual variation and pedigree analysis, as they are highly polymorphic and follow general patterns of codominant Mendelian inheritance (Orrego & King 1990; Bruford & Wayne 1993). Studies of paternity using microsatellite loci typically involve two steps. First, the genotype of an offspring is compared to that of the mother and that of possible sires. Any allele observed in the offspring's genotype that is not shared with the mother must have come from the father (the paternal allele) and, therefore, any male who does not have the necessary paternal allele at every locus can clearly be excluded as the father. The second step of paternity analysis is calculating the paternity exclusion probability for a given set of loci given the allele frequencies in the population. This value is the probability that a nonrelated male drawn at random from the population would be excluded as a potential sire. If the probability of exclusion is high, any male who is *not* excluded is quite likely to be the true sire. Therefore, paternity studies using microsatellite loci require exclusions based on incompatible genotypes, as well as assessments of the statistical significance of paternal assignments.

If enough variable loci are examined, microsatellites can also be used to identify and census individuals within a population because each individual has a unique genotype (e.g. Richard *et al.* 1996). To date, this application in ape species has been limited, but microsatellite genotyping has been used with noninvasive samples from other species to estimate population sizes, monitor population sizes over time, and to estimate the home range of individuals (Waits *et al.* 2001 and references therein).

Y-chromosome

Compared to mtDNA sequencing and autosomal microsatellite studies, analysis of Y-chromosome variability is still a relatively new avenue of research in ape molecular systematics. Y-chromosomes are passed virtually intact from father to son, and polymorphisms are typically due to mutation alone (only a small pseudo-autosomal region recombines, Rappold 1993), although nucleotide diversity on the Y-chromosome may be less than that of autosomes (Shen *et al.* 2000; but see Kayser *et al.* 2000). Y-chromosome sequence and microsatellite analysis can complement the wealth of information already available on the maternal mtDNA lineages by providing parallel information on paternal lineages. For example, comparisons of Y-chromosome data with mitochondrial and autosomal data on humans have provided new insights into sex-specific population sizes and differences in gene-flow during hominin evolution (e.g. Jorde *et al.* 2000; Underhill *et al.* 2000). Assessment of variation on the Y-chromosome typically employs either microsatellite analysis or the sequencing of multiple short segments previously discovered to contain polymorphic nucleotides (single nucleotide polymorphisms or SNPs).

Current genetic research on wild chimpanzees and bonobos is based predominantly on the three classes of molecular markers just described (mitochondrial DNA, autosomal or X-chromosome nuclear DNA, and Y-chromosome DNA). Each type of marker can be used to address a variety of questions (e.g. systematics, social structure, conservation) on several levels (taxon, population, community).

MOLECULAR SYSTEMATICS OF *PAN*

Taxonomy, geographic distribution, and mitochondrial DNA

Much of the early molecular research on wild chimpanzees examined variability in mtDNA sequences, primarily HVR1 (the first hypervariable segment of the mtDNA control region), for the practical reasons discussed above. An early study examined the genetic diversity and population structure of wild chimpanzees by sequencing DNA from shed hairs (Morin *et al.* 1994a). Two mitochondrial loci, HVR1

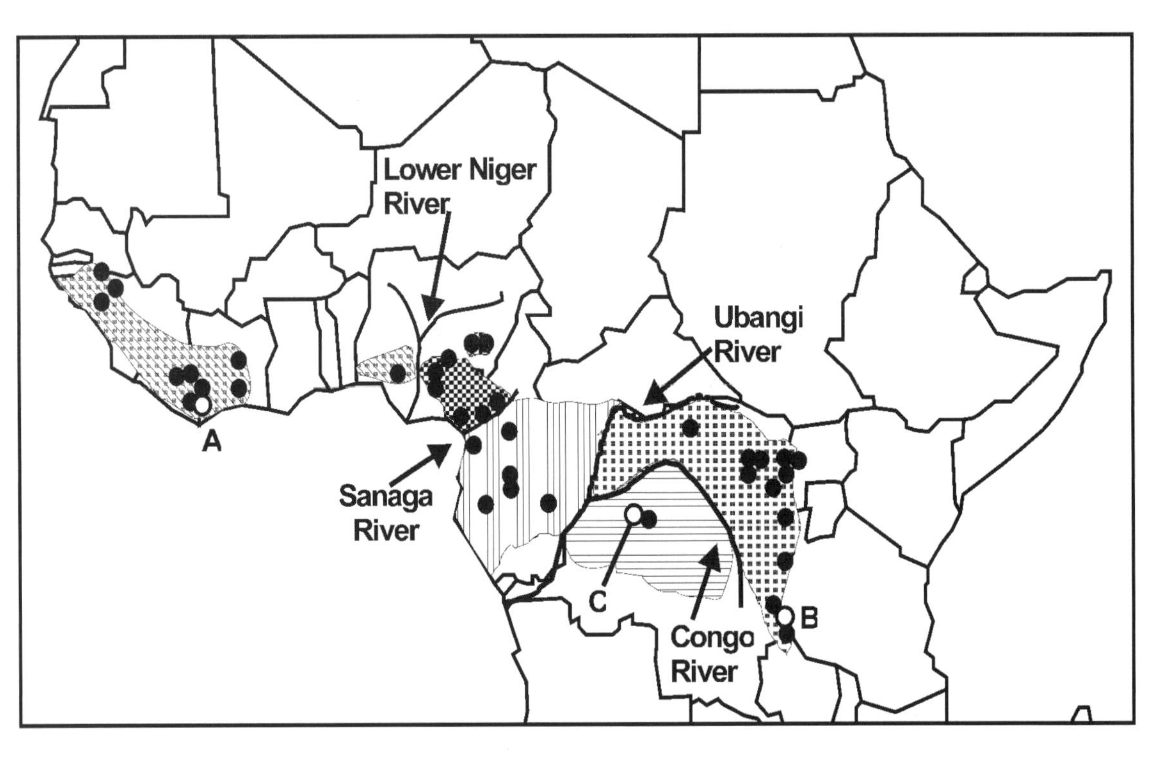

Fig. 19.1. Distribution of chimpanzees and bonobos, and locations of sample collection. Labeled rivers show supposed geographic borders. Although mtDNA studies include samples from a broad geographical range, there remains a large unsampled region on either side of the Ubangui River, which is thought to be the border between the central and eastern chimpanzee subspecies. In contrast to the broad sampling for mtDNA studies, comprehensive microsatellite studies have only been conducted at three sites to date (A, *P. t. verus*: Taï National Park, Côte d'Ivoire; B, *P. t. s. churinfurthii*: Gombe National Park, Tanzania; C = *P. paniscus*: Lomako Forest, Democratic Republic of Congo) (modified from Gagneux *et al.* 2001 and Gonder 2000).

and cytochrome *b*, were analyzed from 66 individuals from 20 different sites across Africa.

The sequence analysis led to several interesting conclusions about chimpanzee systematics and genetic social structure. Parsimony analysis and the overall analysis of genetic variability (percentage nucleotide differences) were consistent with the previously established taxonomy (Hill 1969), which recognized three subspecies of *Pan troglodytes:* West African (*P. t. verus*), Central African (*P. t. troglodytes*), and East African (*P.t. schweinfurthii*) (Figure 19.1). The cytochrome *b* segment of mtDNA was highly conserved across the central and eastern subspecies as there were no fixed differences seen between them. However, the cytochrome *b* sequences from the West African subspecies were significantly different and clearly separated western chimpanzees from the other two subspecies. The more variable and, therefore, more informative HVR1 sequences provided a much more detailed picture in which samples from each subspecies clustered into clear subspecific clades. More recent mitochondrial DNA studies have contributed complementary and contrasting new data on diversity, distribution, and biogeography.

Other workers conducted the first broad genetic survey of a single ape subspecies by investigating the connection between genetic variability and biogeography of eastern chimpanzees (Goldberg & Ruvolo 1997). A large number ($n = 262$) of HVR1 sequences were derived from shed hairs that were collected primarily from populations in the northeastern region of the subspecies distribution. Results confirmed that the control region sequence diversity observed in eastern chimpanzees contained low overall variability relative to that found in other chimpanzee subspecies. Interestingly, the small amount of genetic variability that was observed did not exhibit clear phylogeographic patterning. Given that chimpanzees are a forest-adapted species, Goldberg & Ruvolo (1997) wanted to test the hypothesis that genetic diversity seen today reflects the effects of Pleistocene refugia. If chimpanzee populations expanded from forest refugia, which were relatively isolated, current distributions of genetic variability in eastern chimpanzees should reflect this. Populations living in former refugia areas would be predicted to have higher levels of genetic variability than those chimpanzees living in areas that have been more recently colonized. However, since chimpanzees in areas of former refugia did not in fact show higher levels of variability and there was no clear geographic clustering, the hypothesis was not supported by the data.

While Goldberg & Ruvolo provided new information on the distribution of genetic variability in eastern chimpanzees, Jolly *et al.* (1995) noted that there was an unsampled, 1700-km-wide region between the collection areas of the western (*P. troglodytes verus*) and central chimpanzee samples (*P. troglodytes troglodytes*) analyzed by Morin *et al.* (1994a). This sampling and information gap was quickly filled by Gonder *et al.* (Gonder *et al.* 1997; Gonder 2000), with an analysis of hair samples from chimpanzees in Nigeria and Cameroon, resulting in 78 new HVR1 sequences. These sequences were incorporated into a comprehensive genetic analysis (Gonder 2000), which included the earlier published HVR1 sequences (Figure 19.2) (Morin *et al.* 1994a; Goldberg & Ruvolo 1997; Gagneux *et al.* 1999). The analysis distinguished two major lineages of chimpanzees – one in West Africa and one that includes both Central (western equatorial) and East Africa – that seem to converge at the Sanaga River in Cameroon (Figure 19.1). The western lineage was represented by two monophyletic clades: the West African chimpanzees of the Upper Guinea region (*P. t. verus*), and the chimpanzees of eastern Nigeria and western Cameroon (now termed *P. t. vellerosus*). Sequences from Central (*P. t. troglodytes)* and East African (*P. t. schweinfurthii*) chimpanzees, however, did not show subspecific monophyletic clustering as in the earlier analysis (Morin *et al.* 1994a). The inability of HVR1 sequences to distinguish chimpanzees from the central and eastern subspecies can be plausibly attributed to several factors. One possibility is that of gene flow between the geographically designated subspecies. Alternatively, phylogeographic separation between the subspecies might exist, but analyses of HVR1 sequences do not have the resolution to detect the separation.

Nevertheless, additional geographic boundaries between chimpanzee subspecies have been proposed. For example, the Lower Niger river has been suggested as a barrier between *P. t. verus* and *P. t. vellerosus* (Gonder 2000). However, there are only two mtDNA (HVR1) sequences available from Nigerian chimpanzees west of the Lower Niger and, therefore, the extent to which this river acts as a true barrier to gene flow is still questionable (Gagneux *et al.* 2001). The other potential taxonomic border described by Gonder, the Sanaga River between *P. t. vellerosus* and *P. t. troglodytes*, is known to be incomplete since there are two sequences from individuals north of the Sanaga that cluster with sequences from *P. t. troglodytes* south of the Sanaga (Gonder 2000).

Gagneux *et al.* (2001) recently produced an updated analysis of the mtDNA control region data set, which now includes 340 unique chimpanzee and bonobo HVR1 sequences. This latest analysis reiterated earlier findings that intrasubspecific gene flow is markedly high in chimpanzees because haplotypes are shared across large geographical distributions, sometimes spanning more than 1000 km. The importance of clarifying the distinction (or lack of it) between *P. t. troglodytes* and *P. t. schweinfurthii* and the still incomplete sampling on either side of the Ubangi river, the supposed border between the central and eastern subspecies, was noted. In addition, Gagneux *et al.* (2001) reported that new HVR1 sequences from chimpanzees from the Bondo region in northern Democratic Republic of Congo (i.e. chimpanzees that would by geographic origin be classified as *P. t. schweinfurthii*) appear in a phylogenetic analysis to be intermixed with those of chimpanzees (*P. t. troglodytes*) from west of the Ubangi river. The question of interspecific boundaries and the extent of gene flow between chimpanzee populations are particularly interesting given the ongoing search for viruses related to HIV(human immunodeficiency virus) in *P. troglodytes* populations (Gao *et al.* 1999; Santiago *et al.* 2002).

To date, relatively little analysis of mtDNA from bonobos has been done, although some extensive studies of diversity in bonobo nuclear DNA have been performed

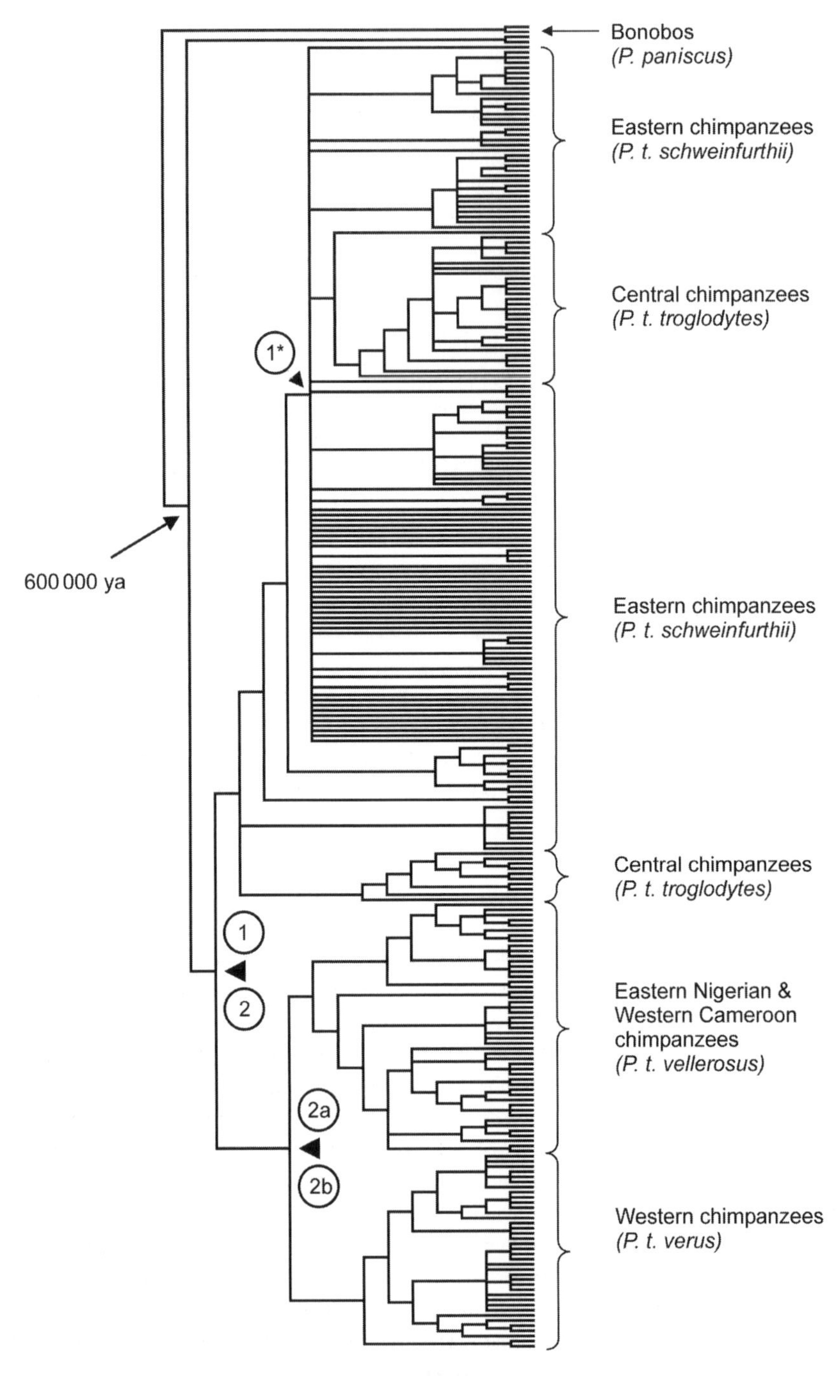

Fig. 19.2. Parsimony consensus tree of chimpanzee phylogeny based on 254 mtDNA (HVR1) sequences (modified from Gonder 2000). Based on this phylogeny the divergence date of all *P. troglodytes* HVR1 sequences is estimated at approximately 600 000 years ago (see Table 19.2). Sequences cluster clearly into an eastern clade (1) and a western clade (2). Within the western clade are two monophyletic clusters: chimpanzees from eastern Nigeria and western Cameroon (2a) and western Africa/Upper Guinea (2b). However, East African and Central African chimpanzees show no such monophyletic clustering (1*). Note that unlike Figure 19.3, branch *length* is not proportional to genetic distance.

Table 19.2. *Comparison of sequence variability and divergence dates of mitochondrial and nuclear DNA. Central African chimpanzees have the highest diversity in both markers*

	Mitochondrial DNA (HVR1)[1]			Nuclear DNA (Xq13.3)[2]		
	n	MPSD[a] (%)	TMRCA[b]	*n*	MPSD (%)	TMRCA
Chimpanzees (*Pan troglodytes*)	254	19.9	600000 (342000–867000)	30	0.13	2100000 ya (1160000–3350000)
Western/Upper Guinea (*P. t. verus*)	38	4.6	140000 (79000–201000)	17	0.05	636000 ya (325000–1090000)
Nigeria/Cameroon (*P. t. vellerosus*)	49	5.2	158000 (88000–228000)	0	nd	nd
Central (*P. t. troglodytes*)	39	6.9	210000 (114000–307000)	12	0.18	2000000 ya (1170000–3030000)
Eastern (*P. t. schweinfurthii*)	128	2.6	78000 (44000–114000)	1	nd	nd
Humans	135	0.07	200000 (342000–867000)	70	0.037%	645000 ya (319000–1150000)

Notes:
n, Sample size; MPSD, mean pairwise sequence differences; TMRCA, time to the most recent common ancestor; range in parenthesis; nd, not determined; [a] Values calculated from Gonder (2000) as average substitution rate divided by number of bp sequenced (346); [b] Calculated following Ward *et al.* (1991) estimate of 0.33 substitutions per site per million years.
[1] Based on compilation of sequences analyzed in Gonder 2000, which includes sequences from Morin *et al.* 1994; Goldberg & Ruvulo 1997; Gagneux *et al.* 1999; Vigilant *et al.* 1991.
[2] From Kaessmann *et al.* 1999.

(Reinartz 2000, see below). However, it is worth noting that the HVR1 sequences available from two sites (Lomako and Wamba) exhibited relatively high variability, roughly equal to that observed within *P. t. verus* (measured as mean transitional differences; *P. paniscus* = 3.43; *P. t. verus* = 3.40; Gagneux *et al.* 1999).

Nuclear DNA variation in *Pan*

Although the analysis of HVR1 sequences provides fundamental information on general levels of genetic diversity, conclusions based on these data must be made with caution, as the mitochondrial genome is a very small fraction of the total genome, and the divergence time-scales of mtDNA and nuclear DNA are different. A full understanding of species-level diversity requires an examination of numerous loci, both nuclear and mitochondrial. As discussed above, analysis of nuclear DNA variability is somewhat complicated by the fact that, unlike mtDNA, nuclear DNA is less abundant in cells, undergoes recombination, and has a much lower mutation rate. As a result, analysis of significantly longer segments of DNA is required before one can detect enough variation to make species-level comparisons. This necessitates the use of high-quality DNA that can only be obtained from blood or tissue samples. Thus, the studies mentioned immediately below are limited to the use of samples from captive animals whose exact geographic origin is unknown; in addition, these studies do not contain representative sample sizes from all of the subspecies under consideration.

Kaessmann *et al.* (1999, 2001) conducted a large-scale study of nuclear DNA variation in humans and the great apes by analyzing DNA from 70 humans, 30 chimpanzees, 5 bonobos, 11 gorillas and 14 orangutans. They sequenced 10 kb of a noncoding region on the X-chromosome (Xq13.3) known to have a low recombination rate. As had been observed in the mtDNA studies, findings showed that chimpanzees (when examining three of the four subspecies; *P. t. vellerosus* was not included as samples were unavailable) have a much higher level of sequence diversity than do

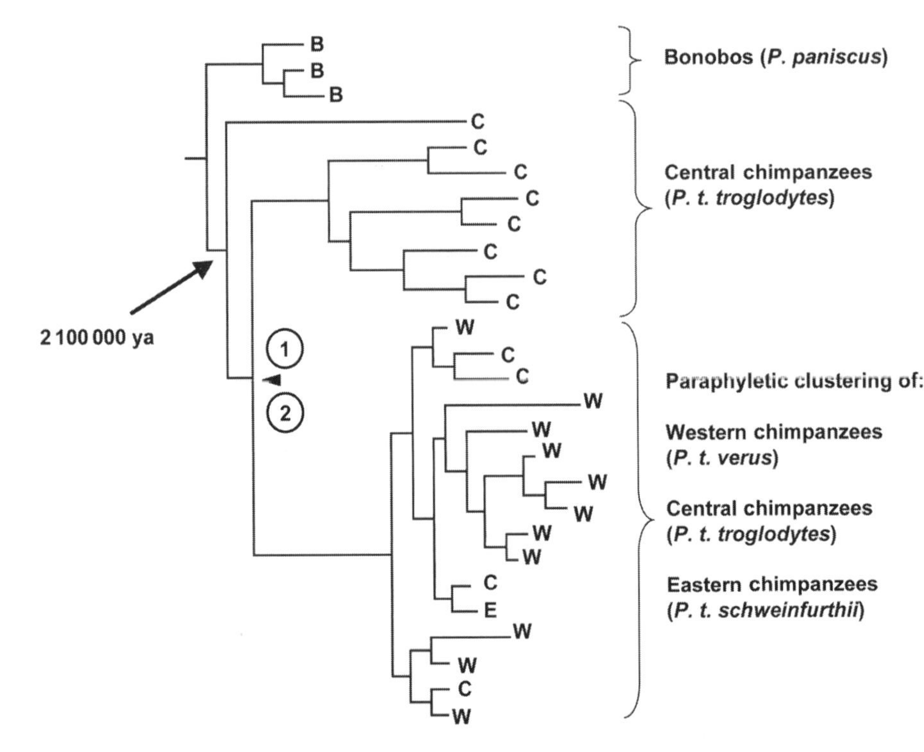

Fig. 19.3. Maximum likelihood phylogenetic tree based on 24 chimpanzee and three bonobo nuclear (Xq13.3) sequences (modified from Kaessmann *et al.* 1999). The divergence date of all *P. troglodytes* Xq13.3 sequences is estimated at approximately 2.1 million years ago (see Table 19.2). Only one sequence is available from the east African subspecies, and no samples from eastern Nigeria and western Cameroon are available. Within chimpanzees, two clades are apparent: one which only includes sequences from central African chimpanzees (1), and a mixed cluster including sequences from the western, central, and eastern subspecies (2). Subspecific clustering is not observed, and the most ancestral sequence is from Central Africa.

humans (Table 19.2) (Kaessmann *et al.* 1999, 2001). However, phylogenetic analysis of the sequences from this nuclear locus showed sequences from the three chimpanzee subspecies as intermixed, with less clustering into eastern and western clades (Figure 19.3), as is the case for the HVR1 sequences (Figure 19.2). This contrast can be attributed to the fact that the effective population size (Ne) of mitochondrial DNA is one-third that of the X-chromosome. As a result, mtDNA types are more likely be lost or become fixed through genetic drift than are genotypes on the X-chromosome or autosomes. Thus, a given set of Xq13 sequences would require three times as long in order to show monophyletic clades (e.g. at the subspecific level) than would mitochondrial sequences from the same set of individuals.

The estimated time to the most recent common ancestor (TMRC) of the chimpanzee sequences is 2100000 years (95% confidence interval of 1160000–3350000) indicating that the genetic history of *Pan troglodytes* is quite deep, especially when compared to that of humans (most recent common ancestor for the Xq13 sequences of humans is estimated at 645000 ya; range = 319000 ya to 1.15 mya). Interestingly, when looking at intrasubspecific variability, greater variability was found within the central (*P. t. troglodytes*) than within the western (*P. t. verus*) subspecies (Kaessmann *et al.* 1999), results consistent with most recent descriptions of the relative levels of diversity in mtDNA (Gonder 2000). Kaessmann *et al.* (1999) also found that, although the bonobo sequences are monophyletic, the distance between some *P. troglodytes* sequences is greater (some *P. troglodytes* sequences differ from each other by as many as 29 base pairs) than the distance between *P. troglodytes* and *P. paniscus* (sequences differ by 13 to 23 positions).

Rather than focusing on analysis of a single large segment of the nuclear genome, Deinard (Deinard 1997; Deinard & Kidd 2000) chose to examine nuclear DNA diversity in chimpanzees ($n = 45$) and bonobos ($n = 18$) by sequencing shorter stretches of five independent, noncoding nuclear loci (*HOXB6*, *APOB*, *DRD2*, *PSUX* and *PSUY*). Problems associated with recombination were precluded by analyzing relatively small fragments, ranging in size from 300 base pairs to just over 1 kilobase. They found that the amount of haplotype polymorphism within and between species varied from locus to locus, reiterating the importance of examining multiple loci when investigating genetic diversity. Overall, Deinard & Kidd (2000) reported findings similar to those presented by Kaessmann *et al.* (1999), with *Pan* having a much higher level of variability than humans. Similarly, phylogenetic analysis among the nuclear loci from each of the three chimpanzee subspecies (due to unavailability of samples, *P. t. vellerosus* was not sequenced and only one East African chimpanzee was sequenced) do not show monophyletic clustering by subspecies, providing further evidence that mtDNA phylogeny does not match the genetic differentiation observed on a nuclear level.

An additional result was that the bonobo, *P. paniscus*, shows much less genetic variability at these five nuclear loci than shown by *P. troglodytes* (Deinard & Kidd 2000). This issue of species-level differences in nuclear DNA diversity was also recently examined by genotype analysis of 17 captive, wild-caught bonobos at 28 nuclear microsatellite loci to evaluate overall levels of allelic diversity (Reinartz *et al.* 2000). Results were compared to those published from analysis of wild chimpanzee populations, and showed that although bonobos currently have an extremely limited range and small population size, they appear to have moderate levels of microsatellite polymorphism. The overall level of allelic diversity in captive, wild-caught bonobos (indicated by a mean observed heterozygosity, H_o, of 0.52) is similar to that observed in eastern chimpanzees ($H_o=0.54$) but less than that observed in chimpanzees as a whole species ($H_o=0.74$) (Reinartz *et al.* 2000). This finding that the current US captive bonobo population does not exhibit comparatively low genetic diversity provides encouraging news for bonobo conservation efforts.

Studies of wild bonobos have been limited to those from two closely situated sites, Lomako and Wamba, in the Democratic Republic of Congo. The focus of these studies was to address questions about relationships of group members, and will be discussed in this chapter's section on molecular ecology. An ongoing study by Eriksson and Hohmann (personal communication) will add to the information on genetic diversity of bonobos from throughout the species range by means of noninvasive samples collected from several previously unsampled regions in the Democratic Republic of Congo.

Y-chromosome variation in *Pan*

Information on Y-chromosome variability in *Pan* is still sparse. Stone *et al.* (2001) examined over 4 kb of sequence on the Y-chromosome in more than 100 unrelated male chimpanzees and several bonobos. Since high-quality DNA was required, the majority of the samples analyzed were from captive chimpanzees. Interestingly, the one sequence available from a *P. t. vellerosus* individual is identical to the majority sequence found in *P. t. verus*, although sequences are not otherwise shared between subspecies. Once again, *P. t. troglodytes* exhibited the highest diversity, while *P. t. verus* showed the lowest. The variable nucleotide positions analyzed by Stone *et al.* (2001) in the Y-chromosome of chimpanzees could potentially be used in applications such as noninvasive genetic studies of male dispersal, paternal lineage history, and male–male relatedness at the population level. However, in order to shed significant light on these questions, it must first be determined whether these markers exhibit sufficient variation in wild populations.

MOLECULAR ECOLOGY

Chimpanzees

A study of chimpanzees (*P. t. schweinfurthii*) belonging to the Kasakela community at Gombe National Park, Tanzania, was one of the first attempts to understand the genetic structure of chimpanzee communities by conducting microsatellite analysis for paternity determination (Morin *et al.* 1994b). This study succeeded in assigning fathers to only two of the offspring examined. The low rate of assignment was attributable in part to the fact that it was not possible to sample some possible sires who had died. The study, none the less, clearly indicated the potential of using molecular-based methods in investigations of wild ape populations. A particularly interesting finding was a suggestion of higher genetic relatedness within community males as compared to within community females, supporting ideas that kin selection may play a role in cooperative behavior among the philopatric males.

A study of a community of chimpanzees (*P. t. verus*) in the Taï National Park, Côte d'Ivoire, included samples from 13 offspring and all males living in the community at the time those offspring were conceived (Gagneux *et al.* 1997).

Although fathers were assigned to seven offspring, all group males were excluded as sires for the remaining six offspring. Analysis of the behavioral records indicated that the mothers of the infants in question were absent from the community for one to several days around the probable time of conception, lending support to the idea that females were specifically choosing to mate and conceive offspring with extra-group males. Extra-group copulations have been observed at a low frequency in chimpanzees (Goodall 1986), but the extent of extra-group paternity was hitherto unknown.

In a more recent assessment of paternity of chimpanzees in the Kasakela community at Gombe, Constable *et al.* (2001) genotyped microsatellite loci using DNA from fecal samples and determined the sires of 14 chimpanzees. By using the almost 40 years of behavioral data on the chimpanzees at Gombe, this study was able to evaluate the relative success of different male reproductive strategies. They found that 50% (7 of 14) of offspring were fathered by high-ranking males, and that consortships by low or middle-ranking males also led to conceptions, as did opportunistic matings by subdominant males. Infants conceived when only one female was cycling were always sired by the alpha male. However, no evidence of extra-group paternity was observed in this study, and it was pointed out that the application of a maximum likelihood method of paternity analysis also succeeded in assigning fathers to the offspring in the Taï study. However, this method of paternity assignment differs from the normal approach of paternity assignment by exclusion, in which only a completely genotypically compatible male is assigned as the father. For most of the reassignments of paternity for the Taï offspring the proposed father and offspring in fact had incompatible genotypes at one or more microsatellite loci and would typically not be considered to represent reliable paternity assignments.

Comparison of the specific results of the two Gombe studies can be made to a limited extent. One of the two paternity assignments made in the earlier study was confirmed in the later study, while the other offspring was not analyzed. In addition, three offspring were reassigned to males that had previously been excluded. The lack of consistency in the results of the two studies can likely be attributed to the presence of undetected errors, including those caused by 'allelic dropout,' in the genotypes (Taberlet *et al.* 1996). These kinds of genotyping errors were virtually unknown when the research of Morin & Woodruff (1992) and Morin *et al.* (1994b) was conducted in the early 1990s.

Allelic dropout describes a situation in which, due to a low starting amount of DNA such as that typically obtained from noninvasive samples, only one of the two alleles at a microsatellite locus is amplified. This result can be falsely interpreted as a finding that the individual is homozygous; that is, has two copies of the same allele, at this locus. Extensive repetition of results is the safeguard against such errors (Taberlet *et al.* 1996), and a newly described method for accurately measuring the amount of DNA in extracts from noninvasive samples allows for the efficient classification of DNAs into those producing reliable results, those requiring extensive repetition, and ones that are unusable (Morin *et al.* 2001).

A new, more extensive study of the Taï chimpanzees used this method to ensure accuracy, and results indicate that extra-group paternity is detected, but at a much lower frequency than originally reported by Gagneux *et al.* (Gagneux *et al.* 1997; Gagneux *et al.* 2001; Vigilant *et al.* 2001).

Several studies using mtDNA sequences (rather than microsatellites) have addressed questions about genetic relatedness and sociality in chimpanzees. Goldberg & Wrangham (1997) tested the hypothesis that affiliate behaviors (e.g. grooming, nesting proximity) among male chimpanzees may have evolved through kin selection. They sequenced 14 individual adult males from the Kanyawara community in the Kibali National Park, Uganda, to determine if maternal relatedness was correlated with affiliative behavior. Their results showed no clear correlation; that is, males did not spend more time in affiliative interactions with males who share their mitochondrial haplotype and thus represent possible maternal relatives. Similarly, Mitani *et al.* (2000) examined six affiliative and cooperative behaviors, including patrolling and meat sharing, among 23 adult males of the Ngogo community in the Kibale National Park, and found no correlation between these variables and maternal relatedness. However, it should be noted that since these studies only examined mtDNA, one cannot exclude the possibility that males affiliate with paternal relatives, an issue that can only be addressed by studies of nuclear DNA.

Bonobos

Bonobos were the subjects of the first comprehensive study of genetic social structure in wild primates using feces, rather than hair, as the source of DNA (Gerloff *et al.* 1995, 1999). The microsatellite genotypes and mtDNA sequences of 36 individuals, including ten offspring, from the Eyengo community and six individuals from neighboring communities in Lomako forest, Democratic Republic of Congo, were analyzed. Clear paternity assignments (99% exclusionary power) were possible for seven of the ten offspring, and the two highest ranking males together sired the majority of the offspring. Since these two males were the sons of high-ranking females, the authors concluded that for bonobos a mother's rank exerts a strong influence on the reproductive success of her sons.

The mtDNA sequence analysis of the Eyengo community, combined with mtDNA sequences from the bonobos at Wamba (Hashimoto *et al.* 1996), provides some additional information about bonobo genetic social structure. Identical haplotypes were found not only among individuals from the same community, but also were shared between communities. This result is not surprising given that the two communities are only separated by approximately 200 km. More extensive sampling across the range of the bonobo is needed to investigate this lack of geographic differentiation, which is presumably an effect of female dispersal.

FUTURE DIRECTIONS

Current results of mitochondrial and nuclear DNA diversity studies in *Pan* demonstrate the need to evaluate genetic diversity within a broad framework, encompassing data from multiple molecular markers. We believe there are several key areas for future research in chimpanzee and bonobo molecular ecology and systematics. First, better sampling is needed, especially in the regions directly west and east of the Ubangi river, to clarify the phylogeography of Central and East African chimpanzees (Gagneux *et al.* 2001). In addition, as the results of comprehensive microsatellite studies focusing on relatedness and genetic social structure from other field sites continue to be published, broader comparisions of microsatellite allele frequences between sites could elucidate the pattern of overall diversity and genetic substructuring in *Pan*. Such studies are currently under way at Mahale Mountains National Park, Tanzania (T. Nishida, M. Sloan, L. Vigilant), Budongo Forest, Uganda (V. Reynolds, U. Immel, L. Vigilant), and the Ngogo community, Kibale National Park, Uganda (J. Mitani, D. Watts, A. Merriweather). However, only if the same markers are used in all studies will a wide interspecific and intersite comparison be possible. In addition, analysis of ancient DNA, using bone, skull, and pelt fragments from museum collections, may provide a new opportunity to sample extinct populations.

Future molecular analysis will need to employ more varied approaches, both in sampling and in molecular techniques. Although our current understanding of chimpanzee molecular ecology and systematics is based on only a few loci, new molecular approaches provide the obvious next step. One of the primary directions for future work is examination of the single nucleotide sites (SNPs) that show high levels of polymorphism in a population. These SNPs are abundant throughout the genome, occurring approximately once every 1000–2000 bases in humans (Li & Sadler 1991). Variation at SNP sites can be detected relatively easily (Kwok 2000), allowing for new comparisons and evaluations of population genetic structures, as well as paternity and individual identification (Krawczak 1999). Most importantly, since they target relatively short stretches of DNA, SNP analyses are especially promising for studies of wild populations, which at this point rely on highly degraded, noninvasive samples. Improvements in sample collection and extraction methods may also result in increases in the quality of the DNA obtained from noninvasive samples, hence expanding the potential uses of such samples.

From a broader perspective, the recent completion of the human genome project (Lander *et al.* 2001; Venter *et al.* 2001) provides new opportunities for wider comparative studies within chimpanzees and between chimpanzees and humans. An international research effort is under way to complete a 'chimpanzee (or primate) genome project' (McConkey & Varki 2000; Normile 2001a), and the obvious next step is examining how the slight, 1–2%, genetic difference between chimpanzees and humans results in such dramatic differences in morphology, cognition, and behavior (Normile 2001b).

Finally, there is a need for greater integration and assimilation of genetic data from the laboratory with behavioral data from the field and vice versa. Genetic diversity and cultural diversity (Whiten *et al.* 1999) may evolve through analogous processes, both dependent on population size and structure. Understanding genetic phenomena, such as gene flow and population structure requires long-term field data on dispersal and mating behavior, and studies of genetic diversity in *Pan* can most interestingly be viewed and discussed within the context of behavioral diversity.

ACKNOWLEDGEMENTS

We thank Phil Morin, Henrik Kaessmann, Diane Doran, Katherine Gonder, and two reviewers for helpful comments, and Gottfried Hohmann, Linda Marchant, and Christophe Boesch for inviting us to contribute to this volume. This chapter is largely a review of research presented at a symposium on chimpanzee and bonobo genetic diversity held 9–10 June, 2000 at the Max Planck Institute for Evolutionary Anthropology in Leipzig, Germany. B. Bradley is supported by the Max Planck Society, National Science Foundation (SBR-9910399), L.S.B. Leakey Foundation, Wenner-Gren Foundation, and Deutscher Akademischer Austauschdienst.

REFERENCES

Beckman, J. S. & Weber, J. L. (1993). Survey of human and rat microsatellites. *Genomics*, 12, 627–31.

Benn, J., Smith, J. M. & Stone, A. C. (2001). Y-chromosome STR analysis in *Pan troglodytes*. *American Journal of Physical Anthropology*, Suppl. 32, 38.

Bradley, B. J., Boesch, C. & Vigilant, L. (2001). Identification and redesign of human microsatellite markers for genotyping wild chimpanzee (*Pan troglodytes verus*) and gorilla (*Gorilla gorilla gorilla*) DNA from feces. *Conservation Genetics*, 3, 289–92.

Brown, E. M., George, M. & Wilson, A. C. (1979). Rapid evolution of animal mitochondrial DNA. *Proceedings of the National Academy of Sciences, USA*, 76, 1967–71.

Brown, W. M., Prager, E. M., Wang, A. & Wilson, A. C. (1982). Mitochondrial DNA sequences of primates: tempo and mode of evolution. *Journal of Molecular Evolution*, 18, 225–39.

Bruford, M. W. & Wayne, R. K. (1993). Microsatellites and their application to population genetic studies. *Current Opinions in Genetic Development*, 3, 939–43.

Cann, R. L., Stoneking, M. & Wilson, A. C. (1987). Mitochondrial DNA and human evolution. *Nature*, 325, 31–6.

Constable, J. L., Ashley, M. V., Goodall, J. & Pusey, A. E. (2001). Noninvasive paternity assignment in Gombe chimpanzees. *Molecular Ecology*, 10, 1279–300.

Deinard, A. S. (1997). The evolutionary genetics of the chimpanzees. PhD Dissertation: Department of Anthropology. Yale, New Haven.

Deinard, A. S. & Kidd, K. (2000). Identifying conservation units within captive chimpanzee populations. *American Journal of Physical Anthropology*, 111, 25–44.

Dowling, T. E., Moritz, C., Palmer, J. & Rieseberg, L. H. (1996). Nucleic Acids III: Analysis of fragments and restriction sites. In *Molecular Systematics*, ed. D. M. Hillis, C. Moritz & B. K. Mable, pp. 249–320. Sunderland, MA: Sinauer.

Edwards, A., Hammond, H. A., Jin, L., Caskey, C. T., Chakraborty, R. (1992). Genetic variation at five trimeric and tetrameric tandem repeat loci in four human population groups. *Genomics*, 12, 241–53.

Ely, J. J., Gonzalez, D. L., Reeves-Daniel, A. & Stone, W. H. (1998). Individual identification and paternity determination in chimpanzees (*Pan troglodytes*) using human short tandem repeat (STR) markers. *International Journal of Primatology*, 19 (2), 255–71.

Enard, W., Khaitovich, P., Klose, J. *et al.* (in press). Intra- and interspecific variation in primate gene expression patterns. *Science*.

Eyre-Walker, A., Smith, N. H. & Smith, J. M. (1999). How clonal are human mitochondria? *Proceedings of the Royal Society of London, Series B, Biological Sciences*, 266, 477–83.

Gagneux, P., Woodruff, D. S. & Boesch, C. (1997). Furtive mating in female chimpanzees. *Nature*, 387, 358–9.

Gagneux, P., Woodruff, D. S. & Boesch C. (2001). Furtive mating in female chimpanzees. *Nature*, 414, 508.

Gagneux, P., Wills, C., Gerloff, U., Tautz, D., Morin, P. A., Boesch, C., Fruth, B., Hohmann, G., Ryder, O. A. & Woodruff, D. S. (1999). Mitochondrial sequences show diverse evolutionary histories of African hominoids. *Proceedings of the National Academy of Sciences, USA*, 96, 5077–82.

Gagneux, P., Gonder, M. K., Goldberg, T. & Morin, P. (2001). Gene flow in wild chimpanzee populations: what genetic data tell us about chimpanzee movement over space and time. *Philosophical Transactions of the Royal Society of London, Series B*, 356, 889–97.

Gao, F., Bailes, E., Robertson, D. L., Chen, Y., Rodenburg, C. M., Michael, S. F., Cummins, L. B., Arthur, L.O., Peeters, M., Shaw, G. M., Sharp, P. M. & Hahn, B. H. (1999). Origin of HIV-1 in the chimpanzee *Pan troglodytes troglodytes*. *Nature*, 397, 436–41.

Gerloff, U., Schlotterer, C., Rassmann, K., Rambold, I., Hohmann, G., Fruth, B. & Tautz, D. (1995). Amplification of hypervariable simple sequence repeats (microsatellites) from excremental DNA of wild living bonobos (*Pan paniscus*). *Molecular Ecology*, 4, 515–18.

Gerloff, U., Hartung, B., Fruth, B., Hohmann, G. & Tautz, D. (1999). Intracommunity relationships, dispersal pattern and paternity success in a wild living community of Bonobos (*Pan paniscus*) determined from DNA analysis of faecal samples. *Proceedings of the Royal Society of London, Series B, Biological Sciences*, 266, 1189–95.

Goldberg, T. L. & Ruvolo, M. (1997). The geographic apportionment of mitochondrial genetic diversity in East African chimpanzees, *Pan troglodytes schweinfurthii*. *Molecular Biology and Evolution*, 14, 976–84.

Goldberg, T. & Wrangham, R. W. (1997). Genetic correlates of social behavior in wild chimpanzees: evidence from mitochondrial DNA. *Animal Behavior*, 54, 559–70.

Gonder, K., Oates, J., Disotell, T., Forstner, M., Morales, J. & Melnick, D. (1997). A new west African chimpanzee subspecies? *Nature*, 388, 337.

Gonder, M. K. (2000). Evolutionary genetics of chimpanzees in Nigeria and Cameroon. PhD Dissertation, Department of Anthropology. CUNY, NY.

Goodall, J. (1986). *The Chimpanzees of Gombe: Patterns of Behavior*. Cambridge, MA: Belknap Press of Harvard University Press.

Goodman, M. (1962). Evolution of the immunologic species specificity of human serum proteins. *Human Biology*, 34, 104–50.

Hamilton, W. D. (1964). The genetical evolution of social behavior, I and II. *Journal of Theoretical Biology*, 7, 477–88.

Hashimoto, C., Furuichi, T. & Takenaka, O. (1996). Matrilineal kin relationship and social behavior of wild bonobos: sequencing the D-loop region of mitochondrial DNA. *Primates*, 37, 305–18.

Hill, W. C. O. (1969). The nomenclature, taxonomy, and distribution of chimpanzees. In *The Chimpanzee*, ed. G. H. Bourne, pp. 22–49. Basel: Karger.

Hofreiter, M., Serre, D., Poinar, H. N., Kuch, M. & Paabo, S. (2001). Ancient DNA. *Nature Reviews*, 2, 353–60.

Jolly, C. J., Oates, J. F. & Disotell, T. R. (1995). Chimpanzee kinship. *Science*, 268, 185–8.

Jorde, L. B., Watkins, W. S., Bamshad, M. J., Dixon, M. E., Ricker, C. E., Seielstad, M. T. & Batzer, M. A. (2000). The distribution of human genetic diversity: a comparison of mitochondrial, autosomal, and Y-chromosome data. *American Journal of Human Genetics*, 66, 979–88.

Kaessmann, H., Wiebe, V. & Paabo, S. (1999). Extensive nuclear DNA sequence diversity among chimpanzees. *Science*, 286, 1159–62.

Kaessmann, H., Wiebe, V., Weiss, G. & Paabo, S. (2001). Great ape DNA sequences reveal a reduced diversity and an expansion in humans. *Nature Genetics*, 27, 155–6.

Kayser, M., Roewer, L., Hedman, M., Henke, L., Henke, J., Brauer, S., Kruger, C., Krawczak, M., Nagy, M., Dobosz, T., Szibor, R., de Knijff, P., Stoneking, M. & Sajantila, A. (2000). Characteristics and frequency of germline mutations at microsatellite loci from the human Y chromosome, as revealed by direct observation in father/son pairs. *American Journal of Human Genetics*, 66, 1580–8.

Kocher, T. D. & Wilson, A. C. (1991). Sequence evolution of mitochondrial DNA in humans and chimpanzees: control region and a protein-coding region. In *Evolution of Life: Fossils, Molecules, and Culture*, ed. S. Osawa & T. Honjo, pp. 391–413. Tokyo: Springer-Verlag.

King, M.-C. & Wilson, A. C. (1975). Evolution at two levels in humans and chimpanzees. *Science*, 188, 107–88.

Krawczak, M. (1999). Informativity assessment for biallelic single nucleotide polymorphisms. *Electrophoresis*, 20, 1676–81.

Kwok, P. Y. (2000). High-throughput genotyping assay approaches. *Pharmacogenomics*, 1, 95–100.

Lander, E. S., Linton, L. M., Birren, B., Nusbaum, C., Zody, M. C., Baldwin, J. *et al.* (2001). Initial sequencing and analysis of the human genome. *Nature*, 409, 860–921.

Li, W. H. & Sadler, L. A. (1991). Low nucleotide diversity in man. *Genetics*, 129, 513–23.

Matova, N. & Cooley, L. (2001). Comparative aspects of animal oogenesis. *Developmental Biology*, 231, 291–320.

McConkey, E. H. & Varki, A. (2000). A primate genome project deserves high priority. *Science*, 289, 1295–6.

Meyer, S., Weiss, G. & von Haeseler, A. (1999). Pattern of nucleotide substitution and rate heterogeneity in the hypervariable regions I and II of human mtDNA. *Genetics*, 152, 1103–10.

Mitani, J. C., Merriwether, D. A. & Zhang, C. (2000). Male affiliation, cooperation and kinship in wild chimpanzees. *Animal Behaviour*, 59, 885–93.

Moore, W. S. (1995). Inferring phylogenies from mtDNA variation: Mitochondrial-gene trees versus nuclear gene trees. *Evolution*, 49, 718–26.

Morin, P. A. & Woodruff, D. S. (1992). Paternity exclusion using multiple hypervariable microsatellite loci amplified from nuclear DNA of hair cells. In *Paternity in Primates: Genetic Tests and Theories*, ed. R. D. Martin, A. F. Dixson & E. J. Wickings, pp. 63–81. Basel: Karger.

Morin, P. A., Moore, J. J., Chakraborty, R., Jin, L., Goodall, J. & Woodruff, D. S. (1994a). Kin selection, social structure, gene flow, and the evolution of chimpanzees. *Science*, 265, 1193–201.

Morin, P. A., Wallis, J., Moore, J. J. & Woodruff, D. S. (1994b). Paternity exclusion in a community of wild chimpanzees using hypervariable simple sequence repeats. *Molecular Ecology*, 3, 467–78.

Morin, P. A., Chambers, K. E., Boesch, C. & Vigilant, L. (2001). Quantitative PCR analysis of DNA from noninvasive samples for accurate microsatellite genotyping of wild chimpanzees (*Pan troglodytes verus*). *Molecular Ecology*, 10, 1835–44.

Mullis, K. B. & Faloona, F. A. (1987). Specific synthesis of DNA *in vitro* via a polymerase-catalyzed chain reaction. *Methods in Enzymology*, 155, 335–50.

Normile, D. (2001a). Chimp sequencing crawls forward. *Science*, 291, 2297.

Normile, D. (2001b). Gene expression differs in human and chimp brains. *Science*, 292, 44–5.

Orrego, C. & King, M. C. (1990). Determination of familial relationships. In *PCR Protocols: A Guide to Methods and Applications*, ed. M. A. Innis, D. H. Gelfand, J. J. Sninsky & T. J. White, pp. 416–26. San Diego: Academic Press.

Rappold, G. A. (1993). The pseudoautosomal regions of the human sex chromosomes. *Human Genetics*, 92, 315–24.

Reinartz, G. E., Karron, J. D., Phillips, R. B. & Weber, J. L. (2000). Patterns of microsatellite polymorphism in the range-restricted bonobo (*Pan paniscus*): considerations for interspecific comparison with chimpanzees (*P. troglodytes*). *Molecular Ecology*, 9, 315–28.

Retief, J. D. & Dixon, G. H. (1993). Evolution of pro-protamine P2 genes in primates. *European Journal of Biochemistry*, 214, 609–15.

Richard, K. R., Whitehead, H. & Wright, J. M. (1996). Polymorphic microsatellites from sperm whales and their use in the genetic identification of individuals from naturally sloughed pieces of skin. *Molecular Ecology*, 5, 313–5.

Robin, E. D. & Wong, R. (1988). Mitochondrial DNA molecules and virtual number of mitochondria per cell in mammalian cells. *Journal of Cell Physiology*, 136, 507–13.

Ruvolo, M. (1997). Molecular phylogeny of the hominoids: inferences from multiple independent DNA sequence data sets. *Molecular Biology and Evolution*, 14, 248–65.

Ruvolo, M., Disotell, T. R., Allard, M. W., Brown, W. M. & Honeycutt, R. L. (1991). Resolution of the African hominoid trichotomy by use of a mitochondrial gene sequence. *Proceedings of the National Academy of Sciences, USA*, 88, 1570–4.

Ruvolo, M., Pan, D., Zehr, S., Goldberg, T., Disotell, T. R. & von Dornum, M. (1994). Gene trees and hominoid phylogeny. *Proceedings of the National Academy of Sciences, USA*, 91, 8900–4.

Saiki, R. K., Scharf, S., Faloona, F., Mullis, K. B., Horn, G. T., Erlich, H. A. & Arnheim, N. (1985). Enzymatic amplification of beta-globin genomic sequences and restriction site analysis for diagnosis of sickle cell anemia. *Science*, 230, 1350–4.

Santiago, M.L., Rodenburg, C.M., Kamenya, S. *et al.* (2002). SIVcpz in wild chimpanzees. *Science* 295, 465.

Sarich, V. & Wilson, A. C. (1967). Immunological time scale for hominid evolution. *Science*, 158, 1200–3.

Satta, Y., Klein, J. & Takahata, N. (2000). DNA archives and our nearest relative: the trichotomy problem revisited. *Molecular Phylogenetics and Evolution*, 14, 259–75.

Shen, P., Wang, F., Underhill, P. A., Franco, C., Yang, W. H., Roxas, A., Sung, R., Lin, A. A., Hyman, R. W., Vollrath, D., Davis, R. W., Cavalli-Sforza, L. L. & Oefner, P. J. (2000). Population genetic implications from sequence variation in four Y chromosome genes. *Proceedings of the National Academy of Sciences, USA*, 97, 7354–9.

Stone, A. C., Griffiths, R. C., Zegura, S. L. & Hammer, M. F. (2002). High levels of Y-chromosome nucleotide diversity in the genus *Pan. Proceedings of the National Academy of Sciences, USA*, 99, 43–8.

Stoneking, M. (1994). Mitochondrial DNA and human evolution. *Journal of Bioenergetics and Biomembranes*, 26, 251–9.

Stoneking, M. (2000). Hypervariable sites in the mtDNA control region are mutational hotspots. *American Journal of Human Genetics*, 67, 1029–32.

Sugiyama, Y., Kawamoto, S., Takenaka, O., Kumazaki, K. & Miwa, N. (1993). Paternity discrimination and intergroup relationships of chimpanzees at Bossou. *Primates* 34, 545–52.

Sunnucks, P. (2000). Efficient genetic markers for population biology. *Trends in Ecology and Evolution*, 15, 199–203.

Taberlet, P., Griffin, S., Goossens, B., Questiau, S., Manceau, V., Escaravage, N., Waits, L. P. & Bouvet, J. (1996). Reliable genotyping of samples with very low DNA quantities using PCR. *Nucleic Acids Research*, 24, 3189–3194.

Tautz, D. (1989). Hypervariability of simple sequences as a general source for polymorphic DNA markers. *Nucleic Acids Research*, 17, 6463–71.

Tautz, D. & Renz, M. (1984). Simple sequences are ubiquitous repetitive components of eukaryotic genomes. *Nucleic Acids Research*, 12, 4127–38.

Underhill, P. A., Shen, P., Lin, A. A., Jin, L., Passarino, G., Yang, W. H., Kauffman, E., Bonne-Tamir, B., Bertranpetit, J., Francalacci, P., Ibrahim, M., Jenkins, T., Kidd, J. R., Mehdi, S. Q., Seielstad, M. T., Wells, R. S., Piazza, A., Davis, R. W., Feldman, M. W., Cavalli-Sforza, L. L. & Oefner, P. J. (2000). Y chromosome sequence variation and the history of human populations. *Nature Genetics*, 26, 358–61.

Venter, J. C., Adams, M. D., Myers, E. W., Li, P. W., Mural, R. J., Sutton, G. G. *et al.* (2001) The sequence of the human genome. *Science*, 291, 1304–51.

Vigilant, L., Pennington, R., Harpending, H., Kocher, T. D. & Wilson, A. C. (1989). Mitochondrial DNA sequences in single hairs from a southern African population. *Proceedings of the National Academy of Sciences, USA*, 86, 9350–4.

Vigilant, L., Stoneking, M., Harpending, H., Hawkes, K. & Wilson, A. C. (1991). African populations and the evolution of human mitochondrial DNA. *Science*, 253, 1503–7.

Vigilant, L., Hofreiter, M., Siedel, H. & Boesch, C. (2001) Paternity and relatedness in wild chimpanzee communities. *Proceedings of the National Academy of Sciences, USA*, 98, 12890–5.

Ward, R. H., Frazier, B. L., Dew-Jager, K. & Pääbo, S. (1991). Extensive mitochondrial diversity within a single Amerindian tribe. *Proceedings of the National Academy of Sciences, USA*, 88, 8720–4.

Waits, L. P., Luikart, G. & Taberlet, P. (2001). Estimating the probability of identity among genotypes in natural populations: cautions and guidelines. *Molecular Ecology*, 10, 249–56.

Weber, J. L. & May, P. E. (1989). Abundant class of human DNA polymorphisms which can be typed using the polymerase chain reaction. *American Journal of Human Genetics*, 44, 388–96.

Whiten, A., Goodall, J., McGrew, W. C., Nishida, T., Reynolds, V., Sugiyama, Y., Tutin, C. E., Wrangham, R. W. & Boesch, C. (1999). Cultures in chimpanzees. *Nature*, 399, 682–5.

Woodruff, D. S. (1993). Noninvasive genotyping of primates. *Primates*, 34, 333–46.

APPENDIX 19.1: GLOSSARY OF GENETIC TERMS AS THEY ARE USED IN CHAPTER 19

allelic dropout a potential source of error in microsatellite genotyping; caused when, for stochastic reasons, only one of the two alleles at a microsatellite locus is amplified due to a low starting amount of DNA.

autosomal refers to the chromosomes that are not involved in sex determination, i.e. chromosomes other than the X and the Y.

base pair (bp) two nucleotides (adenine and thymine or guanine and cytosine), joined by weak bonds that hold the two strands of the double helix together. The term base pair is often, although perhaps incorrectly, used as a synonym for nucleotide; *see* nucleotide.

control region (mtDNA control region) the 1.1 kilobase segment of the mitochondrial DNA genome that contains the origins of replications but does not code for any protein or RNA product.

cytochrome *b* mitochondrial DNA locus coding for the cytochrome *b* protein; the sequence typically exhibits less variability than the HVR1.

D-loop (of mtDNA) the segment of the control region that becomes triple stranded; sometimes misused as a synonym for the control region or HVR1.

DNA–DNA hybridization in vitro annealing of complementary, single-stranded DNAs, typically from different species. DNAs from more closely related species are more similar and disassociate at higher temperatures than those from more distant phylogenetic relatives.

gel electrophoresis a technique for separating charged molecules, such as DNA, by size in an electric field within a gel.

gene flow the movement of genes between populations.

haplotype a particular sequence or combination of alleles; used here as the equivalent to a genotype when referring to a genetic marker that is haploid, such as mitochondrial DNA sequences.

heterozygosity a measure of the genetic variation in a population; with respect to one locus, stated as the frequency of heterozygotes for that locus; H_o = observed heterozygosity.

HVR1 hypervariable region 1; the approximately 400 bp region of the mtDNA control region exhibiting the highest variability.

identity by descent the state of two alleles when they are identical, homologous copies of the same ancestral allele; contrast with identity by state in which identity is not related to common ancestry.

kilobase (kb) 1000 nucleotides of DNA sequence.

marker a genetic element (locus, allele, DNA sequence) that is characterized by molecular techniques.

maximum likelihood tree reconstruction the determination of an optimal phylogenetic tree by comparison of the observed data with that expected under a given model of evolution.

mean pairwise sequence difference the average number of nucleotides that differ between any two individuals within a given stretch of sequence.

microarray technology utilization of a robot to precisely apply tiny droplets containing specific DNA to glass slides (microarrays), which are then exposed to the fluorescently labeled DNA of interest. The labeled DNA is allowed to bind to complementary DNA strands on the slides, and the resulting level of fluorescence reveals how much of a specific DNA fragment is present.

microsatellites polymorphic nuclear loci containing motifs (one to six bases in length) of nucleotide sequences sequentially repeated some 10–30 times; also known as STRs or SSRs.

molecular clock the assumption that mutations are acquired at a consistent rate, so that differences between homologous DNA sequences or proteins can be used to estimate the time since the two molecules shared a common ancestor.

molecular ecology application of DNA analysis to address questions about behavioral ecology, evolution, and conservation through direct measures of relatedness and genetic variability in wild populations.

molecular systematics application of genetic techniques to the identification of phylogenetic relationships among taxa on a molecular level and to estimating the time of their most recent common ancestor.

mitochondrial DNA (mtDNA) the approximately 16 kb, maternally inherited, circular genome found outside the nucleus in the mitochondria.

nucleotide term commonly used to refer to the four subunits (adenine, guanine, thymine, or cytosine) that make up DNA.

parsimony analysis (or maximum parsimony) tree reconstruction inference of an optimal phylogenetic tree from character state data (or sequences) based on the principal of the minimization of character state changes (or mutations).

paternity exclusion probability the probability that an unrelated male drawn at random from the population will be excluded as the sire of an offspring using a given set of loci.

PCR polymerase chain reaction; enzymatic method of replicating a targeted region of DNA in vitro, thus producing several million copies of a specified fragment.

primers short (typically ~20 bases long) synthesized chains of nucleotides used as starting points for enzymatic replication by PCR.

polymorphism a difference in DNA sequence among individuals.

polypeptide a chain of amino acids; a protein.

restriction analysis (or restriction-enzyme analysis) an indirect method of sequence analysis in which DNAs are examined for the presence or absence of restriction sites, that is, recognition sites for enzymes that cut double-stranded DNA. This information can then be used as character states for phylogenetic analysis.

SNP single nucleotide polymorphism; the single nucleotide sites that are polymorphic, or variable, in a population.

SSR simple sequence repeat; *see* microsatellite.

STR short tandem repeat; *see* microsatellite.

trichotomy three-way phylogenetic split among taxa (or other such groups) in which it cannot be determined which two of the three share a more recent common ancestor.

Index

Page references in *italics* refer to figures and tables